FOUNDATIONS OF QUANTUM GRAVITY

Exploring how the subtleties of quantum coherence can be consistently incorporated into Einstein's theory of gravitation, this book is ideal for researchers interested in the foundations of relativity and quantum physics.

The book examines those properties of coherent gravitating systems that are most closely connected to experimental observations. Examples of consistent co-gravitating quantum systems whose overall effects upon the geometry are independent of the coherence state of each constituent are provided, and the properties of the trapping regions of non-singular black objects, black holes, and a dynamic de Sitter cosmology are discussed analytically, numerically, and diagrammatically.

The extensive use of diagrams to summarize the results of the mathematics enables readers to bypass the need for a detailed understanding of the steps involved. Assuming some knowledge of quantum physics and relativity, the book provides textboxes featuring supplementary information for readers particularly interested in the philosophy and foundations of the physics.

JAMES LINDESAY is a Professor of Physics at Howard University, and was the founding Director of the Computational Physics Lab. He has been a visiting professor at Hampton University, Stanford University, and a visiting faculty scientist at MIT.

FOUNDATIONS OF QUANTUM GRAVITY

JAMES LINDESAY
Computational Physics Laboratory,
Howard University

CAMBRIDGE UNIVERSITY PRESS
Cambridge, New York, Melbourne, Madrid, Cape Town,
Singapore, São Paulo, Delhi, Mexico City

Cambridge University Press
The Edinburgh Building, Cambridge CB2 8RU, UK

Published in the United States of America by Cambridge University Press, New York

www.cambridge.org
Information on this title: www.cambridge.org/9781107008403

First published 2013

Printed and bound in the United Kingdom by the MPG Books Group

A catalogue record for this publication is available from the British Library

Library of Congress Cataloguing in Publication data
Lindesay, James.
Foundations of quantum gravity / James Lindesay, Computational Physics Laboratory,
Howard University.
pages cm
Includes bibliographical references and index.
ISBN 978-1-107-00840-3 (hardback)
1. Quantum gravity. I. Title.
QC178.L58 2013
531′.41 – dc23 2013001652

ISBN 978-1-107-00840-3 Hardback

Contents

Preface

This book is the result of decades of efforts by the author to understand physics at its most basic foundations, and to construct models that can address some of the unanswered questions about gravity, quantum mechanics, and the spectrum of fundamental particles. It is felt that foundational explorations should examine the conceptual and philosophical basis of a discipline. As such, less emphasis was placed upon the mathematics of complex calculations, while more emphasis was placed upon the consistencies and critiques of basic premises, in the preparation of this book. As a scientist, there is often a tendency to be drawn towards mathematical and formal pursuits, simply because of the beauty and elegance of mathematics. Such approaches sometimes bypass difficult conceptual approaches and thought experiments, or develop speculative formulations just because of their elegance. This book attempts to establish a balance towards the exploration of *basic* concepts and puzzles.

The prior book on black holes by Lenny Susskind and myself serves as an introduction to quantum physics and relativity for static geometries, as well as information in horizon physics. However, some of my more recent explorations indicate that qualitative modifications in descriptions of the physics occur once dynamics has been incorporated. This then calls into question any intuitions from static geometries that have been used to imply that those very geometries cannot be static. This book incorporates dynamic, spatially coherent geometries, as well as expanding upon the well-established foundations of the prior work. It represents a modest attempt to address some of the questions that were posed to me by Lenny during the writing of our book, as well as an attempt to explore the degree of flexibility in micro-physics that can be consistent with gravitational macro-physics. For instance, Lenny once asked me if an asymptotic Schwarzschild observer is freely falling or fiducial, whether Hawking radiation will be observed by such an observer, and how my answer would be consistent with complementarity. He also had me consider the drastic change in the interior causal structure of a

Schwarzschild space-time that would result from a single, charged electron falling into the center, since the center should change from space-like to time-like as a result of the charge. How does the geometry get modified so drastically?

I also had several fundamental dilemmas of my own. Are the thermal properties of dynamic horizons different than those of static horizons? Can a quantized geometric interaction localize the phase information of a quantum system without breaking its coherence? When the Penrose diagram for the formation of a black hole is constructed, why should its formation modify the scales of a distant Minkowski observer? With regards to microscopic physics, is renormalizability a fundamental requirement for all viable physical theories? After all, some very useful models, like BCS superconductivity, are not analytic in the small coupling limit, and bound states are inherently non-perturbative in the coupling. This book is a compilation of my attempts to answer these and other questions for myself. It is my hope that the reader will find some nugget(s) of interest for further exploration or development out of the various results presented.

I would like to make a few acknowledgments of the many individuals and groups that have been supportive of my efforts, in particular since this is my first manuscript as a single author. As a teacher myself, I have come to even more fully appreciate the dedication that my mentors have exemplified in their support of my development as a scholar. My early general education, science, and math teachers Lee Roy Pitts, Craig Hall, Felton Denham, John Rice, Sharon Belden, Arnold Webb, John Henderson, and Calvin Glasgow, spent hours of evenings and weekends mentoring an immature student within whom they saw some potential, in various projects and activities of pre-college nurturing.

My mentors for undergraduate research at MIT, Francis Low, Ulritch Becker, and Harry Morrison, were truly phenomenal, introducing me to world-class scholarship worthy of attempts at emulation. Harry Morrison, in particular, gave continuing mentorship and support for which I could never express sufficient gratitude. I am likewise grateful for superb mentorship at Stanford by Stan Brodsky, Cliff Will, Roberto Peccei, and my advisor H. Pierre Noyes, who has continued his support throughout my academic career.

Several of my former students and present colleagues have directly contributed to this effort, including Alex Markevich, Ed Jones, Beth Brown, Paul Sheldon, Tehani Finch, Marcus Alfred, and Tepper Gill. In particular, my understanding of black holes and horizons would be rudimentary were it not for the expert tutelage of Lenny Susskind, whose brilliant insights present concepts that are contagious and inspiring. I would also like to acknowledge that my work on linear spinor fields was done partially while I was a Peace Corp Volunteer with the faculty at the University of Dar-es-Salaam, whose hospitality remains unforgettable, and with the support of my mother, Penelope Brown, and grandmother, Elnora Herod, during the intermediate periods of my service in the mid 1980s.

I would like to acknowledge my brother, Crayge Lindesay, for technical expertise in the development of several of the complex computer graphics. I also thank my friend Eileen Johnston, artist, for the design of the book cover. Lastly, I must express my appreciation for the support of friends and family throughout the writing of this book.

Notations and Conventions

Part I

Section

1.1.1 Minkowski metric $((\eta_{\mu\nu})) = \begin{pmatrix} -1 & 0 & 0 & 0 \\ 0 & 1 & 0 & 0 \\ 0 & 0 & 1 & 0 \\ 0 & 0 & 0 & 1 \end{pmatrix}$

1.1.1 τ = proper time $\quad dc\tau^2 = -\eta_{\mu\nu} dx^\mu dx^\nu$

1.1.3 Electrodynamic equations will use cgs-Gaussian units.

1.1.3 Covariant $\mathbf{F} = ((F_{\mu\nu})) = \begin{pmatrix} 0 & -E_x & -E_y & -E_z \\ E_x & 0 & B_z & -B_y \\ E_y & -B_z & 0 & B_x \\ E_z & B_y & -B_x & 0 \end{pmatrix}$.

1.4.2 Compact conformal coordinates:

$$Y_\rightarrow = \left[\tanh\left(\frac{ct_* + r_*}{scale}\right) - \tanh\left(\frac{ct_* - r_*}{scale}\right)\right]/2$$

$$Y_\uparrow = \left[\tanh\left(\frac{ct_* + r_*}{scale}\right) + \tanh\left(\frac{ct_* - r_*}{scale}\right)\right]/2$$

2.1 Relevant fundamental constants:

Kinematic constants: Planck's constant $\hbar \simeq 6.58 \times 10^{-25}$ GeV-s and speed of light $c \simeq 3 \times 10^8$ m/s.

Geometric constant: Newtons gravitational constant $G_N \simeq 6.71 \times 10^{-39} \hbar c^5 / GeV^2$.

Statistical scaling: Boltzmann's constant k_B relates entropy to a temperature scale.

Planck mass $M_P \equiv \sqrt{\frac{\hbar c}{G_N}} \simeq 1.2 \times 10^{19}\, GeV/c^2$.

Planck length $L_P \equiv \sqrt{\frac{\hbar G_N}{c^3}} \simeq 1.6 \times 10^{-35}\, m$.

2.1.1 Momentum state normalization:

$$\langle \mathbf{p}'; m|\mathbf{p}; m\rangle = \frac{\epsilon(p)}{p^0_{(s)}c}\, \delta^3(\mathbf{p}' - \mathbf{p})$$

$$\hat{1} = \int \frac{p^0_{(s)}c}{\epsilon(p)}\, d^3p\, |\mathbf{p}; m\rangle\langle\mathbf{p}; m| = \int d^4p\, |\vec{p}\rangle\langle\vec{p}|\, \delta(\sqrt{-\vec{p}\cdot\vec{p}} - mc).$$

2.3.2 Reduced Compton wavelength of mass m: $\lambda_m \equiv \frac{\hbar}{mc}$.

8.1 Cosmological phenomenology:

Age of universe $t_o \approx 13.7 \times 10^9$ years
Redshift at last scattering (CMB): $z_{LS} \approx 1100$, $k_B T_{LS} \approx 0.3$ eV
Dark energy density ratio to critical density $\Omega_{\Lambda_o} \approx 0.73$

Part II

Locally freely falling coordinate components $\xi^{\tilde{\mu}}$
Curvilinear metric coordinate components x^β

Introduction

This book examines the foundational consistency of quantum mechanics incorporated within relativistic frameworks. Quantum physics remains a perplexing formalism that, although very successful in explaining physical phenomena, poses many philosophical and interpretational questions. Several of the subtleties of quantum physics become more manifest when quantum processes are described using relativistic dynamics. For instance, the successful connection of spin to quantum statistics is a consequence of the consistent incorporation of special relativity into the quantum formalism. There should be similar profound explanations awaiting discovery as gravitating phenomena are successfully incorporated into quantum formulations.

The common theme of this manuscript is the examination of the incorporation of relativistic behaviors upon the foundations of quantum physics. The approach is to keep all formulations as close to observed phenomena as possible, rather than to present a set of speculative models whose primary motivations are internal aesthetics. In the search for the most elegant models of physical phenomena, one must recognize that at its core, physics is an experimental science. The dimensional analysis of fundamental units, taught at the very beginning of introductory physics classes, demonstrates that phenomenology lies at the foundations of physics. Fundamental ideas such as correspondence, the principle of relativity, and complementarity provide direct contact with the physics used to guide this exploration. This manuscript is an elaboration and expansion on previously published work, but also contains some new material.

The target audience of the manuscript includes theoretical physicists and natural philosophers with an interest in the foundational basis of physical models, the experimental consistency of mainstream physics, as well as those internal consistencies required of appropriate models. Readers should have interest in quantum mechanics, general relativity, statistical physics, and the foundations of physics. The content will include concise explanations that should be somewhat self-contained for those with limited interest in a rigorous involvement with the equations. Supplements are provided as asides within the discourse for readers interested in the natural philosophy and foundations of the involved physics. Within the main text, rigorous, concise derivations are included when needed for the development of the subsequent arguments. Appendices have been added for readers interested in more technical details and background than that provided within the various chapters, or when the inclusion of technical rigor would disrupt the flow of the arguments presented in the text.

No attempt has been made to explain the fundamental kinematic/geometric constants $\hbar$, c, G_N, which provide foundational scales of length, time, and mass/energy. Neither has there been an attempt to explain or relate the fundamental thermodynamic constant k_B which relates the temperature convention to microdynamics, or the fundamental couplings, charges, particle masses, and mixing parameters that provide structure to particle interactions and dynamics. There are only a few discussions on the quantization aspects of dynamic couplings.

The approach of this manuscript is to examine the foundations of quantum physics and general relativity using non-perturbative, singularity-free descriptions of the analytic properties of physical measurables consistent with conservation properties (probability, energy-momentum, charge, etc.), cluster decomposability/classical disentanglement, and relativistic covariance. Such an approach has direct correspondence to well-understood physics, including non-relativistic and classical behaviors.

Part I of the manuscript examines those foundations of quantum physics in flat space-time that are relevant to gravitational physics. Most complicated physical models present dilemmas of which aspects are "real", and which are artifacts of the model. In addition, how one interprets a model or theory can affect how one utilizes its results to construct a more elegant model. For instance, the Copenhagen interpretation assigns elements of reality to a wavefunction that "collapses" into a measured state. Some therefore interpret the system as *being* a wavefunction. The approach presented here assigns a wavefunction only as a descriptive tool, describing those aspects of the theory subject to predictive parameterization. Wavefunctions and quantum fields are therefore considered to be merely tools of calculational convenience. To attribute a more substantial "reality" to these devices requires an assertion that the whole of nature is "contained" within models created by humanity up to this stage of our scientific progression. Such assertions have been made in the past but have yet to withstand the tides of increasing knowledge and scrutiny.

Chapter 1 examines classical concepts of space and time. Galilean relativity and special relativity (with accelerations) are discussed from the foundation upwards. Although quantum concepts are later examined in Chapter 2, coherence properties seem most directly examined using the proper dynamics of an interacting subsystem. This motivated the development of the canonical proper time formulation of relativistic dynamics. Such a formulation is particularly useful for gravitational physics, where the coordinates of fiducial observers or metric parameters are generally not those of inertial quantum systems. Since the space-like surfaces defining fixed proper times of gravitating systems are generally different from those defining fixed coordinate times, there are both subtle and non-trivial differences in dynamic descriptions whose interactions are characterized using the disparate temporal parameterizations. Examples of classical proper-time systems are examined in this chapter.

Chapter 2 examines the fundamentals of quantum mechanics in flat space-time. There remain questions about the extent to which "empty" space-time might have quantum properties that affect gravitational curvature. This chapter examines subtle quantum behaviors, like the zero-point motions of sources and fields, the Casimir effect in electromagnetic interactions, and the phenomenon of entanglement. Lifshitz successfully described the Casimir zero-point energies using the van der Waals attractions between sources, in the absence of the singularities associated with the vacuum of quantized field approaches. Therefore, the general approach of this manuscript is to include sources in any analysis of interacting systems, including gravitating systems. Such an approach avoids some of the conceptual complications associated with assigning zero-point energies to fields in the absence of sources.

Statistical physics as relevant to gravity will be developed in this chapter. In addition, one can examine the quantum mechanics of a self-gravitating mass, using canonical proper-time dynamics. Quantum non-locality can indeed prevent the formation of a singularity (regardless of the size of the mass), for much the same reason that the electron in a hydrogen atom does not have a singular wave function. To end the chapter, the thermal properties observed by a system undergoing proper acceleration will be developed. Several of the effects associated with horizons in gravitational physics are shown to be due primarily to the accelerations needed for an observer to remain fixed with respect to the chosen coordinates.

Chapter 3 develops non-perturbative scattering theory and Lagrangian dynamics. Any formulation of physical phenomena that has its insights derived from perturbative considerations must have renormalizability as a crucial tenet of its applicability. Since there are physical systems that have ill-defined low coupling limits (like superconductors), it is somewhat dubious to consider renormalizability as a universal fundamental property of all laws of nature, despite the obvious usefulness and widespread applicability of the tools of renormalization. For this reason, the chapter will develop non-perturbative descriptions of physical principles. Since all kinematic and dynamic analytical behaviors must be contained within any expression describing a physical process, one expects the formulations to be somewhat complicated. Any non-perturbative formulation should have both non-relativistic and classical correspondence limits. In particular, a viable formulation should incorporate the cluster decomposability necessary for disentanglement within a viable unitary relativistic scattering theory.

The successful incorporation of disentangled clusters in relativistic quantum systems involves separating the off-shell (quantum) dynamics from the Lorentz-frame kinematics. Basically, this means that any formulation should allow each coherent cluster of an interacting system to independently maintain or break its coherence within its proper rest frame as the overall system goes off-shell. That proper rest-frame can be most directly parameterized in terms of the *velocities* (rather than *momenta*) characterizing those frames. Such separation of geometric kinematics from off-shell dynamics will prove quite useful in examining quantum behaviors in gravitating systems. A specific example calculating Compton scattering demonstrating the ambiguity expressed by Wheeler and Feynman's approach to electromagnetic interactions between distant sources and sinks provides valuable insights into the fundamental physics of the quanta of interactions. The chapter ends with a development of Lagrangian dynamics and the calculus of unique extrema. This formulation provides an elegant manner for introducing interactions between fundamental constituents.

Chapter 4 examines the formal incorporation of group theory into quantum mechanics. The transformation properties of fundamental particles or constituents under an extended set of operations (like space-time translations, rotations, Lorentz boosts, etc.) are discussed. Both continuous (proper) transformations, as well as discrete (improper) transformations are examined. The concept of a quantum field satisfying a well-defined equation of motion (in configuration space) is developed. There will be special constraints placed upon quantum fields that satisfy microscopic causality, i.e., that cannot develop communication outside of the light cone (for spacelike separations). This condition associates the spin of a particle with the quantum statistics satisfied by that particle.

Linear spinor fields are very convenient for describing cluster decomposable relativistic quantum systems, since these fields satisfy equations of motion that are linear in the space-time derivatives, and therefore in the energy momentum. The linear dispersion relations are ideal for separating dynamic off-shell behaviors from the kinematics. Therefore, the quantum field theory for (non-interacting) general linear spinor fields is developed. The linear spinor field equation has a form that is natural for generalization into gravitational dynamics.

Part II of the manuscript develops fundamental concepts in the geometrodynamics of gravitation. This part of the manuscript utilizes the insights gained from the flat space-time examples in Part I to construct curvilinear geometries that are consistent with quantum sources, or upon which quantum systems can gravitate.

Experiments have demonstrated that the development of spatial-temporal relationships for a gravitating quantum system does not break its coherence, or alternatively, that accelerated motions of an observer do not affect the phase relationships of an inertial system. Such experiments provide verification of the principle of equivalence. One interpretation of experiments that demonstrate the maintenance of quantum coherence of gravitating systems with space-time dependent phases is that space-time coordinates are constructs, i.e., convenient parameters developed by the observer for describing relational aspects of events. A coherence breaking process cannot be involved in localizing the field of a gravitating quantum system. This means that space-time coordinates cannot "bubble up" during such a process in a manner that would break the coherence of a freely falling system being examined in order to establish the phase relationships at differing locations in a gravitational field. Rather, the coordinates of space-time are viewed to be an emergent property of the relational aspects of physical interactions and detections. The very act of creating and measuring temporal and spatial coordinates fixes them, generating a fixed past, but uncertain future.

There is therefore a focus on examining dynamic geometries as expectation-valued constructs of convenience to those choosing a particular coordinate parameterization. In particular, Einstein's equation is taken to describe geometric parameterizations in terms of classical relationships constructed from the expectation values (i.e., averaged behaviors) of quantum constituents $G_{\mu\nu} = \kappa \langle T_{\mu\nu} \rangle$. Those constructs must satisfy geometric consistency and quantum measurement constraints, which can be readily examined on conformal space-time diagrams. Such diagrams were shown to be particularly useful for examining regions of space-time within which space-like coherence can be established, as well as delineating regions of space-time that can have causal influence. One might expect that the construction of conformal space-time diagrams would be more complicated for general curvilinear geometries. A curvilinear geometry is usually described by a metric that describes space-time relationships between events, whether quantum or classical detections. A technique that constructs conformal space-time diagrams using the null trajectories obtained directly from the metric, without needing to solve directly for conformal coordinates, is demonstrated. The construction of such Penrose diagrams is valuable for examining the global causal structures of various geometries. Wherever practical these diagrams have been exhibited for example geometries.

Chapter 5 examines the fundamentals of general relativity, especially as derived from the principle of equivalence. Brief discussions of tensors and curvature are presented, and Einstein's equations are justified and demonstrated to have classical correspondence with Newtonian gravitation. Radially stationary geometries, such as stellar systems and Schwarzschild's geometry, are examined in the classical context, and a notion of gravitational energy is developed. To end the chapter, an axially stationary rotating geometry is briefly examined.

Chapter 6 examines quantum mechanics in curved space-time backgrounds and spatially coherent dynamic geometries. The chapter begins by exploring the quantum dynamics of systems on a background unaffected by those systems. The quantum dynamics of a scalar Klein–Gordon field as a system of considerable familiarity are examined on radially stationary backgrounds. However, by dimensional analysis one can demonstrate that Klein–Gordon fields are not convenient for constructing co-gravitating systems that co-generate the gravitational field. There is a focus upon developing collective disentangled co-gravitating quantum systems consistent with experimental evidence, whose disentangled clusters can generate geometries that correspond to standard general relativity. In particular, substantive gravitating flows can be dimensionally consistent with self-gravitated fields, and can provide well-defined contributory energies to co-gravitating systems. Linear spinor fields are demonstrated to provide a description of micro-physical systems consistent with Dirac spinors that undergo such substantive gravitating flows. To end the chapter, canonical proper-time dynamics are used to develop a self-gravitating system for single and co-gravitating quanta as a tool for demonstrating conformal diagrams for quantum systems.

Chapter 7 examines the physics of horizons and trapping surfaces, especially with regards to dynamic spherically symmetric black holes and black objects. During gravitational collapse, the fundamentals of quantum non-locality are not constrained by the statistical arguments that result in limits upon the quantum degeneracy pressures that can prevent the fall of stellar structures towards a classical singularity. One expects that the quantum measurement constraints associated with microscopic physics should prevent gravitational collapse towards a point singularity much as they prevent the (spatial) collapse of the ground state of electrons in atomic systems, regardless of the strength of attraction. Indeed, using proper-time gravitation with stationary expectation values, one can give credence to such ideas. The energy distribution and global causal structure of a Planck mass–sized stationary non-singular black hole is demonstrated to provide the reader with an intuitive feel for such systems. Ideas involving the temperature and thermodynamics of general stationary systems with finite Planck's constant $\hbar$ are developed, emphasizing expected radiations that would be generally inconsistent with a static geometry. Such considerations suggest that the reader consider utilizing radially dynamic geometries when examining quantum gravitating systems. In particular, the behaviors of quantum fields and co-gravitating quantum clusters on a spatially coherent evaporating black hole are examined in some detail as a means of developing intuitions into the subtleties of co-gravitating quantum systems.

The reader is motivated to examine non-singular black objects with temporally transient trapped surfaces (within which any future-seeking causal trajectory must necessarily have decreasing radial coordinate) that never develop a horizon or a space-like center. Such slowly evolving dynamic black objects should appear to have properties quite similar to those of dynamic black holes in the exterior. However, a non-singular geometry provides an unambiguous background for quantum explorations. A black object free of singularities or space-like boundaries is constructed, and a method of constructing non-singular dynamic black holes is likewise demonstrated. The dynamics of information on non-singular transient black geometries is of particular interest. The formation of a horizon for a transient black hole implies that there must be an interior space-like future boundary (the center) near or upon which any interior information must undergo transmutation (or perhaps enter a causally disjoint region of a parallel space-time, a viewpoint that is not here advocated), whether that boundary is singular or not. However, a temporally transient black object will manifest a *temporary* trapping region in the space-time, within which interior information can eventually escape in its original form (at least in principle). The radiations that leak away the interior energy of a black object carry quanta and information that change the local geometry. This geometry-changing information from an evaporating black object has likely been transmuted by the interior microscopic physics (generating the radiation) that reflects the loss of entanglement information in the exterior (using complementarity). Thus, any information retention associated with the collapsed geometry of a temporally transient black object should mimic that of a transient black hole. However, not all information that has traversed the interior need be transmuted, since the center is time-like. An entangled pair of massless particles produced at the evaporating surface of a black object are propagated until later detection by exterior observers, as an example of the maintenance of coherence information by dynamic geometries.

Chapter 8 examines cosmology and the Big Bang. The equations of standard cosmology are motivated, and in particular, the subtleties of the existence of a persistent dark energy are briefly discussed in regards to the de Sitter cosmology of a positive cosmological constant. One outstanding concern of the standard cosmological model is how the geometry transitions into and away from the thermal descriptions of the Friedmann–Robertson–Walker geometries that model the known behaviors of the Big Bang so well. For this reason, a dynamic de Sitter cosmology is developed that allows a smooth transition from an early inflation or system with coherence of cosmological scale, through a thermal expansion, towards a remnant dark energy consistent with standard cosmology. The description of the dynamics of the cosmology is expressed completely in terms of the physical densities and pressures, irrespective of the behaviors of the geometric scales. The resultant global conformal diagram has one time-like surface (the center) and two space-like surfaces (one past and one future) as its boundaries, consistent with the cosmological principle, and the geometry manifests horizons consistent with a persistent dark energy.

The developed description is particularly useful for parameterizing the early microscopic physics that results in thermalization of the primordial cosmology and "reheating". The spatial coordinates are convenient for describing the scales of microscopic physics, rather than those of the expanding co-moving (geometrically stationary) centers of gravitational clusters. The propagation of cosmological fluctuations are concisely described, and a brief introduction to the acoustic waves of the cosmological energy density is given. Arguments that associate the microphysics of the dark energy with the scale of the fluctuations in the cosmic microwave background radiation are presented. The chapter ends with some discussions on the nature of time in cosmology, especially with regards to the likely recurrences that are associated with any system of finite entropy. Such a finite entropy is a characteristic of the horizon of finite area associated with a remnant dark energy. The consequences of cosmological recurrences are perplexing and intriguing.

The final chapter examines gravitating systems with microscopic interactions. Electromagnetism, which is probably the most understood of interactions, serves as an exemplar of how a micro-physical coupling modifies a geometry. For this reason, Maxwell's equations on a general curvilinear geometry, and in particularly on one with orthogonal coordinates, are explicitly demonstrated. Subsequently, the geometry of a radially stationary charge is developed for radially stationary geometries. The features of a classical charged stationary black hole are briefly discussed. In order to examine the properties of a non-singular stationary charged geometry, the equations for a canonical proper self-gravitating charged system are developed. The modification of the energy density for this system is shown to be as expected due to the electromagnetic self-interaction.

Of particular relevance to this text is the development of gravitating interacting linear spinor fields. The group structure of these fields is particularly well-suited for providing a micro-physical motivation for the behaviors of spinors in gravitational environments described using curvilinear coordinates. This occurs because the algebra of the non-commuting operators of the extended Lorentz group generate the space-time metric. The group algebra of the complete extended Poincaré transformation expands the algebra of a standard Lie transformation group precisely in a manner that incorporates curvilinear coordinate transformations. Special coordinates consistent with the principle of equivalence are derived directly from the group parameters. This is suggestive of some fundamental significance of how the algebra of these fields includes coordinate transformations consistent with the principles of general relativity.

Part I

Galilean and special relativity

1

Classical special relativity

1.1 Foundations of special relativity

The special theory of relativity has had a profound impact upon notions of time and space within the scientific and philosophic communities. This well-established model of local coordinate transformations in the universe is built upon two fundamental postulates:

- *The principle of relativity*: the laws of physics apply in all inertial reference systems;
- *The universality of the speed of light*: the speed of light in a vacuum is the same for all inertial observers, regardless of the motion of the source or observer.

The principle of relativity is not unique to the special theory of relativity; indeed it is assumed within Galilean relativity. However, if the equations of electrodynamics described by Maxwell's equations describe laws of nature, then the second postulate immediately follows from the first, since Maxwell's equations predict a universal speed of propagation of electromagnetic waves in a vacuum. The consequences of these postulates will be developed briefly.

1.1.1 Lorentz transformations

One of the most direct routes towards developing the transformations satisfying the postulates of special relativity involves examining the distance traveled by a propagating light pulse: $(\Delta x)^2 + (\Delta y)^2 + (\Delta z)^2 = (\Delta ct)^2$. One can conveniently define a *space-time interval*, Δs which takes on the invariant value zero for *any* propagating light pulse:

$$\Delta s^2 = (\Delta x)^2 + (\Delta y)^2 + (\Delta z)^2 - (\Delta ct)^2. \tag{1.1}$$

The set of transformations that leave this interval invariant are the *Lorentz transformations*. These transformations are an extension of the subset of transformations that leave the length $(\Delta\ell)^2 = (\Delta x)^2 + (\Delta y)^2 + (\Delta z)^2$ invariant, i.e., the *rotation group* is a subset of general Lorentz transformations at fixed times. Thus, general Lorentz transformations include rotations (described by three parameters) and velocity boosts (described by three additional parameters). The most important aspect of these transformations between different coordinate systems is that observers using *any* set of inertial coordinates will agree upon the speed of light as measured *using those coordinates*.

This implies that this generalized space-time distance traveled by light will be the same for any observer in any frame of reference. The distance form for a geometry defines a distance measure, or *metric*. For special relativity, the space-time metric satisfies:

$$ds^2 = -d(ct)^2 + dx^2 + dy^2 + dz^2 = -c^2 d\tau^2 = \sum_{\mu=0}^{3} \sum_{\nu=0}^{3} dx^\mu \, \eta_{\mu\nu} \, dx^\nu \quad (1.2)$$

where $\eta_{ctct} = -1,\ \eta_{xx} = \eta_{yy} = \eta_{zz} = +1$, and $\eta_{\mu\nu} = 0$ otherwise (temporal components traditionally have the Greek index 0). This is just the usual distance formula expanded to include travel through relative time in a manner that always preserves the speed of light. Henceforth, repeated upper (contravariant) and lower (covariant) indices will be assumed to be summed over, unless specifically stated otherwise (Einstein summation convention). The geometry describing special relativity is known as *Minkowski space-time*. Lorentz transformations will leave the space-time interval invariant:

$$ds^2 = dx^\mu \, \eta_{\mu\nu} \, dx^\nu = dx'^\beta \, \eta_{\beta\lambda} \, dx'^\lambda. \quad (1.3)$$

Proper distance is defined as the distance measured at rest using synchronous locations ($dt = 0$ as measured using that observer's true time coordinate). The proper distance formula is seen to be just that given by the familiar formula described by Pythagoras and prior mathematicians for right-angled triangles: $d\ell^2 = dx^2 + dy^2 + dz^2$. *Proper time* is defined as the true time interval measured by an observer with fixed location ($d\ell = 0$ using the observer's location coordinates). Because of the nature of space-time in special relativity, the space-time distance squared ds^2 is equivalent to the negative value of the proper-time times the speed of light squared $-(c\, d\tau)^2$. Proper time is the time τ measured by the system being described by the Minkowski space-time interval formula. Indeed, time and space are skewed in a very peculiar way, since temporal distance description seems to be an imaginary spatial distance description (because of the square root of -1). The *metric tensor*, η, defines how lengths and distances are described in the space-time. Since the space-time interval is defined to be invariant under Lorentz transformations, both the proper distance and proper time are themselves Lorentz invariants, describing lengths and times in the frame at rest with respect to that proper observer.

Consider general linear transformations between coordinates $\vec{x} \equiv (x^0, x^1, x^2, x^3) = (ct, x, y, z)$ and $\vec{x}' = (ct', x', y', z')$ representing two distinct inertial coordinate frames related by relative speed $\mathbf{v} \equiv \mathbf{u}/\sqrt{1 + u^2/c^2}$. A linear transformation is invertible and single-valued, assuring unique coordinates for a given event. The transformation will be represented as:

$$x'^{\beta} = \Lambda^{\beta}{}_{\mu} x^{\mu}, \tag{1.4}$$

where $\mathbf{\Lambda} = \mathbf{\Lambda}(\mathbf{u})$. This relationship defines the general transformation of (upper index, or *contravariant*) components of *four-vectors*. Substitution into the invariance relation, Eq. 1.3 defines the general constraint on the transformation matrices:

$$\eta_{\mu\lambda} = \Lambda^{\alpha}{}_{\mu} \eta_{\alpha\beta} \Lambda^{\beta}{}_{\lambda}, \tag{1.5}$$

demonstrating that Lorentz transformations leave the Minkowski metric invariant. The inverse of the Minkowski metric tensor will be defined to have components $\eta^{\alpha\beta}$, giving $\eta^{\alpha\beta}\eta_{\beta\lambda} = \delta^{\alpha}_{\lambda}$. This allows the metric tensor and its inverse to be used to lower/raise indices.

By examining the constraint Eq. 1.5, general types of Lorentz transformations can be classified. The determinant satisfies $(\det \mathbf{\Lambda})^2 = 1$, which implies that $\det \mathbf{\Lambda} = \pm 1$. Also, examining the 00 component of that equation $-1 = -(\Lambda^0{}_0)^2 + \sum_j (\Lambda^j{}_0)^2$, the conditions $\Lambda^0{}_0 \geq +1$ or $\Lambda^0{}_0 \leq -1$ must be satisfied. The only types of Lorentz transformations that are connected to the identity transformation by a continuous change of parameters must have $\det \mathbf{\Lambda} = +1$ and $\Lambda^0{}_0 \geq +1$. These transformations are referred to as *proper orthochronous* Lorentz transformations, which will be discussed next.

The proper orthochronous Lorentz group

To express the form of the Lorentz transformation matrices on four-vectors, define $\mathcal{R}^{\nu}{}_{\mu}$ and $\mathcal{L}^{\nu}{}_{\mu}$, which act on (contravariant components of) four-vectors according to $\Lambda^{\mu}{}_{\beta} x^{\beta} = x'^{\mu}$, where $\mathcal{R}$ will rotate coordinates, and $\mathcal{L}$ will velocity boost coordinates. The form of those transformation matrices will be taken as:

$$\begin{aligned}
\mathcal{R}^m{}_k(\boldsymbol{\theta}) &= \cos(\theta)\delta^m_k + (1 - \cos(\theta))\hat{\theta}^k\hat{\theta}^m + \sin(\theta)\hat{\theta}^j\epsilon_{jkm} \\
\mathcal{R}^m{}_0(\boldsymbol{\theta}) &= 0 = \mathcal{R}^0{}_m(\boldsymbol{\theta}) \\
\mathcal{R}^0{}_0(\boldsymbol{\theta}) &= 1 \\
\mathcal{L}^m{}_k(\mathbf{u}) &= \delta^m_k - (1 - u^0)\hat{u}^k\hat{u}^m \\
\mathcal{L}^m{}_0(\mathbf{u}) &= -u^m = \mathcal{L}^0{}_m(\mathbf{u}) \\
\mathcal{L}^0{}_0(\mathbf{u}) &= u^0,
\end{aligned} \tag{1.6}$$

where the hats indicate unit vectors specifying direction only, and ϵ_{jkm} is the anti-symmetric Levi–Civita symbol, defined by:

$$\epsilon_{jkm} \equiv \begin{cases} +1 & (j,k,m) \text{ an even permutation of } (1,2,3) \\ -1 & (j,k,m) \text{ an odd permutation of } (1,2,3) \\ 0 & \text{any two or more indices identical.} \end{cases} \tag{1.7}$$

These transformations represent *passive* transformations, where the system is passively observed from various coordinate systems. An *active* rotating or boosting of the system itself would therefore correspond to an inverse transformation in this convention. A more extensive summary of the techniques of group theory is given in Appendix A.1.1.

For a pure velocity boost in the x-direction, the Lorentz transformation takes the familiar form:

$$\begin{aligned} dct' &= \gamma(dct - \beta\, dx), \\ dx' &= \gamma(dx - \beta\, dct), \\ dy' &= dy, \\ dz' &= dz, \\ \gamma &\equiv \frac{1}{\sqrt{1-\beta^2}}, \qquad \beta \equiv \frac{v}{c}. \end{aligned} \tag{1.8}$$

The proper length ℓ_p of a rod observed moving in a direction tangent to its length is shortened by the *Lorentz contraction*, $L_{observed} = \ell_p/\gamma$. Likewise, a proper-time interval $\Delta\tau$ when observed in relative motion will be expanded through *time dilation*, $\Delta t_{observed} = \gamma\Delta\tau$. These transformations can be expressed as a rotation through a hyperbolic angle ζ:

$$\beta \equiv \tanh\zeta, \qquad \gamma = \cosh\zeta. \tag{1.9}$$

The parameter ζ is sometimes referred to as the *rapidity*. Written in terms of the rapidity, a velocity boost along the x-direction satisfies:

$$\begin{aligned} dct' &= \cosh\zeta\; dct - \sinh\zeta\; dx \\ dx' &= -\sinh\zeta\; dct + \cosh\zeta\; dx \\ dy' &= dy, \quad dz' = dz. \end{aligned} \tag{1.10}$$

The rapidity is particularly useful for describing successive velocity boosts in the same direction to coordinates (ct'', x'', y'', z''), since the resultant rapidity connecting the double-primed coordinates to the unprimed coordinates is obtained by simple addition of the hyperbolic angles, $\zeta'' = \zeta' + \zeta$.

By direct substitution one can verify that the Lorentz space-time transformations leave the interval invariant $ds'^2 = ds^2$. Except for the negative signature associated with the time coordinate (conventionally represented as the 0th space-time component), the invariant interval form is seen to be just that given by the familiar distance formula. By leaving this metric's space-time interval ds^2 invariant for any reference frame, the speed of light is seen to be locally constant if a beam of light always corresponds to zero space-time interval in *any* reference frame,

$$ds^2 = 0 = \left(-c^2 + \left(\frac{dx}{dt}\right)^2 + \left(\frac{dy}{dt}\right)^2 + \left(\frac{dz}{dt}\right)^2\right) dt^2 \Rightarrow v_{light}^2 = c^2.$$

The Lorentz transformations are therefore the most general coordinate transformations that satisfy the postulates of special relativity.

Supplement: Relativistic dynamics

The dispersion relation for electromagnetic radiation $E = pc$ has substantial consequences for relativistic dynamics. Since energy and momentum are conserved, the emission of a photon must involve a recoil of the source. In the well-known *boxcar thought experiment*, a massive boxcar (or free-floating rocket) emits a photon from one interior wall of the boxcar, which is subsequently absorbed by the opposite interior wall. The boxcar is assumed to be completely free to recoil, and does so as the photon traverses the interior. Since there are no external forces acting on the boxcar, the center of mass cannot have moved, and thus mass must have been transfered between the walls by the photon. To preserve the location of the center of mass, the shift in the positions of the walls as viewed by an exterior observer, is given by $\delta x = \frac{\delta M}{M} L$ for a train of length L. The external observer then equates the time and location of the absorption after traversing the train giving $\frac{\delta x}{v} = \frac{L - \delta x}{c}$. (Notice that this requires that the "rigid" boxcar moves in a non-causal manner, since no communication could have arrived at the far end prior to the photon!) During the transit, (non-relativistic) momentum conservation requires that $\frac{E}{c} \simeq (M - \delta M)v$. These equations give the well-known result for mass transfer $E = \delta M c^2$.

More rigorously, particle kinematics defines general forms for the energy and momentum of a particle of rest mass m. To see this, examine the kinematics of Figure 1.1. In the figure, two equal masses m collide inelastically to form a new mass m_F. The lab frame is represented in the lower diagram, while the center-of-momentum system (CMS) is represented in the upper diagram. In both diagrams, the initial particles are colliding on the left, resulting in the final state shown on the right. The CMS is related to the lab system by a Lorentz transformation, $\beta_* = v_*/c = \tanh \zeta_*$. In *both* frames of reference,

Figure 1.1 Kinematics of a completely inelastic collision of two equal masses.

energy and momentum should be conserved. Since the velocity boost is co-linear with all velocities, the rapidity of the lab velocity is directly related to the transformation rapidity ζ_* through $\zeta_* = \zeta_L - \zeta_*$, i.e., $\zeta_L = 2\zeta_*$. The energies of particles of mass M generally depend upon their speed through some functional form $E = \Gamma(\zeta)Mc^2$. Energy conservation in the CMS immediately relates the final mass to the energy forms of the initial masses through $m_F = 2m\Gamma(\zeta_*)$. Energy conservation in the lab frame then defines constraints on the functional form through $[\Gamma(2\zeta_*) + 1]\, m = 2\Gamma(\zeta_*)m$. This "trigonometric" relationship on hyperbolic functions is satisfied by the hyperbolic cosine, $\Gamma(\zeta) = \cosh\zeta = \gamma(v)$, i.e., the Lorentz factor. Thus, the form of the energy consistent with conservation principles is $E = \gamma mc^2$. Likewise, the momentum dependency upon speed can be generally expressed in a form, $p = B(\zeta)mc$. Momentum conservation in the lab frame requires $B(2\zeta_*)m = 2\cosh\zeta_* B(\zeta_*)m$, or $B(\zeta_*) = \sinh\zeta_*$. Thus, the momentum satisfies $p = \sinh\zeta mc = \gamma\beta mc$.

The forms of the energy and momentum in special relativistic kinematics are given by:

$$E = \gamma mc^2, \quad p^j c = \gamma\beta^j mc^2. \tag{1.11}$$

These forms define the energy-momentum dispersion relationship $-E^2 + |\mathbf{p}c|^2 = -\left(mc^2\right)^2$. This relationship motivates a representation of an energy-momentum four-vector. An examination of the inverse Lorentz transformation into the proper frame of a particle of mass m gives:

$$\begin{aligned} dct &= \gamma(dc\tau + \beta\, dx_{p||}) = \gamma\, dc\tau \\ dx_{||} &= \gamma(\beta\, dc\tau + dx_{p||}) = \gamma\beta\, dc\tau. \end{aligned} \tag{1.12}$$

This defines a four-velocity $\vec{u} \equiv \frac{d\vec{x}}{d\tau}$ which transforms as a four-vector under Lorentz transformations, and has components $\vec{u} = (\gamma c, \gamma\boldsymbol{\beta}c)$. From Eq. 1.11, the energy and momentum are components of a four-vector:

$$p^\beta = mu^\beta, \quad \vec{p} = \left(\frac{E}{c}, \mathbf{p}\right), \tag{1.13}$$

where the mass m is a Lorentz invariant.

Discrete Lorentz transformations

Besides the continuous Lorentz transformations previously developed, there are two additional types of discrete transformations of interest. Each of these discrete transformations satisfy $det\,\mathbf{\Lambda} = -1$.

Parity, $\mathcal{P}$ is such a transformation that reflects any vector through the spatial origin, without affecting its temporal component:

$$\mathcal{P} \equiv \begin{pmatrix} +1 & 0 & 0 & 0 \\ 0 & -1 & 0 & 0 \\ 0 & 0 & -1 & 0 \\ 0 & 0 & 0 & -1 \end{pmatrix}. \tag{1.14}$$

Parity reflections are orthochronous $\Lambda^0{}_0 = +1$.

Time reversal, $\mathcal{T}$ is a discrete transformation that reverses the temporal component of a four-vector, without affecting its spatial components:

$$\mathcal{T} \equiv \begin{pmatrix} -1 & 0 & 0 & 0 \\ 0 & +1 & 0 & 0 \\ 0 & 0 & +1 & 0 \\ 0 & 0 & 0 & +1 \end{pmatrix}, \tag{1.15}$$

clearly demonstrating transformations of the type $\Lambda^0{}_0 \leq -1$. Any general Lorentz transformation is either a proper orthochronous transformation, or a product of such a transformation with either $\mathcal{P}$, $\mathcal{T}$, or $\mathcal{PT}$.

1.1.2 Maxwell's equations

Probably the most successful of physical models is the one which describes the electrodynamics of charges. Its success is likely to be related to its importance in describing those phenomena most accessible to scientific exploration, e.g., chemistry, biology, optics, etc. The experimental findings of Coulomb described by Gauss' law, the electromotive inductions incorporated in Faraday's law and the Lorentz force equation, the lack of any experimental evidence for a magnetic monopole or charge, and the generation of magnetic fields using Ampere's law, are all neatly encapsulated, consistent with conservation principles in Maxwell's equations:

$$\begin{aligned} \nabla \cdot \mathbf{E} &= 4\pi\rho \\ \nabla \times \mathbf{E} &= -\frac{1}{c}\frac{\partial \mathbf{B}}{\partial t} \\ \nabla \cdot \mathbf{B} &= 0 \\ \nabla \times \mathbf{B} &= \frac{4\pi}{c}\mathbf{J} + \frac{1}{c}\frac{\partial \mathbf{E}}{\partial t}. \end{aligned} \tag{1.16}$$

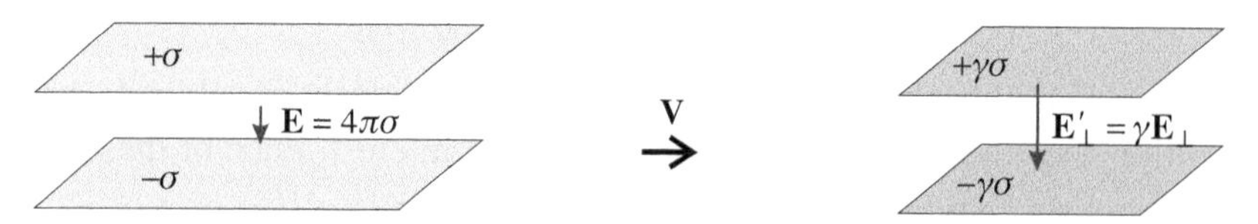

Figure 1.2 Transformation of field normal to motion.

The conservation of charge, or *continuity equation*,

$$\frac{\partial \rho}{\partial t} + \nabla \cdot \mathbf{J} = 0, \tag{1.17}$$

is directly incorporated in these equations. The Lorentz force equation specifies the (mechanical) effect an electromagnetic field has upon a charge q:

$$\left.\frac{d\mathbf{p}}{dt}\right|_{mechanical} = q\left(\mathbf{E} + \frac{\mathbf{v}}{c} \times \mathbf{B}\right). \tag{1.18}$$

These are the equations that motivated the development of the special theory of relativity, as will be briefly discussed. They represent a unification of the actions of electric fields and magnetic fields upon a single coupling, the (electric) charge of a system.

The invariance of the charge of a system under Lorentz transformations is well verified by experiment [1], as well as common experience. Atoms remain precisely neutral despite motions of electrons that give relativistic effects of the order of the fine structure constant, about 1%. Even a tiny variance in charge due to motion would cause the Earth, Sun, and planets to have non-vanishing electric fields due to the large collection of charges with significantly different motions. Thus charge, as well as rest mass, are invariants of a closed system. Fundamental aspects of electromagnetism as it relates to special relativity will be discussed in the next section.

1.1.3 Electrodynamics of moving charges

The approach taken in this development will be to examine how the transformation of perspectives alter the distribution of the sources of a given electromagnetic field configuration, thereby altering the fields [2]. Such an analysis emphasizes the primacy of sources when examining the behaviors of fields.

To begin, consider a parallel plate capacitor with surface charge densities, $\pm\sigma$. Neglecting edge effects, a direct application of Gauss' law yields an electric field given by $\mathbf{E} = 4\pi\sigma\mathbf{n}$ normal to the surface between the plates, and vanishing external to the plates. Next, consider an observer moving in a direction tangent to the plates such that they appear to move as indicated in Figure 1.2. The surface upon

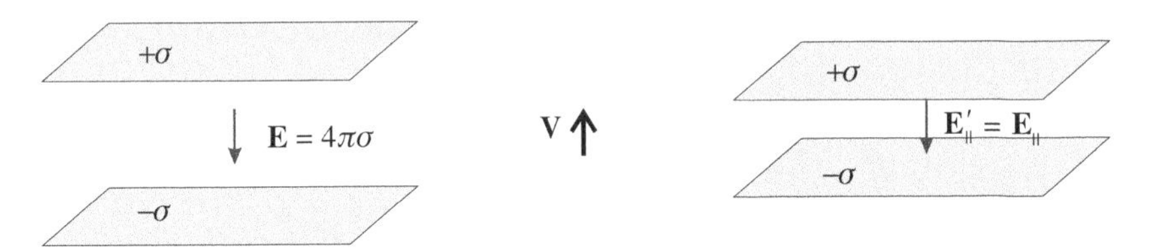

Figure 1.3 Transformation of field parallel to motion.

Figure 1.4 Force on an external charge moving parallel to a wire.

which the source charge densities reside is Lorentz-contracted in the direction of motion, but unchanged in the perpendicular direction, resulting in an increased surface density $\sigma \rightarrow \sigma' = \gamma\sigma$. Therefore, the field generated by this increased charge density (which is normal to the direction of motion) is given by $\mathbf{E'}_\perp = \gamma \mathbf{E}_\perp$.

Likewise, motions normal to the plates can be examined, as demonstrated in Figure 1.3. For such motions, only the separation of the plates is contracted, and the surface charge densities remain unchanged. Since the field between parallel plates does not depend upon separation, the transformed field measured in this configuration is given by $\mathbf{E'}_{||} = \mathbf{E}_{||}$.

The Lorentz force equation 1.18 makes clear that the proper forces exerted on electric charges are due to electric fields only, i.e., in the rest frame of a given charge, the magnetic fields exert no force upon that charge. This means that the magnetic field is just a convenient tool of an inertial observer allowing those inertial coordinates to be utilized in describing the motions of that charge. To explore this point, examine the neutral wire depicted by the left diagram in Figure 1.4. The positive charges are moving to the right, but they have equal but opposite linear density λ to the stationary negative charges. A straightforward application of Gauss' and Ampere's laws generates fields, $\mathbf{E} = 0, \mathbf{B} = \frac{2\lambda v_+}{cr}\hat{\varphi}$. Suppose that an exterior charge q moves coaxially to the right with speed $\mathbf{V}$ a distance r from the axis of the wire. Using this magnetic field, an observer in the frame of reference of the neutral wire calculates a Lorentz force on the charge q given by:

$$\mathbf{f} = q\frac{\mathbf{V}}{c} \times \mathbf{B} = -\frac{2q\lambda v_+ V}{c^2 r}\hat{r}. \tag{1.19}$$

The diagram on the right of Figure 1.4 represents the wire as observed by the moving charge. In the rest frame of the charge, the positive charges have linear density $\lambda'_+ = \gamma(v'_+)\lambda_o = \lambda\gamma(v'_+)/\gamma(v_+)$, and speed $v'_+ = \frac{v_+ - V}{1 - v_+ V/c^2}$. Likewise, the negative charges have linear density $\lambda'_- = -\gamma(V)\lambda$, and speed $-V$. The electric and magnetic fields in this frame can be calculated using Maxwell's equations, giving:

$$
\begin{aligned}
\mathbf{E}' &= \frac{2(\lambda'_+ + \lambda'_-)}{r}\hat{r} = -\gamma(V)\frac{2\lambda v_+ V}{c^2 r}\hat{r} \\
\mathbf{B}' &= \frac{2(\lambda'_+ v'_+ - \lambda'_- V)}{cr}\hat{\varphi} = \gamma(V)\frac{2\lambda_+ v_+}{cr}\hat{\varphi}.
\end{aligned} \tag{1.20}
$$

The force on the charge in its rest frame is given by:

$$
\mathbf{f}' = q\mathbf{E}' = \gamma(V)\mathbf{f}, \tag{1.21}
$$

where $\mathbf{f}$ is the force calculated using the magnetic field in Eq. 1.19, and the factor $\gamma(V)$ is just the time dilation factor connecting the reference frames.

As was the case in Figure 1.3, transverse densities generating linear charge densities and currents in the wire remain unchanged by a velocity boost. Therefore, Lorentz transformations on the charge and current sources modify the fields according to:

$$
\begin{aligned}
E'_x &= E_x & B'_x &= B_x \\
E'_y &= \gamma(E_y - \beta B_z) & B'_y &= \gamma(B_y + \beta E_z) \\
E'_z &= \gamma(E_z + \beta B_y) & B'_z &= \gamma(B_z - \beta E_y)
\end{aligned} \tag{1.22}
$$

where $\beta \equiv \frac{V}{c}\hat{x}$. These transformations clearly mix electric and magnetic fields in differing inertial frames of reference.

The Lorentz transformation properties of an electromagnetic field can be covariantly expressed using the *electromagnetic field tensor*. The contravariant (upper index) components of this tensor can be neatly expressed as follows:

$$
\begin{aligned}
F^{00} &= 0 = F^{11} & F^{22} &= 0 = F^{33} \\
F^{01} &= E_x = -F^{10} & F^{12} &= B_z = -F^{21} \\
F^{02} &= E_y = -F^{20} & F^{23} &= B_x = -F^{32} \\
F^{03} &= E_z = -F^{30} & F^{31} &= B_y = -F^{13},
\end{aligned} \tag{1.23}
$$

or more succinctly:

$$
F^{\mu\nu} = -F^{\nu\mu}, \quad F^{0j} = E_j, \quad F^{jk} = \sum_{m=1}^{3} \epsilon_{jkm} B_m. \tag{1.24}
$$

The electromagnetic field transformations in Eq. 1.22 then guarantee that the field strength $\mathbf{F}$ transforms as a rank 2 tensor $F'^{\alpha\beta} = \Lambda^\alpha{}_\mu \Lambda^\beta{}_\nu F^{\mu\nu}$.

The charge continuity equation 1.17, $\frac{\partial \rho c}{\partial ct} + \frac{\partial}{\partial x^k} J^k = 0$, suggests that the charge density and current density are likewise components of a four-vector. Indeed, by defining the (contravariant) four-current density:

$$\vec{J} \equiv (\rho c, J_x, J_y, J_z) = \rho_* \vec{u}, \tag{1.25}$$

the continuity equation can be written in an invariant form:

$$\frac{\partial}{\partial x^\mu} J^\mu \equiv \partial_\mu J^\mu = 0. \tag{1.26}$$

In the previous expressions, ρ_* is the charge density as measured in the proper frame of the moving charges, whereas ρ is the charge density measured in the inertial laboratory frame. These identifications allow the two inhomogeneous (source-driven) Maxwell equations to be written in the manifestly co-variant form:

$$\partial_\beta F^{\mu\beta} = \frac{4\pi}{c} J^\mu. \tag{1.27}$$

The two homogeneous Maxwell equations typically used to define gauge potentials can likewise be expressed in a manifestly co-variant form. To do this, it is convenient to generalize the rank 3 Levi–Civita symbol from Eq. 1.7 into the rank 4 anti-symmetric form:

$$\epsilon^{\mu\nu\alpha\beta} \equiv \begin{cases} +1 & (\mu, \nu, \alpha, \beta) \text{ an even permutation of } (0, 1, 2, 3) \\ -1 & (\mu, \nu, \alpha, \beta) \text{ an odd permutation of } (0, 1, 2, 3) \\ 0 & \text{any two or more indices identical.} \end{cases} \tag{1.28}$$

The homogeneous equations are then given by the four (dual) equations:

$$\epsilon^{\mu\nu\alpha\beta} \partial_\nu F_{\alpha\beta} = 0, \tag{1.29}$$

where the covariant (lower index) field tensor components are obtained using the Minkowski metric form,

$$F_{\alpha\beta} = \eta_{\alpha\mu} \eta_{\beta\nu} F^{\mu\nu}. \tag{1.30}$$

These manifestly covariant forms are quite convenient for exploring electrodynamics and gauge field properties in various frames of reference, and will henceforth be utilized when describing electromagnetic phenomena.

Using Maxwell's equations, the electromagnetic field can be formally afforded an energy-momentum, which, when added to mechanical energy-momentum densities, gives a locally conserved form. The manifestly covariant form of this tensor is given by:

$$T^{\alpha\beta} = \frac{1}{4\pi} \left(F^{\alpha\mu} F^{\beta\nu} \eta_{\mu\nu} - \frac{1}{4} F^{\mu\nu} F_{\mu\nu} \eta^{\alpha\beta} \right). \tag{1.31}$$

The components of this symmetric tensor include the electromagnetic energy density $T^{00} = u = \frac{1}{8\pi}(\mathbf{E}^2 + \mathbf{B}^2)$, the Poynting energy flux vector $c(T^{0k}) = \mathbf{S} =$

$\frac{c}{4\pi}\mathbf{E} \times \mathbf{B}$, the electromagnetic momentum density $(T^{j0})/c = \mathbf{g} = \frac{1}{4\pi c}\mathbf{E} \times \mathbf{B}$, and the Maxwell stress tensor as its spatial index components T^{jk}. The mechanical energy density flux directly satisfies $\mathbf{J} \cdot \mathbf{E} = -\partial_\beta T^{0\beta}$ (Poynting's theorem), and the local form of the Lorentz force equation describes mechanical momentum density flux $\rho\mathbf{E}^j + \frac{1}{c}(\mathbf{J} \times \mathbf{B})^j = -\partial_\beta T^{j\beta}$. These relationships will be further examined in Section 1.3.2.

1.1.4 Consistency and completeness of Maxwell's equations

As previously mentioned, the special theory of relativity was developed to describe coordinate transformations consistent with the phenomenological equations of electromagnetism. The historical formulas describing the dynamics of charges in electric and magnetic fields were unified using a single coupling (charge) with those fields. Of particular interest are the magnetic properties of a system. Magnetic fields in one frame of reference result in an electric field in the rest frame of a given charge. Through the Lorentz force equation, that charge responds only to electric fields in its proper frame. In this sense, magnetism is a "fictitious" field, i.e., a convenience used by inertial observers to describe the dynamics of mutually interacting electric charges.

There has been some interest in expanding Maxwell's equations in a manner that increases their perceived symmetry by including a new magnetic charge (a magnetic monopole) that modifies the homogeneous Maxwell equations, creating direct, conserved magnetic source terms. Such monopoles imply charge quantization in quantum theories [3]. However, such a model detracts from the unification of electromagnetism in terms of a single coupling, as well as the expression of magnetism *only* as a relativistic re-expression of the proper electric field acting on a charge. Indeed, for the expanded model, one expects the proper electric field on a magnetic monopole to likewise have no effect; only motions of charges generating proper magnetic fields, as well as magnetic monopole generated fields, will exert forces in the rest frame of a monopole.

Supplement: Charge quantization from existence of a magnetic monopole

An argument motivating charge quantization as a consequence of the existence of a magnetic monopole examines the angular momentum of the electromagnetic field generated by an electric charge paired with a magnetic monopole. Suppose the displacement vector between the charge e and monopole q_m is represented by $\mathbf{w}$. The electromagnetic momentum density generated by this configuration is given by $\mathbf{g} = \frac{1}{4\pi c}\mathbf{E} \times \mathbf{B} = \frac{e q_m}{4\pi c}\frac{\mathbf{r}\times(\mathbf{r}-\mathbf{w})}{(|\mathbf{r}||\mathbf{r}-\mathbf{w}|)^3}$. The angular momentum density of the electromagnetic field then satisfies:

$$\mathbf{L}_{EM} = \int \mathbf{r} \times \mathbf{g} d^3 r = \frac{e q_m}{c} \hat{w}.$$

During any collision process, the combination of this angular momentum and the mechanical angular momenta of the particles is conserved. The quantization of the magnitude of this angular momentum in appropriate units of $\hbar$ (or alternatively, periodicity in the phase of a quantum system containing these fields) is the source of the charge quantization condition. However, this condition implies only a quantization of the product eq_m and does *not* address individual quantization of either type of charge.

One should take care not to introduce infinities in developing elegant formulations of electromagnetism. If charges q_s are fundamentally carried by discrete particles, some infinities in classical electromagnetism could result from inappropriately including additional self-energies through a continuum limit of energies such as:

$$U = \frac{1}{2} \sum_{s \neq m} \frac{q_s q_m}{|\mathbf{r}_s - \mathbf{r}_m|},$$

that does not exclude the $s = m$ term. The continuum form of this equation $U = \frac{1}{2} \int \rho(\mathbf{r}')\Phi(\mathbf{r}')\, d^3 r'$, from which expressions such as 1.31 are derived, must involve potentials and fields that do *not* contribute to self-energies already contained in the mechanical terms (such as particle mass).

1.2 General motions in special relativity

One of the most initially unsettling aspects of the theory of special relativity was that time intervals measured by observers in uniform motion relative to one another would be different and not absolute. The time dilation effect mentioned in Section 1.1.1 relates the time intervals measured in the reference frames: $\Delta t = \gamma \Delta \tau$, $\gamma \equiv 1/\sqrt{1 - \frac{v^2}{c^2}}$, where Δt is the "stationary" observer's time interval, $\Delta \tau$ is the time interval measured by the "moving" observer, and v is the speed of the moving observer relative to the stationary observer. Since special relativity demonstrates differing time coordinates associated with different inertial frames of reference, observers in the various frames of reference typically utilize their own *proper time* to describe the dynamics of any other system. Alternatively, the equations that describe the dynamics of a particular system being studied can be expressed in terms of the temporal coordinates of that system (proper time), with which all observers would agree, since that time is parameterized by any clock being carried by the dynamic system itself. This approach has advantages with regards to the exploration of the basic behaviors of interactions [4], since the most relevant perspective for an interacting system is, of course, that system's frame of reference itself, as will be discussed in Section 1.3. However, since typically an observer does not have a directly observable parameterization of an interacting system's proper time, for many descriptions a *covariant* coordinate description of the dynamics involving the space-time relations directly measured by inertial observers is favored, and is the description that has been most utilized in the literature.

Although many of the calculations involved in special relativity primarily deal in inertial, non-accelerating frames of reference (uniform motion), the theory is quite capable of modeling the local behavior of accelerating systems in a flat space-time. This is done by utilizing the laws of physics in instantaneously co-moving inertial frames of reference to determine the dynamics in the accelerating system, then progressively changing inertial frames to construct the global path of the system.

1.2.1 Accelerations in special relativity

For an arbitrarily moving system, the displacements of the system relative to itself vanish, $dx_a = 0$, and the system's proper time represented by τ satisfies $dt_a = d\tau$. The observed displacement of the system x in an inertial time interval, t, can be directly calculated in the case that the local (proper) acceleration measured within the system, a, is held constant. To derive the equations, the (inverse, $\beta \to -\beta$, $\zeta \to -\zeta$) Lorenz transformations in Eq. 1.10 expressed in terms of the *rapidity* ζ will be utilized. This gives:

$$ct = ct' \cosh \zeta' + x' \sinh \zeta', \quad x = ct' \sinh \zeta' + x' \cosh \zeta', \tag{1.32}$$

where successive (active) transformations along a given direction are represented by additive rapidities, i.e., $(ct, x) \leftarrow (ct', x') \leftarrow (ct'', x'')$ is described by the single hyperbolic angle $\zeta = \zeta' + \zeta''$. By considering the primed frame to be the proper frame of the accelerating system (which has vanishing proper velocity and rapidity), the proper acceleration can be described using $\frac{dv_p}{d\tau} = a \Rightarrow \frac{d\zeta_p}{d\tau} = \frac{a}{c}$. Since a does not depend on the proper time, τ, the resultant expressions for the velocity and Lorentz factor are given by:

$$\frac{dx}{dct} = \tanh \frac{a\tau}{c}, \quad \frac{dct}{dc\tau} = \cosh \frac{a\tau}{c}. \tag{1.33}$$

The equations describing the trajectory of a system in flat space-time with a temporally constant proper acceleration are given by:

$$\frac{v}{c} = \beta_a = \tanh\left(\frac{a\tau_a}{c}\right), \qquad \tilde{t} = \frac{c}{a}\left[\sinh\left(\frac{a\tau_a}{c}\right)\right],$$

$$\beta_a = \frac{\frac{a\tilde{t}}{c}}{\sqrt{1+\left(\frac{a\tilde{t}}{c}\right)^2}}, \qquad \gamma_a = \cosh\left(\frac{a\tau_a}{c}\right) = \sqrt{1+\left(\frac{a\tilde{t}}{c}\right)^2},$$

$$\tilde{x} = x_a + \frac{c^2}{a}\left[\cosh\left(\frac{a\tau_a}{c}\right) - 1\right] = x_a + \frac{c^2}{a}\left[\sqrt{1+\left(\frac{a\tilde{t}}{c}\right)^2} - 1\right]. \tag{1.34}$$

The accelerating system is assumed to be momentarily at rest at coordinate $\tilde{x}(\tilde{t}=0) = x_a$ and time $t = 0 = \tau_a$. Only a single local accelerating proper point is described by these equations. If the equations are used to describe two distinct points with initial coordinates x_1 and x_2, each accelerating with the same proper acceleration, then at a future time t, the distance between the points remains $\tilde{x}_2(t) - \tilde{x}_1(t)$ in the inertial frame. Thus, the proper distance $x_{p2} - x_{p1}$ between the two points must have changed. This means that the endpoints of a "rigid" rod used for proper length measurements must have different proper accelerations. This makes sense, because the increasing speeds must result in increasing Lorentz contractions, requiring the endpoints to have differing accelerations.

Therefore, although the proper acceleration used in Eq. 1.34 cannot depend on proper time, it must depend upon the distance coordinate x_a if one is to be able to set up a proper distance grid. Suppose that an accelerating rigid rod becomes inertial. The results of inertial special relativity can be utilized if a single inertial frame of reference can be found such that the speeds of the two endpoints are identical. This can be done only if the trailing endpoint stops accelerating at a time earlier than the leading endpoint, resulting in a final speed of $\beta_{1f} = \beta_{2f} \equiv \beta_f$. From Eq. 1.34, this requires $a_1\tilde{t}_{1f} = a_2\tilde{t}_{2f}$, which relates the accelerations to the (inertial) times that the endpoints each become inertial. The formula for Lorentz contraction $\Delta\tilde{x}(\tilde{t}_{2f}) = \frac{\Delta x_a}{\gamma_f}$ results in the relation:

$$x_{a2} - x_{a1} = \frac{c^2}{a_2} - \frac{c^2}{a_1} \quad \Rightarrow \quad a(x_a) = \frac{c^2}{x_a - x_o}. \tag{1.35}$$

This form demonstrates a surface of infinite proper acceleration at x_o, as will be discussed further shortly.

Figure 1.5 demonstrates the previous discussion for the case that the surface of infinite proper acceleration is chosen to have accelerating coordinate $x_o = 0$.

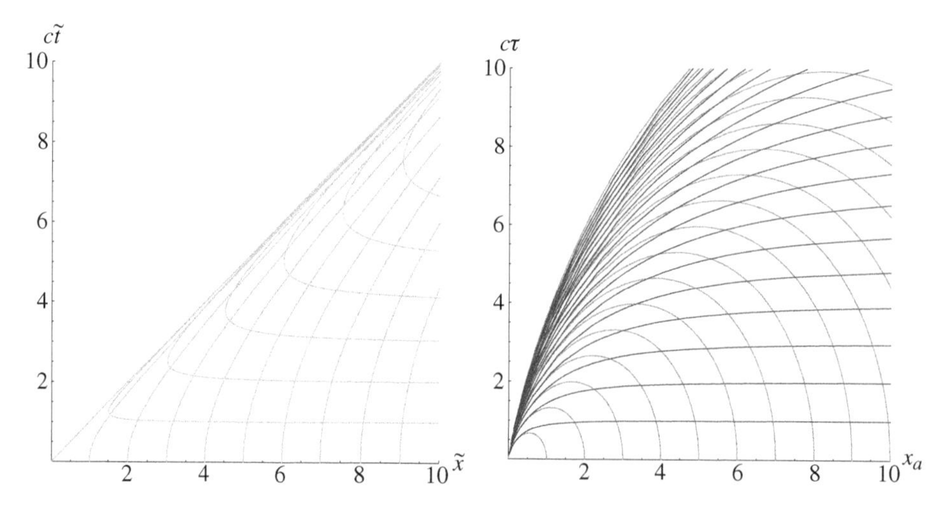

Figure 1.5 Accelerating coordinates in terms of inertial coordinates $(\tilde{ct}, \tilde{x})$, and inertial coordinates in terms of accelerating coordinates $(c\tau_a, x_a)$.

As can be seen from the first diagram, the surface of singular proper acceleration $x_a = 0$ is an outgoing light-like surface defining a *causal patch* in the space-time to the right of this surface. Events to the left of this surface cannot effect those to the right. The functional forms of the coordinate transformations are given by:

$$
\begin{aligned}
c\tilde{t} &= x_a \sinh\left(\frac{c\tau_a}{x_a}\right) & \tilde{x} &= x_a \cosh\left(\frac{c\tau_a}{x_a}\right) \\
c\tau_a &= \frac{1}{2}\sqrt{\tilde{x}^2 - (c\tilde{t})^2}\log\left(\frac{\tilde{x} + c\tilde{t}}{\tilde{x} - c\tilde{t}}\right) & x_a &= \sqrt{\tilde{x}^2 - (c\tilde{t})^2}.
\end{aligned}
\tag{1.36}
$$

Curves of fixed accelerating coordinate x_a originate on the $\tilde{x}$ axis with their corresponding coordinate $(c\tilde{t} = 0 = c\tau_a,\ \tilde{x} = x_a)$. Curves of fixed proper temporal coordinate $c\tau_a$, which are horizontal for large $\tilde{x}$ to the right, are graded in units of the same scale as the distance coordinates. The second diagram in Figure 1.5 plots curves of fixed distance coordinate $\tilde{x}$ (initially vertical), and curves of fixed inertial time $\tilde{ct}$, each graded in units of the same scale. The curves are plotted against axes of accelerating coordinates $(c\tau_a, x_a)$, which will later be shown to generate a non-orthogonal metric form.

To further explore the dynamics of length measurements in the accelerating system, the endpoints of a measuring rod whose trailing endpoint becomes inertial at $\tilde{t}_1$, and whose leading endpoint becomes inertial at $\tilde{t}_2$, is demonstrated in Figure 1.6. As before, the gray coordinate grid represents curves of fixed proper time, τ (horizontal on the far right), and curves of fixed coordinate, x_a (vertical on the $\tilde{x}$-axis at $\tilde{t} = 0 = \tau$). The measuring rod with endpoints located at

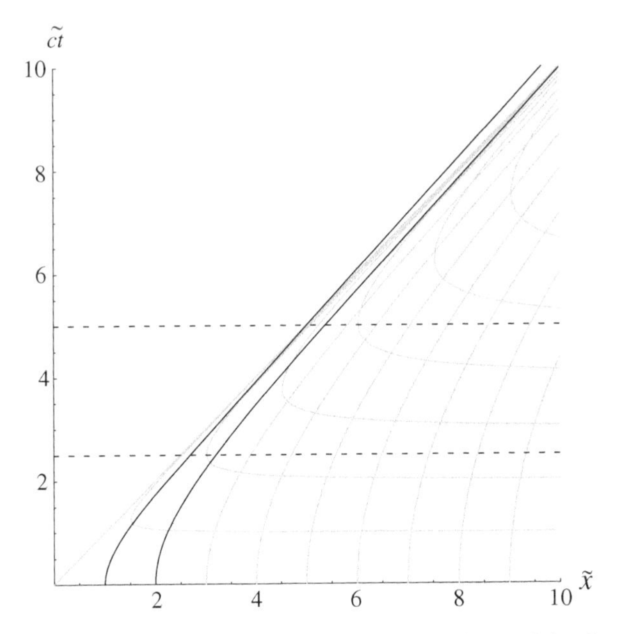

Figure 1.6 Endpoints of an initially acceleration measuring rod that later becomes inertial.

$x_a = 1$ and $x_a = 2$ has its endpoints demonstrated by the black accelerating trajectories. The trailing endpoint ends acceleration at the lower dashed horizontal line, while the leading endpoint ends acceleration at the upper dashed horizontal line. The space-time interval between the end of accelerations of the endpoints is space-like, $(\tilde{x}_{2f} - \tilde{x}_{1f})^2 - (c\tilde{t}_{2f} - c\tilde{t}_{1f})^2 = ((\frac{c}{\beta_f})^2 - 1)(c\tilde{t}_{2f} - c\tilde{t}_{1f})^2 > 0$. Therefore, the cessation of accelerations cannot be causal, and must be prearranged. Causal interactions *cannot* construct a truly rigid rod that maintains its proper length against all accelerations. Once all accelerations cease, the standard Lorentz contraction between the endpoints is subsequently maintained.

The geometry of the perpetually accelerating geometry will be further explored next. The Minkowski space-time interval can be re-expressed in terms of the accelerating coordinates. Excluding y, z displacements, a direct calculation gives:

$$-(dc\tilde{t})^2 + (d\tilde{x}^2) \equiv ds^2 = -(dc\tau_a)^2 + 2\frac{c\tau_a}{x_a}\, dc\tau_a\, dx_a + \left(1 - \left(\frac{c\tau_a}{x_a}\right)^2\right)(dx_a)^2. \tag{1.37}$$

The metric form is clearly non-orthogonal in the coordinates $(c\tau_a, x_a)$, since there is a mixed term of the form $dc\tau_a\, dx_a$. Non-orthogonal metrics often arise in dynamic systems, as will be explored extensively in Part II. In the discussions that follow, it will be useful to construct light-like trajectories in the space-time. In the Minkowski space-time of inertial special relativity, all outgoing light rays have a slope of $+1$, while all ingoing light rays have a slope of -1. Any coordinate system for which all light-like surfaces have slopes of ± 1 are called *conformal coordinates*. Generally, light-like trajectories have a null space-time interval $ds = 0$. Within the causal patch $x_a > 0$ null trajectories satisfy:

$$\frac{dx_{a\gamma}}{dc\tau_a} = \frac{1}{\frac{c\tau_a}{x_{a\gamma}} \pm 1}, \tag{1.38}$$

using the accelerating coordinates, and $\frac{d\tilde{x}_\gamma}{dc\tilde{t}} = \pm 1$ using the inertial coordinates.

Since the familiar Minkowski space-time metric is orthogonal in each of the space-time coordinates, much of the physical insight one has into the meaning of metric coordinates comes from diagonal forms of the metric. It is therefore useful to find a temporal coordinate dt_a that will orthogonalize the metric form 1.37. Such a transformation is given by:

$$dx_a = dx_p, \quad dc\tau_a = \frac{ct_a}{\alpha} dx_p + \frac{x_p}{\alpha} dct_a. \tag{1.39}$$

Analytic integrability of the new coordinate ct_a requires that α be a constant. The transformation will algebraically orthogonalize the metric 1.37 if:

$$ct_a = \alpha \frac{c\tau_a}{x_a}, \tag{1.40}$$

resulting in the *Rindler space* form of the metric:

$$ds^2 = -\left(\frac{x_p}{\alpha}\right)^2 (dct_a)^2 + dx_p^2 + dy^2 + dz^2. \tag{1.41}$$

The metric form 1.41 makes clear that the spatial coordinates $x_p = x_a$ in the accelerating metric 1.37 represents the proper distance coordinate (i.e., simultaneous spatial displacement measurements) in the accelerating frame of reference. As previously discussed, from the metric 1.37, τ_a represents the local ($dx_p = 0$) proper time in the accelerating system. The temporal coordinate t_a is given in Eq. 1.40. For completeness, fixed coordinate curves are respresented in Figure 1.7.

The null trajectories followed by x-moving outgoing/ingoing light rays satisfy:

$$\frac{dx_{p\gamma}}{dct_a} = \pm \frac{x_{p\gamma}}{\alpha}, \tag{1.42}$$

demonstrating the exponential form for light propagation in these coordinates:

$$x_{p\gamma} = x_o e^{\pm ct_a/\alpha}. \tag{1.43}$$

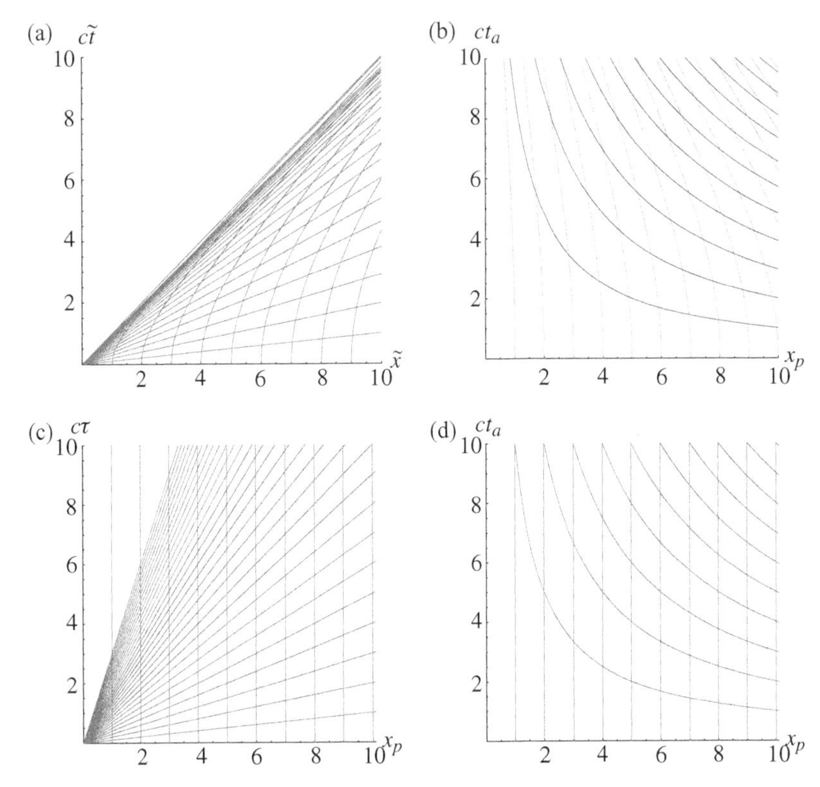

Figure 1.7 Fixed coordinate curves demonstrating transformation into Rindler space-time with $\alpha = 10$. Diagrams (a) and (c) demonstrate fixed t_a curves radially extending from the origins. For all diagrams, curves of constant spatial coordinates are parameterized by their correspondence values on the horizontal axes.

The surface with an infinite proper acceleration previously discussed at $x_p = 0$ represents an outgoing light-like surface which is the *horizon* in these coordinates, as will be discussed both shortly, as well as in Part II.

The temporal aspects of proper clocks will next be examined. Assume that an extended accelerating system that is sufficiently rigid and coordinated can be constructed. As previously discussed, by prior arrangement, the accelerating clocks are set to the time of the nearby inertial clocks when they are momentarily at rest. This means that it can be assumed that the standard clocks at both the rear and the front of the accelerating system read proper-time zero at the time of emission of the photon. Such an accelerating system is depicted in Figure 1.8. Spatially separated accelerating clocks that emit photons at an agreed moment of simultaneity are seen to not detect the companion photons simultaneously. The photon emitted from the front of the measuring rod is detected at the rear detector around the third frame of

Figure 1.8 Accelerating clocks and measuring rod. Left photon travels down from top/front, and right photon travels up from bottom/rear. The far-left frame corresponds to zero time.

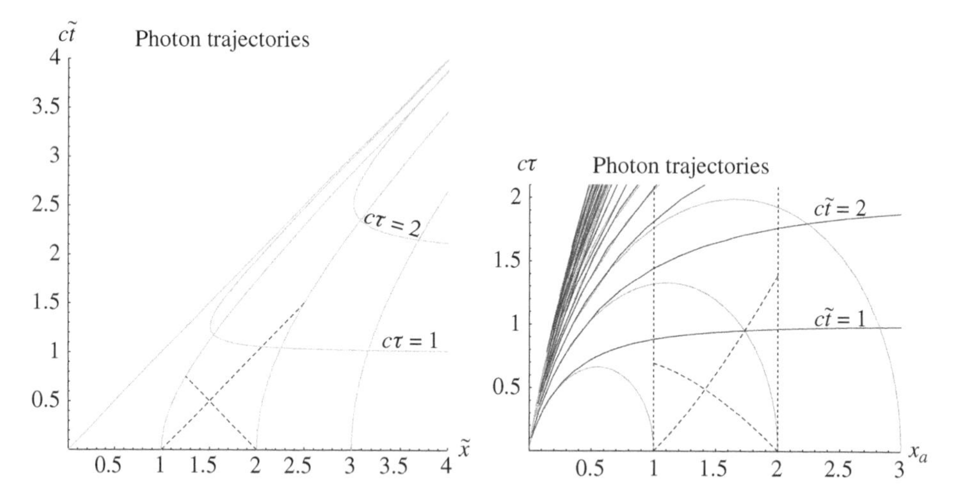

Figure 1.9 Photons emitted from each endpoint of a measuring rod, which propagate towards the opposite endpoint.

the sequence, whereas the photon emitted from the rear is detected at the front after the fifth frame. This also demonstrates why it is difficult to use special relativity to describe extended coordinates for an accelerating system (which motivates the approach of general relativity to describe local accelerations in terms of coordinate-dependent space-time metrics).

Depictions of the photons on space-time diagrams are given in Figure 1.9. The first diagram parameterizes the photon trajectories using inertial coordinate axes, while the second diagram uses accelerating coordinate axes. In the second diagram, the endpoints are depicted as vertical dotted lines. The photons are criss-crossing dashed curves originating on the $\tilde{ct} = 0 = c\tau$ surface. In both instances the photons arrive at the opposite endpoints at differing times, despite having been emitted simultaneously.

The time of interception of the outgoing photon with the front of the measuring rod can be shown to satisfy:

$$\frac{x_{aF}^2 - x_{aR}^2}{2x_{aR}} = ct_F = x_{aF} \sinh\left(\frac{c\tau_F}{x_{aF}}\right).$$

The forward moving photon therefore satisfies:

$$c\tau_F = x_{aF} \log\left(\frac{x_{aF}}{x_{aR}}\right). \tag{1.44}$$

Similarly, the rearward moving photon satisfies:

$$c\tau_R = x_{aR} \log\left(\frac{x_{aF}}{x_{aR}}\right). \tag{1.45}$$

It is crucial to recognize that these proper times represent the *locally* measured times, *not* the time on the separated clock. The parameter τ is the measured time on a local fixed standard clock, whereas t is a mathematical parameter useful in describing the global space-time.

The clock at the front of the measuring rod using Eq. 1.44 can be rewritten in terms of the proper accelerations:

$$\tau_{Front} = -\frac{c}{a_F} \log\left(1 - \frac{a_F \Delta x_a}{c^2}\right) \tag{1.46}$$

at the time of detection of the photon. This formula demonstrates a few points of interest. First, one can determine that in terms of the proper length and time measurements of the observer in the front of the spaceship, the light does not seem to travel at c. The average speed calculated by the front observer is given by $\frac{\Delta x_a}{\tau_F}$, which is always less than c. For small accelerations, this is given by:

$$\bar{v}_{light} \approx \left(1 - \frac{a_F \Delta x_a}{2c^2}\right) c.$$

Furthermore, if the proper distance from the photon emission is greater than:

$$\Delta x_{max} \equiv \frac{c^2}{a_F} = x_{aF},$$

then the photon will *never* reach the receiver. This means that no signal sent from a coordinate $x_a < 0$ will ever be observed by the perpetually accelerating observer. This defines a *causal horizon* at $x_a = 0$ for the observer, beyond which nothing in that region of the universe can affect that observer (assuming of course that the observer continues to accelerate forever). Beyond the horizon, the universe appears *black* to this observer.

This horizon can be seen to represent a breakdown in coordinates, but not a breakdown in the local physics at the horizon. The Minkowski space-time form of the metric within which the accelerations are occurring clearly has no singular points, since it represents the flat space-time of inertial special relativity. Therefore, there can be no singular physical surface within the geometry. However, the metric form 1.37 has a coordinate singularity at $x_a = 0$. Likewise, the metric form 1.41 displays no temporal progression at $x_p = 0$.

Similarly the rearward moving photon reaches the detector in the rear at proper local time:

$$\tau_{first\,photon}^{rear} = \frac{c}{a_R} \log\left(1 + \frac{a_R \Delta x_a}{c^2}\right),$$

which, means that the rear observer averages the speed of the photon (for small accelerations) to be:

$$\bar{v}_{light} \approx \left(1 + \frac{aL}{2c^2}\right) c.$$

These results are qualitatively demonstrated in Figure 1.8.

Next, a clock will be developed for measuring proper time in the accelerating system. The clock will consists of a photon emitter at x_{po} that emits in the forward direction, and a mirror located at $x_{po} + L$ that reflects the photon back to the emitter. Whenever a photon returns, this will be considered to be a tick of the clock, and a new photon will be emitted in the forward direction to initiate the next tick. A single tick of such a construction is represented in Figure 1.10. For the present, this relativistic construction must be macroscopic. There should be a sufficient number of interrelations such that any quantum uncertainties associated with well-defined spatial and temporal separations, as well as delays associated with emissions and reflections, are averaged out to be negligible. It is also important to note from the diagram that the clock ceases to tick if its emitter is placed on the horizon $x_{po} = 0$. Since the horizon itself is a light-like surface, such forward-moving photons never reach the mirror.

The kinematics of actual ticks of this clock as viewed by an inertial observer, as well as an accelerating observer at the rear of the accelerating system, are demonstrated on the space-time diagram in Figure 1.11. Since all measurements are made at $x_p = x_{po}$, these devices can be mass produced and placed at arbitrary positions as standard clocks. Again, the photons travel zero space-time interval, satisfying $\frac{dx_p}{x_p} = \pm \frac{dct_a}{\alpha}$. For the trip out to the mirror, the accelerating time coordinate difference is given by $\frac{c(t_{mirror} - t_{ao})}{\alpha} = \log(\frac{x_{po}+L}{x_{po}})$, while for the trip back to the photon emitter it is given by $\frac{c(t_{af} - t_{mirror})}{\alpha} = -\log(\frac{x_{po}}{x_{po}+L})$. The proper-time interval at the fixed point $x_p = x_{po}$ can be determined from the metric form 1.41 to satisfy $d\tau = \frac{x_p}{\alpha} dt_a$.

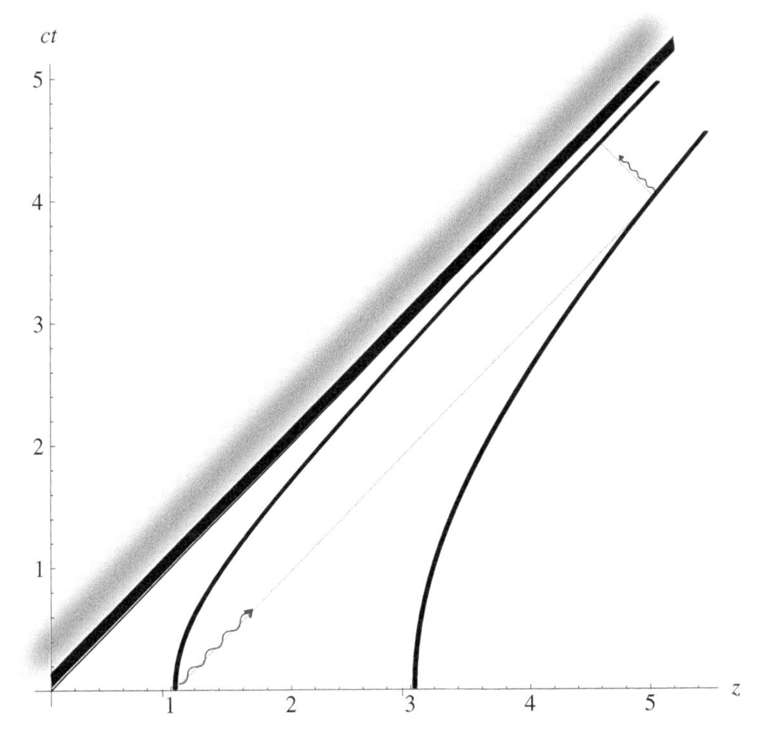

Figure 1.10 Construction of a standard clock using reflected photons.

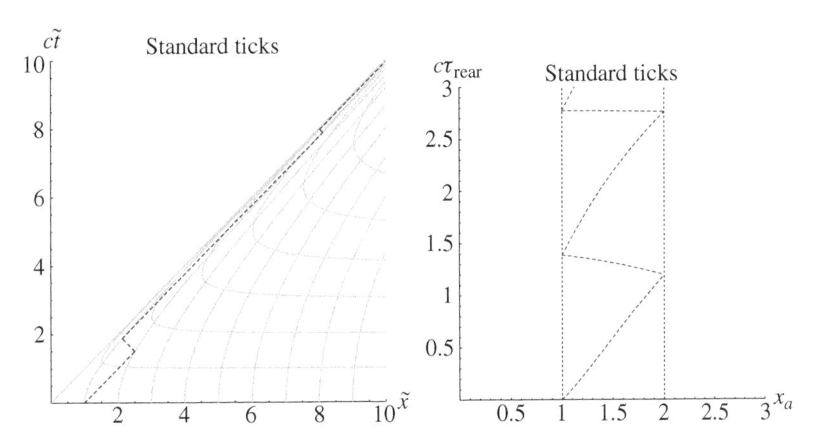

Figure 1.11 Photonic round-trip clock ticks.

Therefore, for the round-trip, the proper-time interval corresponding to a single tick of this clock satisfies:

$$\tau_f - \tau_o = \frac{x_{po}}{\alpha}\,(t_{af} - t_{ao}) = 2\,\frac{x_{po}}{c}\,\log\left(1 + \frac{L}{x_{po}}\right). \tag{1.47}$$

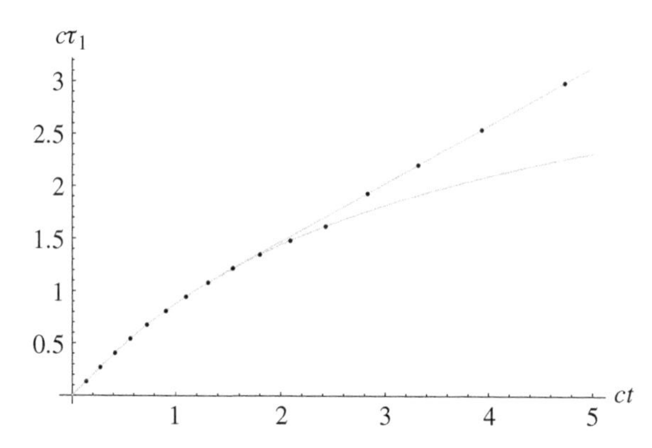

Figure 1.12 Proper ticks of accelerating standard clock that becomes inertial. Lower curve continues uniform acceleration.

This clock can be made as compact as one desires by making L small. The diagram on the right of Figure 1.11 shows the intervals between ticks are indeed uniform, despite apparent alterations in the photon trajectories as parameterized using this observer's proper time.

In this accelerating frame of reference, it is clear the rate that the clock ticks explicitly depends on the location x_{po}. One of the key assumptions of general relativity is the equivalence of an acceleration in the forward direction as an artificial gravitational field oriented opposite the acceleration (more is said about this *principle of equivalence* in Chapter 5 on general relativity in Part II). Although a similar analysis in the uniformly moving (inertial) system gives tick rates independent of position, this analysis shows that in general the temporal tick rate is relative to the speed *and* the position of the observer for accelerating and gravitating systems. A single Lorentz transformation from an inertial system that brings spatially separated accelerating clocks to rest can only be found at $t = 0 = \tau$. One should also note that the transition from accelerating to inertial motion for the extended measuring rod previously discussed is space-like, and thus affects, at most, two ticks of this standard clock. Ticks of an accelerating clock that becomes inertial are demonstrated in Figure 1.12. The tick intervals of such a standard clock that occur *during* any change in the acceleration will be slightly anomalous, but ticks during periods of steady acceleration (or inertial motion) will be uniformly spaced.

1.2.2 Coordinate transformations and proper accelerations

General space-time metrics

The previous section demonstrated the usefulness of using coordinate transformations to construct convenient representations that parameterize the locations of events in a given space-time. Given the inertial Minkowski coordinates, $\vec{\xi} \equiv (c\tilde{t}, \tilde{x}, \tilde{y}, \tilde{z})$, the invariant metric form generates a locally varying tensor **g** satisfying:

$$ds^2 = d\xi^\mu \, \eta_{\mu\nu} \, d\xi^\nu = dx^\alpha \frac{\partial \xi^\mu}{\partial x^\alpha} \eta_{\mu\nu} \frac{\partial \xi^\nu}{\partial x^\beta} dx^\beta \equiv dx^\alpha g_{\alpha\beta}(\vec{x}) dx^\beta$$
$$\text{where } g_{\alpha\beta}(\vec{x}) \equiv \frac{\partial \xi^\mu}{\partial x^\alpha} \eta_{\mu\nu} \frac{\partial \xi^\nu}{\partial x^\beta}. \tag{1.48}$$

This space-time metric will be quite useful in describing events in general curvilinear coordinates in Part II.

General integrals over forms $d\xi^0 \, d\xi^1 \, d\xi^2 \, d\xi^3 = \left| \frac{\partial(\xi^0 \xi^1 \xi^2 \xi^3)}{\partial(x^0 x^1 x^2 x^3)} \right| dx^0 dx^1 dx^2 dx^3$ define the standard *Jacobian* transformation determinant for variable changes on multi-variable surfaces. The Jacobian form can be related to the general space-time metric by taking the determinant of Eq. 1.48, noting that the determinant of a product matrix is the product of the determinants:

$$\det \, \mathbf{g} = \left| \frac{\partial \vec{\xi}}{\partial \vec{x}} \right| \det \, \boldsymbol{\eta} \left| \frac{\partial \vec{\xi}}{\partial \vec{x}} \right|. \tag{1.49}$$

Thus, the Jacobian satisfies:

$$\left| \frac{\partial \vec{\xi}}{\partial \vec{x}} \right| = \sqrt{- \det \, \mathbf{g}}. \tag{1.50}$$

The determinant of the metric is often written as $g \equiv det \, \mathbf{g}$, and thus the Jacobian form for variable transformations in integrals over space-time is just $\sqrt{-g}$.

General proper accelerations

An acceleration described using local Minkowski coordinates momentarily at rest with a system can also be described in terms of the general curvilinear coordinates. The velocities and accelerations are related via the chain rule,

$$\frac{d\xi^\lambda}{d\tau} = \frac{\partial \xi^\lambda}{\partial x^\beta} \frac{dx^\beta}{d\tau},$$
$$\frac{d^2\xi^\lambda}{d\tau^2} = \frac{\partial \xi^\lambda}{\partial x^\beta} \frac{d^2 x^\beta}{d\tau^2} + \frac{\partial^2 \xi^\lambda}{\partial x^\alpha \partial x^\beta} \frac{dx^\alpha}{d\tau} \frac{dx^\beta}{d\tau}, \tag{1.51}$$

where τ represents the proper time of the accelerating system. Thus, a proper acceleration from an instantaneously co-moving Minkowski frame of reference is described in terms of the general coordinates using:

$$\frac{d^2\xi^\lambda}{d\tau^2} = \frac{\partial \xi^\lambda}{\partial x^\beta} \left(\frac{d^2 x^\beta}{d\tau^2} + \Gamma^\beta_{\mu\nu} \frac{dx^\mu}{d\tau} \frac{dx^\nu}{d\tau} \right), \tag{1.52}$$

where the *connections* are defined by,

$$\Gamma^{\beta}_{\mu\nu} \equiv \frac{\partial x^{\beta}}{\partial \xi^{\lambda}} \frac{\partial^2 \xi^{\lambda}}{\partial x^{\mu} \partial x^{\nu}}. \tag{1.53}$$

These connections are also sometimes referred to as *Christoffel symbols*.

Substituting the coordinates of the uniformly accelerating system in Eq. 1.34, the space-like four-vector defined by $\tilde{a}^{\beta} \equiv \frac{d^2 \tilde{x}^{\beta}}{d\tau_a^2}$ gives the invariant proper acceleration $\tilde{a}^{\alpha} \eta_{\alpha\beta} \tilde{a}^{\beta} = a^2$. The first factor on the right hand side of Eq. 1.52 simply transforms this four-vector from the general curvilinear coordinate system to that inertial co-moving system. Thus, the four-acceleration is usually defined as:

$$A^{\beta} \equiv \frac{d^2 x^{\beta}}{d\tau^2} + \Gamma^{\beta}_{\mu\nu} \frac{dx^{\mu}}{d\tau} \frac{dx^{\nu}}{d\tau}. \tag{1.54}$$

When the proper acceleration of a system vanishes, the relation $A^{\beta} = 0$ defines what will later be referred to as the *geodesic equation*, and the system undergoes geodesic motion in the space-time.

1.3 Canonical proper-time dynamics

The canonical proper-time formulation [4] of relativistic dynamics provides a framework from which one can describe the dynamics of classical and quantum systems using the clock of those very systems. This formulation has at its core philosophy a presumption that the effects of external interactions upon a given system should be expressible in fundamental form when parameterized in terms of the proper time of that system. The framework utilizes a canonical transformation on the inertial time variable that is used to describe the dynamics, but does not transform other dynamical variables such as momenta or positions. This means that the time scales of the dynamics are described in terms of the natural local time coordinate, which is the most meaningful parameterization of the dynamics of the interacting particle or system itself, while other physical parameters can be expressed in forms which are most convenient for calculations. This type of formulation should give insights into the fundamental form of an interaction, since the response and back-response of an interacting system is most likely defined by its proper-scaled dynamics.

1.3.1 Proper-time Hamilton's equations

From classical physics, Hamilton's equations describe the dynamics of an observable $W(q, p, t)$ in terms of the Poisson bracket of that observable with the Hamiltonian:

$$\frac{dW(q,p,t)}{dt} = \{H, W(q,p,t)\} + \left(\frac{\partial W}{\partial t}\right)_{q,p}, \tag{1.55}$$

where $\{A, B\} \equiv \frac{\partial A}{\partial p}\frac{\partial B}{\partial q} - \frac{\partial A}{\partial q}\frac{\partial B}{\partial p}$. In particular, the temporal dynamics of the generalized coordinate and momentum satisfies:

$$\frac{dq}{dt} = \frac{\partial H}{\partial p}, \quad \frac{dp}{dt} = -\frac{\partial H}{\partial q}. \tag{1.56}$$

The inertial time, t is related to the proper time, τ through the standard Lorentz factor γ using:

$$dt = \gamma\, d\tau \doteq \frac{H}{Mc^2} d\tau. \tag{1.57}$$

The second form in Eq. 1.57 follows from the relationship between the energy of a system compared to its rest energy. The canonical proper-energy form K is defined to generate dynamical changes with regards to the *proper time* of that system:

$$\frac{dW}{d\tau} = \frac{dW}{dt}\frac{dt}{d\tau} \equiv \{K, W\} + \left(\frac{\partial W}{\partial \tau}\right)_{q,p}. \tag{1.58}$$

From this equation, along with Hamilton's equation, it then follows that $\{K, W\} = \frac{H}{Mc^2}\{H, W\}$. The dynamics of the coordinates from Eqs. 1.56 then satisfy:

$$\frac{dq}{d\tau} = \frac{\partial K}{\partial p}, \quad \frac{dp}{d\tau} = -\frac{\partial K}{\partial q}. \tag{1.59}$$

Thus the canonical proper energy form generates the dynamics of the system in terms of the proper time of that system.

The proper-energy form K is expected to correspond to the Hamiltonian when the Hamiltonian itself corresponds to the rest energy, $K|_{H=Mc^2} = H = Mc^2$. Holding the system mass M fixed during the "boost" from H to K results in the form:

$$K[H] = \frac{H^2}{2Mc^2} + \frac{Mc^2}{2}. \tag{1.60}$$

A direct substitution of the relativistic form $H = \sqrt{(pc)^2 + (Mc^2)^2}$ into this equation yields:

$$K = \frac{p^2}{2M} + Mc^2. \tag{1.61}$$

In this non-interacting case, both systems are inertial. There are a few points of interest in this example:

- The equation for K is the same as that of a non-relativistic free particle, despite the system being completely relativistic;
- The momentum in K is the momentum of the particle as parameterized by the external observer, not the proper momentum (which always vanishes). This is clearly *not* a Lorentz transformation of the dynamical parameters of the system;
- The sometimes troublesome square root does not appear in the expression for K.

Usefulness in quantum dynamics

Proper-time quantum dynamics can be formulated using standard correspondence techniques connecting classical dynamics to quantum behaviors: $\{H, W\} \Rightarrow [H, W]/i\hbar$. The temporal evolution of (Heisenberg representation) physical parameters is then expressed in terms of the commutator $i\hbar \frac{dW}{dt} = [H, W]$. The parameter K conjugate to the proper time then generates infinitesimal proper-time translations $i\hbar \frac{dW}{d\tau} == [K, W]$. Eigenstates and eigenvalues of K describe the spectrum of states of the system at a given time as relevant to the internal dynamics of that system.

Usefulness in gravitational dynamics

The canonical proper-time formulation is quite convenient for examining fiducial (fixed coordinate, thus generally accelerating) systems in a gravitational field, since the internal dynamics is parameterized by the proper time, τ of that system, while the coordinates x^k, p_k remain in terms of a convenient external (often inertial or distant) observer. One expects that the time scale of the dynamics of a local gravitating system is clearly most directly expressed in terms of that system's proper time, rather than the temporal coordinates of some external observer. For instance, a laboratory on the surface of the Earth is fiducial and accelerating, in that it is held at a fixed radial distance from the center of the Earth. As will be explored in Part II, a direct coordinate transformation is useful for implicitly incorporating proper-time dynamics through the system four-velocity via:

$$\frac{dW}{d\tau} = \frac{dx^\beta}{d\tau} \frac{\partial}{\partial x^\beta} W \equiv u^\beta \partial_\beta W, \tag{1.62}$$

where the system four-velocity is defined by $u^\beta \equiv \frac{dx^\beta}{d\tau}$.

1.3.2 Proper-time interactions

Next, the inclusion of interactions in a canonical proper-time formulation will be explored. This will be done by examining the dynamics of a few relevant interacting systems.

Uniform proper acceleration

The solutions for a system with a constant proper acceleration given in Section 1.2.1 provide functional forms that can be used to determine the appropriate canonical proper-energy form. From Eq. 1.34

$$\frac{dq}{d\tau} = c \sinh\left(\frac{a\tau}{c}\right) = \frac{\partial K}{\partial p}, \quad \frac{dp}{d\tau} = m\,a\,\cosh\left(\frac{a\tau}{c}\right) = -\frac{\partial K}{\partial q}. \tag{1.63}$$

In Eqs. 1.63, the momentum is assumed to take the standard relativistic form proportional to four-velocity $\mathbf{p} = m\mathbf{u} = m\frac{dq}{d\tau}$. The form of K describing this dynamics is given by:

$$K(q, p) = \frac{p^2}{2m} - ma\left(\frac{a}{2c^2}q^2 + q\right) + mc^2. \tag{1.64}$$

For completeness, the Hamiltonian forms will also be demonstrated. Hamilton's equations give,

$$\frac{dq}{dt} = \frac{pc}{\sqrt{(mc)^2 + p^2}} = \frac{\partial H}{\partial p}, \quad \frac{dp}{dt} = m\,a = -\frac{\partial H}{\partial q}, \tag{1.65}$$

which are generated by a Hamiltonian of the form:

$$H = \sqrt{(mc^2)^2 + (pc)^2} - maq. \tag{1.66}$$

There is no obvious relationship between the canonical proper-energy form and Hamiltonian form for this interacting system.

Proper-time harmonic oscillator

A simple harmonic oscillator undergoes the following motion:

$$q(\tau) = q_o \cos\omega\tau, \quad \frac{dq}{d\tau} = -q_o\omega \sin\omega\tau \equiv \frac{p}{m}. \tag{1.67}$$

The canonical Hamilton's equations are given by:

$$\frac{dq}{d\tau} = \frac{p}{m} = \frac{\partial K}{\partial p}, \quad \frac{dp}{d\tau} = -m\omega^2 q = -\frac{\partial K}{\partial q}, \tag{1.68}$$

which defines a canonical proper-energy form given by:

$$K = \frac{p^2}{2m} + \frac{1}{2}m\omega^2 q^2 + mc^2. \tag{1.69}$$

This form is completely relativistic, despite its appearance. The Hamilton's equations for this system are given by:

$$\frac{dq}{dt} = \frac{pc}{\sqrt{(mc)^2 + p^2}} = \frac{\partial H}{\partial p}, \quad \frac{dp}{dt} = \frac{m\omega^2 q}{\sqrt{1 + (\frac{\omega}{c})^2(q_o^2 - q^2)}} = -\frac{\partial H}{\partial q}, \tag{1.70}$$

which gives a Hamiltonian of the form:

$$H = \sqrt{(mc^2)^2 + (pc)^2} - mc^2 \left(\sqrt{1 + \left(\frac{\omega}{c}\right)^2 (q_o^2 - q^2)} - \sqrt{1 + \left(\frac{\omega}{c}\right)^2 q_o^2} \right). \tag{1.71}$$

Because of the relativistic Lorentz factors involved, the interaction form in the Hamiltonian representation is considerably more complicated for this system.

Accelerating harmonic oscillator

Next, consider the motion generated by combining the uniform acceleration interaction form with the harmonic oscillator interaction:

$$K = \frac{p^2}{2m} + \frac{1}{2} m\omega^2 q^2 - ma\left(\frac{a}{2c^2} q^2 + q\right) + mc^2. \tag{1.72}$$

By completing the square, the proper-energy form can be rewritten as:

$$K = \frac{p^2}{2m} + \frac{1}{2} m \left(\omega^2 - \left(\frac{a}{c}\right)^2\right) \left(q - \frac{a}{\omega^2 - \left(\frac{a}{c}\right)^2}\right)^2 + m\left(c^2 - \frac{a^2}{2\left(\omega^2 - \left(\frac{a}{c}\right)^2\right)}\right). \tag{1.73}$$

This second form elucidates a few characteristics of this system. Unless the oscillator has sufficient "stiffness", with $\omega^2 > \left(\frac{a}{c}\right)^2$, it will not oscillate. Thus, the *effective* stiffness of the oscillator is decreased by the acceleration a. Likewise, if $\omega^2 < \left(\frac{a}{c}\right)^2$, the system undergoes uniform acceleration with a decreased effective acceleration.

The solutions are given by:

$$\begin{aligned}
\omega &> \left|\frac{a}{c}\right|, \quad q(\tau) = \frac{a}{\omega^2 - \left(\frac{a}{c}\right)^2}\left(1 - \cos\left(\sqrt{\omega^2 - \left(\frac{a}{c}\right)^2}\,\tau\right)\right) \\
\omega &< \left|\frac{a}{c}\right|, \quad q(\tau) = \frac{a}{\omega^2 - \left(\frac{a}{c}\right)^2}\left(1 - \cosh\left(\sqrt{\left(\frac{a}{c}\right)^2 - \omega^2}\,\tau\right)\right) \\
\omega &= \left|\frac{a}{c}\right|, \quad q(\tau) = \frac{1}{2} a\tau^2.
\end{aligned} \tag{1.74}$$

The ground state of the oscillator, as well as its frequency, are clearly modified by the acceleration.

Proper-time gravitation

Consider a canonical proper-energy form for a gravitating mass m given by:

$$K = \frac{\mathbf{p}^2}{2m} - m\frac{G_N M}{|\mathbf{r}|} + mc^2. \tag{1.75}$$

The equations of motion for this mass are:

$$\frac{\partial K}{\partial p_k} = \frac{p^k}{m} = \dot{r}^k, \quad \frac{\partial K}{\partial r^k} = m\frac{G_N M}{|\mathbf{r}|^2}\hat{r}_k = -\dot{p}_k. \tag{1.76}$$

It should be emphasized that the time derivatives in these expressions are derivatives with respect to the proper time, τ of the gravitating mass, *not* the Newtonian time, t. Thus, despite their appearance, these are relativistic expressions.

Electrodynamics and minimal coupling

The Lorentz force equation describes how charges respond to external fields. The mechanical three-momentum $\gamma m \mathbf{v}^j = m u^j$ and mechanical power $\gamma m c^2 = m c u^0$ satisfy:

$$m\frac{d\mathbf{u}}{dt} = q\left(\mathbf{E} + \frac{\mathbf{v}}{c} \times \mathbf{B}\right), \quad m\frac{du^0}{dt} = \frac{q}{c}\mathbf{v}\cdot\mathbf{E}. \tag{1.77}$$

Multiplying by the Lorentz factor of the charge u^0, this equation can be re-expressed as:

$$m\frac{du_\beta}{d\tau} = \frac{q}{c}F_{\beta\lambda}\, u^\lambda = \frac{q}{c}F_{\beta\lambda}\, \dot{r}^\lambda, \tag{1.78}$$

where the dot represents derivatives with respect to proper time, τ. The proper-time canonical equations are developed by defining the *canonical momentum* $\vec{p}$ using minimal coupling to the gauge fields:

$$m\dot{r}^\beta = m u^\beta \equiv p^\beta - \frac{q}{c}A^\beta. \tag{1.79}$$

By adding the term $\frac{q}{c}\frac{d}{d\tau}A^\beta = u^\lambda \partial_\lambda A^\beta$ to Eq. 1.78, canonical forms for equations describing the dynamics are obtained:

$$\begin{aligned} \frac{\partial K}{\partial p_\beta} &= \dot{r}^\beta = \frac{p^\beta - \frac{q}{c}A^\beta}{m} \\ -\frac{\partial K}{\partial r^\beta} &= \dot{p}_\beta = \frac{q}{c}\left(F_{\beta\lambda} + \partial_\lambda A_\beta\right)\left(\frac{p^\lambda - \frac{q}{c}A^\lambda}{m}\right). \end{aligned} \tag{1.80}$$

Minimal coupling ensures that the integrability condition is satisfied:

$$\frac{\partial^2 K}{\partial p^\beta \partial r^\lambda} = \frac{\partial^2 K}{\partial r^\lambda \partial p^\beta} \quad \Rightarrow \quad F_{\lambda\beta} = \partial_\lambda A_\beta - \partial_\beta A_\lambda, \tag{1.81}$$

which is a form that automatically satisfies the homogeneous Maxwell equations 1.29. Thus, the Eqs. 1.80 can be directly integrated to obtain the canonical proper-energy form:

$$K = \frac{\left(p^\beta - \frac{q}{c}A^\beta\right)\eta_{\beta\lambda}\left(p^\lambda - \frac{q}{c}A^\lambda\right)}{2m} + \frac{3}{2}mc^2, \tag{1.82}$$

where $\eta_{\beta\lambda}$ is the Minkowski metric, and K takes the value mc^2 when three-velocity vanishes, $\dot{r}^j = 0$ in the absence of external fields.

As a brief summary of the advantages of the canonical proper-energy form:

- The form generates the dynamics parameterized by the proper time of a system, which is a Lorentz invariant. However, its form need not be manifestly invariant, since canonical positions and momenta are parameterized by an inertial observer;
- The form allows the use of non-relativistic techniques to solve fully relativistic problems;
- The description of the dynamics of the reaction of a system to external influences is straightforward.

The formulation is particularly useful for describing electrodynamics in terms of the proper time of the impacted charge itself.

1.4 Conformal space-time diagrams

Diagrammatic techniques are often very useful for visualizing complicated physical concepts. Graphical techniques in perturbative quantum mechanics have guided the understanding of such subtle phenomena as anomalous magnetic moments and electronic band structures. Likewise, space-time diagrams can enhance one's intuitive feel for processes in geometrodynamics. In this section, the utility of such diagrams will be briefly demonstrated.

1.4.1 Traditional space-time diagrams

The space-time diagrams previously developed demonstrate the traditional use of the temporal coordinate ct (in units of length) as the label of the vertical axis, with Cartesian or radial distance labeling the horizontal axis. Minkowski space-time was developed such that light beams maintain a speed c as measured by any inertial observer, satisfying $\mathbf{r} = \mathbf{r}_o + ct\,\hat{\mathbf{k}}$. Therefore, an outgoing light beam has a slope of $+1$ on the Minkowski space-time diagram, while an ingoing light beam has a slope of -1. Space-time coordinates for which this is true are generally known as *conformal coordinates*.

As an example, consider the observers depicted in Figure 1.13. In the diagram, an observer that remains at the origin emits an outgoing light beam at $t = 0$. Since this observer remains at the origin, he/she moves in a time-like (vertical) manner,

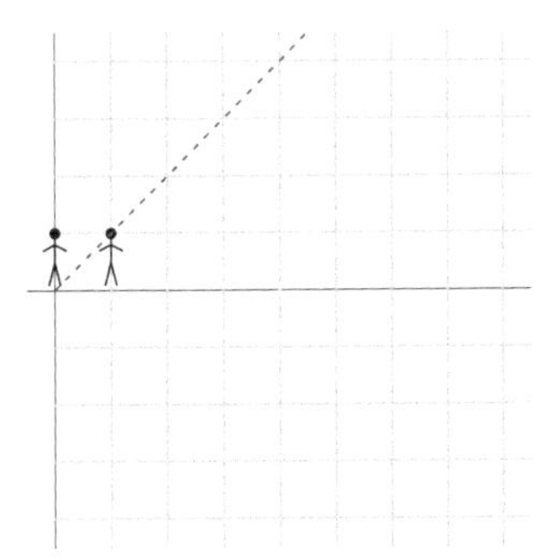

Figure 1.13 Space-time diagram of flat Minkowski geometry.

remaining on the time axis, $r = 0$. Any past or future event that might *possibly* be coincident with this observer (involving arbitrary motions) is defined to have *time-like* separation from this observer. The outgoing light beam is represented by the dashed line with slope $+1$. If an event at the origin is to have causal impact upon any future event, the communication cannot have propagated at a speed greater than that of light. The *future light cone* of an event at the origin consists of all points on an outgoing spherical surface centered at the origin propagating at the speed of light. Points on the light cone have *light-like* separations. Only future events with time-like or light-like separations can have been causally affected by the event at the origin. Such events must lie upon or within the future light cone and have unit slope or greater when connected to the event at the origin. The classical trajectory of a massless particle is *always* a line with slope ± 1, while the classical trajectory of any massive system is a time-like curve whose slope cannot anywhere have a value of $\pm$unity. Using similar reasoning, any past event that could have affected the event at the origin must lie upon or within the *past light cone* of the origin, which is an ingoing light-like trajectory. Thus, a conformal diagram is quite useful for establishing causal relationships on the space-time.

The second observer in Figure 1.13 located at unit longitudinal distance from the origin at time $t = 0$, has *space-like* separation from the event at the origin. Only at the future time $ct = 1$ unit will the origin enter the past light cone of this observer, at which time the light beam can be detected. The event at the origin can have *no* effect on this second observer prior to that time. However, as will be discussed in the next chapter, there can be space-like coherence between quantum processes detected at the two locations. Generally, quantum boundary conditions are coherent constraints placed upon the system on an acausal surface. However, these constraints cannot allow *communications* at a speed greater than that of light. The conditions constrain only correlations consistent with dynamic conservation principles upon any space-like separated measurements on the quantum coherent system.

Formally, it should be noted that events with time-like separation have a negative metric interval $ds^2 < 0$, while space-like separated events have positive interval $ds^2 > 0$. Light-like events always fall on surfaces with the null interval $ds^2 = 0$. Since the interval is invariant, these relationships are preserved under arbitrary coordinate transformations. Also, the diagrams developed in this section will always be two-dimensional. The vertical axis will always be temporal, but the horizontal axis has been represented as an arbitrary longitudinal spatial direction. In previous sections, this direction has been chosen to be a single Cartesian direction. However, it can likewise be a radial coordinate. Generally, a single point on a space-time diagram represents a fixed-time surface area transverse to the longitudinal coordinate being displayed.

1.4.2 Penrose diagrams

One of the advantages of a traditional space-time diagram is the intuitive representation of distance, time, and scale given in terms of separation of events represented on the diagram. In Figure 1.13, the two spatially separate observers of equal height are represented with equivalent scale on the diagram. However, arbitrarily distant observers cannot be displayed on a finite traditional space-time diagram. One can map all of space-time onto a finite region by performing a coordinate transformation that maps infinity onto a finite number, and if the transformation is conformal, light rays continue to be represented by lines with $\pm$ unit slope. A Penrose diagram of a given geometry is a two-dimensional space-time diagram that has the following properties:

- The coordinates are conformal, i.e., outgoing light-like surfaces have slope $+1$, and ingoing light-like surfaces have slope -1;
- The range and domain of the diagram are compact, i.e., all of space-time are represented in a finite region.

Thus, Penrose diagrams are compact conformal space-time diagrams. Any function can be used to compactify conformal coordinates once they are found. For example, in treatments in this manuscript, conformal coordinates (ct_*, r_*) will be transformed into compact coordinates using:

$$\begin{aligned} Y_{\rightarrow} &= \left[\tanh\left(\frac{ct_* + r_*}{scale}\right) - \tanh\left(\frac{ct_* - r_*}{scale}\right)\right] \Big/ 2 \\ Y_{\uparrow} &= \left[\tanh\left(\frac{ct_* + r_*}{scale}\right) + \tanh\left(\frac{ct_* - r_*}{scale}\right)\right] \Big/ 2. \end{aligned} \tag{1.83}$$

Here, $Y_{\rightarrow}$ labels the horizontal axis, while $Y_{\uparrow}$ labels the vertical axes, and *scale* is an arbitrary (length) scale.

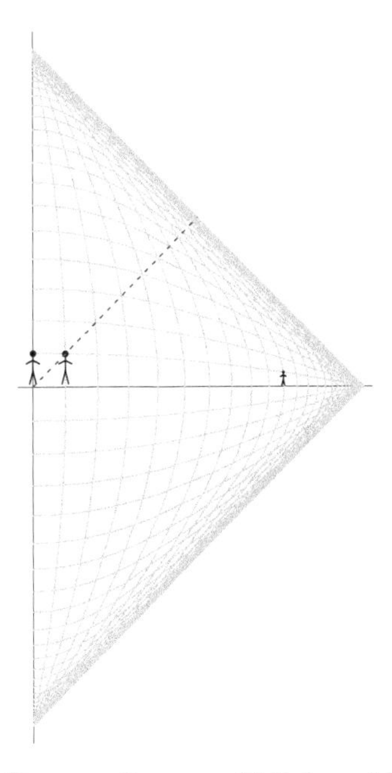

Figure 1.14 Penrose diagram of Minkowski space-time.

The compact conformal space-time diagram representing Figure 1.13 is displayed in Figure 1.14. The advantages of this type of diagram are apparent. Since the coordinates are conformal, causal relationships can be immediately ascertained. The outgoing light beam preserves its slope of $+1$. In addition, the observer at (0, 10 units) is represented on this diagram but not on the traditional diagram. Using the given convention to generate the compact coordinates, an event with longitudinal distance zero in the distant past has coordinates $(ct, r) = (-\infty, 0)$, $(Y_\rightarrow, Y_\uparrow) = (0, -1)$. If an outgoing light beam is emitted from this point, it forms the lower right boundary of the Minkowski space-time Penrose diagram giving past light-like infinity, which is traditionally referred to as *skri minus* ($\mathcal{I}^-$). Skri minus is the originating surface of all ingoing light-like trajectories on this conformal diagram. This outgoing light-like surface reaches $t = 0$ at longitudinal coordinate infinity, having coordinates $(ct, r) = (0, \infty)$, $(Y_\rightarrow, Y_\uparrow) = (1, 0)$. An ingoing light beam emitted from this point forms the upper-right boundary of this diagram, giving future light-like infinity, traditionally referred to as *skri plus* ($\mathcal{I}^+$). Skri plus is the terminating surface of all outgoing light-like trajectories on this conformal diagram. This surface reaches the "center" with vanishing longitudinal coordinates after infinite time, at the coordinates $(ct, r) = (+\infty, 0)$, $(Y_\rightarrow, Y_\uparrow) = (0, +1)$. The left boundary of the diagram represents the longitudinal center of spatial coordinates $r = 0$, $Y_\rightarrow = 0$.

One should note that the intuitive scale of distance is *not* preserved in the compact conformal diagram, as demonstrated by the sizes of the spatially separated observers of identical height represented on the diagram. Distant coordinates are compressed into regions near the boundaries of the diagram, distorting relative scale. However, all observers are indeed represented on the diagram. Therefore, these diagrams display the global causal structure of the given space-time. They will turn out to be quite useful for exhibiting dynamics in geometries with horizons, which in the case of black holes, have regions within which outgoing light beams *do not* terminate on skri plus. Such geometries will be explored extensively in Part II of this book.

The next chapter will examine the foundations of the quantum mechanics of relativistic systems. Quantum measurement constraints will considerably effect interpretations of particle trajectories and momenta, and introduce new constraints upon the types of quantum systems that can satisfy causal relationships.

2

Quantum mechanics, classical, and special relativity

2.1 Fundamentals of quantum mechanics

Quantum mechanics remains one of the most successful, yet enigmatic, formulations of physics. Uncertainty and measurement constraints are incorporated at the core of this fundamental description of micro-physical dynamics. Quantum mechanics successfully describes the observed structures in chemistry and materials science. In particular, the impenetrability of matter due to Pauli exclusion is a consequence of incorporating special relativity into quantum mechanics. *Microscopic causality*, or the requirement that communications cannot propagate at greater than the speed of light, relates a particle's spin to its quantum statistics but does not exclude the space-like correlations associated with a coherent quantum state.

This section will examine some of the foundations of quantum physics. An emphasis will be placed upon how the microscopic fundamentals of quantum mechanics relate to the geometric fundamentals of gravitational mechanics.

2.1.1 Quantum formalism

There are several interpretations of quantum mechanics offering models of the underlying (often hidden) dynamics that generate the observed quantum phenomenology (for example, the Copenhagen interpretation, or the many-worlds interpretation). Since any interpretation consistent with quantum physics cannot be proved or disproved by experiment, only those aspects of quantum formalism directly connected to experimental observables will be developed, with minimal interpretation imposed. To establish the conventions utilized in what follows, a brief overview of the formalism that describes those calculable observables modeled by quantum physics will be given.

Quantum state vectors

An arbitrary quantum state vector represented by the *ket*, $|\psi\rangle$, is an element in a complex valued vector space. The *ket* symbol, $|\xi\rangle$, encompassing the label ξ is analogous to the arrow $\vec{A}$ designating a standard vector with a denumerable number of dimensions over the label A. The state vector represents what information can be ascertained about a coherent quantum state prior to any measurement. The results of operations upon quantum state vectors using *operators*, $\hat{\mathcal{O}}$, which generally transform or modify the vectors, allow state vectors to be parameterized and described relative to one another. Vectors are generally elements in a linear space, which means that linear combinations and transformations of vectors result in other vectors. The inner product of two state vectors $\langle\phi|\psi\rangle$ generates a scalar complex number, where the *bra* represents the dual form of the complex vector, referred to as the *hermitian conjugate* or *hermitian adjoint* of the ket. The operation of hermitian conjugation will be indicated by a superscript dagger (†), and is defined from complex conjugation of the scalar $\langle\phi|\hat{\mathcal{O}}|\psi\rangle$ using $\langle\phi|\hat{\mathcal{O}}|\psi\rangle^* \equiv \langle\psi|\hat{\mathcal{O}}^\dagger|\phi\rangle$. The hermitian conjugate thus involves transposing kets and bras, while taking the complex conjugate. This means that the inner product of a (non-trivial) state vector with itself is a positive real number associated with that state vector analogous to the length of a standard vector. This property allows the positive semi-definite modulus square of quantum state vectors to be associated with probability metrics from multiple measurements on identical systems.

General vectors are conveniently described in terms of component projections along *basis vectors* which have useful or elegant properties. A *complete* set of basis vectors is a minimal collection of vectors that are *linearly independent*, which means that no vector in the collection can be represented in terms of a weighted sum over the other vectors. A complete basis is particularly convenient if each basis vector is *orthogonal* to the other basis vectors, which means that the inner product between any different basis vectors vanishes. An *orthonormal* basis has the further property of normalization for each vector. These properties can be formally represented. For a discrete (denumerable) set of basis vectors,

$$\begin{aligned} \langle\xi_m|\xi_n\rangle &= \delta_{mn} && \text{orthonormality} \\ \hat{1} &= \sum_m |\xi_m\rangle\langle\xi_m| && \text{completeness.} \end{aligned} \tag{2.1}$$

A continuous complete set of orthonormal basis vectors satisfies:

$$\begin{aligned} \langle\zeta'|\zeta\rangle &= \delta(\zeta' - \zeta) && \text{orthonormality} \\ \hat{1} &= \int d\zeta\, |\zeta\rangle\langle\zeta| && \text{completeness.} \end{aligned} \tag{2.2}$$

Such continuous vectors can form the basis of objects in a *Hilbert space*, which is a space of complex functions $f(\zeta)$ satisfying $\int_{all\ \zeta} d\zeta |f(\zeta)|^2 < \infty$.

Therefore, any arbitrary state vector can be described in terms of a weighted sum over the well-understood basis vectors, with each weight (component) given by the inner product of the basis vector with the general state vector:

$$|\phi\rangle = \hat{1}|\phi\rangle = \sum_m |\xi_m\rangle\langle\xi_m|\phi\rangle = \sum_m |\xi_m\rangle\,\phi_m$$

$$|\psi\rangle = \hat{1}|\psi\rangle = \int d\zeta|\zeta\rangle\langle\zeta|\psi\rangle = \int d\zeta|\zeta\rangle\,\psi(\zeta). \tag{2.3}$$

The functions $\phi_m \equiv \langle\xi_m|\phi\rangle$ (for a discrete basis) or $\psi(\zeta) \equiv \langle\zeta|\psi\rangle$ (for a continuous basis) are the component of the state vector relative to the given basis ($|\xi_m\rangle$ or $|\zeta\rangle$), and each component is called the *wavefunction* of the state in that basis. Inner products of quantum states can be expressed in terms of the components:

$$\langle\phi|\psi\rangle = \langle\phi|\hat{1}|\psi\rangle = \begin{cases} \sum_m \phi_m^* \psi_m, & \text{discrete states} \\ \int d\zeta\ \phi^*(\zeta)\,\psi(\zeta), & \text{continuous states.} \end{cases} \tag{2.4}$$

In particular, for a normalized quantum state vector $\langle\psi|\psi\rangle = 1 = \sum_m |\psi_m|^2$, the projection $\hat{\mathcal{P}}_{\{m_s\}}$ into a subset of discrete states $\{m_s\}$ will give the probability that the quantum state will be measured within that subset of states:

$$\begin{aligned}\langle\psi|\hat{\mathcal{P}}_{\{m_s\}}|\psi\rangle &= \langle\psi|\sum_{m\subset\{m_s\}} |\xi_m\rangle\langle\xi_m|\,|\psi\rangle = \sum_{m\subset\{m_s\}} |\psi_m|^2 \\ &= \text{Prob}(\psi\ \textit{measured in}\ \{m_s\}).\end{aligned} \tag{2.5}$$

Similarly for continuum systems,

$$\langle\psi|\hat{\mathcal{P}}_{\delta\Omega}|\psi\rangle = \int_{\zeta\ \subset\ \delta\Omega} d\zeta\ |\psi(\zeta)|^2 = \text{Prob}(\psi(\zeta)\ \textit{measured in}\ \delta\Omega). \tag{2.6}$$

The subset probabilities $|\psi_{m_s}|^2$ or $d\zeta\ \ |\psi(\zeta)|^2$ are clearly positive (or zero) contributors to the overall probability that the system will be measured in *some* state. In particular, for continuous systems, the *probability density* $|\psi(\zeta)|^2$ is a density which, integrated over some interval of the continuous parameter, gives the likelihood the system will be measured to be in that interval. The quantum state vectors are not interpreted to describe a "reality" associated with the evolution of a quantum system. Rather, through these probabilities, what information can be ascertained about the general behaviors of many such systems is contained within this tool of quantum theory.

A system with independent quantum components is represented in terms of a product state of those independent components. If the component states are labeled using parameters $\{\phi_1, \phi_2, \ldots, \phi_s\}$, then the composite system $|\Psi_{\{\phi_1,\ldots,\phi_s\}}\rangle$ is constructed as:

$$|\Psi_{\{\phi_1,\ldots,\phi_s\}}\rangle \equiv |\phi_1\rangle_1 \cdots |\phi_S\rangle_S. \tag{2.7}$$

Probabilities developed from such composite states will be the product of the independent probabilities of the component states.

Operators

Generally, an operation on a quantum state vector $\hat{O}|\psi\rangle$ generates another vector in the space,

$$\hat{O}|\psi\rangle = \sum_{m\,n} |\xi_m\rangle\, \mathcal{O}_{mn}\, \psi_n, \tag{2.8}$$

where the *matrix element* of the operator is defined by $\mathcal{O}_{mn} \equiv \langle\xi_m|\hat{O}|\xi_n\rangle$. A *linear operation* $\hat{O}$ on state vectors is one that satisfies:

$$\hat{O}(a_1\,|\psi_1\rangle + a_2\,|\psi_2\rangle) = a_1\,\hat{O}|\psi_1\rangle + a_2\,\hat{O}|\psi_2\rangle, \tag{2.9}$$

where a_1 and a_2 are arbitrary complex numbers. Elementary behaviors of quantum systems are linear, and operations upon the state vectors representing those systems that describe their characteristics and evolutions are linear operators. (However, non-linear behaviors *can* result from interactions.) The *expectation value* of the operator $\hat{O}$ is the average value of the operator relative to the state:

$$\langle\hat{O}\rangle_\psi \equiv \langle\psi|\hat{O}|\psi\rangle = \sum_{m\,n} \psi_m^* \mathcal{O}_{mn} \psi_n, \tag{2.10}$$

where the right-most form is replaced by a double integral for continuous basis states.

If the quantum state is an *eigenstate* of the operator, then the resultant state is proportional to the original state, and the constant of proportionality is the *eigenvalue*. Typically, that eigenvalue is then used to label the eigenstate $\hat{Q}\,|\lambda\rangle = \lambda\,|\lambda\rangle$. Such special eigenstates are often used as the basis vectors upon which general quantum states are decomposed. Physical observables are real parameters, and such observables must be represented by *hermitian operators*, which are operators whose hermitian adjoint is the same as the operator itself. For instance, the momentum operator is hermitian $\hat{P}^\dagger = \hat{P}$, with eigenstates labeled by the physical momentum $\hat{P}\,|p\rangle = p\,|p\rangle$. By examining the relationships $\langle\lambda'|\hat{O}|\lambda\rangle^* = \langle\lambda|\hat{O}^\dagger|\lambda'\rangle$ one can directly demonstrate that the eigenstates associated with physical observables with different eigenvalues are always orthogonal.

Functions of quantum operators are defined in terms of the series expansion of that function:

$$f(\hat{\mathcal{O}}) \equiv \sum_{s=0}^{\infty} \frac{f^{(s)}(0)}{s!} \hat{\mathcal{O}}^s. \tag{2.11}$$

Such relationships are most conveniently evaluated using basis states that are eigenstates of the operator, $\hat{\mathcal{O}}$. If two operators $\hat{\mathcal{O}}$ and $\hat{Q}$ are independent, then the order of the operations in a product does not matter, and the operators *commute*, $\hat{\mathcal{O}}\hat{Q} = \hat{Q}\hat{\mathcal{O}}$. The *commutator* of the two operators is defined as $[\hat{\mathcal{O}}, \hat{Q}] \equiv \hat{\mathcal{O}}\hat{Q} - \hat{Q}\hat{\mathcal{O}}$, while the *anti-commutator* of the two operators is defined as $\{\hat{\mathcal{O}}, \hat{Q}\} \equiv \hat{\mathcal{O}}\hat{Q} + \hat{Q}\hat{\mathcal{O}}$. Independent operators have vanishing commutators. The independent eigenstates of the commuting operators can be written as direct products as previously discussed.

A *unitary* transformation upon a quantum state is a transformation that preserves the normalization of the state vector. If the transformation is represented by $|\psi'\rangle \equiv \hat{U}|\psi\rangle$, this means that $\langle\psi'|\psi'\rangle = \langle\psi|\hat{U}^\dagger\hat{U}|\psi\rangle = \langle\psi|\psi\rangle$. Therefore, unitary operators satisfy $\hat{U}^\dagger\hat{U} = \hat{1}$. Quantum states carrying conserved internal quantum numbers (like mass, charge, lepton number, baryon number, etc.) must evolve unitarily to preserve their normalizations.

Generators for infinitesimal transformations

As previously stated, physical observables are represented by hermitian operators in quantum mechanics. Certain operators are associated with a *group* of transformations characterized by continuous parameters. A group $\mathcal{G}$ is a set of elements with a group operation that has the following properties:

- *closure*: if A and B are elements of $\mathcal{G}$, then there is a group operation $A \cdot B$ that combines these two elements into an element of $\mathcal{G}$;
- *identity*: there is an element of the group I which leaves any other element unchanged under the group operation $I \cdot A = A = A \cdot I$;
- *inverse*: for any element in the group A, there is another element in the group A^{-1} (its inverse) that combines with that element to give the identity, $A^{-1} \cdot A = I$;
- *associativity*: for any three elements in the group, $(A \cdot B) \cdot C = A \cdot (B \cdot C)$.

Group theory for quantum systems will be discussed in more detail in Chapter 4.

Of particular interest is the set of operators representing observables that are canonically conjugate to corresponding continuous parameters of a group. For quantum systems, *generators* are complex forms that operate on kets whose hermitian adjoints operate on bras. Consider, for instance, the set of all possible translations from a particular Minkowski space-time coordinate $\vec{x} \equiv (ct, x, y, z)$.

Under an infinitesimal translation $\vec{dx}$, a state vector will transform according to:

$$|x^\mu + dx^\mu\rangle \simeq |x^\mu\rangle + dx^\beta \frac{\partial}{\partial x^\beta}|x^\mu\rangle \equiv |x^\mu\rangle + \frac{1}{i\hbar} dx^\beta \hat{\mathrm{P}}_\beta |x^\mu\rangle, \tag{2.12}$$

which defines the covariant four-momentum operator $\hat{\mathrm{P}}_\beta$ as the generator for infinitesimal space-time translations. When operating on arbitrary quantum state vectors, the four-momentum operator takes the conjugate form when acting on the associated wave functions:

$$\langle x^\mu|\hat{\mathrm{P}}_\beta|\psi\rangle = \langle\psi|i\hbar\frac{\partial}{\partial x^\beta}|x^\mu\rangle^* = \frac{\hbar}{i}\frac{\partial}{\partial x^\beta}\psi(x^\mu) \equiv \hat{p}_\beta\,\psi(x^\mu). \tag{2.13}$$

This allows one to develop the form of the wave function associated with four-momentum eigenstates $|\underline{p}\rangle$. Replacing the state vector in Eq. 2.13, this function must satisfy:

$$\langle\vec{x}|\hat{\mathrm{P}}_\beta|\underline{p}\rangle = \frac{\hbar}{i}\frac{\partial}{\partial x^\beta}\langle\vec{x}|\underline{p}\rangle = p_\beta\,\langle\vec{x}|\underline{p}\rangle. \tag{2.14}$$

The normalization criteria of generic configuration and four-momentum eigenstates will be chosen to be $\langle\vec{x}'|\vec{x}\rangle = \delta^4(\vec{x}' - \vec{x})$, $\langle\underline{p}'|\underline{p}\rangle = \delta^4(\underline{p}' - \underline{p})$. The solution satisfying Eq. 2.14 is given by:

$$\langle\vec{x}|\underline{p}\rangle = \frac{e^{\frac{i}{\hbar}p_\beta x^\beta}}{(2\pi\hbar)^2}. \tag{2.15}$$

Thus, configuration and momentum basis states are directly related through Fourier transformation.

Energy-momentum and space-time eigenstates transform through a unitary representation of the ten-dimensional *Poincaré group* (or the inhomogeneous Lorentz group) $U(\mathbf{\Lambda}, \vec{a})$, where the parameters describe the three rotations and three velocity boosts of Lorentz transformations $\mathbf{\Lambda}$, and space-time translations $\vec{a}$:

$$U(\mathbf{\Lambda}_2, \vec{a}_2)\, U(\mathbf{\Lambda}_1, \vec{a}_1) = U(\mathbf{\Lambda}_2\, \mathbf{\Lambda}_1, \mathbf{\Lambda}_2\vec{a}_1 + \vec{a}_2). \tag{2.16}$$

The *passive* transformations involve the group of coordinate and reference frame transformations on a given quantum state parameterized by contravariant components:

$$U(\mathbf{\Lambda}, \vec{a})|\vec{p}\rangle = |\mathbf{\Lambda}\vec{p}\rangle, \quad U(\mathbf{\Lambda}, \vec{a})|\vec{x}\rangle = |\mathbf{\Lambda}\vec{x} + \vec{a}\rangle. \tag{2.17}$$

For transformations under the Poincaré group, the states can be characterized in terms of *standard state* vectors, which remain invariant under the little group of transformations for the particle type. For massive particles, the standard four-momentum state (invariant under rotations) is parameterized by $\vec{p}_{(s)} \equiv (mc, 0, 0, 0)$. Massless states will have a standard state (invariant under rotations in along the standard direction of motion) given by $\vec{p}_{(s)} \equiv (1, 0, 0, 1)$, with the momenta expressed in the appropriate standard units.

Fixed mass particle states $|\mathbf{p}; m\rangle$ with energy $\epsilon(p) \equiv \sqrt{(mc^2)^2 + |\mathbf{p}c|^2}$ will be chosen to satisfy the normalization condition:

$$\langle \mathbf{p}'; m|\mathbf{p}; m\rangle = \frac{\epsilon(p)}{p^0_{(s)}c}\, \delta^3(\mathbf{p}' - \mathbf{p}), \tag{2.18}$$

and the completeness condition:

$$\hat{1} = \int \frac{p^0_{(s)}c}{\epsilon(p)}\, d^3p \; |\mathbf{p}; m\rangle\langle \mathbf{p}; m| = \int d^4p \; |\vec{p}\rangle\langle\vec{p}|\; \delta(\sqrt{-\vec{p}\cdot\vec{p}} - mc). \tag{2.19}$$

This normalization will be chosen so that all derived probabilities, amplitudes, etc., will have direct correspondence with those of non-relativistic quantum physics. It should be noted that the mass-shell constraint need not be a precondition for a viable physical model.

2.1.2 Uncertainty and measurement

In quantum mechanics, an observer is unable to ascertain the behavior of a system *between* interactions with a probe designed to measure that behavior, and those very interactions alter the system into a new state. Interactions that couple only weakly to a given quantum number parameterizing the state need not break the coherence labeled by that particular quantum number. However, a probe specific to a given state will break the coherence of that state if the measurement gives a result. The observer is only able to predict the likelihood of a particular outcome to probing the system (including the likelihood of not having *any* interaction whatsoever), but the actual particular outcome to a given probing is unknown (and usually unknowable) to the observer (and the observed) until the experiment is done. The predictive information about the quantum system is contained in the quantum state vector $|\Psi\rangle$, and like all vectors, its length (the probability measure) satisfies the triangle inequality, as demonstrated in Figure 2.1. Information about the state of a quantum system can be ascertained using operators on the state vector which describes that system. In quantum physics, unless the system is an *eigenstate* of an operator, its operation upon the state vector changes that vector. A particular measurement momentarily projects the quantum system into the eigenstate associated with that measurement.

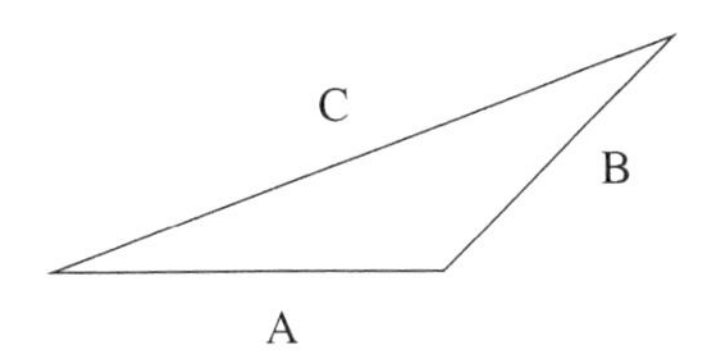

Figure 2.1 The triangle inequality. (Length of side A) + (Length of side B) $\geq$ (Length of side C).

A quantum prediction involving a physical observable is a projective length in this quantum vector space, which is assumed to be the average value that would result from many, many experiments. As with many types of averages, any particular observation might *never* measure the average value itself. For example, if the average number of people in any given room in a building is 2.3, one can be sure to never actually obtain this number from one given measurement. Of course, with any averaging exercise, there is usually a scatter of values about that average, and the square root of the average distance squared of the scatter from that average value is the root-mean-squared deviation, or *uncertainty*, of the individual values relative to that average. In most classical instances of averaging, by doing more careful and systematic experiments, the uncertainty can be made arbitrarily small. However, in quantum mechanics, certain physical observables (like momentum and position, or energy and time), are inherently skewed relative to one another such that they cannot be simultaneously measured by a single measurement.

The triangle inequality manifests as quantum uncertainty and quantum measurability constraints. As previously stated, in quantum mechanics the state of a system is represented by a quantum state vector $|\psi\rangle$ that contains what information that *can* be extracted for predicting the subsequent behavior of that quantum system (as long as it maintains its coherence). After a measurement is taken, the state momentarily has a known parameterization, and the prior state vector becomes obsolete, losing further predictive information. This loss of *coherence*, in some interpretations known as the *collapse of the wave function*, results whenever any definitive interrelation (involving measurement or communication) occurs. Since the quantum state vector behaves fundamentally as a (complex) vector, it has both *magnitude* (or length) and direction/{*phase*}. For standard vectors, $\vec{A}$, $\vec{B}$, and $\vec{C} = \vec{A} + \vec{B}$, the law of cosines is the generalization of the length of the hypotenuse of a triangle that need not be a right-angled triangle:

$$|\vec{C}|^2 = |\vec{A}|^2 + |\vec{B}|^2 + 2\vec{A} \cdot \vec{B} \leq (|\vec{A}| + |\vec{B}|)^2 = |\vec{A}|^2 + |\vec{B}|^2 + 2|\vec{A}||\vec{B}|$$

or, simplifying:

$$|\vec{A} \cdot \vec{B}| \leq |\vec{A}||\vec{B}| = \sqrt{|\vec{A}|^2|\vec{B}|^2}.$$

For quantum state vectors, this relation becomes:

$$|\langle\phi_1|\phi_2\rangle| \leq \sqrt{\langle\phi_1|\phi_1\rangle\langle\phi_2|\phi_2\rangle}. \tag{2.20}$$

For two quantum states, their inner product is generally a complex scalar.

This result is related to the quantum uncertainty inherent in a set of careful measurements. Defining the root mean-squared (RMS) deviation from the expectation value of a physical parameter Q by:

$$\Delta Q \equiv \left[\langle\psi|(\hat{Q} - \langle\psi|\hat{Q}|\psi\rangle)^2|\psi\rangle\right]^{1/2},$$

one can ascertain the average scatter of measurements from the average value $\langle\psi|\hat{Q}|\psi\rangle$, (i.e., how close any given measurement can be expected to be to the average measurement). For this reason, ΔQ is called the *uncertainty* of physical parameter Q relative to state $|\psi\rangle$. If one examines quantum states that are defined in terms of the difference in measurement of two hermitian measurables $\hat{Q}_1$ and $\hat{Q}_2$ from the expected average values:

$$|\phi_j\rangle \equiv \left(\hat{Q}_j - \langle\psi|\hat{Q}_j|\psi\rangle\right)|\psi\rangle \equiv \Delta\hat{Q}_j|\psi\rangle,$$

a measure of the uncertainties of these physical parameters relative to an arbitrary quantum state $|\psi\rangle$ can be developed. Substitution of these forms into Eq. 2.20 gives a quantum form of the triangle inequality:

$$\begin{gathered}(\Delta Q_1 \cdot \Delta Q_2)^2 \geq \tfrac{1}{4}|\langle\psi|[\hat{Q}_1, \hat{Q}_2]|\psi\rangle + \langle\psi|\{\Delta\hat{Q}_1, \Delta\hat{Q}_2\}|\psi\rangle|^2, \\ \text{where } [\hat{A}, \hat{B}] \equiv \hat{A}\hat{B} - \hat{B}\hat{A} \quad \text{and} \quad \{\hat{A}, \hat{B}\} \equiv \hat{A}\hat{B} + \hat{B}\hat{A}\end{gathered} \tag{2.21}$$

define the *commutator* $[\hat{A}, \hat{B}]$ and *anti-commutator* $\{\hat{A}, \hat{B}\}$ of the two operators. For hermitian operators representing observables, the commutators must be pure imaginary $[\hat{A}, \hat{B}]^\dagger = -[\hat{A}, \hat{B}]$, and the anti-commutators must be pure real. Therefore, the quantum form of the triangle inequality is given by:

$$\Delta Q_1 \cdot \Delta Q_2 \geq \frac{1}{2}|\langle\psi|[\hat{Q}_1, \hat{Q}_2]|\psi\rangle|. \tag{2.22}$$

For quantum mechanical operations, the order of operation is important, so this difference in order often does not vanish.

If one examines quantum states defined in terms of the difference in measurement of momentum p and position x from the expected average values, one obtains a measure of the product of uncertainties of these physical parameters. Substitution of these forms into Equation 2.22 gives a quantum form of the triangle inequality:

$$\Delta p \cdot \Delta x \geq \frac{1}{2}|\langle\psi|[\hat{P}, \hat{X}]|\psi\rangle|. \tag{2.23}$$

Since the commutator does not vanish, the momentum p and the position x cannot be known simultaneously. Furthermore, even after many measurements on identical systems, there remains an inherent uncertainty in the measurements given by:

$$\Delta p \cdot \Delta x \geq \frac{\hbar}{2}.$$

This equation sets a fundamental lower limit on how accurately momentum and position can both be measured in an averaged way for a given system. They cannot be simultaneously measured *at all* by a single experiment (a single interaction). A similar relationship holds for energy and time, despite it being problematic to describe time in terms of a quantum operator. Since the Hamiltonian operator, whose eigenvalues give energies, is the generator for infinitesimal time translations, any temporal wave packet must satisfy these quantum measurement constraints. Of course, macroscopic experience consists of a multitude of ongoing, rapid, disparate probings from the scattering of light, collisions of molecules, and so on, so this fundamental limitation goes unnoticed. Macroscopic measurements are indistinguishably close to the expectation values describing the averages of many measurements.

Thus, by generically describing quantum states in terms of vectors with well-defined phase coherence properties, the triangle inequality places a fundamental limit on the possibility of using multiple quantum measurements to precisely measure non-commuting observables. More so, quantum systems completely exclude the simultaneous measurement of non-commuting observables in a single experiment. Due to the fundamental limits placed on measurements by quantum mechanics, care will be taken never to refer to particles as *point* particles. Rather, the structure of particles will be referred to within the context of their (or their components') behaviors under the relevant group of transformations.

Supplement: Classical-quantum correspondence

Centuries ago, Hamilton showed that the rate that a physical parameter changes in time is governed by well-defined relationships with the energy. The quantum mechanical expression of this principle is illustrated by the relation $\frac{d\langle\hat{Q}\rangle}{dt} = \langle[\hat{H},\hat{Q}]\rangle/i\hbar$, where $\langle\hat{Q}\rangle$ represents the value of the average of many measurements of the physical parameter Q, and $\hat{H}$ is the energy operator (called the *Hamiltonian*). This implies a form for an "energy operator", $\hat{H} = i\hbar\frac{d}{dt}$. If the Hamilton equations for the speed and force:

$$v = \frac{d\langle\hat{X}\rangle}{dt}, \quad F = \frac{d\langle\hat{p}\rangle}{dt},$$

are to apply for the non-relativistic Hamiltonian made up of the sum of the kinetic and potential energies $\hat{H} = \hat{KE} + U(\hat{x}) = \frac{\hat{p}^2}{2M} + U(\hat{x})$, then the form of the momentum operator must be $\hat{p} = -i\hbar\frac{d}{dx}$. Substitution of this form into Hamilton's equations gives the usual form of the momentum $v = \frac{\langle\hat{p}\rangle}{M}$ and Newton's second law of motion $F = -\left\langle\frac{dU(x)}{dx}\right\rangle$ connecting force to negative potential energy gradient for conservative force fields. Therefore, quantum processes can be directly interpreted to correspond to the classical physical behaviors of the expectation (averaged) values of given quantum parameters.

2.1.3 Zero-point motions

The discussion in the previous section demonstrates that a physical system in its stable ground state cannot be described in terms of components at specified coordinate locations with similarly specified momenta. The *zero-point* motions in locations of real systems are inherent within the finite parameterizations (energies, momenta, etc.) of those systems. If those components also couple to quanta (fields) of interactions, the zero-point motions are associated with vacuum energies of those quanta. The standard vacuum is defined as the state corresponding to the minimum energy of the system. In this treatment, vacuum energy will *not* be considered to be due to the background fluctuations of the basis of a particular perturbative expansion (like vacuum polarization or mass renormalization of *bare* particles). The actual physical manifestations must be independent of any basis of expansion of the physical states. For this reason, vacuum energies will be examined only relative to the zero-point motions of the quantum correlated sources. The vacuum energies of the quanta of interaction are useful constructs only insofar as they provide a convenient mechanism to calculate the effects of the zero-point motions of the sources of those quanta. Any infinities associated with vacuum energies of interacting quanta must therefore be nothing more than calculational artifacts of that particular method of computation.

There are several physical systems that directly exhibit zero-point behaviors. For instance, the van der Waals interaction between non-polar molecules, which is found to fall off like $1/r^6$ well beyond the scales of the individual molecules, results from subtle correlations of zero-point motions within those molecules. To demonstrate this behavior, an instructive system consists of collections of electrically neutral composite particles that can polarize due to external fields or quantum motions. The interaction Hamiltonian for an electric dipole, $\mathbf{d}$, in an electric field is, $H = -\mathbf{d} \cdot \mathbf{E}$. This gives an interaction energy between two dipoles of the form:

$$H_{int} = \frac{(\mathbf{d}_1 \cdot \mathbf{d}_2) r^2 - 3(\mathbf{d}_1 \cdot \mathbf{r})(\mathbf{d}_2 \cdot \mathbf{r})}{r^5}, \tag{2.24}$$

where $\mathbf{r} \equiv \mathbf{r}_2 - \mathbf{r}_1$ is the relative displacement between the dipoles. Since the average polarization vanishes, there is no first-order perturbative effect upon the overall system due to this interaction. However, this weakly interacting system does provide a non-vanishing contribution to second order in perturbation theory,

$$\Delta E^{(2)} = \sum_{m \neq 0} \frac{\langle 0|H_{int}|m\rangle\langle m|H_{int}|0\rangle}{E_0 - E_m}, \tag{2.25}$$

which naively generates an r^{-6} behavior (due to the r^{-3} behavior of H_{int}).

Zero-point behaviors can be directly examined using the canonical motions of equal mass identical oscillators polarized parallel to the z-axis separated along the x-axis by a distance $D = |x_2 - x_1|$. Consider collective modes of the canonical system:

$$K_P = \frac{p_{1z}^2}{2m} + \frac{1}{2} m\omega_o^2 z_1^2 + mc^2 + \frac{p_{2z}^2}{2m} + \frac{1}{2} m\omega_o^2 z_2^2 + mc^2 + \frac{q z_1 q z_2}{D^3}. \tag{2.26}$$

The canonical equations of motion of this equation has normal modes given by $\omega_\pm = \sqrt{\omega_o^2 \pm \frac{q^2}{mD^3}}$, resulting in a zero-point energy of $K_{\text{zero-pt}} = \frac{1}{2}\hbar\omega_+ + \frac{1}{2}\hbar\omega_- + 2mc^2$. Thus, for weakly interacting dipoles $\frac{q^2}{m\omega_o^2 D^3} \ll 1$, zero-point motions result in an interaction of the form:

$$K_{\text{zero-pt}} \simeq 2mc^2 + \hbar\omega_o - \frac{\hbar\omega}{8}\left(\frac{q^2}{m\omega_o^2}\right)^2 \frac{1}{D^6}. \tag{2.27}$$

This means that the long-range form of the interaction between two oscillating electric dipoles is attractive and falls off with the inverse sixth power of the separation, consistent with the form of a van der Waals interaction between neutral systems.

However, once retardation (i.e., causal interaction) is included, the form of the interaction between the molecules for intermediate/long distances is found to behave as r^{-7} at temperatures low enough for coherent polarizations to be significant. Such fluctuations are the source of the well-known *Casimir effect*, which will be briefly explored in what follows.

Quantized proper-time oscillators

The canonical proper-time oscillator examined in Eq. 1.68 can be developed for quantized oscillators. The operator form of the canonical proper energy is given by:

$$\hat{K} = \frac{\hat{p}^2}{2m} + \frac{1}{2} m\omega_o^2 \hat{x}^2 + m\,c^2. \tag{2.28}$$

A normalized solution of the proper-time dependent equation $\hat{K}\psi(x,\tau) = i\hbar\frac{\partial}{\partial\tau}\psi(x,\tau)$ is given by:

$$\psi(x,\tau) = \left(\frac{m\omega_o}{\pi\hbar}\right)^{1/4} exp\left\{ -\frac{m\omega_o}{2\hbar}\left[\left(x - \bar{x}e^{-i\omega_o\tau}\right)^2 + i\bar{x}^2 e^{-i\omega_o\tau} sin\omega_o\tau + i\left(\frac{\hbar}{m} + \frac{2c^2}{\omega_o}\right)\tau\right]\right\}. \tag{2.29}$$

However, it is quite useful to develop energy eigenstates for the oscillator.

Algebraic techniques can be utilized by introducing a non-hermitian operator $\hat{b}$ related to the momentum and position operators by:

$$\begin{aligned}\hat{p} &= \frac{1}{i}\sqrt{\frac{m\hbar\omega_o}{2}}\,(\hat{b} - \hat{b}^\dagger),\\ \hat{x} &= \sqrt{\frac{\hbar}{2m\omega_o}}\,(\hat{b} + \hat{b}^\dagger).\end{aligned} \tag{2.30}$$

The momentum-position commutation relation $[\hat{p}, \hat{x}] = \frac{\hbar}{i}$ implies that the bosonic operators satisfy $[\hat{b}, \hat{b}^\dagger] = 1$. By substituting the Eqs. 2.30 into Eq. 2.28, the positive semi-definite number operator $\hat{N} \equiv \hat{b}^\dagger \hat{b}$ define eigenstates of the proper energy form $\hat{K} = (\hat{N} + \frac{1}{2})\hbar\omega_o + mc^2$. The number operator satisfies the commutation relation $[\hat{N}, \hat{b}] = -\hat{b}$, defining $\hat{b}$ as a lowering/annihilation operator, and the relation $[\hat{N}, \hat{b}^\dagger] = +\hat{b}^\dagger$, defining $\hat{b}^\dagger$ as a raising/creation operator. One should note that non-hermitian operators $\hat{b}$, $\hat{b}^\dagger$ cannot represent physical observables. These commutation relations allow one algebraically to construct eigenstates of the number operator:

$$\begin{gathered}\hat{N}|n\rangle = n|n\rangle, \quad \hat{b}|n\rangle = \sqrt{n}\,|n-1\rangle, \quad \hat{b}^\dagger|n\rangle = \sqrt{n+1}\,|n+1\rangle,\\ \langle n'|n\rangle = \delta_{n'n}, \quad \hat{1} = \sum_{n=0}^{\infty} |n\rangle\langle n|, \quad |n\rangle = \frac{(\hat{b}^\dagger)^n}{\sqrt{n!}}|0\rangle.\end{gathered} \tag{2.31}$$

Stationary eigenstates then define wavefunctions $\psi_n(x) \equiv \langle x|n\rangle$ that satisfy the differential equation form of $\hat{K}\,\psi_n(x) = [(n + \frac{1}{2})\hbar\omega_o + mc^2]\psi_n(x)$. It should be noted that, despite the customary denotation of this procedure as *second quantization*, no additional quantization of quantum operators has occurred. Rather, multi-quanta states are represented using actions by convenient non-hermitian operators upon a unique vacuum or ground state. For this reason, these states describe the system in the so-called *operator product representation.*

The state $|0\rangle$ has the lowest energy corresponding to zero quanta of excitation, representing the vacuum state (or ground state) of the oscillator. The zero-point energy of the oscillator beyond rest energy of the mass is given by the term $\frac{1}{2}\hbar\omega_o$. The average momentum and position of the oscillator both vanish, so that a straightforward calculation of the uncertainties for zero-point motions yields $\frac{(\Delta p)^2}{2m} = \frac{1}{4}\hbar\omega_o = \frac{1}{2}m\omega_o^2(\Delta x)^2$, demonstrating that the zero-point energy is equally partitioned between kinetic and potential energy terms. For the zero-point motions, the uncertain condition is optimized:

$$\Delta p_0\,\Delta x_0 = \frac{\hbar}{2}, \tag{2.32}$$

while excited states have uncertainties satisfying the general form $\Delta p_n\,\Delta x_n = \frac{\hbar}{2}(2n + 1)$.

Thermal correlations

A harmonic oscillator in a thermal bath will undergo fluctuations in time. Statistical thermodynamics will be discussed generally in Section 2.2.3. However, it is noteworthy to briefly mention a characteristic of thermal fluctuations relevant to the present discussion.

It is convenient to define a function that measures the extent to which the behavior of a given system at a later time is correlated to its prior behaviors. The temporal thermal correlation function with temperature parameterized by $\beta = 1/k_B T$ is defined by the symmetrized form:

$$C_x(\tau - \tau_o) \equiv \frac{1}{2} \langle\{\hat{x}(\tau), \hat{x}(\tau_o)\}\rangle_T = \frac{1}{2}\, Tr\, e^{\beta(F_K - \hat{K})} \{\hat{x}(\tau), \hat{x}(\tau_o)\}, \qquad (2.33)$$

where $\hat{x}(\tau) \equiv e^{\frac{i}{\hbar}\tau\hat{K}}\, \hat{x}\, e^{-\frac{i}{\hbar}\tau\hat{K}}$, and $Tr\, e^{\beta(F_K - \hat{K})} = 1$ defines the proper free energy F_K. The operator $\hat{x}(\tau)$ represents the time dependent Heisenberg form of the position of the oscillator, and the density operator $\hat{\rho} \equiv e^{\beta(F_K - \hat{K})}$ gives the proper thermal weights of occupation of states in the equilibrium system. The frequency-dependent Fourier transform of correlations for *positive* frequencies can then be shown by direct algebra to satisfy:

$$\tilde{C}_x(\omega) \equiv \int_{-\infty}^{\infty} \frac{d\tau}{\sqrt{2\pi}} C_x(\tau)\, e^{i\omega\tau} = \sqrt{\frac{\pi}{2}} \frac{\hbar}{m\omega_o}\, \delta(\omega - \omega_o) \left(\langle\hat{N}\rangle_T + \frac{1}{2} \right), \qquad (2.34)$$

where the thermal average of the number of excitations is given by $\langle\hat{N}\rangle_T = \frac{1}{e^{\beta\omega_o} - 1}$. It is clear that the zero-point correlations are therefore independent of thermal effects.

The Casimir effect

Collective systems that are absent from other interactions that do not overwhelm the van der Waals forces can also be constructed. A category of such systems include those that exhibit the Casimir effect [5]. Casimir considered the change in the photon vacuum energy due to the placement of two parallel conducting plates separated by a distance, a. He calculated the (photon) vacuum energy per unit plate area to be of the form:

$$\frac{1}{2} \left(\sum_{modes} \hbar c k_{interior} - \sum_{modes} \hbar c k_{exterior} \right) / A = -\frac{\pi^2\, \hbar c}{720\, a^3}, \qquad (2.35)$$

resulting in an attractive force given by:

$$F/A = -\frac{\pi^2\, \hbar c}{240\, a^4} \approx -0.013\, dynes/cm^2 \left(\frac{a}{microns} \right)^{-4}, \qquad (2.36)$$

independent of the charges of the sources. This result depends *only* on the geometric factors corresponding to the area of the plates, A, and their separation, a, along with the fundamental constants, $\hbar$ and c. For the instructive example of interacting dipoles resulting in Eq. 2.27, a lack of dependence upon the charges of the sources implies that the spring constant $m\omega_o^2$ associated with the restoring force should be due to the electrostatic coupling between the dipole charges $\pm q$, eliminating the scale of electromagnetic coupling $e^2/\hbar c$. It should be noted that the Casimir energy density 2.35 and pressure 2.36 obey the equation of state $P_{vacuum} = 3\rho_{vacuum}$ (in the region *excluding* the sources, which has negative energy density and negative pressure).

Although the effect does not depend on the electromagnetic coupling strength of the sources, it does depend on the nature of the interaction (monopole forces can be neutralized, leaving only polarization forces) and the configuration. For instance, a spherical geometry will result in a repulsive Casimir force. Boyer [6] and others subsequently derived the repulsive force for a spherical geometry with energy of the form:

$$\frac{1}{2}\left(\sum_{modes} \hbar c k_{interior} - \sum_{modes} \hbar c k_{exterior}\right) = 0.92353 \frac{\hbar c}{2a}. \tag{2.37}$$

This shows that the change in electromagnetic vacuum energy is dependent upon the geometry of the boundary conditions, although it does not depend on the detailed couplings of the involved interactions. Both predictions have been confirmed experimentally. It is interesting to note that this energy grows inversely with the geometric scale $E_{Casimir} \sim \frac{\hbar c}{a}$.

Supplement: Vacuum energies of quanta

For many statistical problems, the sums needed for proper counting can be most easily calculated by conversion into an integral. For photons $E = |\mathbf{p}|c$ in a large volume V that contains many modes, the *density of states* factor that allows this conversion is of the form:

$$\sum_{n_x,n_y,n_z} f_{n_x,n_y,n_z} \doteq \frac{V}{\pi^2(\hbar c)^3} \int \epsilon^2 d\epsilon f(\epsilon).$$

Thus, average energies will involve integrals of the type $\int \epsilon^3 \mathcal{P}(\epsilon) d\epsilon$. In particular, defining the average energy density $\langle\rho\rangle = \langle E\rangle / V$ of a thermal system of photons, including the vacuum mode $\langle\rho\rangle = \langle\rho_{black\,body}\rangle + \langle\rho_{vacuum}\rangle$, in terms of any ultraviolet cutoff, ϵ_{UV}, the vacuum energy density:

$$\langle\rho_{vacuum}\rangle = \frac{1}{2}\left(\frac{\epsilon_{UV}^4}{4\pi^2(\hbar c)^3}\right),$$

is clearly quite large.

For the Casimir effect, the vacuum energy interior to plates of area A separated by distance a requires a calculation of the sum:

$$E^{interior}_{vacuum} = \sum_{\{n\}} \frac{1}{2}\hbar|\mathbf{k}_{\{n\}}|c \Rightarrow \sum_{n} \frac{A}{2} \int \frac{d^2k_\perp}{(2\pi)^2}\sqrt{k_\perp^2 + \left(\frac{n\pi}{a}\right)^2}.$$

The sum generates an infinite outward pressure on the plates, but there is also an infinite inward pressure due to the exterior (non-denumerable) modes. Therefore, to calculate the Casimir energy, one needs to examine mathematical forms of the type:[1]

$$\Delta_N \equiv \sum_{n=0}^{N} f(n) - \int_0^N f(x)\,dx.$$

Euler developed a formula to approximate such sums by integrating by parts and partitioning the form:

$$\Delta_N = \frac{1}{2}[f(N) + f(0)] + \sum_{n=0}^{N-1} \int_n^{n+1} \left(x - n - \frac{1}{2}\right) f'(x)\,dx,$$

then using Euler polynomials (or alternatively Bernoulli numbers B_k) to obtain:

$$\Delta_N = \frac{1}{2}[f(N) + f(0)] + \sum_{s}^{[\frac{K}{2}]} \frac{B_{2s}}{(2s)!}\left[f^{(2s-1)}(N) - f^{(2s-1)}(0)\right] + O\left[\frac{N B_K}{K!} f^{(K)}(\bar{x})\right],$$

where K is an integer, and the mean value $\bar{x}$ satisfies $0 \le \bar{x} \le N$. Consistent with the counting of photon states, Casimir chose the regularized form $f(x) = x^3 e^{-\lambda x}|_{\lambda\to 0}$ to evaluate the finite energy difference, giving $\Delta_\infty = -1/120$.

It should be noted that the vacuum energy density in the absence of some cutoff is infinite and poorly defined. However, the Casimir energy density is well-defined and independent of any cutoff or regularization technique.

Lifshitz and his collaborators [7] demonstrated that the Casimir force can be thought of as the superposition of the van der Waals attractions between individual molecules that make up the attracting media. This allows the Casimir effect to be interpreted directly in terms of the zero-point motions of the sources as an

[1] See, for instance J.P. Dowling,"The Mathematics of the Casimir Effect", *Mathematics Magazine* **62**, 324 (December 1989).

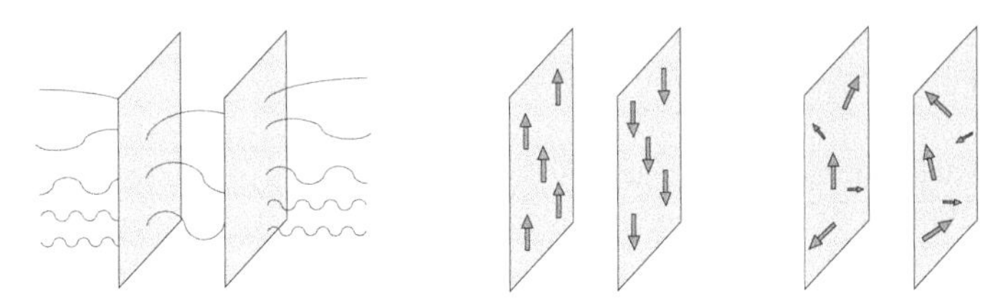

Figure 2.2 Casimir attraction: The first diagram explains the Casimir pressure in terms of a fewer number of modes (denumerably infinite) between the plates as compared to the number of vacuum modes in the exterior region. The second diagram demonstrates an attraction due to space-like correlated van der Waals-induced polarizations resulting in net attraction between the sources. Uncorrelated polarizations demonstrated in the third diagram result in no net attraction.

alternative to vacuum energy of the associated quanta. At zero temperature, the coherent zero-point motions of source currents on opposing plates correlate in a manner resulting in a net attraction, whereas if the motions were independently random, there would be no net attraction, as shown in Figure 2.2. On dimensional grounds, one can argue that the number of source particles per unit area undergoing zero-point motions that contribute to the Casimir result vary as a^{-2}.

In calculating a general form for molecular forces between dielectric bodies, Lifshitz utilized Maxwell's equations driven only by random fluctuating currents $\langle \mathbf{J} \rangle_{time\ average} = 0$:

$$\nabla \times \mathbf{E} = -\frac{1}{c}\frac{\partial \mathbf{B}}{\partial t}, \quad \nabla \times \mathbf{B} = \frac{4\pi}{c}\mathbf{J} + \frac{\epsilon}{c}\frac{\partial \mathbf{E}}{\partial t}. \tag{2.38}$$

Despite having vanishing time averages, the current fluctuations have non-vanishing temporal self correlations similar to those previously demonstrated in Eq. 2.33. He calculated the flux of the Maxwell stress tensor T_{jk} (from standard macroscopic electromagnetism) over the area of the parallel plate to obtain the mechanical force between the plates, consistent with Casimir [8]. The Maxwell stress tensor is related to the electromagnetic fields through $T_{jk} \equiv -\frac{1}{4\pi}[E_j E_k + B_j B_k - \frac{1}{2}\delta_{jk}(E^2 + B^2)]$, where the electromagnetic momentum density is given by $\mathbf{g} = \frac{1}{4\pi c}\mathbf{E} \times \mathbf{B}$. The momentum conservation relationship is $\frac{\partial \mathbf{g}_k}{\partial t} + \nabla_j T_{jk} + (\rho \mathbf{E} + \frac{1}{c}\mathbf{J} \times \mathbf{B})_k = 0$, relating electromagnetic and mechanical forces.

Schwinger also developed the mathematics of the Casimir effect using language that "makes no reference to quantum oscillators and their associated zero point energy." [9]. He and his collaborators demonstrated that for tenuous dielectrics ($\epsilon - 1 \ll 1$), the Casimir force can be thought of as a superposition of the van der Waals attraction between the individual molecules that make up the media [10]. Others have likewise noted that correlated zero-point motions can be used to describe vacuum energy behavior in the Casimir effect. As expressed by Daniel Kleppner [11],

> The van der Waals interaction is generally described in terms of a correlation between the instantaneous dipoles of two atoms or molecules. However, it is evident that one can just as easily portray it as the result of a change in vacuum energy due to an alteration in the mode structure of the system. The two descriptions, though they appear to have nothing in common, are both correct.

Additionally, as pointed out by Wheeler and Feynman [12], and others [13], one cannot unambiguously separate the properties of fields from the interaction of those fields with their sources and sinks, as will be demonstrated in the next chapter.

That *vacuum energy* can be thought of as resulting from the zero-point motions of the sources is also supported by the calculation of Bohr and Rosenfeld [14] in their examination of measurability in perturbative quantum electrodynamics. In their (*near-field*) analysis, the authors minimized the effect of a *classical* measurement of an electric field averaged over a finite volume on the value of a magnetic field at right angles averaged over (1) a non-overlapping volume, and (2) an overlapping volume, and vice versa (with respect to the fields). When this minimum disturbance of sources was put equal to the *minimum* uncertainty ($\frac{\hbar}{2}$) allowed for the measurement by the uncertainty principle, they reproduce the result of averaging the quantum mechanical commutation relations of the electromagnetic fields over the corresponding volumes. Such arguments can be extended to the corresponding case when the sources and detectors are interacting gravitationally.

Another system that manifests physically measurable effects due to zero-point energy is liquid ^{4}He. One sees that this is the case by noting that atomic radii are related to atomic volume V_a (which can be measured) by $R_a \sim V_a^{1/3}$. The uncertainty relation gives momenta of the order $\Delta p \sim \frac{\hbar}{V_a^{1/3}}$. Since the system is non-relativistic, one can estimate the zero-point kinetic energy to be of the order, $E_{0,KE} \sim \frac{(\Delta p)^2}{2m_{He}} \sim \frac{\hbar^2}{2m_{He}V_a^{2/3}}$. The minimum in the inter-atomic potential energy is located around the atomic separation, R_a, and because of the low mass of ^{4}He, the value of the small attractive (van der Waals) potential is comparable to the zero-point kinetic energy. Therefore, this bosonic system forms a low density liquid at densities much less than those associated with the nuclear masses involved, or even usual molecular liquids. The lattice spacing for solid helium is expected to be even smaller than the average spacing for the liquid, which means that a large external pressure is necessary to overcome the zero-point energy in order to form solid helium.

For thermal quantum systems, typical thermal energies $k_B T_{crit}$ are given by averaged energies for constituent particles of mass m, which define a thermal distance scale $R_{thermal} \approx \frac{\hbar c}{k_B T_{crit}}$ that satisfies $R_{thermal}\lambda_m \sim (\Delta x)^2$, in terms of the reduced Compton wavelength of the mass scale $\lambda_m \equiv \frac{\hbar}{mc}$, and the scale of zero-point motions of those masses Δx. This relationship just follows from the momentum-position uncertainty principle. For example, for a degenerate free Fermi gas, the number density relationship $n = \frac{g_m}{6\pi^2}(2m\epsilon_{thermal})^{3/2}$ implies $R_\epsilon \lambda_m = \frac{1}{2}(\frac{6\pi^2}{g_m})^{2/3}(\Delta x)^2$, where g_m is the degeneracy of the fermion. For a non-relativistic simple harmonic oscillator, the zero-point energy satisfies $\frac{(\Delta p)^2}{2m} + \frac{1}{2}m\omega^2(\Delta x)^2 = \frac{1}{2}\hbar\omega$. The zero-point kinetic and potential energies each partition half of the vacuum energy $\frac{1}{4}\hbar\omega$. The resulting uncertainties $(\Delta x)^2 = \frac{\hbar}{2m\omega}$ and $(\Delta p)^2 = \frac{\hbar m\omega}{2}$ saturate the quantum measurement condition $\Delta x \, \Delta p = \frac{\hbar}{2}$. Thus, for simple harmonic motions $R_{\hbar\omega}\lambda_m = 2(\Delta x)^2$.

Therefore, it is apparent that the zero-point motions of sources correspond to the vacuum energies of the associated quanta. Physical effects associated with the zero-point motions need not directly depend on the coupling constant of the involved interactions but do depend on the nature of the interaction (for instance, whether there exists neutralizing charges, whether it is a long-range interaction). In physical systems, zero-point motions correspond to a saturation of the triangle inequality. The application of these results to cosmological dark energy will be discussed in Part II.

2.1.4 Quantum entanglement

Formulations of quantum mechanics incorporate correlated wave-like properties over space-like (acausal) regions. Generally, quantum coherence refers to the maintenance of physically conserved parameters in the unmeasured nature of quantum states at space-like separations, for which a system is sometimes interpreted as developing supraluminal correlations (without the exchange of signals) in the observable behavior of such quantum states. The term *entanglement* typically refers to a multi-component system whose components have been prepared in a coherent state but can be individually measured while the other component(s) remain coherent. Quantum phenomena generally exhibit properties of space-like phase coherence within seemingly otherwise stochastic processes. Once it is recognized that the microscopic universe is fundamentally quantum mechanical, it then becomes puzzling as to why classical behaviors are predictive in a broad variety of circumstances. Classical physics is sensible only when describing the dynamics and kinematics between the *dis*entangled components of a system. This means that components (clusters) that interrelate classically cannot have (exterior) coherence properties. These behaviors describe what is meant by *cluster decomposition*, or classically disentangled clusters.

From its beginnings, many have questioned whether the non-local, non-classical aspects of quantum theory might be due to hidden variables obeying an algorithm that only give the illusion of quantum behaviors. As an illustration, consider the

Einstein–Podolsky–Rosen (EPR) [15] analysis. Einstein was quite uneasy about quantum mechanics. In a well-known paper, the EPR authors stated that, "We are forced to conclude that the quantum-mechanical description of physical reality given by wave functions is not complete." This statement asserts that wave functions do not satisfy *local realism*. Local realism maintains that each "element of reality" can be influenced only by events within its past light cone. Realism presumes that objects have all properties definite whether they are observed/interact or not. Since the experimental basis of quantum mechanics was well-established by the time of this analysis, the question posed was not whether quantum theory was *correct*, but rather whether quantum theory gives only an *incomplete* description of physical phenomena. Of particular concern was the principle of *locality*, i.e., that influences between spatially separated regions of a quantum coherent or entangled system must travel in a way that they continuously pass through space-time. If the postulates of relativity are incorporated, any communications or causations/effects cannot be conveyed faster than the speed of light. However, the quantum behaviors of certain entangled systems appear paradoxical when interpreted in terms of an evolving local algorithm.

Consider, for example, the decay of a pseudo-scalar π^0 meson. The meson decays to an entangled pair of photons combining in a manner consistent with overall angular momentum and parity $J^P = 0^-$. The photons travel in opposite directions to conserve momentum and have opposite polarizations to conserve angular momentum, in the *entangled* state:

$$|\Psi(\mathbf{P}_{total} = \mathbf{0}, J = 0)\rangle = \frac{1}{\sqrt{2}} \left(| + \mathbf{p}, R\rangle \, | - \mathbf{p}, R\rangle - | + \mathbf{p}, L\rangle \, | - \mathbf{p}, L\rangle \right), \tag{2.39}$$

where R refers to right-handed polarization, and L refers to left-handed polarization. A measurement on the photon with momentum $+\mathbf{p}$ of polarization R requires that the polarization measured on the photon with momentum $-\mathbf{p}$ must likewise be R, despite this state being unknown prior to either measurement. Seemingly, a measurement on one component of the coherent system affects the measurement upon another space-like separated component.

In order to revive local realism, the EPR authors postulated some currently unknown properties of the physical universe that should describe the apparent acausal quantum behaviors, which might be parameterized through *hidden variables* carried by the entangled components. The authors proposed an experiment that could disprove their beliefs. Their proposition suggested that a measurement of position for one particle of an entangled pair and a measurement of momentum for the other would allow violation of the uncertainty principle.

In the early 1960s, J.S. Bell [16] performed an analysis of the thought experiment of EPR. Using a construction that was very favorable for describing a consistent quantum system using a local hidden variable, a straightforward experiment that could distinguish between quantum non-locality and a hidden statistical parameter that should give similar, but sometimes differing, results was suggested.

2.1.5 Bell's Inequality

The formal *unknown* and *unknowable* character of quantum mechanics was parametrically categorized by Bell. He addressed the inconsistency of the predictions of quantum mechanics with those of a theory of local "statistical ignorance", where a hidden variable describes the behavior of a given physical system, being just unknown to the observers. For instance, if a spin zero, negative parity particle decays to an electron and a positron while conserving angular momentum and parity, angular momentum conservation requires that any eventual measurements of the spins of the products of the decays must be anti-correlated in order to cancel, even if those measurements at spatially separated positions cannot allow any possible communication between those separate measurements. If the positron is measured to have a spin directed along direction $\hat{n}_1$, and the electron is non-locally measured to have a spin directed along $\hat{n}_2$, then quantum mechanics predicts that the normalized averaged value of the product of the measured spins Π_{QM} be correlated according to:

$$\Pi_{QM}(\hat{n}_1, \hat{n}_2) \equiv \left(\frac{2}{\hbar}\right)^2 \langle pair|\mathbf{S}_{positron}(\hat{n}_1)\,\mathbf{S}_{electron}(\hat{n}_2)|pair\rangle = -\hat{n}_1 \cdot \hat{n}_2 \,.$$

This results from the quantum requirement that the overall spin state of the correlated particles $|pair\rangle = \frac{1}{\sqrt{2}}(|\uparrow\rangle|\downarrow\rangle - |\downarrow\rangle|\uparrow\rangle)$ be of a singlet form to conserve angular momentum.

However, Bell showed that if there is actually a hidden local variable in each measured system that describes the distant correlations in a classical statistical manner, the average value of the product of the measured spin directions Π_{HV} must satisfy the inequality:

$$|\Pi_{HV}(\hat{n}_1, \hat{n}_2) - \Pi_{HV}(\hat{n}_1, \hat{n}_3)| \leq 1 + \Pi_{HV}(\hat{n}_2, \hat{n}_3)\,.$$

It is straightforward to choose angles that give contradictory predictions (for instance, $\hat{n}_1$ and $\hat{n}_2$ perpendicular to each other and bisected by $\hat{n}_3$). Experiments [17] have subsequently supported the predictions of quantum mechanics. No violations of quantum theory on energy scales of tens of GeVs have been observed [18]. Other formulations that sacrifice local realism for non-local realism [19] predict results that are likewise contrary to experiment [20]. These experiments generally demonstrate a stronger statistical correlation of measured results than have been predicted by any hidden variable theory. At this time, all experimental evidence is consistent with quantum theory. In particular, it would be quite difficult to model even the local behavior of identical fermions without using the quantum paradigm. The absence of the previously described forms of realism is what is meant by the statement that quantum mechanics has both *unknown* and *unknowable* aspects.

Supplement: Goedel and Incompleteness

There has been considerable work done on whether or not a formal system that explains all phenomena can be developed from a closed set of principles. In the early 1930s, Kurt Goedel[2] examined whether logical systems can be both self-consistent and complete. A theory was defined to be a set of symbols that can be manipulated using a set of rules. There are assumed to be procedures by which a truth value can be assigned to a given assertion. The theory is *self-consistent* if it contains no internally contradictory proofs, i.e., the theory cannot be used to prove a false statement. A theory is *complete* if any true theorem produced by the theory can be proven to be true using that theory.

Goedel was able to map any such logical system onto number theory, and show that a given non-trivial, formal (enumerable) theory can construct a statement that *is* true, but cannot be proved to be true using that theory. This was accomplished by developing a self-referential statement; a statement that refers to its own truth. He represented statements as numbers in a theory about numbers. The self-referential statement about $\mathcal{S}_G$ of the form:

The statement $\mathcal{S}_G$ cannot be proved true in this theory,

can then be added to the theory as a property of a certain number.[3] This then either incorporates an inconsistency in the theory or demonstrates that the theory is incomplete.

This *Incompleteness Theorem* furthermore implies that no consistent enumerable theory can contain a statement of its own consistency. A complete theory cannot be based purely on logic or closed, self-consistent mathematical formulas. However, it is clear that physics is an experimental science that cannot be ascertained purely via logic or mathematics. The expression of those dimensional parameters (length, mass, time, etc.) developed during dimensional analysis at the very beginning of introductory physics classes define entities that cannot be arguments of mathematical functions. Theoretical physics is thus an endeavor that attempts to model experimental phenomenology in the most elegant manner possible.

[2] J. van Heijenoort, *From Frege to Godel*, Harvard University Press (1967).

[3] C. Chall, *Kuhn, Godel, and Bell: Paradigm, Incompleteness, and Indeterminism*, Senior Thesis, Dept. of Physics, Howard University (2009).

2.1.6 Quantum fields

Multi-particle states of identical elementary particles always obey the quantum statistics of Fermi–Dirac, or Bose–Einstein. Proximal identical fermions are always in entangled states that are completely anti-symmetric under the exchange of any two of those fermions, resulting in the Pauli exclusion principle giving atomic structure, the impenetrability of most forms of matter, and increased quantum degeneracy pressures. Similarly, entangled identical bosons are always in states that are completely symmetric under the exchange of any two of those bosons, resulting in phenomena such as superfluidity, superconductivity, and the lasing of photons.

The operator product representation of quantum states is quite convenient for representing the quantum statistics of identical particles. This is achieved by defining anti-commutation/commutation relations for momentum-space field operators consistent with the quantum statistics of the particles. Fermion annihilation operators $\hat{f}(\mathbf{p}, \lambda; m)$ and creation operators $\hat{f}^{\dagger}(\mathbf{p}, \lambda; m)$ incorporate Fermi statistics through the anti-commutation relations:

$$\begin{aligned}\{\hat{f}(\mathbf{p}', \lambda'; m), \hat{f}(\mathbf{p}, \lambda; m)\} &= 0 = \{\hat{f}^{\dagger}(\mathbf{p}', \lambda'; m), \hat{f}^{\dagger}(\mathbf{p}, \lambda; m)\}, \\ \{\hat{f}(\mathbf{p}', \lambda'; m), \hat{f}^{\dagger}(\mathbf{p}, \lambda; m)\} &= \delta_{\lambda'\lambda} \frac{\sqrt{|\mathbf{p}|^2+(mc)^2}}{p^0_{(s)}(m)} \delta^3(\mathbf{p}' - \mathbf{p}),\end{aligned} \tag{2.40}$$

where the λs represent the discrete internal quantum numbers of the particles, and the standard state component p^0_s is unity for massless particles and mc for massive particles. Multi-particle momentum eigenstates satisfying Fermi statistics can then be constructed using products of the creation operators acting on the unique vacuum state defined for the system $\langle vac|vac\rangle = 1$, which satisfies $\hat{f}(\mathbf{p}, \lambda; m)|vac\rangle = 0$:

$$\begin{aligned}&|\mathbf{p}_1, \lambda_1 : \mathbf{p}_2, \lambda_2 : \ldots : \mathbf{p}_N, \lambda_N\rangle \\ &\quad \equiv \hat{f}^{\dagger}(\mathbf{p}_1, \lambda_1; m)\hat{f}^{\dagger}(\mathbf{p}_2, \lambda_2; m) \cdots \hat{f}^{\dagger}(\mathbf{p}_N, \lambda_N; m)|vac\rangle.\end{aligned} \tag{2.41}$$

This appropriately normalized form vanishes if any two sets of quantum numbers are identical, due to the anti-commutativity of the operators. Similarly, boson annihilation operators $\hat{b}(\mathbf{p}, \lambda; m)$ and creation operators $\hat{b}^{\dagger}(\mathbf{p}, \lambda; m)$ incorporate Bose statistics through the commutation relations:

$$\begin{aligned}[\hat{b}(\mathbf{p}', \lambda'; m), \hat{b}(\mathbf{p}, \lambda; m)] &= 0 = [\hat{b}^{\dagger}(\mathbf{p}', \lambda'; m), \hat{b}^{\dagger}(\mathbf{p}, \lambda; m)] = 0, \\ [\hat{b}(\mathbf{p}', \lambda'; m), \hat{b}^{\dagger}(\mathbf{p}, \lambda; m)] &= \delta_{\lambda'\lambda} \frac{\sqrt{|\mathbf{p}|^2+(mc)^2}}{p^0_{(s)}(m)} \delta^3(\mathbf{p}' - \mathbf{p}),\end{aligned} \tag{2.42}$$

Multi-particle momentum eigenstates satisfying Bose statistics are constructed using products of the creation operators acting on the vacuum state that satisfies

$\hat{b}(\mathbf{p}, \lambda; m)|vac\rangle = 0$:

$$|\mathbf{p}_1, \lambda_1 : \mathbf{p}_2, \lambda_2 : \ldots : \mathbf{p}_N, \lambda_N\rangle \\ \equiv \hat{b}^\dagger(\mathbf{p}_1, \lambda_1; m)\hat{b}^\dagger(\mathbf{p}_2, \lambda_2; m) \cdots \hat{b}^\dagger(\mathbf{p}_N, \lambda_N; m)|vac\rangle. \quad (2.43)$$

Unlike fermions, there is no exclusion of bosons with identical quantum numbers. The general normalization form for fermions/bosons is given by:

$$\langle \mathbf{p}'_1, \lambda'_1 : \mathbf{p}'_2, \lambda'_2 : \ldots : \mathbf{p}'_N, \lambda'_N | \mathbf{p}_1, \lambda_1 : \mathbf{p}_2, \lambda_2 : \ldots : \mathbf{p}_N, \lambda_N\rangle \\ = \sum_{\{\mathcal{P}\}} (\mp)^{\mathcal{P}} \frac{\epsilon(p_1)}{p^0_{(s)}c} \delta^3(\mathbf{p}_{\mathcal{P}_1} - \mathbf{p}_1)\delta_{\lambda_{\mathcal{P}_1},\lambda_1} \cdots \frac{\epsilon(p_N)}{p^0_{(s)}c} \delta^3(\mathbf{p}_{\mathcal{P}_N} - \mathbf{p}_N)\delta_{\lambda_{\mathcal{P}_N},\lambda_N}, \quad (2.44)$$

where the permutation index $\mathcal{P}$ is the number of particle exchanges necessary to bring the indices $\{\mathcal{P}_1, \mathcal{P}_2, \ldots, \mathcal{P}_N\}$ into standard order $\{1, 2, \ldots, N\}$.

Besides being a convenient incorporation of the quantum statistics of particles, the quantum field operators provide a convenient mechanism for including particle creation in models consistent with physical principles. The configuration space-form of the operator product that will generate a given particle state is the *quantum field* representing that particle. The fields must incorporate the transformation properties appropriate for the particle type (i.e., scalar, vector, etc.). General transformation properties of quantum fields will be explored in Chapter 4. For the present, fields for asymptotic electromagnetic radiations (photons) and scalar invariant particles will be developed next.

Photon field

The electromagnetic field, represented by the vector potential $\hat{\mathbf{A}}(\mathbf{r}, t)$ will be a hermitian boson field satisfying the source-free form of Maxwell's equations. The vector field ($J^P = 1^-$) has two possible polarization states that are transverse to the direction of propagation $\mathbf{k}$ (from Gauss' law), as well as transverse to each other (from Faraday's law). A massive vector particle $J = 1$ will generally have $2J + 1 = 3$ components of angular momentum in its rest frame. This means that in different Lorentz frames of reference, the massive particle will have different helicities. However, since the electromagnetic field is massless, there can be no Lorentz transformation that can bring it to rest. The component of polarization along the direction of motion, defined as the *helicity*:

$$\hat{\mathrm{h}} \equiv \frac{\hat{\mathbf{J}} \cdot \hat{\mathbf{P}}}{|\hat{\mathbf{P}}|}, \quad (2.45)$$

takes the value J_z for the standard state vector $p^0_{(s)} = (1, 0, 0, 1)$. A Lorentz boost along the z-direction leaves J_z invariant (since the boost generator in this direction commutes with the angular momentum component in this direction). A general

Lorentz boost into an arbitrary state can be expressed as a boost along the z-axis (or general standard direction), which leaves the helicity unchanged, followed by a rotation to its final orientation. Therefore, the helicity of the massless particle is invariant under (continuous) Lorentz transformations and labels the state of the particle. However, since angular momentum is a pseudo-vector under parity transformations, h $\rightarrow$ -h, i.e., helicity is a pseudo-scalar. A parity transformation therefore inverts both the momentum and helicity of a massless particle. Photons will be labeled by the possible helicities h$= \pm 1$ for right-/left-handed polarization states, as well as by momenta, $\mathbf{p} \equiv \hbar \mathbf{k}$.

The polarization vectors and wave vector will be assumed to form an orthonormal triad $(\mathbf{e}(\mathbf{k}, 1), \mathbf{e}(\mathbf{k}, 2), \frac{\mathbf{k}}{|\mathbf{k}|})$. The photon field is expected to take the form:

$$\hat{\mathbf{A}}(\mathbf{r}, t) = \sum_{\lambda} \int d^3k \, \eta(k, \lambda) \left[\mathbf{e}(\mathbf{k}, \lambda) \, \hat{b}(\mathbf{k}, \lambda) \, e^{i\vec{k}\cdot\vec{x}} + \mathbf{e}^*(\mathbf{k}, \lambda) \, \hat{b}^\dagger(\mathbf{k}, \lambda) \, e^{-i\vec{k}\cdot\vec{x}} \right], \tag{2.46}$$

where $k^0 = |\mathbf{k}|$, and $\mathbf{e}(\mathbf{k}, \lambda) \cdot \mathbf{e}^*(\mathbf{k}, \lambda') = \delta_{\lambda,\lambda'}$. The normalization factor $\eta(k, \lambda)$ can be directly determined by examining the form of the energy operator:

$$\hat{H} = \frac{1}{8\pi} \int d^3r \left\{ \vec{E}(\mathbf{r}, t) \cdot \vec{E}(\mathbf{r}, t) + \vec{B}(\mathbf{r}, t) \cdot \vec{B}(\mathbf{r}, t) \right\}$$

$$\Rightarrow \hat{H} = \sum_{\lambda} \int d^3k \, (2\pi k \, \eta(k, \lambda))^2 \hat{b}^\dagger(\mathbf{k}, \lambda) \hat{b}(\mathbf{k}, \lambda) + \text{constant}, \tag{2.47}$$

where the infinite constant is formally eliminated by requiring :*normal ordering*: (all annihilation operators to the right of creation operators) when evaluating products of fields $\langle vac| : \hat{H} : |vac\rangle < \infty$. Thus, the properly normalized photon field is of the form:

$$\hat{\mathbf{A}}(\mathbf{r}, t) = \sum_{\lambda} \int d^3k \, \frac{\hbar^2}{2\pi} \sqrt{\frac{c}{|\mathbf{k}|}} \left[\mathbf{e}(\mathbf{k}, \lambda) \, \hat{b}(\mathbf{k}, \lambda) \, e^{i\vec{k}\cdot\vec{x}} + \mathbf{e}^*(\mathbf{k}, \lambda) \, \hat{b}^\dagger(\mathbf{k}, \lambda) \, e^{-i\vec{k}\cdot\vec{x}} \right]. \tag{2.48}$$

The operator product form of the photon field is seen to either create or annihilate an appropriate massless particle with consistent momentum and helicity.

Scalar fields

A scalar field will likewise obey Bose statistics and will be chosen to satisfy the normalization:

$$\hat{\Phi}(\vec{x}) \equiv \frac{\hbar}{\sqrt{(2\pi)^3 2 p^0_{(s)} c}} \int \frac{p^0_{(s)} c \, d^3p}{\epsilon(\mathbf{p})} \left[\hat{b}(\mathbf{p}) \, e^{\frac{i}{\hbar}\vec{p}\cdot\vec{x}} + \hat{b}^\dagger(\mathbf{p}) \, e^{-\frac{i}{\hbar}\vec{p}\cdot\vec{x}} \right]. \tag{2.49}$$

This field is directly seen to satisfy the Klein–Gordon equation:

$$\left(\nabla^2 - \frac{1}{c^2}\frac{\partial^2}{\partial t^2}\right)\hat{\Phi}(\vec{x}) = \left(\frac{mc}{\hbar}\right)^2 \hat{\Phi}(\vec{x}). \tag{2.50}$$

The normalization has been chosen such that the canonical momentum is given by $\hat{\Pi}(\vec{x}) = \frac{\partial}{\partial t}\hat{\Phi}(\vec{x})$ and satisfies the equal-times commutation relation:

$$[\hat{\Phi}(\mathbf{r}_1, t), \hat{\Pi}(\mathbf{r}_2, t)] = i\hbar\delta^3(\mathbf{r}_1 - \mathbf{r}_2). \tag{2.51}$$

This demonstrates why the operator product representation is sometimes referred to as second quantization, since the field satisfies an equivalent Klein–Gordon wave equation as does a standard one-particle wave function.

Condensed matter fields

Quantum fields are also convenient for describing the dynamics of fluids. For the present discussion, the bosonic excitation spectrum will be assumed to be labeled by discrete quantum numbers λ. Field operators are described by:

$$\hat{\psi}(\mathbf{r}, t) = \sum_\lambda \phi_\lambda(\mathbf{r}, t)\,\hat{a}_\lambda. \tag{2.52}$$

The connection of the quantum field to the single-particle wavefunction is given by:

$$\phi_\lambda(\mathbf{r}, t) = \langle vac|\hat{\psi}(\mathbf{r}, t)|\lambda\rangle. \tag{2.53}$$

As long as the wavefunction is orthonormal $\int \phi_\lambda^*(\mathbf{r}, t)\,\phi_{\lambda'}(\mathbf{r}, t)\,d^3r = \delta_{\lambda\lambda'}$, the number operator can be expressed as:

$$\int \hat{\psi}^\dagger(\mathbf{r}, t)\,\hat{\psi}(\mathbf{r}, t)\,d^3r = \sum_\lambda \hat{a}_\lambda^\dagger \hat{a}_\lambda, \tag{2.54}$$

which defines a number density for the excitations $\hat{n}(\mathbf{r}, t) \equiv \hat{\psi}^\dagger(\mathbf{r}, t)\,\hat{\psi}(\mathbf{r}, t)$. The canonical commutation relation $[\hat{\psi}(\mathbf{r}, t), \hat{\psi}^\dagger(\mathbf{r}', t)] = \sum_\lambda \phi_\lambda(\mathbf{r}, t)\phi_\lambda^*(\mathbf{r}', t)$ is directly related to the completeness of the wavefunctions. This example demonstrates direct connections of quantum fields to wave mechanics.

2.1.7 Space-like correlations

The state vectors that describe quantum systems have the property of maintaining coherence over distances beyond what would be allowed by classical communications between the disparate spatially separated regions within which that coherence is maintained. One is able to calculate how the description of one of the simplest systems, a massless scalar field, alters when its spatial parameterization is examined at two different positions in different orders of operation. When comparing different orders of description of the field at the space-time positions $\vec{x}_1 = (ct_1, \mathbf{r}_1)$ and $\vec{x}_2 = (ct_2, \mathbf{r}_2)$ from Eq. 2.50, the vacuum expectation value satisfies:

$$\langle vac|\hat{\Phi}(\vec{x}_1)\hat{\Phi}(\vec{x}_2) + \hat{\Phi}(\vec{x}_2)\hat{\Phi}(\vec{x}_1)|vac\rangle$$
$$= \frac{\hbar^2}{4\pi^2 \Delta s^2} = \frac{\hbar^2}{4\pi^2[(x_1 - x_2)^2 + (y_1 - y_2)^2 + (z_1 - z_2)^2 - c^2(t_1 - t_2)^2]}.$$

Therefore, even if the times of measurement are the same $t_1 = t_2$, this value (the equal time correlation function) does not vanish if the spatial positions differ, but rather falls off with the inverse square of the distance of separation. The form directly manifests consequences of the space-like phase coherence of the quantum system.

Since the fields are expected to average to zero value in a vacuum, $\langle vac|\hat{\Phi}(\vec{x})|vac\rangle = 0$. If the behaviors at different positions were not correlated, one would expect the average values to be independent:

$$\langle vac|\hat{\Phi}_{independent}(\vec{x}_1)\,\hat{\Phi}_{independent}(\vec{x}_2)|vac\rangle$$
$$= \langle vac|\hat{\Phi}_{independent}(\vec{x}_1)|vac\rangle\langle vac|\hat{\Phi}_{independent}(\vec{x}_2)|vac\rangle.$$

Clearly, since:

$$\langle vac|\hat{\Phi}(\vec{x}_1)\hat{\Phi}(\vec{x}_2) + \hat{\Phi}(\vec{x}_2)\hat{\Phi}(\vec{x}_1)|vac\rangle \neq 2\langle vac|\hat{\Phi}(\vec{x}_1)|vac\rangle\langle vac|\hat{\Phi}(\vec{x}_2)|vac\rangle,$$

for the boson field, this quantum system maintains correlated behavior over space-like separations.

However, a measurement at $\mathbf{r}_1$ that affects one at $\mathbf{r}_2$ involves a *difference* in the behavior of the boson field at these points. Indeed, the spin-statistics theorem [21,22] requires that the commutator between product field descriptions at different space-time position must vanish if the distance involved does not allow a communication at the speed of light:

$$\langle vac|\hat{\Phi}(\vec{x}_1)\hat{\Phi}(\vec{x}_2) - \hat{\Phi}(\vec{x}_2)\hat{\Phi}(\vec{x}_1)|vac\rangle = 0$$

if $(x_1 - x_2)^2 + (y_1 - y_2)^2 + (z_1 - z_2)^2 > c^2(t_1 - t_2)^2$, (a space-like separation). Therefore, quantum coherent systems manifest space-like correlations without allowing supra-luminal (faster than light) communications. This means that causally related decoherences cannot be communicated faster than light.

2.2 Quantum mechanics and statistics

Quantum systems often consist of components with varying degrees of entanglement. The density matrix formulation provides a convenient framework for describing both pure quantum states and incoherent sums of such pure states. A *pure* state is a coherent quantum system that can be represented in terms of the quantum state vector $|\psi\rangle$. *Entangled* states are constituents of a multi-component pure state. For instance, two particles that have been created by a single decay maintain their (relative) coherence until either particle has that coherence broken by a measurement. That composite state remains a pure quantum state until such a measurement occurs.

A *mixed* state consists of a set of pure states that are independent of each other (with regards to quantum coherence properties). The constituent components maintain defined proportions in a mixed states. For example, a partially polarized beam of independent particles represents a mixed state, with the degree of polarization parameterizing the proportions of the constituent particles with specified spin orientations. An elegant mechanism for describing arbitrarily mixed quantum system will be developed in the next section.

2.2.1 Density matrix formalism

A normalized quantum state vector provides a convenient device for comparing the relative probabilities that a quantum system will be measured to be in a given state, or set of states. Examining Eq. 2.1, the normalization and completeness conditions infer that $1 = \langle\psi|\psi\rangle = \sum_n \langle\psi|n\rangle\langle n|\psi\rangle$. Thus, a partial sum of this expression over a subset of states $\{\tilde{n}\}$ provides the proportion of possible measurements consistent with that subset. Thus, $\delta\mathcal{P}_n \equiv \langle\psi|n\rangle\langle n|\psi\rangle = |\psi_n|^2$ represents the probability that the coherent quantum state will be measured in the state parameterized by n. Likewise, for continuous systems, the probability that the quantum state will be measured within an interval dq of a state parameterized by q is given by $d\mathcal{P}(q) \equiv |\Psi(q)|^2\, dq$.

The average value of many measurements of a given observable represented by operator $\hat{A}$ on a quantum state ψ (the *expectation value*) is given by:

$$\langle\hat{A}\rangle_\psi = \langle\psi|\hat{A}|\psi\rangle = \sum_n\sum_{n'}\langle\psi|n'\rangle\langle n'|\hat{A}|n\rangle\langle n|\psi\rangle = \sum_{n\,n'}\langle n|\psi\rangle\langle\psi|n'\rangle\langle n'|\hat{A}|n\rangle, \tag{2.55}$$

where the values have been expressed relative to a complete set of basis states $\{|n\rangle\}$. Thus, the expectation value of this quantum operator can be expressed in terms of a matrix representation:

$$\begin{gathered}\langle\hat{A}\rangle = \textstyle\sum_{n\,n'} \rho_{n\,n'} A_{n'\,n} = Tr\,\hat{\rho}\,\hat{A} \\ \hat{\rho} \equiv |\psi\rangle\langle\psi|, \quad \rho_{n\,n'} = \langle n|\psi\rangle\langle\psi|n'\rangle, \quad A_{n'\,n} = \langle n'|\hat{A}|n\rangle.\end{gathered} \tag{2.56}$$

Matrix elements of the operator $\hat{\rho}$ define the *density matrix* of the pure state. A diagonal element of the density matrix $\rho_{n\,n} = |\psi_n|^2$ represents the probability that the system would be measured to be in the state labeled $|n\rangle$. One should note that since the trace of a matrix is representation independent $Tr\,(\mathbf{S}^{-1}\mathbf{Q}\mathbf{S}) = Tr\,\mathbf{Q}$, *any* convenient basis of states can be utilized to evaluate the expectation value. If the state is normalized, the density matrix is likewise normalized:

$$Tr\,\hat{\rho} = \sum_n \langle n|\psi\rangle\langle\psi|n\rangle = \sum_n \langle\psi|n\rangle\langle n|\psi\rangle = \langle\psi|\psi\rangle = 1. \tag{2.57}$$

As previously stated, the density matrix formalism is particularly useful for describing systems made up of disentangled components. Such a *mixed state* is a system of independent quantum states $|\psi_a\rangle$, each of proportion w_a. For mixed states, the average value of a given observable is given by $\langle\hat{A}\rangle_{\{a\}} = \sum_a w_a \langle\psi_a|\hat{A}|\psi_a\rangle$, where the weights w_a satisfy $\sum_a w_a = 1$. The density operator of a mixed state is thus given by:

$$\hat{\rho} = \sum_a w_a\,|\psi_a\rangle\langle\psi_a|, \quad Tr\,\hat{\rho} = 1, \quad \langle\hat{A}\rangle_{\{a\}} = Tr\,\hat{\rho}\hat{A}. \tag{2.58}$$

Since the weights satisfy $0 \leq w_a \leq 1$, the relationship $w_a^2 \leq w_a$ implies that:

$$\hat{\rho}^2 \leq \hat{\rho}, \tag{2.59}$$

where the equality holds only for a pure state, with a single weight of value unity and all others vanishing. Thus, a pure state is a projection, $\hat{\rho}^2_{\text{pure}} = \hat{\rho}_{\text{pure}}$.

2.2.2 Properties of a density matrix

- A density matrix is hermitian, $\hat{\rho}^\dagger = \hat{\rho}$. This implies that its eigenvalues are real. For pure states, if λ_m is an eigenvalue of ρ_{pure}, then $\hat{\rho}^2_{\text{pure}} - \hat{\rho}_{\text{pure}} = \hat{0}$ implies that $\lambda_m(\lambda_m - 1) = 0$. Thus, the eigenvalues of pure state density matrices must take the values $\lambda_m = 0$ or $\lambda_m = 1$. Since $Tr\,\hat{\rho} = 1$, this means that all eigenvalues vanish for a pure state density matrix except for a single eigenvalue of unity.
- A pure state density matrix does *not* represent an ensemble of consistent microstates; rather it is an alternative depiction of a single coherent state. A composite entangled state will not be measured as a product (factored) state. For instance, for systems of fixed angular momentum, such a measurement might violate angular momentum conservation (for instance, a Clebsche–Gordon decomposition of a state involves a coherent sum of composite states).
- Since $\hat{\rho}^2 \leq \hat{\rho}$, the eigenvalues of a density matrix are positive semi-definite $0 \leq \lambda_m \leq 1$.
- Expectation values involve the trace of the density matrix with given operators. Since a trace is a representation independent operation, one can evaluate the trace using *any* convenient set of basis states.
- For independent (disentangled) systems A and B, the joint density operator factors $\hat{\rho}_{AB} = \hat{\rho}_A \times \hat{\rho}_B$.

Supplement: Comparison of quantum processes to Markov processes

As previously shown, the hermitian density operator associated with a pure quantum state is a projector $\mathcal{P}^2 = \mathcal{P}$ with real eigenvalues. Projective operators can also be constructed from Markov processes.[4] A future/past directed Markov process is stochastic in that its next state is correlated only to its most recent state, not to any prior/subsequent states. If the process is represented by a matrix $((\mathbf{M}))_{jk} \equiv m_{jk}$, each column c of a future directed Markov process satisfies $\sum_{j=1}^{N} m_{jc} = 1$, while each row r of a past directed Markov process satisfies $\sum_{k=1}^{N} m_{rk} = 1$. Quite generally, the matrix elements in Markov processes are real probabilities satisfying $0 \leq m_{jk} \leq 1$. Thus, it can be shown that future-directed Markov matrices satisfy $\mathbf{M}_{F2}\,\mathbf{M}_{F1} = \mathbf{M}_{F2}$, while past directed Markov matrices satisfy $\mathbf{M}_{P2}\,\mathbf{M}_{P1} = \mathbf{M}_{P1}$.

One is able to demonstrate that the resulting matrix from a infinite chain of a given Markov process is a projector, known as a *Perron* projection,[5] $\mathbf{P_M} \equiv \mathbf{M}^{\infty}$. Thus, one might construct chains of Markov processes analogous to path integrals that have endpoints associated with a given Perron projector. For the chain of a future directed Markov process, a real density matrix can be constructed from the Perron projector (whose columns are identical) using:

$$\rho_P \equiv \frac{\mathbf{P_M}^T \mathbf{P_M}}{Tr(\mathbf{P_M}^T \mathbf{P_M})}.$$

This density matrix can be shown likewise to be a projector, which means that it can represent a pure state. Conversely, operators satisfying $\langle \hat{A} \rangle = Tr(\rho_P\, \mathbf{A})$ can be mapped onto matrices satisfying $\langle \hat{A}_P \rangle = Tr(\mathbf{P_M}\, \mathbf{A_P})$ defined by $\mathbf{A_P} \equiv \frac{\mathbf{A}\mathbf{P}^T}{Tr(\mathbf{P_M}^T \mathbf{P_M})}$.

However, complex density matrix forms whose components have negative or complex phase relationships *cannot* be constructed using this procedure, since the Markov elements are real probabilities. Markov stochastic processes do not model all interference behaviors inherent in quantum systems.

2.2.3 Quantum statistical ensembles

Thermal systems are most likely to be measured in configurations near those describing thermal equilibrium, the most probable configuration. Even very different systems sharing minimal coherent exchanges can reach mutual thermodynamic equilibrium. Therefore, statistical physics offers an arena in which descriptions of

[4] Thomas Etter, private communication (2005).
[5] Philip Kurian, private communication (2010).

incoherent mixtures of fundamentally quantum systems can elegantly be described using density matrices.

At the core of statistical physics are the probabilities associated with a given configuration. Since measures of disconnected independent probabilities are multiplicative, the logarithm of such probabilities serves as an additive state variable that can be related to other state variables like energy. Boltzmann related the thermodynamic entropy to statistical probabilities in this manner and was able to connect this statistical entropy to the parameters describing the dynamics of an ideal gas. Since entropy was shown to be related to the counting of statistically relevant states, quantum mechanics ultimately provided the ideal foundation of countable states.

Although thermodynamic systems parameterize only the equilibrium (average) distribution of statistical processes, they are also quite accurate for describing quasi-equilibrium processes for which the system reaches equilibrium on a time scale much more rapid than the quasi-static changes being made. Therefore, the statistical physics described by quasi-static ensembles related to particular thermodynamic state variables has broad applicability to many problems of physical interest. A few familiar ensembles will be briefly discussed.

Microcanonical ensemble

For an isolated system, the postulate of equal *a priori* probabilities presumes that any of the Ω distinct states accessible to the system are equally likely to occur, where Ω is the thermodynamic weight. For such systems, the weights w_a (and probabilities $\mathcal{P}_a$) of the incoherent sum satisfy:

$$w_a = \begin{cases} \frac{1}{\Omega} & \text{for any of the } \Omega \text{ accessible states} \\ 0 & \text{for inaccessible states.} \end{cases}$$

Using Boltzmann's identification, one can define a statistical entropy using:

$$S_{thermal} = k_B \log \Omega = -k_B \, log \, w_a = -k_B \sum_a w_a \log w_a = -k_B \sum_a \mathcal{P}_a \log \mathcal{P}_a. \tag{2.60}$$

The thermal entropy is the form that appears in the first and second laws of thermodynamics. Using the orthogonality (independence) of the states $|\psi_a\rangle$, the log term can be expressed as $\log w_a = \langle\psi_a| \log \left(\sum_b w_b \, |\psi_b\rangle\langle\psi_b|\right) |\psi_a\rangle = \langle\psi_a| \log \hat{\rho} |\psi_a\rangle$. Thus, the statistical entropy takes the von Neumann form:

$$S_v = -k_B \sum_a \mathcal{P}_a \log \mathcal{P}_a = \langle -k_B \log \hat{\rho} \rangle = -k_B \, Tr \, \hat{\rho} \log \hat{\rho}. \tag{2.61}$$

The general entropy form in Eq. 2.61 is maximum when the weights $w_a = \frac{1}{\Omega}$ are all equal. Thus, the thermal entropy represents the maximum possible statistical entropy for this ensemble. The density matrix limits its operations onto the *accessible* states (which is a subset of all possible states).

Other ensembles

The *canonical ensemble* in statistical physics describes the dynamics of macroscopic parameters in a "small" subsystem embedded in a much larger system which is considered to provide a thermal bath of fixed temperature. For energy eigenstates, the thermal weights are obtained by developing the thermodynamic weight associated with the thermal bath contributing a partition of energy E_a to the small subsystem, and are given by:

$$w_a = \frac{e^{-E_a/k_B T}}{\sum_b e^{-E_b/k_B T}}. \tag{2.62}$$

This gives a thermal density operator of the form:

$$\hat{\rho}_{\text{canonical}} = \sum_a \frac{e^{-E_a/k_B T}}{\sum_b e^{-E_b/k_B T}} |\psi_a\rangle\langle\psi_a| = \frac{e^{-\hat{H}/k_B T}}{Tr e^{-\hat{H}/k_B T}} \sum_a |\psi_a\rangle\langle\psi_a|$$
$$= \frac{e^{-\hat{H}/k_B T}}{Tr e^{-\hat{H}/k_B T}}. \tag{2.63}$$

Thus, the canonical ensemble thermal density operator can be expressed in terms of the Helmholtz free energy, $F = -k_B T \log Tr\, e^{-\hat{H}/k_B T}$, in the form $\hat{\rho}_{\text{canonical}} = e^{(F-\hat{H})/k_B T}$. Similarly, the *grand canonical ensemble* optimizes a subsystem open to both energy *and* number fluctuations, yielding:

$$\hat{\rho}_{\text{grand canonical}} = \frac{e^{-(\hat{H}-\mu\hat{N})/k_B T}}{Tr e^{-(\hat{H}-\mu\hat{N})/k_B T}}. \tag{2.64}$$

Ensembles that describe fixed pressures, or other thermal baths, have also been developed.

2.2.4 Information content

The idea of *information* as a basic conserved property of the universe has been quite successful in fields such as cosmology [23–25] and communications [26]. In both classical and quantum physics there is a sense that information cannot be lost from an isolated system. For classical systems, this principle is described by Liouville's theorem [27], which asserts the conservation of the volume of the phase space of the system. The phase space consists of the positions and momenta of the constituents of the system, and as the system evolves, the shapes of fixed volumes contort due to the motions of the constituents. However, the phase space volume itself remains fixed. Information is apparently lost only due to one's inability to follow the fine details of the contorting volume as it evolves. Some information obtained from coarse-grained measurements on the system indeed becomes lost, giving rise to the second law of thermodynamics, whereas classical fine-grained measurements can recover the detailed histories of the various trajectories. The unitarity of the time evolution of an isolated quantum system ensures that the number of states describing the system remains conserved with time. Thus, the idea of information conservation derives from the number of states of a system being conserved, even if some of them become dynamically degenerate. For both classical and quantum systems, a measure of the information content of a system can be quantified using the thermodynamic parameters describing that system.

Most accepted descriptions of information relate it to the statistical description of entropy $S = k_B \log \Omega$, where Ω counts the number of accessible microstates corresponding to the macrostate being described (a measure of disorder introduced by coarse graining). Thus, the entropies of statistically independent systems add up. If the macrostate fluctuates in a thermal bath, the statistical entropy fluctuates (both up and down), while the average value corresponds to the thermodynamic entropy, which satisfies the well-supported laws of thermodynamics. In particular, the second law of thermodynamics states that the overall thermally averaged entropy of a system and its environment must be a non-decreasing function. However, due to thermal contact with the temperature bath, the order/disorder does fluctuate. Also, the second law does not assert that the thermal entropy of a subsystem cannot decrease, as occurs, for example, in reverse osmosis and biological processes. Since the thermal entropy is a state variable, it is one of the parameters that can be used to describe the state of the system, which is independent of the history of that system. Therefore, an equilibrium system gives no information about how the particular state of that system has been achieved, or the history of that state. This means that any information derived from equilibrium entropy informs about only the present macrostate of the system.

Several of the original successes of quantum physics involved its ability to explain thermodynamic phenomena such as black body radiation. Quantum physics inherently incorporates discrete, countable quantized processes. The von Neumann entropy demonstrated in Section 2.2.3 describes a basis-independent form for a general quantum system. This form can be used to examine the entropy of entangled quantum coherent systems. Generally, one can establish measures of information as a difference between the maximum entropy associated with equivalent occupation of all available microstates versus the entropy defined by a coherently maintained probability distribution with unequal occupation of microstates. The maximum entropy state as described is common among all systems, assuming that only the distribution of probabilities varies amongst the individual systems.

It is quite useful to develop a dimensionless measure of the relative information expressed by general statistically mixed systems. As mentioned, the von Neumann entropy S_υ in Eq. 2.61 provides a mechanism to develop a relative measure of the order maintained in a non-thermal system. A system in a single state can be described as having maximum information content (and zero entropy). Likewise, a thermal system is in a maximum entropy state, the macrostate with the largest number of possible microstates, which cannot reveal any history, since state variables in thermodynamics are path independent. A thermal state will be chosen to have zero information. Therefore, the dimensionless information measure will be defined by:

$$\mathcal{I} \equiv \left(S_{max} - S_{\upsilon}\right) / S_{bit}, \tag{2.65}$$

where S_{bit} is the statistical entropy associated with a single bit of information, and $S_{max} = S_{thermal}$ is the entropy associated with the state of least order.

There are some circumstances for which an information comparison between different types of systems would be convenient. The normalized information content [28] has been developed to compare these systems. The normalized information content, which is defined:

$$\mathcal{N}_{\mathcal{I}} \equiv \left(S_{max} - S_{\upsilon}\right) / S_{max} \tag{2.66}$$

provides a normalized relative measure $0 \leq \mathcal{N}_{\mathcal{I}} \leq 1$. Since the measure is normalized, it can be used to "compare apples to oranges". Systems with normalized information content near unity are highly constrained, while those with vanishing $\mathcal{N}_{\mathcal{I}}$ are informationally gray.

Properties of the information content

- For a thermal system, the statistical entropy is equal to the thermal entropy, which is the maximum entropy of the system (the most probable configuration). Thus, $\mathcal{I}_{thermal} = 0 = \mathcal{N}_{\mathcal{I}_{thermal}}$.
- For a pure state, the statistical entropy vanishes (the system is in a single state). Thus, $\mathcal{N}_{\mathcal{I}_{pure}} = 1$; i.e., all information is expressed in the single accessible state, and $\mathcal{I}_{pure}$ has its maximum value.
- For a homogeneous state, $w_{a_o} = 1$ and all others vanish, which presents the same information content as that particular pure state, a_o.
- Intermediate values of normalized information content express the degree to which the state is constrained or organized towards a particular bias.

2.2.5 *Statistics of quantum entanglement*

One of the most intriguing aspects of quantum processes is that of space-like coherence. If a multi-component system remains coherent as components separate, the system maintains its quantum nature over space-like separations, as previously discussed. Such components are considered to be *entangled* until a measurement breaks the coherence of any given component. If the information from a causally disjoint region of space-time is unavailable, one must develop tools to describe the physics of an entangled component using available measurements.

The state vector describing an entangled system cannot be decomposed as a direct product of the component states. As an example, consider a two-component system $|\Psi_{entangled}\rangle$ with components $\Psi_{\alpha\beta} \equiv \langle \alpha\, \beta | \Psi_{entangled}\rangle$ that is entangled across space-like separated regions A and B. The pure state density operator of the composite system takes the form:

$$\hat{\rho}_{entangled} = |\Psi_{entangled}\rangle\langle\Psi_{entangled}|. \tag{2.67}$$

This pure state density operator is a projector and has vanishing statistical entropy. If only information on measurements in region A can be used to model the dynamics, the density operator describing this physics must sum over the states for which no information is available:

$$\langle\alpha|\hat{\rho}_A|\alpha'\rangle \equiv ((\rho_A))_{\alpha\alpha'} = \sum_\beta \langle\alpha\,\beta|\hat{\rho}_{entangled}|\alpha'\,\beta\rangle = \sum_\beta \Psi_{\alpha\beta}\Psi^*_{\alpha'\beta}. \tag{2.68}$$

The density operator describing the physics in region B can likewise be modeled:

$$\langle\beta|\hat{\rho}_B|\beta'\rangle \equiv ((\rho_B))_{\beta\beta'} = \sum_\alpha \langle\alpha\,\beta|\hat{\rho}_{entangled}|\alpha\,\beta'\rangle = \sum_\alpha \Psi_{\alpha\beta}\Psi^*_{\alpha\beta'}. \tag{2.69}$$

The *entanglement entropies*, S_A for system A and S_B for system B, are defined using these density matrices in the entropy form in Eq. 2.61:

$$S_{A,B} = -k_B\, Tr\left[((\rho_{A,B}))\cdot\log((\rho_{A,B}))\right]. \tag{2.70}$$

These density matrices and entropies indeed satisfy all of the required properties appropriate for describing physical parameters.

A quite interesting property of the entanglement entropies associated with the separated components is that they have identical values. This can be shown by examining the eigenvalues of the density matrices, which can then be used to calculate the (representation independent) entropies. The density matrices satisfy:

$$((\rho_A))\cdot\mathbf{u}^{(A)} = \lambda_{(A)}\mathbf{u}^{(A)} \Rightarrow \sum_{\beta'\alpha'} \Psi_{\alpha\beta'}\Psi^*_{\alpha'\beta'}\, u^{(A)}_{\alpha'} = \lambda_A u^{(A)}_\alpha, \tag{2.71}$$

$$((\rho_B))\cdot\mathbf{u}^{(B)} = \lambda_{(B)}\mathbf{u}^{(B)} \Rightarrow \sum_{\alpha'\beta'} \Psi_{\alpha'\beta}\Psi^*_{\alpha'\beta'}\, u^{(B)}_{\beta'} = \lambda_B u^{(B)}_\beta. \tag{2.72}$$

If one multiplies $\Psi_{\alpha\beta}$ by the complex conjugate of Eq. 2.71, then sums over the index α, one obtains:

$$\sum_{\beta'}\left(\sum_\alpha \Psi_{\alpha\beta}\Psi^*_{\alpha\beta'}\right)\left(\sum_{\alpha'} \Psi_{\alpha'\beta'}\, u^{(A)*}_{\alpha'}\right) = \lambda_A\left(\sum_\alpha \Psi_{\alpha\beta} u^{(A)*}_\alpha\right). \tag{2.73}$$

The quantity in the first parenthesis on the left-hand side of Eq. 2.73 is $((\rho_B))$, the density matrix describing measurements in region B. Therefore, the form $(\sum_\alpha \Psi_{\alpha\beta} u_\alpha^{(A)*})$ is an eigenvector of this density matrix with eigenvalue λ_A. From Eq. 2.72, this indicates that the eigenvalues of $((\rho_B))$ are the same as those of $((\rho_A))$. Therefore:

$$S_{entanglement}^{(A)} = S_{entanglement}^{(B)}, \tag{2.74}$$

i.e., entanglement entropies are always equal!

Supplement: An example of entanglement

As an example of entangled statistics, consider the state of zero angular momentum represented by:

$$|\Psi_{entangled}\rangle \equiv \frac{1}{\sqrt{2}}(|\uparrow\downarrow\rangle - |\downarrow\uparrow\rangle).$$

The entangled (pure state) density matrix takes the form of a 4×4 matrix:

$$((\rho_{entangled})) = \begin{pmatrix} 0 & 0 & 0 & 0 \\ 0 & \frac{1}{2} & -\frac{1}{2} & 0 \\ 0 & -\frac{1}{2} & \frac{1}{2} & 0 \\ 0 & 0 & 0 & 0 \end{pmatrix}.$$

The density matrix $((\rho_A))$ then takes the form of the 2×2 matrix given by:

$$((\rho_A)) = \begin{pmatrix} \frac{1}{2} & 0 \\ 0 & \frac{1}{2} \end{pmatrix},$$

which is clearly *not* a pure state. The entanglement entropy is given by:

$$S_{entanglement}^{(A)} = k_B \log 2 = S_{entanglement}^{(B)}.$$

Since this is the maximum entropy associated with the accessible states, the entangled information contents vanish, despite having been generated from a pure entangled state.

As a concrete example, consider two otherwise isolated quantum entangled subsystems which have a slow, reversible thermal leak that passes thermal entropy from one system to the other. Since the systems are entangled, not all information about one subsystem is completely available to the other subsystem. Since the two entangled systems are also undergoing thermal exchanges, one is able to describe the information content of one of the systems in terms of a difference between its additive thermal entropy and its entanglement entropy (see e.g., Ref. [23],

Chapter 8). Any information available to one of the subsystems must sum over all possible states of the other that are consistent with the given measurements in the original subsystem. Because of the nature of quantum entanglement, the entanglement entropy $S^{(A)}_{entanglement}$ of subsystem A is equal to the entanglement entropy $S^{(B)}_{entanglement}$ of entangled subsystem B. Since the overall entangled system must of course be in a pure quantum state, the overall entanglement entropy must vanish, $S^{(A+B)}_{entanglement} = 0$ (a pure state has probability one). This clearly demonstrates that the entanglement entropy is not additive. However, when the density matrices represent thermal probabilities, the resulting thermal entropies $S^{(A)}_{thermal}$ and $S^{(B)}_{thermal}$ are indeed additive, $S^{(A+B)}_{thermal} = S^{(A)}_{thermal} + S^{(B)}_{thermal}$. The information in a subsystem can generally be defined in terms of the difference between the coarse-grained entropy and the defined form for the degree of observed fine-graining of its entropy (in this case, the entanglement entropy), $\mathcal{I}^{(A)} = (S^{(A)}_{coarse-grained} - S^{(A)}_{fine-grained})/S_{bit}$. This indeed serves as a measure of information loss due to the coarse graining of the macroscopic observation. Since the thermal entropy typically parameterizes the maximum value entropy state, a subsystem with measured fine-graining the same as that of a thermal system gives zero information, whereas a system observed in a pure state with vanishing entropy gives maximum information. For the overall, isolated entangled system, this information measure coincides with the system's total thermal entropy, which is indeed conserved for the reversible leak.

2.2.6 No-cloning theorem

If a mechanism for faithfully copying quantum states can be found, measurements on various copies could be used to ultimately violate quantum measurement constraints. The no-cloning theorem [29,30] demonstrates that quantum states that evolve using linear dynamics cannot be cloned in this manner.

Cloning will be defined as a quantum state replication process that results in a new, identical state that (1) is not entangled with the original state, and (2) evolves in a parallel manner to the original state. A separate (normalized) *generic* blank state $|blank\rangle$ will provide the raw medium upon which the replication occurs. Quantum states can be generally represented in terms of state vectors. Vector spaces are linear (i.e., sums of vectors produce other vectors). Quantum states evolve using linear (matrix) operators $\hat{U}$ that preserve probabilities. The cloning process of the state $|\psi\rangle$ is expected to preserve independent probabilities and thus should be represented through a unitary operation $\hat{U}_C$:

$$\hat{U}_C\ (|\psi\rangle \times |blank\rangle) = |\psi\rangle \times |\psi\rangle_{copy}. \tag{2.75}$$

In general, another state $|\phi\rangle$ can likewise be cloned:

$$\hat{U}_C\ (|\phi\rangle \times |blank\rangle) = |\phi\rangle \times |\phi\rangle_{copy}. \tag{2.76}$$

The states subscripted with *copy* represent the cloned states.

If the unitary operation is to preserve normalization, the norms of the states on the left-hand sides should be identical to the norms of the states on the right-hand sides:

$$\langle blank|\langle\phi|\,\hat{U}_C^\dagger\hat{U}_C\,|\psi\rangle|blank\rangle =_{copy} \langle\phi|\langle\phi|\psi\rangle|\psi\rangle_{copy}. \tag{2.77}$$

This can be satisfied only if:

$$\langle\phi|\psi\rangle = \langle\phi|\psi\rangle^2, \tag{2.78}$$

which is generally not true (for generic blank states). This is only true if the states are identical or orthogonal and never true if their inner product has a non-vanishing complex phase. Therefore, a linear operation $\hat{U}_C$ that clones a *general* quantum state and preserves normalization cannot be found.

2.3 Quantum mechanics and gravity

Gravitation is an interaction of considerable familiarity, yet mathematical subtlety. The interaction is quite weak, with, for instance, the left-over polarizing force generating static cling easily overcoming the attractive force of an entire planet. Objects as disparate as satellites, cannon balls, and moons all follow identical trajectories when given identical initial conditions. This characteristic provides the basis of the geometrization of gravitation that will be discussed in Part II.

Much of the complexity in understanding the fundamental nature of gravity comes from the irreducible dimensions of its coupling constant G_N. For electrodynamics, the coupling can be parameterized in terms of the fine structure constant $\alpha \equiv \frac{e^2}{\hbar c}$, which is a dimensionless number of the order of 1/137 at low energies. This allows one to examine solutions describing physical systems that are perturbative in orders of the dimensionless coupling α. Straightforward mathematical functions describing quantum probabilities can functionally include this dimensionless form. However, no such dimensionless form can be constructed using the gravitational coupling. Whether its the *Planck length*, $L_P \equiv \sqrt{\frac{\hbar G_N}{c^3}} \simeq 1.6 \times 10^{-35}\, m$ or the Planck mass, $M_P \equiv \sqrt{\frac{\hbar c}{G_N}} \simeq 1.2 \times 10^{19}\, GeV/c^2$, dimensional reductions of the gravitational coupling (the fundamental geometric constant G_N) using the fundamental kinematic constants $\hbar, c$ result in dimensional quantities. This means that any mathematical forms generated by quantum processes involving the gravitational coupling must involve dimensionless forms that include intermediate energies that can be infinite. Therefore, despite being a weak force, it is quite problematic to construct perturbative forms in a dimensionless coupling, which must be of the form $\frac{G_N}{\hbar c}\frac{energy^2}{c^4}$.

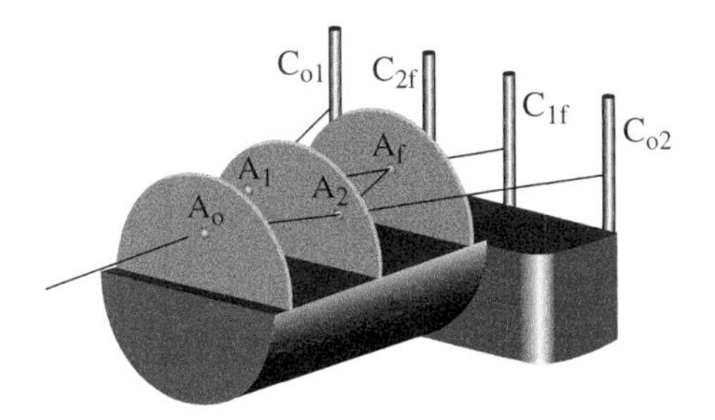

Figure 2.3 Diagram of apparatus used by Collela, et al. to demonstrate coherence of gravitating neutrons. Apertures are labeled A, while counters are labeled C.

This section will examine the fundamental behaviors of quantum systems propagating in a gravitational field. The discussion will attempt to maintain close consistency with geometric arguments that will be developed later. To begin, a brief exposition of experimental findings will be presented.

2.3.1 Coherence of gravitating systems

There have been remarkably few direct experimental observations of the quantum behaviors of gravitating systems. There is experimental evidence that quantum coherence is maintained by the nearly static gravitational field near Earth's surface. During the early and mid 1970s, experiments performed by Collela, et al. [31] demonstrated the gravitation of coherent neutrons diffracting from an apparatus whose orientation could be changed relative to the Earth's gravitational field. The type of apparatus utilized is illustrated in Figure 2.3. In the diagram, a collimated beam of neutrons is incident upon the aperture labeled A_o. The apparatus can be rotated such that the apertures A_1 and A_2 can be oriented to have differing gravitational potentials. Interfering neutrons are detected after having passed through aperture A_f by counters C_{2f} and C_{1f}, to which there are no classical paths for the incident neutrons. The experimental results demonstrate an interference pattern in the count difference $N_{C_{2f}} - N_{C_{1f}}$ measured as a function of the gravitational potential difference between the apertures A_1 and A_2. Gravitating quantum systems were therefore seen to maintain spatial coherence.

Another experiment measured the small difference between the ticks of two interfering quantum clocks [32]. In that experiment, very cold caesium atoms gravitated across a laser beam that superimposed the atoms into states across different gravitational potentials. The resultant difference in phase generated interference in the (very rapid) quantum oscillations (Compton frequencies) associated with the mass-energies of the atoms. Therefore, gravitating quantum systems have also been shown to maintain temporal coherence.

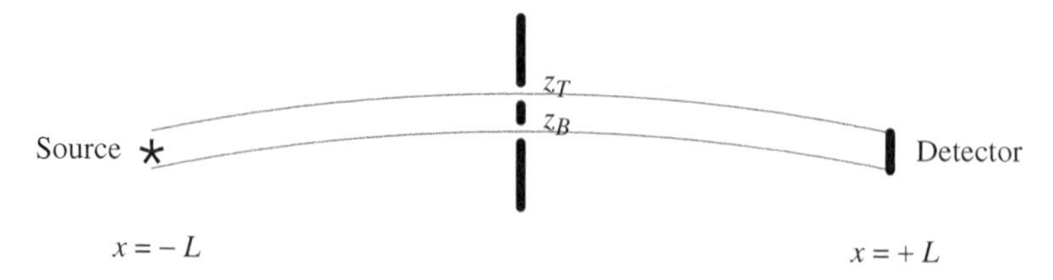

Figure 2.4 Example of coherent gravitation. Absorbing screen is located on the plane $x = 0$. Gravitational acceleration is downward.

Additional experiments demonstrating coherence maintained during gravitation examined the oscillation between values of the momenta of fermions occupying different rungs on an optical ladder [33] in a gravitational potential. The energy spectrum is known as a Wannier–Stark ladder of states, and a constant energy difference of $mg\lambda/2$ separates the successive rungs, which are separated by distance λ. The phase difference $\Delta\phi = \Delta E\, t/\hbar$ is a function of time, resulting in oscillations that can be used as precise gravimeters [34].

These experimental results provide illustrative examples of the usefulness of the canonical proper-time formalism developed in Section 1.3 in describing gravitating systems. A gravitating quantum system should exhibit coherence on surfaces defined by its proper time. The frame of reference of a given coherent or entangled system provides the only unique set of coordinates describing that system. The various regions across a coherent state might propagate with differing speeds through varying gravitational potentials, which means that space-like surfaces of simultaneity defined by fixed coordinate time, t, are generally *not* coincident with the space-like surfaces of simultaneity defined by fixed proper time, τ. In special relativity, surfaces of simultaneity are relative to the observers' motions.

Equations will be developed for the example illustrated in Figure 2.4. An absorbing screen placed on the plane $x = 0$ has small openings at heights z_T and z_B, and a coherent particle propagates with a perpendicular component of momentum $p_\perp$ in the x direction. For the accelerating particle, the surface corresponding to the absorbing screen is chosen so that projectiles through the surface will have vanishing $\dot{z}$ (where *dot* represents a derivative with respect to *proper* time) at that surface, allowing temporal synchronization with coordinate time on that surface $\tau = 0 = t$. The particle, therefore, has a well-defined temporal phase for interference.

Because of the different path lengths, there will be the usual phases associated with the geometric paths (parameterized by proper time) followed by the propagating waves:

$$\Delta\phi_{geometric} = \frac{i}{\hbar}\left(\int_{path\ T} p_k\, \dot{x}^k d\tau - \int_{path\ B} p_k\, \dot{x}^k d\tau\right). \qquad (2.79)$$

However, there are additional phase factors associated with the gravitational field. For the coherent system, the generator for infinitesimal proper-time translations is given by the canonical proper energy K. Therefore, the phase difference between the two paths due to gravitation will be given by:

$$\Delta\phi_{gravitational} = -\frac{i}{\hbar}\left(\int_{path\,T} K\,d\tau - \int_{path\,B} K\,d\tau\right). \tag{2.80}$$

The sum of the geometric and dynamic terms generates the canonical proper Lagrangian form, $L_K(x,\dot{x}) \equiv p_k\,\dot{x}^k - K(x,p)$ given in a path integral formulation over the allowed paths, $\Delta\phi_{total} = \frac{i}{\hbar}\int_{path} L_K(x,\dot{x})\,d\tau$.

The equations of motion generated using the canonical proper-time formulation ensures that the canonical proper energy is conserved, $\frac{dK}{d\tau} = 0$, which means that the gravitational phase difference satisfies $\Delta\phi_{gravitational} = -\frac{i}{\hbar}(K_T - K_B)\Delta\tau$. For motions near the Earth's surface, the canonical proper energy takes the form:

$$K = \frac{\mathbf{p}\cdot\mathbf{p}}{2m} + U(\mathbf{r}) + mc^2 \simeq \frac{\mathbf{p}\cdot\mathbf{p}}{2m} + mg(z - R_{Earth}) + mc^2. \tag{2.81}$$

Since the canonical energy form is independent of the coordinates perpendicular to the gravitational field, the perpendicular components of momentum $\mathbf{p}_\perp$ are constant. For the different paths, the constant canonical energy forms take the values given on the screen s:

$$K_s = \frac{\mathbf{p}_\perp\cdot\mathbf{p}_\perp}{2m} + mg(z_s - R_{Earth}) + mc^2. \tag{2.82}$$

Therefore, the gravitational phase difference is given by $\Delta\phi_{gravitational} \simeq -\frac{i}{\hbar}mg(z_T - z_B)\Delta\tau$.

The geometric trajectory can be solved using the form Eq. 2.81 for a constant canonical proper energy. The solution satisfies:

$$z(\tau) = z_s - \frac{1}{2}g\tau^2, \tag{2.83}$$

so that the total proper time is given by $\Delta\tau/2 = \sqrt{\frac{2z_B}{g}} = \frac{mL}{p_\perp}$. The transformation from proper- to coordinate-time follows from the Minkowski metric form, which satisfies:

$$\frac{dct}{dc\tau} = \sqrt{1 + \left(\frac{\mathbf{p}_\perp^2}{mc}\right)^2 + \left(\frac{\dot{z}}{c}\right)^2} = \sqrt{1 + \left(\frac{\mathbf{p}_\perp^2}{mc}\right)^2 + \left(\frac{g\,\tau}{c}\right)^2}. \tag{2.84}$$

This result integrates into a closed form solution connecting the temporal coordinates.

The experimental results discussed imply that (at least for stationary sources) gravitating systems maintain their quantum behavior (i.e., that the quasi-static near

Earth gravitational field does not break the phase coherence of neutrons or atoms as needed for interference). The experiments were also a test of the principle of equivalence (i.e., that the motions of the observer do not break the coherence of an inertial system, in these cases, the gravitating particles), as will be discussed later when examining the principles of general relativity. The experimental results involve both Newton's gravitational constant, G_N and Planck's constant, $\hbar$ in a single equation form. Such results, along with other observed phenomena, generally require that gravitating sub-clusters can maintain quantum coherence while having their internal dynamics influenced by disentangled clusters (e.g., the rest of the planet) co-contributing to the local gravity. Such a configuration of co-gravitating quantum systems will be constructed in the next subsection.

Note that the exhibition of quantum coherent behavior for gravitating systems need not require the quantization of the gravitation field itself (much as Planck used quantization of the source resonators to derive the properties of black body radiation without assuming quantization of the electromagnetic radiation). Although justifications of the redshift of gravitating photons need not make use of Planck's constant, those quanta likewise maintain their coherence during extended interactions with *dynamic* gravitational fields (e.g., lensing of the cosmic microwave background, gravitational redshift measurements of sources due to traversing astrophysical objects, etc.). Of particular interest, interacting gravitating systems continue to gravitate after disentanglement, which motivates a formulation that incorporates straightforward cluster decomposability within macroscopic gravitational environments.

Macroscopic quantum fluids (like superfluids and superconductors) also maintain persistent quantum flows that satisfy quantization conditions in the non-inertial environments of most laboratory measurements. For instance, vortices of superflow with quantized circulation maintain the angular momentum of a rotating vessel of liquid helium cooled below the superfluid transition temperature. Such angular momenta should individually and collectively precess in a gravitational field. Since any expression of the precession frequency involves both Planck's constant and Newton's gravitational constant, such phenomena give additional insights on the coherence properties of mixed gravitating systems. Any gravitationally induced precession of the distributive motions that maintain coherent relational aspects in space-time make a strong statement about the fundamental nature of the gravitating behaviors of the disparate constituents of such a macroscopic quantum system.

2.3.2 *Static field quantum co-gravitation*

The fields generated by static and quasi-static sources are the best understood gravitating systems. In particular, spherically symmetric systems are particularly

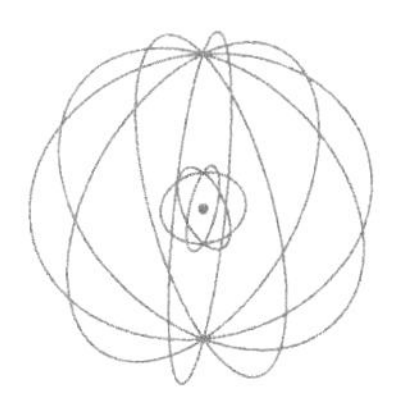

Figure 2.5 Thought experiment of an incoherent sum of coherent gravitating particles acting as sources for exterior particles.

relevant and illustrative. As previously stated, experiments [31] demonstrate that static gravitational sources do not break the coherence of gravitating quantum systems. Orbiting gravitating particles should be able to maintain their coherence in a quantum stationary state. Using these observations, various thought experiments can be examined.

Consider a set of freely co-orbiting quantum coherent (neutral) particles that are in composite spherically symmetric states. Since the gravitation does not break that coherence, the local energy densities of those quanta should likewise serve as gravitational sources generating the background field. By Birkhoff's theorem (or Gauss' law for gravitational fields), only the interior energy density effects the local gravitation. Such spherically symmetric shells can be incoherently combined to construct a co-gravitating system with quantum components. A semi-classical depiction of this thought experiment is shown in Figure 2.5. If other particles are measured interior to a given gravitating particle, those measured particles serve as gravitational sources for the given particle, even if it is coherent.

In the following section, such self-gravitating and co-gravitating systems will be developed using this reasoning. Both semi-classical *and* quantum systems will be developed. In particular, the canonical proper-energy form will be used to generate quantum behaviors, ensuring that relativistic energies are properly incorporated.

Proper-time quantum gravitating particles

Consider the stationary gravitation of mass m due to a source mass M_ℓ which depends only on ℓ, which is the angular momentum quantum number associated with the stationary quantum state. The proper-energy form from Eq. 1.75 will be used to examine the dynamics of this mass. The normalized wave function that satisfies the stationary canonical proper-energy equation for this mass is given by:

$$\begin{gathered}\left(\frac{\hat{p}^2}{2m} - \frac{G_N m M_\ell(r)}{r} + mc^2\right)\psi_{\ell\ell_z}(r,\theta,\phi) = K_\ell\ \psi_{\ell\ell_z}(r,\theta,\phi),\\ \psi_{\ell\ell_z}(r,\theta,\phi) = R_\ell(r)Y_\ell^{\ell_z}(\theta,\phi).\end{gathered} \tag{2.85}$$

The proper-energy eigenvalues K_ℓ include relativistic velocities.

Spherically symmetric forms can be developed by summing over the $2\ell + 1$ values of the index ℓ_z using the angular addition theorem:

$$\sum_{\ell_z=-\ell}^{\ell} \left| Y_\ell^{\ell_z}(\theta, \phi) \right|^2 = \frac{2\ell + 1}{4\pi} P_\ell(1).$$

For such forms, equations analogous to those for hydrogenic systems can be developed. The radial scale of the solutions is given by:

$$a \equiv \frac{\hbar^2}{G_N m^3} = \left(\frac{\lambda_m}{L_P} \right)^2 \lambda_m = \left(\frac{M_P}{m} \right)^2 \lambda_m, \tag{2.86}$$

where $\lambda_m \equiv \frac{\hbar}{mc}$. Defining the reduced radial wavefunctions $u_\ell(r/a) \equiv r R_\ell(r)$ parameterized by dimensionless variable $\zeta \equiv r/a$, the Eq. 2.39 can be re-written:

$$\frac{d^2 u_\ell(\zeta)}{d\zeta^2} - \frac{\ell(\ell+1)}{\zeta^2} u_\ell(\zeta) + \left(\frac{2}{\zeta} \right) \left(\frac{M_\ell(\zeta)}{m} \right) u_\ell(\zeta) = \epsilon_\ell \, u_\ell(\zeta), \tag{2.87}$$

where the dimensionless form $\epsilon_\ell = -2 \left(\frac{\lambda_m}{L_P} \right)^4 \frac{K_\ell - mc^2}{m c^2}$ parameterizes the gravitational binding energy of the mass. The interior mass will be presumed to be composed of particles of mass m_ℓ, including self-sourcing:

$$M_\ell(\zeta) = \int_0^\zeta \rho_{mass}(\zeta') \, d\zeta' = \int_0^\zeta \sum_{\text{all } \ell'} n_{\ell'} \, m_{\ell'} \, u_{\ell'}^2(\zeta') \, d\zeta'. \tag{2.88}$$

This term represents an incoherent sum of the co-gravitating interior masses $m_{\ell'}$. Generally, the integrand is a *mass* density matrix for which the weights are given by the masses constituting the incoherent sum. In the following treatment, all co-gravitating particles will be assumed to have the same mass m.

Supplement: Semi-classical co-gravitation

The analysis of the present discussion can be alternatively motivated using a semi-classical Bohr model for the gravitating masses. Assume that the orbital angular momentum of a circularly orbiting mass m must be quantized using Bohr's criterion, $L_n = m r_n v_n = n\hbar$. If the interior gravitational source mass M_n must generate a gravitational force that provides the necessary centripetal acceleration, the speed of the mass and radius parameterizing this orbital must satisfy:

$$v_n = \frac{G_N m M_n}{n\hbar}, \qquad r_n = \frac{n^2 \hbar^2}{G_N m^2 M_n}.$$

The radial scale a_n and energy E_n will depend only on the principle quantum number n:

$$a_n \equiv \frac{\hbar^2}{G_N m^2 M_n}, \qquad E_n = -\frac{m(G_N m M_n)^2}{2\hbar^2 n^2}.$$

As long as the dependence M_n assures that the radius increases with principle quantum number n, one can develop a model where each interior mass contributes as a gravitational source for the subsequent masses. A physical co-gravitating system can be developed by assuming that incoherently summed symmetric shells of mass m combine to give interior mass satisfying $M_n = n\, m$ for shell n. The previously defined forms can then be re-written in terms of the reduced Compton wavelength $\lambda_m \equiv \frac{\hbar}{m\, c}$ of the mass m and the Planck length, giving:

$$a_n = \left(\frac{\lambda_m}{L_P}\right)^2 \frac{\lambda_m}{n}, \qquad E_n = -\frac{1}{2}\left(\frac{L_P}{\lambda_m}\right)^4 m\, c^2,$$
$$v_n = \left(\frac{L_P}{\lambda_m}\right)^2 c, \qquad r_n = n\, a_1.$$

In contrast to the behaviors of electrons in atomic physics, each successive shell experiences increased attraction due to the interior masses. It is interesting to note that the semi-classical speeds of the masses, as well as their gravitational binding energies, are independent of the orbital in this model.

A mass whose interior probability density provides its local source gravitational field will be referred to as a *self-gravitating* mass. If additionally the gravitational energy generates the mass itself, the self-gravitating mass will be referred to as a *self-generating* mass. The small r behavior of Eq. 2.87 is dominated by angular momentum, giving a form $u_\ell(\zeta) \overset{\zeta\to 0}{\Longrightarrow} \zeta^{\ell+1}$. However, a self-gravitating central mass can have non-vanishing probability density at the center. Such a self-gravitating single mass satisfies:

$$\frac{du_C(\zeta)}{d\zeta^2} + \left(\frac{2}{\zeta}\right)\left(\int_0^\zeta u_C^2(\zeta')\, d\zeta'\right) u_C(\zeta) = \epsilon_C\, u_C(\zeta). \tag{2.89}$$

This form is clearly non-linear, so that initial conditions and eigenvalues are non-trivially related to the solution. Figure 2.6 demonstrates a self-generating solution to this equation, with $\epsilon_C = 0$. The diagram on the left demonstrates the probability density $|u_C(r/a)|^2$, while the diagram on the right is a density plot of the self-gravitating mass density consistent with vanishing overall gravitational binding energy. The mass interior to a given radius is demonstrated in Figure 2.7. For the system, the gravitational source mass at a given radial coordinate gravitates in a field generated by the summed mass density within that radial coordinate. A

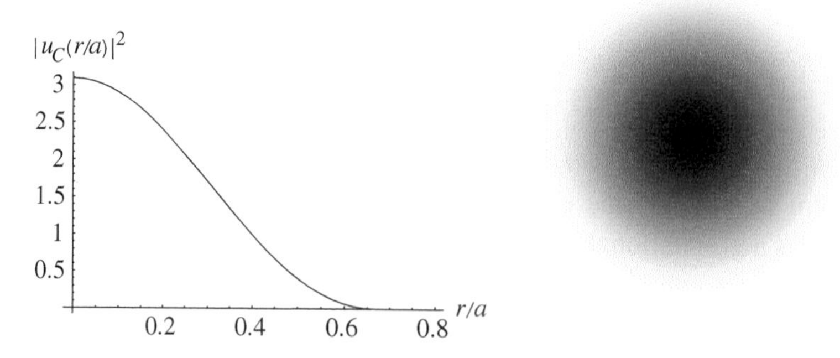

Figure 2.6 Self-gravitating, self-generating central mass. *Left*: Probability distribution, *right*: mass density distribution.

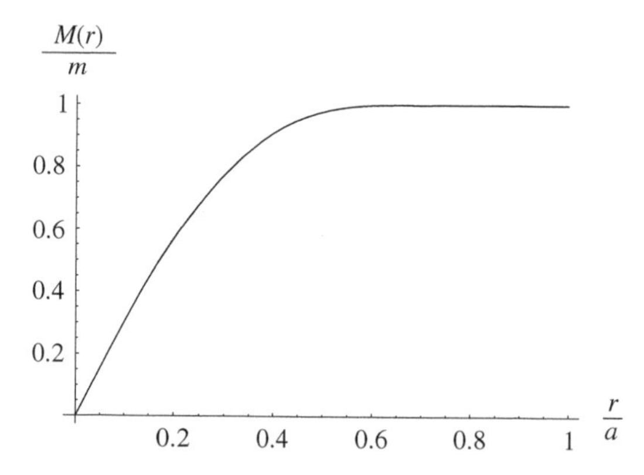

Figure 2.7 Interior mass serving as gravitational source for self-generating mass.

comparison is made with a system that is self-gravitating but has non-vanishing binding energy eigenvalue in Figure 2.8. The energy eigenvalue for the normalized probability density was obtained by examining the small r behavior of Eq. 2.89, yielding a form $\epsilon_C = \frac{4}{3}|u_C(0)|^2$. The self-generating mass density is seen to be more concentrated at the center.

Additional masses can be added to the system in a spherically symmetric manner (modifying the solution of the central mass). A set of masses whose interior probability densities provide their local source gravitational field will be referred to as *co-gravitating* masses. As an example, consider a system consisting of a central mass m with $n_\ell = 2\ell + 1$ co-gravitating masses each of mass m, summed over the eigenvalues of ℓ_z as shown previously. Figure 2.9 demonstrates such a set of co-gravitating masses. The system consists of a central mass and seven equal masses with an angular momentum of $\ell = 3$. The diagram on the left demonstrates

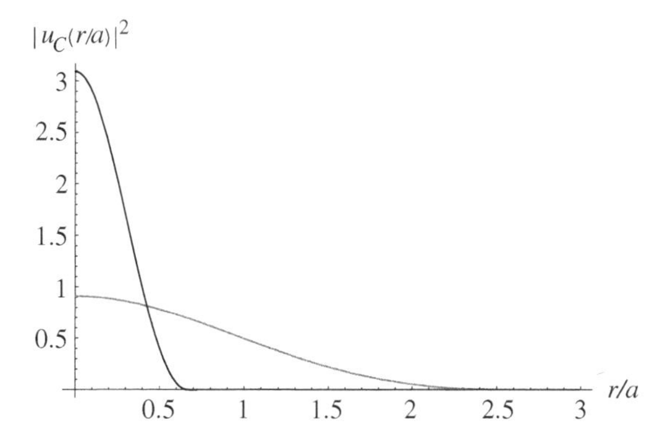

Figure 2.8 Self-gravitating mass density (gray), and self-gravitating, self-generating mass density (black).

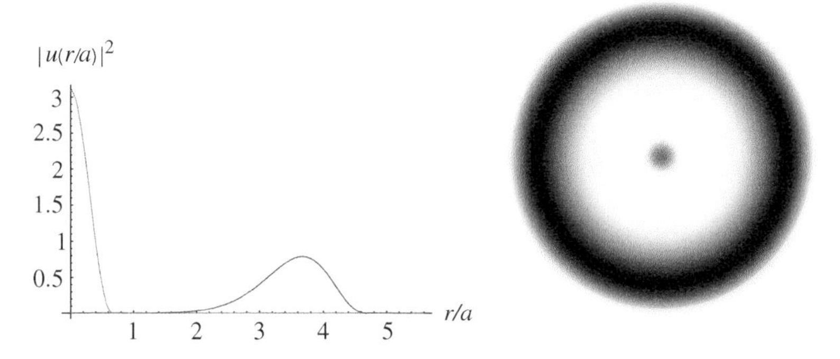

Figure 2.9 Central mass and a spherically symmetric arrangement of $n_\ell = 7$ co-gravitating masses with $\ell = 3$. *Left*: Probability distributions, *right*: mass density distribution.

the normalized co-gravitating probability densities of the central mass and any of the orbiting masses. The diagram on the right demonstrates the co-gravitating mass density of the system. The mass interior to a given radius is demonstrated in Figure 2.10. The form of the central mass is very slightly shifted from that in Figure 2.6. The binding energies of the orbiting masses were chosen to be self-generating $\epsilon_3 = 0$. Each of the co-gravitating masses with non-vanishing angular momenta satisfy the phase coherence experimentally observed [31].

For completeness, the mass density forms will be demonstrated on a compact conformal space-time diagram, exhibiting the global causal structure of the geometry. The conformal density plot is demonstrated in Figure 2.11. The apparent compression of the densities for early and late times is due only to the mathematical transformation into finite spatial and temporal extent on the Penrose diagram of the essentially Minkowski space-time. These diagrams will be quite useful in later

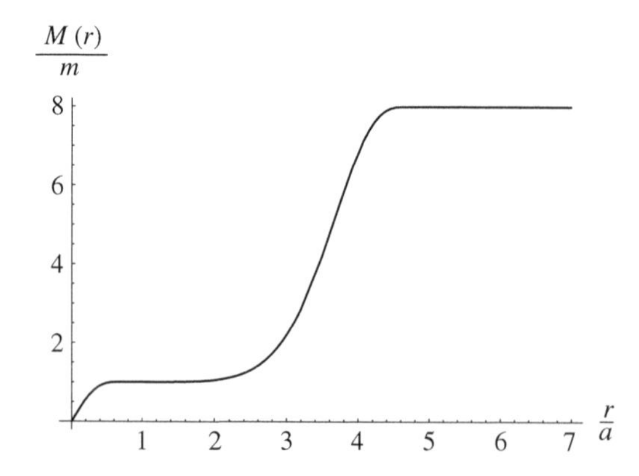

Figure 2.10 Interior mass serving as a gravitational source for co-generating masses.

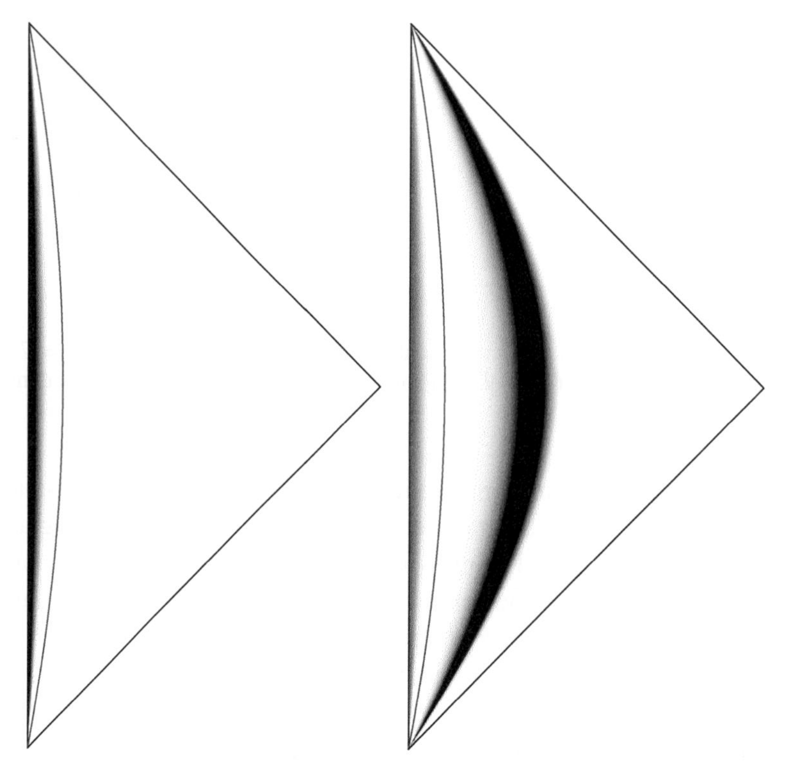

Figure 2.11 Conformal stationary mass density plots of self-gravitating central mass (*left*) and co-gravitating central mass (*right*) with seven $\ell = 3$ masses. The solid, fixed radial curve represents the scale $r = a$.

examinations of gravitating quantum systems. It should be emphasized that these calculations *have not* discussed space-time curvature effects. Such curvature effects (which are not significant in weakly gravitating environments) will be examined in Part II.

2.3.3 Energy conditions

Energy conditions refer to the physical expectation that mass-energy distributions are expected to have local velocities that are everywhere less than (or in the case of massless energies, equal to) that of light. For distributions of classical particles, this certainly holds true, since for any region of the distribution, the enclosed four-momentum must be time-like $\vec{p} \cdot \vec{p} < 0$. However, space-like coherence properties of quantum systems make such expectations less obvious for a coherent state. In this subsection, the expectation of local four-momentum densities of coherent systems will be briefly explored.

To begin, consider bound states of a square well described by:

$$K = \frac{p^2}{2m} + V + mc^2, \quad K < mc^2, \tag{2.90}$$

where V is constant. The form of this interaction is chosen such that $V = -V_B$ in region I with $x < a$. In this region, the spatial solutions are oscillatory, and the wave number can be obtained using:

$$K = \frac{\hbar^2 k^2}{2m} - V_B + mc^2 \Rightarrow \hbar k = \pm\sqrt{2m(K + V_B - mc^2)}. \tag{2.91}$$

The interaction is chosen to vanish $V = 0$ in region II with $x > a$, giving an exponential spatial solution $e^{-\kappa x}$, where κ satisfies:

$$K = -\frac{\hbar^2 \kappa^2}{2m} + mc^2 \Rightarrow \hbar\kappa = \pm\sqrt{2m(mc^2 - K)}. \tag{2.92}$$

In region II with vanishing interaction, the energy can be directly related to the canonical proper energy by noting $E^2 + (\hbar\kappa c)^2 = (mc^2)^2$. This allows one to solve for the conserved energy in terms of the conserved canonical proper energy:

$$E^2 = mc^2(2K - mc^2). \tag{2.93}$$

Clearly, the canonical proper-energy spectrum is different from the standard energy spectrum. Relative motions for given energies connect proper times to differing inertial times.

Substituting Eq. 2.93 and Eq. 2.91 from region I in the invariant $\vec{p} \cdot \vec{p}$ gives:

$$\vec{p} \cdot \vec{p} = (\hbar k c)^2 - E^2 = mc^2(2V_B - mc^2). \tag{2.94}$$

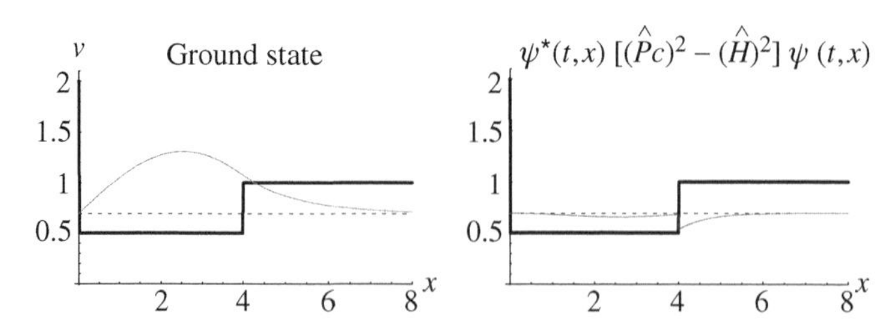

Figure 2.12 Slightly bound ground state probability amplitude (*left*) and local energy conditions (*right*).

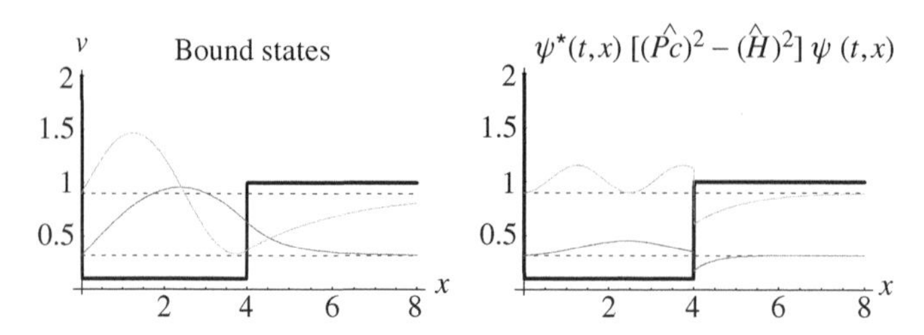

Figure 2.13 Deeply bound states probability amplitudes (*left*) and local energy conditions (*right*).

This means that if the binding is greater than half of the rest energy of the free particle, energy conditions will not be satisfied within that region.

Figure 2.12 demonstrates a slightly bound system where classical energy conditions are everywhere satisfied. The probability density is shown in the left diagram, and the diagram on the right demonstrates that the square of the local four-momentum density is everywhere negative.

On the other hand, Figure 2.13 demonstrates a deeply bound quantum system. The deeply bound states have momenta that dominate over the energy, thus violating energy conditions within the bound regions. This is why the diagram on the right demonstrates space-like behaviors within the bound region, violating classical energy conditions. Therefore, possible violations of classical energy conditions due to the introduction of quantum behaviors should not be surprising.

2.4 Thermal properties of acceleration

Observers in accelerating frames of reference experience "fictitious" forces manifested in the non-inertial local coordinates. Detectors undergoing the non-inertial motions are in modified ground states when compared to inertial detectors (see Appendix B.1.2). The modified states involve non-vanishing occupation of excited states of inertial detectors. The distribution of these excitations for detectors undergoing uniform acceleration will be briefly discussed in this section and Appendix B.1.3.

Consider massless scalar radiations with energy $\epsilon = pc$ propagating in the direction of the accelerating detector. Since the phase of a waveform can be used to count the wavefronts, it is an invariant. The rate at which the wavefronts pass any observer is Doppler shifted due to motions of that observer. For an accelerating observer following the trajectory given in Eq. 1.36, the wavefunction has a phase of the form [35]

$$\begin{aligned}\psi_\epsilon^{\pm}(\tau_a) &= \psi_o exp\left[\frac{i}{\hbar}(p\,\tilde{x}(\tau_a) \mp \epsilon\,\tilde{t}(\tau_a))\right] \\ &= \psi_o exp\left[i\frac{\epsilon}{\hbar c}(x_a - \frac{c^2}{a})\right] exp\left[i\frac{\epsilon c}{\hbar a}e^{\mp\frac{a\tau_a}{c}}\right].\end{aligned} \tag{2.95}$$

The frequency distribution of the wavefronts will be ascertained by examining the Fourier transform $\bar{\psi}^{\pm}(\omega) = \int_{-\infty}^{\infty} d\tau_a e^{i\omega\tau_a}\psi_\epsilon^{\pm}(\tau_a)$. This integral requires evaluation of the form:

$$\mathcal{I}_\omega = \int_{-\infty}^{\infty} d\tau\, exp\left[-\left(\omega\tau + \frac{\epsilon c}{\hbar a}e^{\mp\frac{a\tau}{c}}\right)\right]. \tag{2.96}$$

By making the substitution $\zeta = \frac{\epsilon c}{\hbar a}e^{\frac{a\tau}{c}}$, and noting that:

$$\int_0^{\infty} \zeta^{i(w-i0^+)-1}\, e^{iz\zeta}\, d\zeta = z^{-iw}\,\Gamma(iw)\, e^{-\pi w/2}, \tag{2.97}$$

this integral can be evaluated as:

$$\begin{aligned}\mathcal{I}_\omega &= \frac{c}{a}\left(\frac{\hbar a}{\epsilon c}\right)^{i\omega\frac{c}{a}}\Gamma\left(i\omega\frac{c}{a}\right)e^{-\frac{\pi}{2}\omega\frac{c}{a}}, \\ |\mathcal{I}_\omega|^2 &= \frac{2\pi}{\omega}\frac{c}{a}\frac{1}{e^{2\pi\omega\frac{c}{a}} - 1}.\end{aligned} \tag{2.98}$$

Notice that the energy of the radiation does not appear in this result. Therefore, the frequency spectrum of the wave density observed by the accelerating observer has a time-dependent Doppler shift with a Planck factor if the temperature associated with the acceleration is given by:

$$k_B\,T = \frac{\hbar a}{2\pi c}, \tag{2.99}$$

yielding:

$$\frac{\hbar\,\omega}{k_B T}\,|\mathcal{I}_\omega|^2 = \frac{1}{e^{\frac{\hbar\,\omega}{k_B T}} - 1}. \tag{2.100}$$

The thermal effects are generated by the forces accelerating the detector. They are established given sufficient equilibration time after the establishment of local acceleration and vanish after relaxation following the cessation of local acceleration. Appendix B.1.3 demonstrates a related derivation of thermal distributions resulting from examining the relative normalizations of outgoing massless radiations.

Supplement: Evolution of the vacuum concept

Concepts of the meaning of the *vacuum* have evolved as various models of the physical universe have developed.[6] Its common meaning is a region devoid of material content. As an example, consider a sealed cylinder with a piston that has all gas evacuated at room temperature. If a force is applied to partially withdraw the piston, the evacuated space within the cylinder constitutes a common vacuum. If the force is released, the piston will retract to zero enclosed volume. However, if the piston is withdrawn for a period of time that allows the system to come to thermal equilibrium, once the applied force is removed, the piston *does not* retract to zero volume. Thermal electromagnetic radiation fills the equilibrium volume, exerting radiation pressure that matches the exterior atmospheric pressure. Because of the couplings of the bounding materials with electromagnetic radiation (due to constituent charges), the region devoid of material particles mediates thermal exchanges between those constituents via photons.

One should note that unless source/sink particles that couple to the radiations were present on the boundaries, those radiations would not be present to mediate the thermal exchanges between those particles. In particular, consider the zero-point motions generating the Casimir effect considered in Section 2.1.3. The van der Waals interactions mediate the statistical exchanges between the boundary molecules. If one wishes to attribute dynamic properties to the region interior to the boundary sources/sinks, for conducting plates separated by distance a, the energy density is attractive, $\rho_v \sim -\frac{\hbar c}{a^4}$, while that contained in a conducting spherical boundary of radius a is repulsive, $\rho_v \sim +\frac{\hbar c}{a^4}$. Indeed, since the attributes of the dynamics of this materially evacuated region is extremely sensitive to the boundary conditions of the sources, one must take care when referring to such radiations resulting from zero-point motions as "vacuum energy".

When the radiations resulting from zero-point motions ρ_v behave as do thermal radiations, they must be homogeneous and isotropic. An accelerating

[6] See for instance T.H. Boyer, "The Classical Vacuum", *Scientific American*, pp. 70–8 (Aug. 1985).

observer measures radiations associated with zero-point motions, and additional radiations with a thermal spectrum. This thermal spectrum has some intriguing consequences. For instance, a uniformly accelerating charge radiates, yet it experiences no classical radiative reaction, since this force is proportional to the change in acceleration $F_{RR} = \frac{2}{3}\frac{e^2}{c^3}\frac{d^2 v}{dt^2}$. This becomes more sensible by noting that the charge is immersed in a co-moving thermal bath of temperature $k_B T = \frac{\hbar a}{2\pi c}$. Thus, there is no transfer of energy between the charge and the thermal bath. The field of the charge is non-radiative as seen in the accelerating frame.[7] Radiated energies are provided by the force maintaining the uniform acceleration, which undergoes the fluctuations needed to hold the acceleration constant.

Consistent with previous discussions, instances for which the dynamics of evacuated regions can be given defined attributes, always involve interactions between boundary sources and sinks of the radiations. In the present text, all such radiations will be attributed to zero-point motions of sources, rather than as a form of vacuum energy, to avoid possible confusion in source-free regions of space-time.

One might conclude that the thermal properties of an accelerating system are a consequence of the existence of a horizon and the consequential loss of information/attribution of entropy. However, the previous discussion (local accelerations $\rightarrow$ associated temperatures) demonstrates that the thermal behavior is rather a result of the proper acceleration itself (which is a local property). Generally, a horizon is a global construct (since it is a light-like surface separating a region forever causally separated), and if the acceleration ceases, a horizon never existed. Yet "thermal" excitations of detectors *do* occur during accelerations. The energy sources of these excitations are the forces responsible for the acceleration, as well as exchanges with inertial systems.

[7] See, for instance, P. Candelas and D.W. Sciama, "Is there a quantum equivalence principle?", *Phys. Rev. D* **27**, 1715 (1983).

3

Microscopic formulations of particle interactions

The fundamental "players" in the cosmological arena are microscopic particles and the interactions by which they exchange well-defined quantum numbers. Many of the critical properties of micro-physics can be determined by their behaviors in nearly flat space-time, as described by Minkowski. There are several requirements that a successful model of fundamental processes should fulfill. Among these characteristics are:

- *Describes quantum phenomenology*: Quantum mechanics successfully describes the subtle behaviors of matter and energies undergoing microscopic exchanges. Quantum behaviors inherently have aspects beyond measurement.
- *Conservation properties*: Most particles have internal quantum numbers, like charge, lepton number, baryon number, etc., that are carried undiminished throughout complicated interactions with other particles. In addition, they carry properties like mass and spin, whose kinematic transformations under space-time transformations are well-defined, and which satisfy composite conservation laws (energy, momentum, angular momentum) for sufficiently isolated homogeneous and isotropic systems.
- *Unitarity*: Despite being unable to follow quantum coherent particle properties while that coherence is maintained, the evolution of those properties is described in a manner that ensures conservation of probability. This means that these properties do not just "pop" into or out of existence between the detections and interrelations of the particles.
- *Cluster decomposability and classical correspondence*: Classical physics is quite successful in describing much of common phenomenology. Classical models exploit those characteristics of a system that can be isolated, studied, and parameterized independent of observation. Fundamental descriptions of quantum systems should provide mechanisms for both establishing and maintaining long-range coherence, as well as describing clusters whose internal dynamics

are essentially isolated from the internal dynamics of other disentangled clusters.
- *Poincaré covariance*: Relativistic kinematics describes the exchanges of energy-momentum and angular momentum among interacting constituents. Despite the complicated quadratic form of the energy-momentum dispersion relation, the resulting kinematics and space-time symmetries must be properly incorporated. Since the kinematics of special relativity has well-understood non-relativistic correspondence, a fundamental description should be relativistically consistent.

One should note that fundamental physical processes need not be analytic for small coupling, or renormalizable relative to specific perturbative procedures. For instance, the interaction between the Cooper pairs of superconducting electrons has an essential singularity at vanishing coupling [36]. Generally, phase transitions and bound states cannot be analytically described using a perturbative series describing a differing phase or scattering states. The Efimov effect [37] describes a profound influence that short-range forces between a pair of interacting particles can have on a distant, weakly interacting third particle. This chapter will develop formulations that generally incorporate the fundamental properties and subtle effects of relativistic quantum phenomenology.

Particles (usually fermions) have conservation properties that are quite different from those of the carriers of interaction (so-called quanta). Internal quantum numbers like electric charge, mass, total spin, lepton number, etc., of a particle that breaks its coherence remain unchanged and conserved, whereas quanta of interactions are absorbed when coherence is broken during detection. The quanta act as carriers of conserved space-time quantum numbers (and in the case of gluons, color charge) that are exchanged between particles. For this reason, it is problematic to discuss the carriers of interaction in the absence of sources or sinks. Vacuum states of such quanta are physically ill-defined, since modes of those states must be defined relative to sources/sinks on the boundaries. Vacuum modes often formally introduce infinities into calculations that depend on mode densities, requiring the development of complementary formal techniques to eliminate or ignore those infinities. The approach developed here always associates the quanta of interactions with their sources and sinks.

Perturbative relativistic quantum field theory has provided a robust framework for describing elementary particle physics and successfully describing many quantitative predictions of quantum electrodynamics (QED), weak-electromagnetic unification, and quantum chromodynamics. The general form of the interactions between systems is quite often so complicated that meaningful descriptions can most directly be obtained only through comparisons of the interacting systems to more solvable systems that are not substantially distinct. A technology for successively improving the approximations then provides a mechanism for describing the modeled interactions to arbitrary precision, in principle. However, as previously explained, a perturbative formulation of a fundamental interaction remains flawed at the foundational level, as well as its adequacy to describe the complete spectrum of phenomena associated with the interaction.

One such flaw was noted by Dyson [38] during the development of renormalization methods in QED, when he showed that the renormalized perturbation series is not uniformly convergent when extended beyond 137 terms. This convinced him that renormalized QED could never be a fundamental theory [39].

Supplement: Convergence of perturbative QED

The argument given by Freeman Dyson, that the power series generated by renormalized perturbative quantum electrodynamics is not uniformly convergent, was as follows. Most problems in QED give results in the form of a power series in e^2. The renormalization procedure occurs term by term to finite order in the series. Examine the interaction between two like charges e, and suppose that a physical quantity $F(e^2)$ can be represented by the convergent series:

$$F(e^2) = \sum_{s=0}^{\infty} a_s e^{2s}.$$

This implies that $F(0)$ is analytic, and thus that $F(-e^2)$ is convergent for sufficiently small $-e^2$ near 0 (by the definition of analyticity). Such a theory would produce Coulomb-like potentials that are attractive for *like* charges, rather than unlike charges.

Within calculations of quantum scattering processes between the like charges, there are intermediate states with a large number N of $e^+ e^-$ pairs. Some such states will have the electrons spatially separated from the positrons in a manner that has the like-particle Coulomb attractions sufficiently overcoming the rest energies of the source charges. Such a state is then energetically unstable to further creation of $e^+ e^-$ pairs, cascading into collapse of the ground state. This therefore means that $F(0)$ cannot be analytic, and the series could not be uniformly convergent.

Dyson estimated that the minimum number of pairs needed to initiate the pathology is about $N \sim \frac{1}{\alpha} \approx 137$, which indicates the perturbative order where the series is expected to become anomalous. It should be noted that similar arguments should be applicable to a perturbative formulation of gravitational force, for which the interaction between masses is already attractive.

Another fundamental flaw is the so called *correspondence* problem, that one cannot develop an unambiguous "potential" that can be used in a non-relativistic Hamiltonian from a perturbative, relativistic quantum field theory. Among other things, this prevents the development of a rigorous model for low-energy nuclear physics [40].

A perturbative description of scattering states cannot predict low-lying bound states, since (as will be shown) those states correspond to singularities in the corresponding scattering amplitudes. This problem is particularly exacerbated in the theory of strong interactions, since the conventional mass parameters of the quarks cannot be directly asymptotically measured because of confinement. Furthermore, those mass parameters cannot be indirectly inferred from measurements of free and bound state masses below hadron production threshold because of "infrared slavery".

In contrast, the full analytic structure of a system's behavior at all energies is manifest in the amplitudes of non-perturbative scattering theory. However, a scattering theoretical approach has often been stifled by the sheer complexity such complete solutions entail. This complexity is magnified for relativistic scattering theory due to the nature of the energy-momentum dispersion relation, the expectation of particle-anti-particle production thresholds in the kinematics, and the difficulty of ensuring cluster decomposability in a covariant relativistic scattering formalism.

3.1 Non-perturbative scattering theory

Formal scattering theory describes a set of fully interacting scattering states in terms of boundary states that are asymptotically indistinguishable from those scattering states. The boundary states correspond to self-interacting boundary conditions for the scattering. The boundary states are presumed to represent a complete set of states, which allows even near-zone expansion of the scattering states in terms of these states. The Lippmann–Schwinger equation describes the connection between the two sets of eigenstates, as will be briefly discussed.

3.1.1 The Lippmann–Schwinger equation

A fully-interacting system can be represented using eigenstates $|\Psi(M)\rangle$ of the local energy-momentum of the system. The invariant energy of the system Mc^2 will be represented in terms of an invariant mass operator $\hat{\mathcal{M}}$. The fully-interacting system has constituent clusters that asymptotically do not interact but can undergo scattering with other clusters. The boundary states $|\Phi(M)\rangle$ do not have interactions between the asymptotic clusters but maintain all self-interactions within the clusters, having physical masses and charges. The boundary states *can* include sub-cluster bound states. Sub-cluster bound states *can* appear at the boundary if there are three or more interacting particles. Boundary states are eigenstates of the invariant self-interacting mass operator $\hat{\mathcal{M}}_o$ and serve as boundary conditions for the scattering system:

$$\begin{aligned} \hat{\mathcal{M}}\,|\Psi(M)\rangle &= M\,|\Psi(M)\rangle, \\ \hat{\mathcal{M}}_o|\Phi(M)\rangle &= M\,|\Phi(M)\rangle, \\ |\Psi(M)\rangle &\overset{\text{boundary}}{\Longrightarrow} |\Phi(M)\rangle, \end{aligned} \tag{3.1}$$

where the "interaction potential" is defined by $\hat{\Delta} \equiv \hat{\mathcal{M}} - \hat{\mathcal{M}}_o$. Therefore, the fully interacting states satisfy the equation:

$$(\hat{\mathcal{M}}_o - M)\,|\Psi(M)\rangle = -\hat{\Delta}\,|\Psi(M)\rangle. \tag{3.2}$$

Eigenvalues of fully interacting states and boundary states overlap in the continuum for "scattering" states with energy Mc^2, which is presumed to be the (conserved) energy of the scattering state at the (asymptotic) boundary. Since there is an overlap of the eigenvalues of $\hat{\mathcal{M}}$ and $\hat{\mathcal{M}}_o$ in the continuum, there is a need to go *off-shell* with general complex eigenvalues, Z. This allows an inversion of the operator $(\hat{\mathcal{M}}_o - Z)$, resulting in a solution of the form:

$$\begin{aligned} |\Psi(Z)\rangle &= |\Phi(M)\rangle - (\hat{\mathcal{M}}_o - Z)^{-1}\hat{\Delta}\,|\Psi(Z)\rangle, \text{ or} \\ |\Psi(Z)\rangle &= \left(\hat{1} - (\hat{\mathcal{M}}_o - Z)^{-1}\hat{T}(Z)\right)|\Phi(M)\rangle \end{aligned} \tag{3.3}$$

where $\hat{\Delta}\,|\Psi(Z)\rangle \equiv \hat{T}(Z)\,|\Phi(M)\rangle$ defines the transition operator $\hat{T}(Z)$.

The integral equations for the transition operator are given by:

$$\hat{T}(Z) = \hat{\Delta} - \hat{\Delta}\,(\hat{\mathcal{M}} - Z)^{-1}\hat{\Delta}, \text{ and} \tag{3.4}$$

$$\hat{T}(Z) = \hat{\Delta} - \hat{\Delta}\,(\hat{\mathcal{M}}_o - Z)^{-1}\hat{T}(Z). \tag{3.5}$$

As defined, the transition operator is most naturally evaluated using the boundary states, defining a transition amplitude T-matrix $T(M_f|M_o; Z) \equiv \langle\Phi(M_f)|\hat{T}(Z)|\Phi(M_o)\rangle$. Energy-momentum values corresponding to $M_f \neq M_o$ define an *off-diagonal* transition matrix. These equations describe the full analytic form of the transition amplitudes for arbitrary scattering processes.

Bound states are eigenstates of the fully interacting system with discrete energies $\{M_r c^2\}$. The bound-state masses M_r are less than the sum of the constituent boundary state masses. Because of the singularity in Z apparent from 3.4, bound states cannot be obtained perturbatively but are indeed described in the equations for the transition operator. The bound states are seen to be represented as poles in the fully interacting resolvant, or propagator, as a function of the off-shell parameter Z. By inserting a complete set of fully-interacting states, including all bound states, between the operators in Eq. 3.4, the transition matrix is singular when the off-shell parameter takes on a value precisely the same as that of a discrete bound state. Note that $\langle \Phi(M)|\hat{\Delta}|\Psi(M_r)\rangle = (M_r - M)\langle \Phi(M)|\Psi(M_r)\rangle$ and that the discrete spectrum never overlaps continuum eigenvalues. The non-perturbative description of the discrete spectrum of bound states can be obtained using the observation that:

$$\lim_{Z \to M_r} (Z - M_r)\, T(M|M_o; Z) = (M_r - M)\, \langle \Phi(M)|\Psi_r\rangle\langle \Psi_r|\Phi(M_o)\rangle\, (M_r - M_o)\,, \tag{3.6}$$

which extracts (momentum space) bound-state wavefunctions from the poles in the scattering/transition amplitude. The integral equation defining bound states is obtained by noting that only the singular term in the transition amplitude contributes to the eigenvalue equation:

$$T(M_f|M_o; M_r) = -\int dM\, \langle \Phi(M_f)|\hat{\Delta}\, [M - M_r]^{-1}|\Phi(M)\rangle T(M|M_o; M_r). \tag{3.7}$$

The transition amplitudes will be related to physical observables in Section 3.3.4.

3.2 Faddeev formulation of scattering theory

Faddeev [41] successfully developed a formal mechanism for obtaining non-relativistic scattering amplitudes in a few-particle sector using inputs from the various scattering sub-clusters involved. The method defines channels that characterize all possible physical clusters in the initial and final states, and guarantees unitarity and cluster decomposability of the physical amplitudes, as long as the input interactions within the sub-clusters are unitary. For present purposes, the most useful property of the Faddeev channel decomposition is that it eliminates those singularities in kernels that would occur in calculating the scattering amplitudes due to disconnected diagrams. There is no need for a renormalization procedure during the calculation, and the input amplitudes are parameterized using physical masses, charges, etc.

3.2.1 Channel decomposition

Faddeev channels are defined by all potential clusters in interacting systems. For three-particle scattering, the channels can be specified by the *spectator* index,

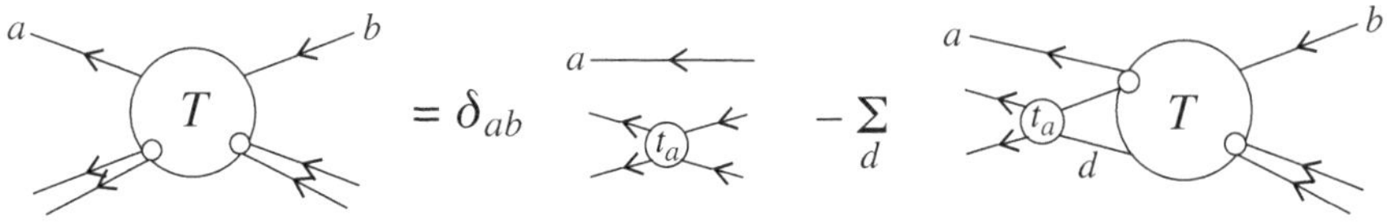

Figure 3.1 Faddeev channel T matrix.

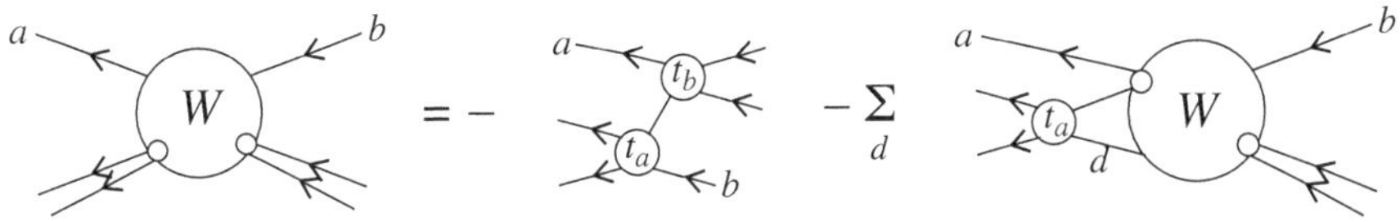

Figure 3.2 Faddeev channel W matrix.

which labels the non-interacting particle. This means that the interaction potential, V_3, represents the potential between the pair 1 and 2. Assuming only pairwise interactions, the Hamiltonian and channel transition operators of the fully-interacting system are given by:

$$H = H_o + V_1 + V_2 + V_3 = H_o + \sum_a V_a,$$

$$T_a(Z) = V_a - V_a(H_o - Z)^{-1} T_a(Z) \text{ for pairwise scattering,}$$
$$T_{ab}(Z) = \delta_{ab} T_a(Z) - \sum_d \bar{\delta}_{ad} T_a(Z)(H_o - Z)^{-1} T_{db}(Z). \tag{3.8}$$

The factor $\bar{\delta}_{ad} \equiv 1 - \delta_{ad}$ ensures that the second-term interaction mixes channels and excludes self-interactions. The transition matrix for the system is constructed by summing over the Faddeev channels $T(Z) = \sum_{ab} T_{ab}(Z)$, where the channel amplitudes are diagrammatically demonstrated in Figure 3.1. The full transition matrix $T(Z)$ can directly be shown to be unitary as long as the $T_a(Z)$ satisfy pairwise unitarity [42].

The fully connected amplitudes W_{ab}, defined by $T_{ab}(Z) \equiv \delta_{ab} T_a(Z) + W_{ab}(Z)$, are particularly convenient for calculating unitary amplitudes. These amplitudes mix interacting clusters and satisfy the set of equations given by:

$$W_{ab}(Z) = -\bar{\delta}_{ab} T_a(Z)(H_o - Z)^{-1} T_b(Z) - \sum_d \bar{\delta}_{ad} T_a(Z)(H_o - Z)^{-1} W_{db}(Z), \tag{3.9}$$

which are diagrammatically demonstrated in Figure 3.2. It is important to note that the $\bar{\delta}_{ad}$ in Eq. 3.9 excludes any contributions from self-energy bubbles, vacuum polarizations, etc., that do not involve a coupling between channels.

The amplitudes $T_a(Z)$ in Eq. 3.8 embed the pairwise interactions into the three-particle space. Faddeev achieves this proper unitary embedding by specifying this input to take the form:

$$\langle \mathbf{p}'_1, \mathbf{p}'_2, \mathbf{p}'_3 | T_a(Z) | \mathbf{p}_1, \mathbf{p}_2, \mathbf{p}_3 \rangle \equiv \delta^3(\mathbf{p}'_a - \mathbf{p}_a) \langle \mathbf{p}'_{a+}, \mathbf{p}'_{a-} | t_a \left(Z - \frac{p_a^2}{2n_a} \right) | \mathbf{p}_{a+}, \mathbf{p}_{a-} \rangle \tag{3.10}$$

where $t_a(z_a)$, is a transition operator that is unitary in the $(a+, a-)$ pair subspace, $n_a = (\frac{1}{m_a} + \frac{1}{m_{a+}+m_{a-}})^{-1}$ is the reduced mass between spectator a and the $(a+, a-)$ pair, and the momenta are specified in the three-particle center-of-momentum system. This input form achieves cluster decomposability since, if particle a does not interact with either other particle, it remains a spectator with conserved momentum, and $T_{a+}(Z) = 0 = T_{a-}(Z)$. Therefore, only the scattering channel $T_a(Z)$ will contribute to the three-particle scattering. The formulation has been generalized to four or more particles [43].

3.2.2 Long-range coherence and the Efimov effect

A subtle effect of quantum scattering of interest to the foundations of quantum theory becomes evident when the interacting system involves three or more particles. Through an analysis of wavefunctions, Noyes [44] noted that long-range effects occur in the three-particle system, even if all pairwise interactions are short range. This ("eternal triangle") effect changes the interaction between a given pair, if a third interacting particle is brought into the system *anywhere*, regardless of the range of the forces involved.

By examining the scaling behavior of the Hamiltonian eigenstates in the case of resonantly interacting particles, Efimov [37] determined that the number of three-particle bound state solutions grows as the magnitude of the scattering length for a pair becomes large relative to the scale of forces (weak pairwise interactions). The three-particle binding energy of such a state is small. The scattering length for the pairwise interaction can be parameterized most directly in terms of the *phase shift* for the s-wave ($\ell = 0$) low-energy scattering δ_0, from which a partial wave cross-section is given by:

$$\sigma_0 = \frac{4\pi}{|\mathbf{k}|^2} \sin^2 \delta_0. \tag{3.11}$$

The wave number $\mathbf{k}$ is related to the internal pair momentum by $\mathbf{q} \equiv \hbar \mathbf{k}$. The *scattering length*, a, and *effective range*, r_o, are defined in terms of the low energy behavior of the phase shift,

$$|\mathbf{k}| \cot \delta_0 = -\frac{1}{a} + \frac{1}{2} r_o^2 |\mathbf{k}|^2 + O\left[\frac{(r_o|\mathbf{k}|)^4}{r_o}\right]. \tag{3.12}$$

Thus, the low-energy s-wave partial cross-section can be parameterized:

$$\sigma_0(|\mathbf{k}|) = \frac{4\pi a^2}{1 + a(a - r_o)|\mathbf{k}|^2 + \left(\frac{1}{2} a\, r_o\right)^2 |\mathbf{k}|^4}. \tag{3.13}$$

The scattering length directly provides the length/area scale for the cross-section at zero energy $|\mathbf{k}| \to 0$.

The partial wave scattering amplitudes $S_\ell = e^{2i\delta_\ell}$ and transition amplitudes $\tau_\ell = \frac{S_\ell - 1}{2i} = e^{i\delta_\ell} \sin \delta_\ell$ are directly related to the phase shift for a given partial wave. These amplitudes directly measure the probability amplitude for scattering into a given angular momentum state ℓ. If the potential is repulsive, the scattering length is positive $a > 0$. If the potential is attractive but does not bind, the scattering length is negative $a < 0$. As the strength of the potential increases to a sufficient form such that a new bound state occurs at zero binding energy, the phase shift takes the form $\delta_0(0) = (n + \frac{1}{2})\pi$ just at the transition, giving maximal cross-section. From Eq. 3.12, this means that the scattering length becomes large at these resonant values. Thus, the scattering length changes sign as each new bound state appears. Generically, the phase shift at zero energy is given by $n\pi$, where n is the number of pairwise bound states [45]. This can be proven by counting the number of poles in the analytic form of the on-shell transition amplitude in Eq. 3.4 within the appropriate closed contour.

Using the integral equation counting methods described in Appendix C.1, an inclusion of low energy pairwise scattering into three-particle dynamics can be shown to result in a logarithmic growth in the number of three-particle bound states as the two-particle binding energy decreases to zero (i.e., for a very weak pairwise interaction) [46]:

$$N^{(3)}_{\text{bound states}} \approx \frac{1}{\pi} \log\left(\frac{|a|}{r_o}\right). \tag{3.14}$$

It should be emphasized that such effects cannot be explored using perturbative methods, demonstrating potentially significant inconsistencies in any attempt to fundamentally model an attractive interaction using a perturbative approach. The range of the three-particle bound states is large compared to the effective range r_o. Even if the attractive interaction is short-range but resonant (not quite able to form a bound pair), the addition to the system of the third attracting particle can be sufficient to produce bound, long-range, three-particle bound states, significantly altering the analytic structure of the system.

3.3 Unitarity, Poincaré covariance, and cluster decomposability

Despite the success of Faddeev's approach in describing non-relativistic few-particle quantum dynamics, attempts to generalize this approach to relativistic systems met with several difficulties. Foremost amongst these difficulties were analytic complications introduced by the relativistic energy-momentum form, $E_{Total} = \sum \epsilon_a = \sqrt{|\sum \mathbf{p}_a c|^2 + (M^* c^2)^2}$. If the overall momentum is conserved during intermediate processes as the energy goes off-diagonal, it becomes difficult to insure cluster decomposability of the kinematics. The Faddeev cluster transition amplitude, $T_{(a)}$, contains a spectator momentum conserving delta function. This then non-trivially inserts that spectator's momentum into integrations over intermediate momenta. In addition, the off-shell parameter $Z - \epsilon_a$ that enters into the two-particle embedded amplitude in effect samples differing regions of the pair off-shell energy depending upon the energy of the non-interacting spectator, thereby potentially modifying the energy spectrum of the pair just due to the relativistic kinematics of the spectator, independent of its dynamics within the system. This means that, for instance, even a distant relativistic electron would have an effect upon the bound-state spectrum of a local hydrogen atom. This is clearly not physical. The solution obtained involves ensuring that the Lorentz frame is maintained in all intermediate calculations. Off-diagonal and off-shell quantum dynamics in gravitating systems should likewise be described within well-defined coordinate frames of reference. However, the resultant conservation of intermediate frame *velocities* will introduce somewhat complicated kinematic factors that must be properly incorporated into a consistent relativistic formulation. On-shell, the conservation of Lorentz frame velocities will directly correspond with the appropriate conservation of four-momentum.

3.3.1 Relativistic few-particle scattering theory

Generally, the disentanglement of relativistic dynamic quantum clusters that is necessary for correspondence with classical dynamics requires that the geometric aspects of the kinematics associated with the Lorentz and Poincaré transformation properties of a given cluster must be separate and distinct from the internal coherent descriptions and off-shell analytic behaviors of disparate clusters. The methods described in this section demonstrate the techniques developed in the establishment of cluster decomposable formulations in relativistic few-particle scattering [47,48]. These formulations exhibit the expected disentanglement of interacting quantum scattering states needed for classical correspondence properties, in spite of the kinematic complications introduced by the non-linear nature of the relativistic energy-momentum dispersion relation for massive systems. The solution requires proper cluster independence of the geometric parameters describing the kinematics between subsystems, from the internal quantum dynamics associated with the description of an interacting system in terms of the boundary (i.e., only *self*-interacting) states. Cluster-decomposable relativistic scattering theory is most directly realized as follows:

- the clusters should be characterized using the channel decomposition that Faddeev [41] developed for describing cluster-decomposable unitary non-relativistic systems;
- the kinematics between external and intermediate quantum states (*off-diagonal* dynamics) should ensure Lorentz frame (three-velocity) conservation rather than three-momentum conservation. Velocity conservation has been referred to as the *point form* representation of the dynamics by Dirac [49] and others, whereas intermediate momentum conservation has been referred to as the *contact form* representation [50];
- the internal dynamics of the intermediate quantum states (or descriptions of the interacting state in terms of the complete basis of boundary states) for the various non-interacting clusters should be independent;
- the complex analytic extension of the dynamic invariant energy of a given cluster (the *off-shell* behavior) should only parametrically affect the kinematics of the other clusters, i.e., the internal dynamics of one cluster should not alter the energy spectrum of another;
- formulations of scattering theory with these properties should have appropriate non-relativistic behaviors [47,48] as well as give the expected results from perturbative representations such as quantum electrodynamics [13];
- a linear dispersion relation between energy and momentum allows single parameter descriptions of off-diagonal and off-shell behaviors, with proportionate energy-momentum properties. Lorentz-frame conservation is most directly realized by going off-shell in invariant center-of-momentum energy (mass), maintaining kinematic energy-momentum relationships.

3.3.2 Normalization of states and relativistic kinematics

The relativistic energy of a single particle (on arbitrary mass shell) takes the usual form, $\epsilon(\mathbf{p}) \equiv \sqrt{(mc^2)^2 + |\mathbf{p}c|^2}$. As previously defined, the standard state of a particle defines a unique reference frame from which all other particle motions can be obtained through a Lorentz transformation. The standard state contravariant momentum for massive particles is taken to be $\vec{p}_{(s)} = \{mc, 0, 0, 0\}$, where the four-momentum of that particle in an arbitrary frame boosted by four-velocity $\vec{u}$ is given

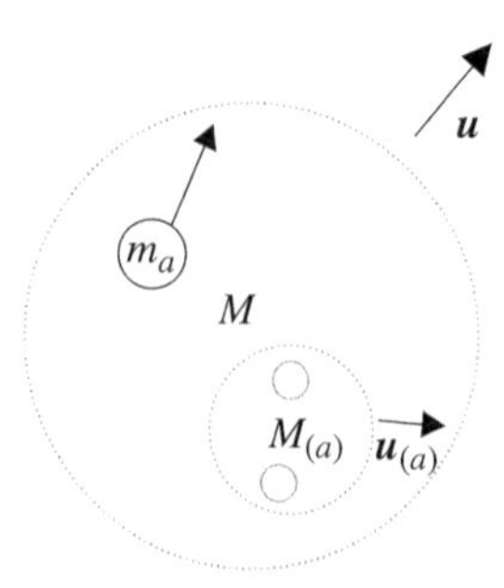

Figure 3.3 Kinematic parameters for clusters with internal degrees of freedom.

by $\vec{p} = m\vec{u}$. The standard state contravariant momentum for massless particles will be taken to be $\vec{p}_{(s)} = \{1, 0, 0, 1\}$.

As previously discussed, the normalization of momentum states has been chosen to have direct correspondence with non-relativistic normalizations. The mass shell orthonormalization is given by $\langle \mathbf{p}' | \mathbf{p} \rangle = \frac{\epsilon(\mathbf{p})}{\vec{p}^0_{(s)}} \delta^3(\mathbf{p}' - \mathbf{p})$, and the completeness relation consistent with this normalization is given by $\hat{1} = \int \frac{\vec{p}^0_{(s)}}{\epsilon(\mathbf{p})} d^3p \, |\mathbf{p}\rangle\langle\mathbf{p}| = \int d^4p \, \delta(\sqrt{-\vec{p} \cdot \vec{p}} - mc) \, |\vec{p}\rangle\langle\vec{p}|$. It should be noted that the mass-shell constraint is not necessary. General scatterings need not preserve particle mass and often do not preserve sub-cluster invariant energies. These states will describe *fully dressed* particles, properly normalized and labeled by their asymptotically defined parameters (mass, charge, etc.).

The cluster kinematics will be parameterized in terms of the overall invariant energy Mc^2 and the Lorentz four-velocity that boosts the overall system out of its rest frame $\vec{u}$, and the interacting cluster (a) invariant energy $M_{(a)}c^2$ and Lorentz frame velocity $\vec{u}_{(a)}$, as demonstrated in Figure 3.3. The invariant dynamic cluster energy can be shown to satisfy the condition:

$$M_{(a)} = M u^0_{(a)} - \sqrt{m_a^2 + M^2 \, |\mathbf{u}_{(a)}|^2}. \tag{3.15}$$

Using this invariant, the internal momentum of the particles in the two-particle center-of-momentum system can be directly calculated as:

$$|\mathbf{q}_{(a)}(M^2_{(a)}, m_{a+}, m_{a-})|^2 = \frac{\left[M^2_{(a)} - (m_{a+} + m_{a-})^2\right]\left[M^2_{(a)} - (m_{a+} - m_{a-})^2\right]}{4M^2_{(a)}} c^2 \tag{3.16}$$

Therefore, the kinematic parameters describing the dynamics of the scattering process can be described in terms of the invariant mass that goes off-diagonal M, the three-velocity components defining the overall frame of reference $\mathbf{u}$, and the three-velocity components that describe the frame of reference of the interacting cluster $\mathbf{u}_{(a)}$.

3.3.3 Cluster-decomposable scattering amplitudes

The relativistic form of the transition operators are taken from the Lippman–Schwinger equation, with the relativistic resolvants (propagators) given in terms of the invariant energy $\hat{R}_o(Z) = \frac{\mathcal{P}_{E+}}{(\hat{\mathcal{M}}_o - Z)}$ where the positive energy projector $\mathcal{P}_{E+}$ ensures that within complete intermediate integrations, only positive energy states are included. This ensures that the propagators will be properly causal. The transition operator $\hat{T}_{(a)}(Z)$ that embeds the pair interactions into the three-particle space, and the channel transition operators $\hat{T}_{ab}(Z)$ take the form:

$$\hat{T}_{(a)}(Z) = \hat{\Delta}_a - \hat{\Delta}_a \frac{\mathcal{P}_{E+}}{(\hat{\mathcal{M}}_o - Z)} \hat{T}_{(a)}(Z) \text{ pairwise scattering,} \tag{3.17}$$

$$\hat{T}_{ab}(Z) = \delta_{ab} \hat{T}_{(a)}(Z) - \sum_d \bar{\delta}_{ad} \hat{T}_{(a)}(Z) \frac{\mathcal{P}_{E+}}{(\hat{\mathcal{M}}_o - Z)} \hat{T}_{db}(Z). \tag{3.18}$$

The fully-connected relativistic operator $W_{ab}(Z)$ is similarly given by:

$$W_{ab}(Z) = -\bar{\delta}_{ab} T_{(a)}(Z) \frac{\mathcal{P}_{E+}}{(\hat{\mathcal{M}}_o - Z)} T_{(b)}(Z) - \sum_d \bar{\delta}_{ad} T_{(a)}(Z) \frac{\mathcal{P}_{E+}}{(\hat{\mathcal{M}}_o - Z)} W_{db}(Z). \tag{3.19}$$

Scattering amplitudes are defined by appropriate matrix elements of these operators between the boundary states.

The boundary states describing three self-interacting-only clusters are most naturally expressed in terms of the individual invariant energies and momenta of those clusters. However, the scattering equations (3.19) require intermediate states that maintain the frame of reference. The Jacobian factors that change the individual particle momenta into appropriate Lorentz-frame-conserving parameters are (neglecting factors of c for the present):

$$\rho^{(3)}_{(a)}(M, u^0_{(a)}) \equiv u^0 u^0_{(a)} \frac{m_a m_{a+} m_{a-}}{\epsilon_a \epsilon_{a+} \epsilon_{a-}} \frac{\partial(\mathbf{p}_a, \mathbf{p}_{a+}, \mathbf{p}_{a-})}{\partial(M, \mathbf{u}, \mathbf{u}_{(a)}, \hat{\mathbf{q}}_{(a)})},$$

$$\rho^{(3)}_{(a)}(M, u^0_{(a)}) = \frac{m_a m_{a+} m_{a-} M^3 M^2_{(a)} \, |\mathbf{q}_{(a)}(M^2_{(a)}, m_{a+}, m_{a-})|}{M u^0_{(a)} - M_{(a)}}. \tag{3.20}$$

The amplitudes $T_{(a)}(Z)$ in Eq. 3.17 must again embed the pairwise interactions into the three-particle space. Unitary embedding can be achieved by specifying this input to take the form:

$$\langle \mathbf{p}'_1, \mathbf{p}'_2, \mathbf{p}'_3 | T_{(a)}(Z) | \mathbf{p}_1, \mathbf{p}_2, \mathbf{p}_3 \rangle$$
$$\equiv \frac{u^0 \delta^3(\mathbf{u}' - \mathbf{u})\, u^0_{(a)} \delta^3(\mathbf{u}'_{(a)} - \mathbf{u}_{(a)})}{\sqrt{\rho^{(3)}_{(a)}(M', u^0_{(a)})\, \rho^{(3)}_{(a)}(M, u^0_{(a)})}}$$
$$\times\, \tau_{(a)} \left(\frac{M' - e_a}{-\vec{u} \cdot \vec{u}_{(a)}}, \hat{\mathbf{q}}'_{(a)} \Big| \frac{M - e_a}{-\vec{u} \cdot \vec{u}_{(a)}}, \hat{\mathbf{q}}_{(a)}; \frac{Z - e_a}{-\vec{u} \cdot \vec{u}_{(a)}} \right), \tag{3.21}$$

where e will be discussed in Eq. 3.23. The input transition amplitude $\tau_{(a)}$ is obtained from the two-particle transition operator $t_{(a)}(z_{(a)})$, which is unitary in the $(a+, a-)$ pair subspace. This transition operator has matrix elements in the two-particle space given by:

$$\langle \mathbf{p}_{a+}, \mathbf{p}_{a-} | t_{(a)}(\zeta) | \mathbf{p}_{a+o}, \mathbf{p}_{a-o} \rangle$$
$$\equiv u^0_{(a)} \delta^3(\mathbf{u}_{(a)} - \mathbf{u}_{(a)o}) \frac{\tau_a \left(M_{(a)}, \hat{\mathbf{q}}_{(a)} | M_{(a)o}, \hat{\mathbf{q}}_{(a)o}; \zeta \right)}{\sqrt{\rho^{(2)}_{(a)}(M_{(a)})\, \rho^{(2)}_{(a)}(M_{(a)o})}} \tag{3.22}$$

where $\rho^{(2)}_{(a)}(M_{(a)}) = m_{a+} m_{a-} M^2_{(a)} |\mathbf{q}_{(a)}(M^2_{(a)}, m_{a+}, m_{a-})|$ are the analogous two-particle Jacobian-transforming kinematic parameters involving the momenta, $\mathbf{p}_{a+}, \mathbf{p}_{a-}$ to $M_{(a)}, \mathbf{u}_{(a)}, \hat{\mathbf{q}}_{(a)}$. This input form achieves cluster decomposability since, if particle a does not interact with either other particle, it remains a spectator with conserved momentum, and $T_{(a+)}(Z) = 0 = T_{(a-)}(Z)$. Therefore, only the scattering channel $T_{(a)}(Z)$ contributes to the three-particle scattering.

Another criterion that needs to be satisfied in order to maintain cluster decomposed energy spectra is that the off-shell behavior of the scattering must be parametric, and thus purely kinematic, with regards to the energy of the spectator e_a in Eq. 3.21 for off-diagonal intermediate states. This means that this parameter must be defined in terms of the on-shell, on-diagonal scattering kinematics. The kinematic form of the energy of spectator a for a final-state scattering pair is given by:

$$e_a(M_o, u^{*0}_{(a)}) = u^0_{(a)} \sqrt{m_a^2 c^2 + M_o^2 |\mathbf{u}^*_{(a)}|^2} - M_o\, |\mathbf{u}^*_{(a)}|^2, \tag{3.23}$$

where $u^{*0}_{(a)} = -\vec{u}_{(a)} \cdot \vec{u}$, while that for a final-state bound pair of mass $\mu_{(a)}$ is given by:

$$e_a(M_o, \mu_{(a)}) = \frac{M_o^2 + m_a^2 - \mu^2_{(a)}}{2M_o} c^2. \tag{3.24}$$

The development of appropriate amplitudes on the boundaries will be examined in the next section.

3.3.4 Physical amplitudes for scattering and bound states on the boundary

Physical transition amplitudes correspond to a restricted sector of the fully off-diagonal, off-shell transition amplitudes. The scattering formalism drives the off-diagonal amplitudes to conserve overall energy-momentum as the off-shell parameter Z is assigned the physical value for the invariant energy. It should be noted that there is a branch cut in the Z plane extending from the scattering threshold energy to unbounded positive energies. This branch cut is due to the overlap of the spectrum of energies of the boundary states with the spectrum of scattering energies of the fully interacting states, giving a non-analytic structure to the resolvant $(\hat{\mathcal{M}}_o - Z)^{-1}$. The direction of approach to real eigenvalues above scattering threshold, characterizes *in* versus *out* states, according to whether the real axis is approached from below versus above.

On-shell condition

The amplitudes are placed on-shell by taking the limit of $Z = M_o \pm i\eta$ as $\eta \to 0^+$, i.e., as Z approaches the real physical value associated with the invariant energy of the system M_o. The off-shell parameter appears in resolvants, which satisfy:

$$\lim_{\eta \to 0^+} \frac{1}{M - (M_o \pm i\eta)} = \frac{\mathcal{P}}{M - M_o} \pm i\pi\, \delta(M - M_o), \tag{3.25}$$

where $\mathcal{P}$ here represents the Cauchy principal value. The invariant energy delta function in this relationship, along with overall Lorentz-frame velocity conservation, results in energy-momentum conservation in scattering theoretic formulations.

Two-particle cross-section

Most practical scattering experiments involve two initial particles scattering into an arbitrary final state. The differential cross-section into a set of final state particles with four-momenta $\{\vec{p}_{af}\}$ is given by:

$$d\sigma = \frac{1}{\mathcal{F}(\vec{p}_{1o}, \vec{p}_{2o})} \prod_a \left(\frac{m_a c^2}{\epsilon_{af}} d^3 p_{af} \right) (2\pi)^4 \delta^4 \left(\sum_a \vec{p}_{af} - \vec{p}_{1o} - \vec{p}_{2o} \right) |A_{fo}|^2, \tag{3.26}$$

where the incoming flux factor is given by $\mathcal{F}(\vec{p}_{1o}, \vec{p}_{2o}) \equiv \frac{\sqrt{(\vec{p}_{1o}\cdot\vec{p}_{2o})^2 + (m_{1o} m_{2o} c^2)^2}}{(m_{1o}c)(m_{2o}c)}$, the physical transition probability density into the set of final state of particles with masses $\{m_a\}$ is specified by $|A_{fo}|^2$, and the relativistic phase space is given by the remaining factors of differential momenta and the delta function. The various physical transition amplitudes are obtained from the calculated transition amplitudes by appropriately extracting poles in the incoming and outgoing propagators from the transition amplitudes, as will be demonstrated next.

Three free to three free scattering

The transition operator for three free particles scattering into three free particles is obtained from the on-shell form of the sum of the channel amplitudes over all channels, $\hat{T}^{(+)}(M_o) = \sum_{ab} \hat{T}^{(+)}_{ab}(M_o)$. The invariant energy-conserving delta function generated by the on-shell resolvant generates a four-momentum-conserving delta function multiplied by the physical transition amplitude as follows:

$$\langle \Phi_{(o)} : \vec{p}_1; \vec{p}_2; \vec{p}_3 \left| \sum_{ab} \hat{T}^{(+)}_{ab}(M_o) \right| \Phi_{(o)} : \vec{p}_{1o}; \vec{p}_{2o}; \vec{p}_{3o} \rangle \, \delta(M - M_o)$$
$$= -A^{(+)}_{oo} (\vec{p}_1; \vec{p}_2; \vec{p}_3 | \vec{p}_{1o}; \vec{p}_{2o}; \vec{p}_{3o}) \, \delta^4(\vec{p}_1 + \vec{p}_2 + \vec{p}_3 - \vec{p}_{1o} - \vec{p}_{2o} - \vec{p}_{3o}). \tag{3.27}$$

Coalescence and breakup

Coalescence refers to a scattering process in which three initially free particles scatter into an outgoing bound pair along with a single spectator particle. Conversely, *breakup* refers to a scattering process involving an initial spectator particle scattering off of a bound pair, resulting in three final state free particles. The three-particle kinematics allows such processes to proceed.

The initial or final bound state propagator on the boundary can be extracted from the calculated analytic transition amplitudes by defining the operators $\hat{K}_{ab}$ and $\hat{\hat{K}}_{ab}$, given by:

$$\begin{aligned} \hat{R}_o(Z)\hat{W}_{ab}(Z)\hat{R}_o(Z) &\equiv \hat{R}_{(a)}(Z)\hat{K}_{ab}(Z)\hat{R}_o(Z) \quad \text{breakup}, \\ \hat{R}_o(Z)\hat{W}_{ab}(Z)\hat{R}_o(Z) &\equiv \hat{R}_o(Z)\hat{\hat{K}}_{ab}(Z)\hat{R}_{(b)}(Z) \quad \text{coalescence}. \end{aligned} \tag{3.28}$$

The physical breakup amplitude is then obtained from the expression:

$$\left\langle \Phi_{(o)} : \vec{p}_1; \vec{p}_2; \vec{p}_3 \left| \sum_{a} \hat{K}^{(+)}_{ab}(M_o) \right| \Phi_{(b)} : \vec{p}_{bo}; \psi_{(b)}(\vec{p}_{(b)o}) \right\rangle \delta(M - M_o)$$
$$= -A^{(+)}_{ob} \left(\vec{p}_1; \vec{p}_2; \vec{p}_3 | \vec{p}_{bo}; \psi_{(b)}(\vec{p}_{(b)o})\right) \, \delta^4(\vec{p}_1 + \vec{p}_2 + \vec{p}_3 - \vec{p}_{bo} - \vec{p}_{(b)o}), \tag{3.29}$$

while the physical coalescence amplitude is obtained from:

$$\left\langle \Phi_{(a)} : \vec{p}_a; \psi_{(a)}(\vec{p}_{(a)}) \left| \sum_{b} \hat{\hat{K}}^{(+)}_{ab}(M_o) \right| \Phi_{(o)} : \vec{p}_{bo}; \psi_{(b)}(\vec{p}_{(b)o}) \right\rangle \delta(M - M_o)$$
$$= -A^{(+)}_{ao} \left(\vec{p}_a; \psi_{(a)}(\vec{p}_{(a)}) | \vec{p}_{1o}; \vec{p}_{2o}; \vec{p}_{3o}\right) \delta^4(\vec{p}_a + \vec{p}_{(a)} - \vec{p}_{1o} - \vec{p}_{2o} - \vec{p}_{3o}). \tag{3.30}$$

Elastic, inelastic, and rearrangement scattering

Finally, an initial particle + bound pair can scatter into a final state particle + bound pair. If the bound pair remains unchanged in its initial state, the scattering is *elastic*. If the same bound pair changes its rest energy, the scattering is *inelastic*. If there is a particle exchange, yielding a different final state spectator and bound pair, this is referred to as *rearrangement* scattering. If the interaction produces confined bound states, these are the only physical amplitudes accessible to the system. The physical amplitudes are obtained from the operators $\hat{Q}_{ab}$, defined by:

$$\hat{R}_o(Z)\hat{W}_{ab}(Z)\hat{R}_o(Z) \equiv \hat{R}_{(a)}(Z)\hat{Q}_{ab}(Z)\hat{R}_{(b)}(Z). \tag{3.31}$$

Physical amplitudes A_{ab} are obtained from these operators using:

$$\begin{aligned}&\left\langle \Phi_{(a)} : \vec{p}_a; \psi_{(a)}(\vec{p}_{(a)}) \middle| \hat{Q}_{ab}^{(+)}(M_o) \middle| \Phi_{(b)} : \vec{p}_b; \psi_{(b)o}(\vec{p}_{(b)o}) \right\rangle \delta(M - M_o) \\ &\quad = -A_{ab}^{(+)}\left(\vec{p}_a; \psi_{(a)}(\vec{p}_{(a)}) | \vec{p}_b; \psi_{(b)}(\vec{p}_{(b)})\right) \delta^4(\vec{p}_a + \vec{p}_{(a)} - \vec{p}_{bo} - \vec{p}_{(b)o}). \end{aligned} \tag{3.32}$$

3.3.5 Non-relativistic correspondence

The non-relativistic behavior of the scattering amplitudes will next be demonstrated. In the limit that all velocities are small compared to that of light, the pair internal momentum takes the form $\mathbf{q}_{(a)} \Rightarrow \frac{m_{a+}\mathbf{p}^*_{a-} - m_{a-}\mathbf{p}^*_{a+}}{m_{a+}+m_{a-}}$, which is the pair-reduced momentum that is shared (equal but opposite) by each particle in the two-particle center-of-momentum system with reduced mass, $\frac{1}{m^{reduced}_{(a)}} = \frac{1}{m_{a+}} + \frac{1}{m_{a-}}$. The asterisks label the momentum values in the three-particle center-of-momentum system. The invariant pair energy takes the form $M_{(a)}c^2 \Rightarrow (m_{a+} + m_{a-})c^2 + \frac{|\mathbf{q}_{(a)}|^2}{m^{reduced}_{(a)}} \equiv (m_{a+} + m_{a-})c^2 + E_{(a)}$. These kinematic identifications relate the invariant scattering amplitude to the non-relativistic scattering amplitude through the identifications:

$$\begin{aligned}&\tau_{(a)}(m_{a+} + m_{a-} + E_{(a)}, \hat{\mathbf{q}}_{(a)} | m_{a+} + m_{a-} + E_{(a)o}, \hat{\mathbf{q}}_{(a)o}; m_{a+} + m_{a-} + z_{(a)}) \\ &\quad \Rightarrow \tau^{NR}_{(a)}(E_{(a)}, \hat{\mathbf{q}}_{(a)} | E_{(a)o}, \hat{\mathbf{q}}_{(a)o}; z_{(a)}) \\ &\quad \Rightarrow \frac{m_{a+}m_{a-}\sqrt{|\mathbf{q}_{(a)}||\mathbf{q}_{(a)o}|}}{m_{a+} + m_{a-}} \tau^{NR}_{(a)}(\mathbf{q}_{(a)} | \mathbf{q}_{(a)o}; z_{(a)}). \end{aligned} \tag{3.33}$$

The kinematic factor in the last step results from the change in variables from energy to momentum. The relativistic off-shell parameter can be expanded in the form $\zeta_{(a)} \Rightarrow m_{a+} + m_{a-} + z - \frac{|\mathbf{p}^*_a|^2}{2n_a}$, where the spectator reduced mass n_a is given by $\frac{1}{n_a} = \frac{1}{m_a} + \frac{1}{m_{a+}+m_{a-}}$, and $\mathbf{p}^*_a$ is the reduced momentum of the spectator in the three-particle center-of-momentum system. Therefore, the non-relativistic off-shell dependency in the pair interaction input is given by $z_{(a)} = z - \frac{|\mathbf{p}_a^*|^2}{2n_a}$.

The non-relatistic form of the Jacobian factor in Eq. C.17 that relates Lorentz velocities to particle momenta is given by:

$$F_{ad}(u_{(a)}, u'_{(d)}, u_{(a)}, u'_{(d)}) \Rightarrow \left(\frac{m_a m_d}{q'_{(a)} q'_{(d)}} \right)^{1/2} \frac{(m_{a+} + m_{a-})^2 (m_{d+} + m_{d-})^2}{m_1 m_2 m_3}. \quad (3.34)$$

Substituting these non-relativistic forms into the equation for the fully connected transition amplitude expressed in Eq. C.13 generated by Eq. 3.19 yields the Faddeev equation:

$$W_{ab}^{NR}\left(\mathbf{p}_a^*, \mathbf{q}_{(a)} | \mathbf{p}_{ao}^*, \mathbf{q}_{(a)o}; z\right) = W_{ab}^{D-NR}\left(\mathbf{p}_a^*, \mathbf{q}_{(a)} | \mathbf{p}_{ao}^*, \mathbf{q}_{(a)o}; z\right) +$$
$$- \sum_d \bar{\delta}_{ad} \int d^3 p'_d d^3 q'_{(d)} \delta(\mathbf{p}_{(a)}^* - \mathbf{p}'_{(a)})\, \tau_{(a)}^{NR}(\mathbf{q}_{(a)} | \mathbf{q}'_{(a)}; z - \frac{|\mathbf{p}_a^*|^2}{2n_a})$$
$$\times \frac{1}{E' - z} W_{db}^{NR}\left(\mathbf{p}'_d, \mathbf{q}'_{(d)} | \mathbf{p}_{ao}^*, \mathbf{q}_{(a)o}; z\right), \quad (3.35)$$

where the driving term is the non-relativistic limit of that W_{ab}^D in Eq. C.18 given in Appendix C.2. Therefore, the relativistic few-particle scattering formulation has a direct, well-defined correspondence with the standard Faddeev few-particle formalism. This means that all subtle non-relativistic effects that are predicted using the Faddeev equation (like, for instance, the Efimov effect) are also contained within this relativistic formulation for non-perturbative scattering.

3.3.6 Non-perturbative scattering theory

The general properties of few-particle scattering theory will be summarized here. The approach is to construct relativistic, cluster decomposable, and unitary amplitudes in the fixed-particle number sector. In order to calculate any arbitrary N particle scattering processes, one needs to specify all of the clusters with $n < N$ particles so as to identify the channels in the expansion of the transition matrix.

The formulation requires conservation of the velocity of the Lorentz frame of each cluster $\mathbf{v} = \frac{\mathbf{P}}{E}$ among any matrix element needed in the calculation. This differs from non-relativistic scattering theory, which conserves the net momenta of the clusters. If the momentum of a cluster is conserved $(\mathbf{P} = \mathbf{P}')$ as it goes off-energy-diagonal, for intermediate states $E \neq E'$, which then implies that the bra and ket are not in same Lorentz frame. However, the conservation of Lorentz-frame velocity means that both energy and momentum go off-diagonal together for intermediate states.

This velocity conserving form is formally related to Dirac's point form of relativistic dynamics [49]. The commutation relation $[\hat{K}_j, \hat{P}_k] = i\delta_{jk}\hat{H}$ demonstrates that interactions present in the Hamiltonian generator $\hat{H}$ must also manifest in either the boost generator $\hat{K}$ (the *instant form* of dynamics) or the generator for spatial displacements $\hat{P}$ (the *point form* of relativistic dynamics). The use of point form formulations was shown to be equivalent to the instant form and the light-front form by Sokolov [50], but the point form is demonstrated here to have the convenience of manifest Lorentz invariance.

The quantum dynamics of an interacting cluster is, therefore, most naturally described in the proper rest-frame of that cluster. This makes the formulation completely consistent with a proper-time formulation of the coherent process, since the proper time is associated with that unique Lorentz frame. The coherence of *each* self-interacting sub-cluster maintains the Lorentz frame as the quantum dynamics is described by off-diagonal matrix elements.

Another key property is that the spectator energies enter into the off-shell, off-diagonal kinematics of an interacting cluster only parametrically, using the spectator energy form, $e_a(M_o, -\vec{u}_{(a)} \cdot \vec{u})$. This guarantees that the proper-energy spectra of the self-interacting clusters are independent of the kinematics of all other clusters that are spectators to those internal interactions. This property is crucial for classical correspondence, since spectators can undergo measurements that de-cohere the spectators without affecting another cluster undergoing coherent internal interactions. Therefore, the amplitudes have direct correspondence limits with both non-relativistic and classical physics, including standard configuration-space wavefunctions with proper normalizations.

The off-shell amplitudes not only describe scattering states but also include bound states. For instance, the poles of the non-relativistic transition amplitude for Coulomb interactions have been shown to correspond to the bound states of hydrogenic atoms, and the extracted amplitude describes the bound state wavefunctions [51–53]. Also, finite-particle-number amplitudes for which the number of particles changes consistent with global conservation properties can be constructed from the fixed-particle-number amplitudes by including appropriate amplitudes involving particle-type exchanges through particle-antiparticle channels defined from those amplitudes. The antiparticles can be obtained using Feynman's assertion that "... one man's virtual particle is another man's virtual antiparticle" [54]. For momentum-space amplitudes, one can associate an incoming/outgoing particle with four-momentum $\vec{p}$ with an outgoing/incoming anti-particle with four-momentum $-\vec{p}$.

All of these properties will be exploited in the following example of Compton scattering. This example has been chosen to demonstrate that the standard result for the scattering of asymptotic free photons is physically indistinguishable from

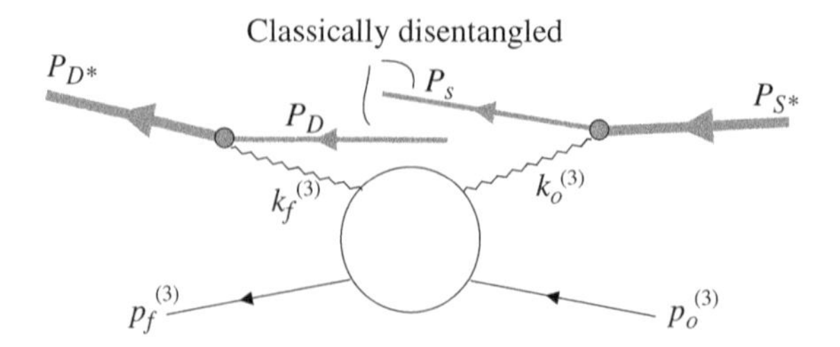

Figure 3.4 Classical disentanglement of a detector and source.

that of a distant source and sink that produce and absorb those photons. If photons are always associated with sources and sinks, the vacuum modes are defined by those particles, not by supposed empty space-time. This then eliminates certain vacuum energies that might later be expected to contribute to the energy-momentum densities generating space-time curvatures independent of the sources. The local energy-momentum densities would thus be due to particle dynamics and the fields directly associated with those dynamics.

3.3.7 Compton scattering

As an example of connecting an analytic scattering formalism to perturbative approaches, quanta as boundary states will be extracted from non-perturbative few-particle amplitudes [13]. The system to be examined will consist of a photon source, the spinor particle of interest, and a photon detector. The source cluster will interact with the particle via a two-body unitary scattering amplitude from which the incoming quantum will be extracted. Since the source is a transitioning boundary state, this scattering must be *anelastic* in the sense that the source changes its mass when the quantum is emitted. Likewise, the detector must engage in a unitary anelastic two-body scattering with the particle. The scattering of asymptotic (massless, transverse) photons from the spinor particle will be seen to yield the usual manifestly covariant (QED) result in the weak coupling limit.

The achievement of this result requires several properties of the non-perturbative formulation. A crucial property is the *cluster decomposition* of the source and detector, i.e., these boundary states are *disentangled* from each other in the usual quantum mechanical sense. These states can be complex subsystems, self-interacting and "fully dressed" (described by physically observable masses, charges, etc.), which only enter kinematically into the dynamics described by the overall integral equations.

Explicitly, the requirement that the source and detector be disentangled from each other, except through the interaction with the particle, is illustrated in Figure 3.4: Thus, one can separate out the specific process of interest by an

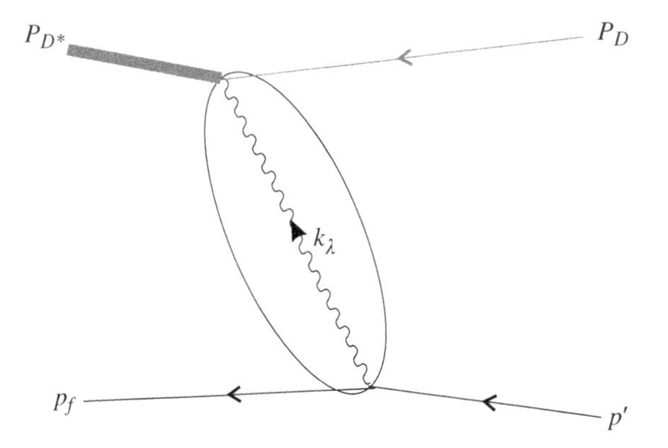

Figure 3.5 Extraction of an outgoing photon absorbed by a boundary detector.

appropriate choice of kinematic boundary conditions. This approach is related to the Wheeler–Feynman [12] point of view that includes both retarded and advanced components in source emission and photon detection or absorption, which is complementary to the point of view of the photons as asymptotic particles. The equivalence of these two points of view becomes manifest in that, beyond the specified kinematic properties, the amplitude derived is independent of the structure of the source and the detector.

The inclusion of the source and detector in Figure 3.4 converts the Compton scattering process into a three-particle scattering process, which incorporate the long-range photons as pairwise interactions between either the source and the charged target particle, or the detector and the target particle. The photon detection process involves the channel (*S*) for which the source is a spectator. Diagrammatically, one must calculate the process indicated in Figure 3.5. The on-shell kinematics to be extracted from the transition amplitude satisfy $\vec{k} = \vec{P}_{D*} - \vec{P}_D$. The on-shell invariant energy condition is given by $z_{(S)} \to M_{(S)}$, where $M_{(S)}c^2 \equiv -\vec{u}_{(S)} \cdot (\vec{k} + \vec{P}_{(D)} + \vec{p}_f)$, and $M_{(S)f}c^2 \equiv -\vec{u}_{(S)} \cdot (\vec{P}_{(D^*)} + \vec{p}_f)$. The dynamics of the asymptotic photon from the on-shell Lippmann–Schwinger equation yields:

$$\lim_{z_{(S)} \to M_{(S)} \pm i0} \langle \vec{P}_{D*}, \vec{p}_f | \hat{t}_{(S)}(z_{(S)}) | \vec{P}_D, \vec{p}' \rangle$$
$$= \pm 2\pi i \delta(M_{(S)f} - M_{(S)}) \langle \vec{P}_{D*}, \vec{p}_f | \hat{\Delta}_{(S)} | \vec{P}_D, \vec{k}, \vec{p}_f \rangle \langle \vec{P}_D, \vec{k}, \vec{p}_f | \hat{\Delta}_{(S)} | \vec{P}_D, \vec{p}' \rangle. \tag{3.36}$$

It remains only to evaluate the vertex functions, $\Delta_{(S)}$.

For Dirac spinors [55], the coupling depends upon the spinor current and the photon gauge potential, $\Delta = \frac{q}{c}\gamma^\mu A_\mu$. Its matrix element is given by:

$$\langle \vec{P}_D; \vec{k}, \lambda; \vec{p}_f, s_f | \hat{\Delta}_{(S)} | \vec{P}'_D; \vec{p}', s' \rangle \doteq \frac{u^0_{(S)f} \delta^3(\mathbf{u}_{(S)f} - \mathbf{u}'_{(S)})}{(M_{(S)f} \, M'_{(S)})^{3/2}} \frac{\mathcal{J}^\mu(\vec{p}_f, s_f | \vec{p}', s')}{2\pi} e_\mu(\vec{k}, \lambda). \tag{3.37}$$

The product of the polarization vectors of a photon with momentum $\vec{k}$ and polarization state λ define a tensor given by $e_\mu(\vec{k}, \lambda) e^*_\nu(\vec{k}, \lambda) \equiv \Pi_{\mu\nu}(\vec{k}, \lambda)$. Using this definition and defining the final state invariant energy using $M_f c^2 \equiv -\vec{u} \cdot (\vec{p}_f + \vec{P}_{D^*} + \vec{P}_S)$, the on-shell form of the pairwise transition amplitude having the source as spectator satisfies:

$$\lim_{Z \to M_f \pm i0} \langle \vec{P}_S, \sigma_S; \vec{p}_f, s; \vec{P}_{D*}, \sigma_{D*} | \hat{T}_{(S)}(Z) | \vec{P}'_S, \sigma'_S; \vec{p}', s'; \vec{P}_D, \sigma_D \rangle$$
$$= \pm 2\pi i \delta^4(\vec{p}_f + \vec{k}_f - \vec{p}') \frac{u^0_f \, \delta^3(\mathbf{u}_f - \mathbf{u}') \delta_{\sigma_S, \sigma'_S}}{(M_f \, M')^{3/2}}$$
$$\frac{\mathcal{J}^{\mu *}_D(\vec{P}_{D*}, \sigma_{D*} | \vec{P}_D, \sigma_D)}{2\pi i} \Pi_{\mu\nu}(\vec{k}, \lambda) \frac{\mathcal{J}^\nu(\vec{p}, s | \vec{p}', s')}{2\pi i}. \tag{3.38}$$

There is a similar form for the pairwise transition amplitude having the detector as spectator. These amplitudes are inputs into the full three-particle amplitude given in Eq. 3.18. Because of the $\bar{\delta}_{ab}$ excluding self-interactions in the transition amplitudes, and the classical disentanglement of the source and the detector (other than due to the Compton scattering), the unitary three-particle amplitude for this scattering is given by only the three terms:

$$T = \delta_{PP} T_{(P)} + W_{DS} + W_{SD}, \tag{3.39}$$

where the $T_{(P)}$ amplitude corresponds to a single photon that is emitted by the source and directly absorbed by the detector without interacting with the particle.

First, examine the amplitude W_{DS}. For the intermediate states, an integration over negative energies will be associated with positive energy antiparticles with the corresponding momenta assigned using Feynman's identifications mentioned previously [54]. This is diagrammatically demonstrated in Figure 3.6. The on-shell condition will set the off-shell parameter to equivalent values given by $Zc^2 \to -\vec{u}_f \cdot (\vec{p}_f + \vec{P}_{D*} + \vec{P}_S) = -\vec{u}_f \cdot (\vec{p}_o + \vec{P}_D + \vec{P}_{S*})$. The

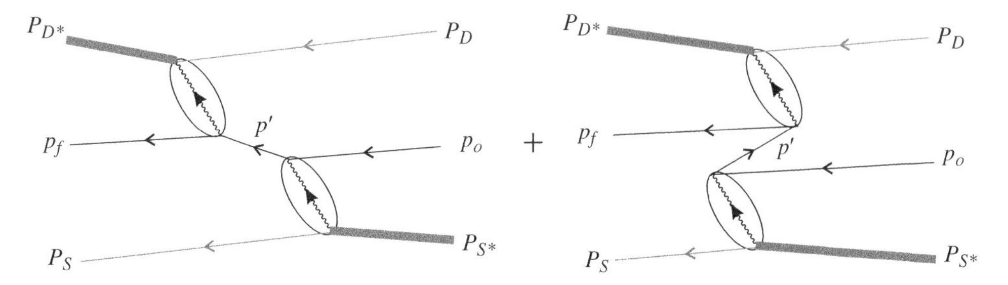

Figure 3.6 Causal propagator for intermediate particle states.

resolvant then takes the form:

$$\frac{\mathcal{P}_{E+}}{(\hat{\mathcal{M}}_o - Z)c^2} \doteq \frac{\mathcal{P}_{E+}(\vec{p}')}{(\sqrt{|\mathbf{p}'c|^2 + (mc^2)^2} + E_D + E_S) - (\epsilon_f + E_{D*} + E_S) \mp i0} + \frac{\mathcal{P}_{E+}(\vec{p}')}{(\sqrt{|\mathbf{p}'c|^2 + (mc^2)^2} + \epsilon_f + \epsilon_o + E_{D*} + E_{S*}) - (\epsilon_o + E_D + E_{S*}) \mp i0}, \tag{3.40}$$

where the energies are specified in the three-particle rest-frame. Simplifying this expression gives:

$$\frac{\mathcal{P}_{E+}(\vec{p}')}{(\hat{\mathcal{M}}_o - Z)c^2} = \frac{2\sqrt{|\mathbf{p}'c|^2 + (mc^2)^2}\mathcal{P}_{E+}}{(mc^2)^2 + \vec{p}c \cdot \vec{p}c \mp i0}. \tag{3.41}$$

For Dirac particles, the positive energy projector takes the form $\mathcal{P}_{E+}(\vec{p}') = \left(\frac{\gamma^\mu p'_\mu - mc\mathbf{1}}{2mc}\right)$, yielding the causal propagator form:

$$\frac{\mathcal{P}_{E+}}{(\hat{\mathcal{M}}_o - Z)c^2} \doteq -\frac{\sqrt{|\mathbf{p}'c|^2 + (mc^2)^2}}{mc^2} \frac{1}{(\gamma^\mu p'_\mu c + mc^2\mathbf{1} \pm i0)}. \tag{3.42}$$

The first factor precisely matches the invariant normalization factor for the momentum eigenstates. The resolvant in the calculation for the amplitude W_{SD} takes a similar form.

By dividing out the factors describing the interaction of the photons with the source and detector, a photon-spinor transition amplitude A can be written in terms

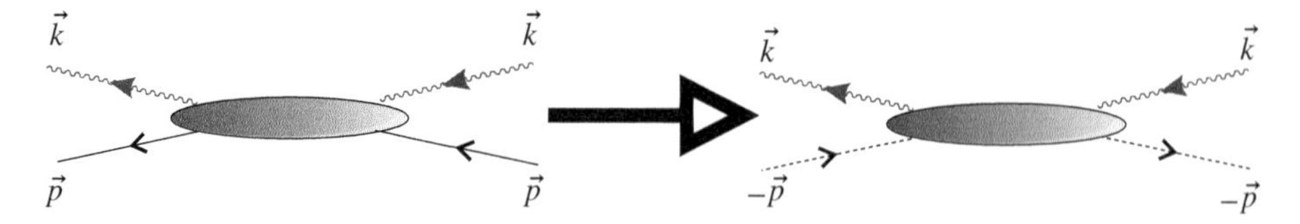

Figure 3.7 u-channel amplitude, where time increases towards the left.

of kinematic parameters only:

$$A(\vec{k}_f, \lambda_f; \vec{p}_f, s_f | \vec{k}_o, \lambda_o; \vec{p}_o, s_o) = \sum_{s'} \left(\frac{1}{2\pi i}\right)^2$$

$$\times \left[\mathcal{J}^{\mu *}(\vec{p}_f, s_f | \vec{p}_o + \vec{k}_o, s') e^*_\mu(\vec{k}_f, \lambda_f) \frac{2mc\mathcal{P}_{E+}(\vec{p}_o + \vec{k}_o)}{(mc)^2 + (\vec{p}_o + \vec{k}_o)^2} e_\nu(\vec{k}_o, \lambda_o) \mathcal{J}^\nu(\vec{p}_o, s_o | \vec{p}_o + \vec{k}_o, s') \right.$$

$$\left. + \mathcal{J}^\nu(\vec{p}_o, s_o | \vec{p}_o - \vec{k}_f, s') e_\nu(\vec{k}_o, \lambda_o) \frac{2mc\mathcal{P}_{E+}(\vec{p}_o - \vec{k}_f)}{(mc)^2 + (\vec{p}_o - \vec{k}_f)^2} e^*_\mu(\vec{k}_f, \lambda_f) \mathcal{J}^{\mu *}(\vec{p}_f, s_f | \vec{p}_o - \vec{k}_f, s') \right] \tag{3.43}$$

The parameters in this amplitude are the *physical* mass, charge, and four-momenta of the scattering constituents. In the weak coupling limit, the vertex factors $\mathcal{J}^\mu$ are directly related to the current of the free Dirac spinors:

$$\mathcal{J}^\mu(\vec{p}', s' | \vec{p}, s) \approx \frac{q}{c} \, \bar{\mathbf{u}}(\vec{p}', s') \gamma^\mu \mathbf{u}(\vec{p}, s), \tag{3.44}$$

yielding the second-order amplitude calculated from perturbative quantum electrodynamics [57].

The appropriate contributions of the anti-particle to the Compton scattering amplitude have already been included. Once the unitary, fully off-shell, standard amplitude has been calculated, one can construct both the anti-particle Compton scattering amplitude (the *u-channel* amplitude) and the particle-anti-particle annihilation amplitude (the *t-channel* amplitude) by the appropriate Feynman identification and substitution process [53]. This substitution procedure (crossing) involves replacing ingoing/outgoing particle legs with outgoing/ingoing anti-particle legs on the scattering diagram as appropriate. The u-channel amplitude is obtained from the standard amplitude by changing both the incoming and outgoing particle states into anti-particle states as indicated in Figure 3.7. Similarly, the annihilation or t-channel amplitude that calculates the annihilation of a particle-anti-particle pair directly into two photons is indicated in Figure 3.8. This annihilation amplitude can be properly embedded to construct a unitary amplitude that changes particle

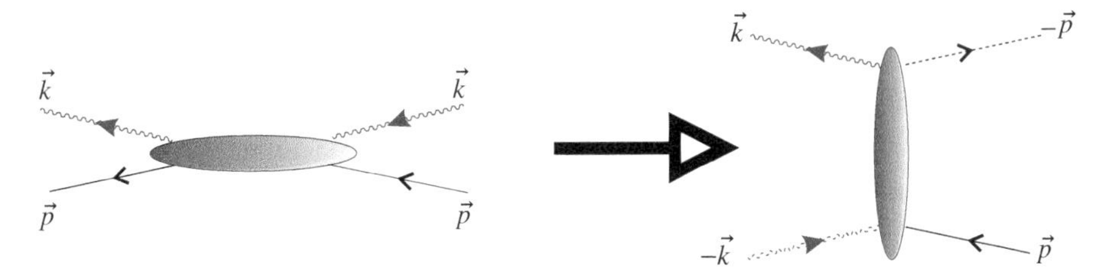

Figure 3.8 Annihilation channel amplitude, where time increases towards the left.

number. Therefore, a finite particle number scattering formulation *does not* imply that the number of particles remains fixed in all scattering processes. Annihilations and pair productions preserve the appropriate conserved quantum numbers and can be directly calculated from properly embedded channels obtained from the non-perturbative amplitudes.

Supplement: Do quantized carriers of interactions exist devoid of sources?

There is a distinction between the particles that exchange kinematic and dynamic parameters while maintaining certain defining characteristics, and the carriers of the interactions by which those exchanges occur. For instance, whereas spinor particles have some associated conserved quantum number, photons are "consumed" by detection. Photon number is generally not conserved during interactions. All interactions are characterized using bosonic representations, which can usually be represented using semi-classical models. However, the fundamental behavior of the antisymmetry of two fermions under exchange is a non-trivial property to represent using classical models.

As discussed in Section 2.1.3, Bohr and Rosenfeld demonstrated that the quantization properties of the electric and magnetic fields follow directly from the behaviors of quantized charged sources producing classical electromagnetic fields. Therefore, it might be argued that, whereas quantization of the sources and sinks of photons is a necessity, photon quantization itself seems merely a convenience. In particular, a consideration of the vacuum modes of photons in the absence of boundary conditions (with boundaries composed of sources and sinks) is ill-defined. A brief history of some concepts of the quantum characteristics associated with photons will be presented here.

One of the first successes of early quantum theory was the explanation of the spectrum of *radiations emitted by a black body* in thermal equilibrium. Planck asserted that the boundaries of a black body are composed of *resonators* whose energy levels are quantized.[1] The resonators are the sources and sinks of the electromagnetic modes in the black body cavity, requiring that the radiations are emitted and absorbed during quantized transitions of resonators. Using the statistical techniques discussed in Section 2.2, the average number density $\bar{n}(\omega)$ and energy density $\bar{u}_{thermal}(\omega)$ of the electromagnetic radiations are determined to satisfy:

$$\bar{n}(\omega) = \frac{1}{e^{\hbar\omega/k_B T} - 1}, \qquad \bar{u}_{thermal}(\omega) = \frac{\hbar\omega^3}{\pi^2 c^3}\, \bar{n}(\omega).$$

Unlike classical formulations, these formulas correctly describe the thermal radiance of bodies.

Later, Einstein successfully explained the lack of emissions of electrons from metals irradiated with low frequency light, regardless of the intensity of that light, and the onset of emissions of photoelectrons whenever the incident light exceeds a threshold frequency.[2] That threshold frequency is associated with the work function binding that electron, $h\nu_{threshold} = w$. This *photoelectric effect* demonstrates that the most likely way to have an electron that is bound within a metal transition into a free electron, is to have the incident light composed of radiations generated by external transitions of frequency greater than w/h.

Einstein eventually developed a semi-classical formulation of the interaction of electromagnetic radiation with atoms that predicted the amplification of light beams by lasers.[3] For a stationary state of a collection of systems with two (quantized) energy levels coupled via the absorption or emission of a quantum of radiation (with frequency $h\nu = \hbar\omega = E_2 - E_1$), he developed the transition rate equation describing the number of systems in each energy state in the presence of electromagnetic energy density $\bar{u}(\omega)$:

$$\frac{dN_1}{dt} = -\frac{dN_2}{dt} = N_2\, A_{21} - N_1\, B_{12}\, \bar{u} + N_2\, B_{21}\, \bar{u}.$$

The various factors, A_{21} (the coefficient for spontaneous emission), B_{12} (the coefficient for stimulated absorption), and B_{21} (the coefficient for stimulated emission), are independent of the electromagnetic energy density $\bar{u}$. For a system in thermal equilibrium, the average occupancy number for either level does not change in time. Algebraically solving the equilibrium equation, and

[1] M. Planck, *Verh. dt. Phys. Gex.* **2**, 202 (1900).
[2] A. Einstein, *Annln. Phys.* **17**, 132 (1905).
[3] A. Einstein, *Phys. Z.* **18**, 121 (1917). See also, D. ter Haar, *The Old Quantum Theory*, Pergamon, Oxford (1967).

using Boltzmann statistics to determine the ratio of thermal occupancy of the levels,

$$\bar{u}_{thermal} = \frac{A_{21}}{\frac{N_1}{N_2} B_{12} - B_{21}}, \quad \frac{N_1}{N_2} = \frac{g_1}{g_2} e^{\hbar\omega/k_B T},$$

one can establish the relationships between the coefficients needed to maintain consistency with Planck's black body formula: $\frac{g_1}{g_2} B_{12} = B_{21}$ and $\frac{\hbar\omega^3}{\pi^2 c^3} B_{21} = A_{21}$. This then relates the rate of thermally stimulated emissions to the rate of spontaneous emission and the average number of photons in the mode:

$$B_{21}\, \bar{u}_{thermal}(\omega) = A_{21}\, \bar{n}(\omega).$$

Thus, only semi-classical considerations are needed to demonstrate mode amplification proportional to the number of quanta already present in that mode.

One should note that the assertion of the *photon* as a quantum of radiation was not presented in the scientific literature until 1926.[4] Modern attempts to ascertain the quantum nature of the photon focus upon rth order coherence of the photon field.[5] The degree of rth order coherence (or an rth order correlation function) determines the averaged response in experiments that measures the properties of r coherent photons. Generally, photo-detectors give a response only if photons are present and do not measure zero-point energies of the quantized photon field. First-order correlation functions for the second quantized photon field yield the same results as the classical functions. In general, for any eigenstate of the photon field, or any chaotic state, the rth order correlation functions of the second quantized photon field yield the same results as the classical functions. However, it is informative to examine the second-order coherence for a single-mode excitation, $\frac{\langle \hat{a}^\dagger \hat{a}^\dagger \hat{a}\hat{a}\rangle}{\langle \hat{a}^\dagger \hat{a}\rangle^2} = \frac{\langle \hat{n}(\hat{n}-1)\rangle}{\langle \hat{n}\rangle^2}$. For quantum systems, the act of measurement interferes with the measured system. In this case, a measurement of the number of photons decreases a subsequent measurement by one photon, which is non-classical. However, such experiments require eventual interaction with quantized sources, sinks, or media, making it problematic to separate the quantum properties of the particles from those of the radiations.

[4] G.N. Lewis, *Nature*, **118**, 874 (1926).
[5] R. Loudon, *The Quantum Theory of Light*, Oxford University Press (1983).

3.4 Lagrangian dynamics

Much of physical phenomenology can be modeled in terms of extrema of functional forms, whether the straight lines of inertial Newtonian motions, formulations of accelerating classical trajectories and quantum dynamics through Lagrangian dynamics, the minimization of the free energies of thermodynamic or hydrodynamic systems, the optical path lengths of electromagnetic rays through macroscopic media, or the description of gravitating bodies through geodesic equations. The calculus of variations provides a formalism for determining the unique functions that extremize a specified functional, which will generally be referred to as the *action*, whose value is determined not by a local value of the function, but rather by the global form of the function. For all problems of interest here, the action W will be expressed as an invariant generated by an integral of a *Lagrangian density* over a bounded space-time volume:

$$W = \int_{\Omega_{(4)}} \mathcal{L} d^4 x, \tag{3.45}$$

where $\mathcal{L}$ has units of energy density, or energy per unit volume. Typically, the Lagrangian density depends upon a set of functions or fields $\boldsymbol{\Psi} = \{\psi_a\}$ and their (first) derivatives, as a functional of the type $\mathcal{L} = \mathcal{L}(\boldsymbol{\Psi}(\vec{x}), \vec{\nabla}\boldsymbol{\Psi}(\vec{x}))$. The equations of dynamics for systems that can be described using such Lagrangian densities will be developed.

Euler–Lagrange equations

Variations of the action under arbitrary functional variations $\delta\psi_a(\vec{x})$ satisfy:

$$\delta W = \int_{\Omega_{(4)}} d^4 x \sum_a \left(\frac{\partial \mathcal{L}}{\partial \psi_a} \delta\psi_a + \frac{\partial \mathcal{L}}{\partial(\partial_\mu \psi_a)} \partial_\mu \delta\psi_a \right), \tag{3.46}$$

where space-time partial derivatives are denoted $\partial_\mu \equiv \frac{\partial}{\partial x^\mu}$, and $\delta\partial_\mu \psi_a = \partial_\mu \delta\psi_a$. The functional variations are required to vanish on the boundary/endpoints $\delta\psi(\vec{x})|_{\vec{x} \in \partial\Omega_{(4)}} = 0$, describing a known state on that surface. The second term in the integral can be written:

$$\int_{\Omega_{(4)}} d^4 x \frac{\partial \mathcal{L}}{\partial(\partial_\mu \boldsymbol{\Psi})} \partial_\mu \delta\boldsymbol{\Psi} = \int_{\Omega_{(4)}} d^4 x \left[\partial_\mu \left(\frac{\partial \mathcal{L}}{\partial(\partial_\mu \boldsymbol{\Psi})} \delta\boldsymbol{\Psi} \right) - \partial_\mu \left(\frac{\partial \mathcal{L}}{\partial(\partial_\mu \boldsymbol{\Psi})} \right) \delta\boldsymbol{\Psi} \right]. \tag{3.47}$$

Since the functional variations are presumed to vanish on the boundary,

$$\int_{\Omega_{(4)}} d^4 x \, \partial_\mu \left(\frac{\partial \mathcal{L}}{\partial(\partial_\mu \boldsymbol{\Psi})} \delta\boldsymbol{\Psi} \right) = \oint_{\partial\Omega_{(4)}} d^3 \Sigma_\mu \frac{\partial \mathcal{L}}{\partial(\partial_\mu \boldsymbol{\Psi})} \delta\boldsymbol{\Psi} = 0. \tag{3.48}$$

This means that an extremal functional variation of the action must satisfy:

$$\delta W = \int_{\Omega_{(4)}} d^4x \left[\frac{\partial \mathcal{L}}{\partial \boldsymbol{\Psi}} - \partial_\mu \left(\frac{\partial \mathcal{L}}{\partial(\partial_\mu \boldsymbol{\Psi})} \right) \right] \delta \boldsymbol{\Psi}(\vec{x}) = 0, \tag{3.49}$$

for arbitrary functional variations $\delta \boldsymbol{\Psi}(\vec{x})$. Therefore the functions $\boldsymbol{\Psi}(\vec{x})$ must satisfy the *Euler–Lagrange equations*:

$$\frac{\partial \mathcal{L}}{\partial \psi_a} - \partial_\mu \left(\frac{\partial \mathcal{L}}{\partial(\partial_\mu \psi_a)} \right) = 0. \tag{3.50}$$

These differential equations (one for each component $\psi_a(\vec{x})$) represent the equations of motions that can be solved for the functional forms that extremize the action.

Local continuity and conservation equations

The Euler–Lagrange equations provide a mechanism for developing a conserved energy-momentum form. If one directly examines how the Lagrangian density varies in space-time $\partial_\beta \mathcal{L} = \frac{\partial \mathcal{L}}{\partial \psi} \partial_\beta \psi + \frac{\partial \mathcal{L}}{\partial(\partial_\mu \psi)} \partial_\beta \partial_\mu \psi$, a substitution into the first term on the right-hand side of this equation from Eq. 3.50 yields a total derivative:

$$\partial_\mu \mathcal{T}^\mu_\beta = 0, \qquad \text{where } \mathcal{T}^\mu_\beta \equiv \sum_a \frac{\partial \mathcal{L}}{\partial(\partial_\mu \psi_a)} \partial_\beta \psi_a - \delta^\mu_\beta \mathcal{L}. \tag{3.51}$$

This conserved energy-momentum form can also be derived as a special case of a general property of Lagrangians that are invariant under transformations of the functions $\boldsymbol{\Psi}(\vec{x})$.

Consider the variation of the Lagrangian density under a transformation that mixes the components $\delta \boldsymbol{\Psi}(\vec{x})$;

$$\delta \mathcal{L} = \frac{\partial \mathcal{L}}{\partial \boldsymbol{\Psi}} \delta \boldsymbol{\Psi} + \frac{\partial \mathcal{L}}{\partial(\partial_\mu \boldsymbol{\Psi})} \delta \partial_\mu \boldsymbol{\Psi}, \tag{3.52}$$

where $\delta \boldsymbol{\Psi}(\vec{x}) \equiv \boldsymbol{\Psi}'(\vec{x}) - \boldsymbol{\Psi}(\vec{x})$ is a field variation at a single point. The Lagrangian will be assumed to be invariant under the symmetry transformation $\delta \mathcal{L} = 0$. Using the Euler–Lagrange equations, the variation can be expressed:

$$\delta \mathcal{L} = \partial_\mu \left[\frac{\partial \mathcal{L}}{\partial(\partial_\mu \boldsymbol{\Psi})} \delta \boldsymbol{\Psi} \right] + \frac{\partial \mathcal{L}}{\partial(\partial_\mu \boldsymbol{\Psi})} [\delta, \partial_\mu] \boldsymbol{\Psi} = 0, \tag{3.53}$$

for a general symmetry transformation δ.

If the set of internal symmetry transformations form a group, the variation of $\boldsymbol{\Psi}(\vec{x})$ can be expressed in terms of the group parameters and the representation of the group generators for infinitesimal transformations of the same dimension as

$\mathbf{\Psi}(\vec{x})$, given by:

$$\delta\mathbf{\Psi} = i\,\delta\alpha^r \mathbf{G}_r \mathbf{\Psi}. \tag{3.54}$$

For a global transformation, $\delta\partial_\mu \mathbf{\Psi} = \partial_\mu \delta\mathbf{\Psi}$, so that:

$$\delta\mathcal{L} = i\,\delta\alpha^r \partial_\mu \left[\frac{\partial\mathcal{L}}{\partial(\partial_\mu \mathbf{\Psi})} \mathbf{G}_r \mathbf{\Psi} \right] = 0, \tag{3.55}$$

which defines the internal group symmetry. This equation defines a continuity equation for a current density associated with each symmetry $\delta\alpha^r$;

$$J_r^\mu \equiv \frac{i}{\hbar} \frac{\partial\mathcal{L}}{\partial(\partial_\mu \mathbf{\Psi})} \mathbf{G}_r \mathbf{\Psi} + \text{hermitian conjugate}, \quad \partial_\mu J_r^\mu = 0. \tag{3.56}$$

This current density carries units of probability or number current density. The local conservation property furthermore defines a globally conserved charge. If one integrates the current continuity equation over a space-like volume $\Omega_{(3)}$, the result is:

$$0 = \int_{\Omega_{(3)}} d^3x \partial_\mu J_r^\mu = \frac{d}{dct} \int_{\Omega_{(3)}} d^3x J_r^0 + \oint_{\Sigma=\partial\Omega_{(3)}} d^2\mathbf{\Sigma} \cdot \mathbf{J}_r. \tag{3.57}$$

Assuming that the net flux of current through the closed surface vanishes (which is usually the case for localized currents through asymptotic closed surfaces), the conserved charge associated with the symmetry transformation $\delta\alpha^r$ is given by:

$$Q_r \equiv \int_{\Omega_{(3)}} d^3x J_r^0, \quad \frac{d}{dt} Q_r = 0. \tag{3.58}$$

The association of a conserved charge for each global symmetry of the system is known as *Noether's theorem.*

Angular momentum and Lorentz transformations

Functions describing the angular momentum of systems that can be described using Lagrangian densities will be developed next. The Lorentz group of transformations consists of the three angles describing an arbitrary rotation relating the relative orientations of the frames of reference, and three components of a velocity boost between frames of reference. Examining infinitesimal transformations for the present, these six parameters will conveniently specify the elements of a real, antisymmetric, dimensionless 4×4 matrix $\delta\omega^{\mu\nu}$, whose components will be defined by:

$$\frac{\delta v^j}{c} \equiv \delta\omega^{j0}, \quad \delta\theta^m \equiv \epsilon^m_{jk} \delta\omega^{jk}. \tag{3.59}$$

Infinitesimal parameters specified this way define commensurate generators related to angular momentum and Lorentz boost generators (up to the dimensional factor, $\hbar$). These generators for infinitesimal transformations parameterized by $\delta\omega^{\mu\nu}$ will be defined by matrices $\mathcal{J}_{\mu\nu}$ that transform the components of $\mathbf{\Psi}$ and are anti-symmetric under interchange of the space-time indices. The dynamic functions then transform under Lorentz transformations according to:

$$\begin{aligned}\delta\mathbf{\Psi} &= \tfrac{i}{2}\,\delta\omega^{\mu\nu}\mathcal{J}_{\mu\nu}\,\mathbf{\Psi},\\ \delta(\partial_\beta\mathbf{\Psi}) &= \tfrac{i}{2}\,\delta\omega^{\mu\nu}\mathcal{J}_{\mu\nu}\,(\partial_\beta\mathbf{\Psi}) + \delta\omega_\beta^\alpha\,(\partial_\alpha\mathbf{\Psi}).\end{aligned} \tag{3.60}$$

Lorentz transformation of the derivatives is seen to modify both the (spinor) components of the dynamic functions $\mathbf{\Psi}$, as well as the space-time index associated with the derivative operation.

For a system that is invariant under Lorentz transformations, the transformation properties described in Eq. 3.60 can be inserted into Eq. 3.53 to yield:

$$\delta\mathcal{L} = 0 = \frac{i}{2}\delta\omega^{\mu\nu}\partial_\beta\left[\frac{\partial\mathcal{L}}{\partial(\partial_\beta\mathbf{\Psi})}\mathcal{J}_{\mu\nu}\,\mathbf{\Psi}\right] + \frac{\partial\mathcal{L}}{\partial(\partial^\mu\mathbf{\Psi})}\,\delta\omega^{\mu\nu}\,\partial_\nu\mathbf{\Psi}. \tag{3.61}$$

The last term in Eq. 3.61 gives the antisymmetric part of the tensor $\mathcal{T}^{\mu\nu}$:

$$\frac{\partial\mathcal{L}}{\partial(\partial_\mu\mathbf{\Psi})}\partial^\nu\mathbf{\Psi} - \frac{\partial\mathcal{L}}{\partial(\partial_\nu\mathbf{\Psi})}\partial^\mu\mathbf{\Psi} = \mathcal{T}^{\mu\nu} - \mathcal{T}^{\nu\mu} = -i\partial_\beta\left[\frac{\partial\mathcal{L}}{\partial(\partial_\beta\mathbf{\Psi})}\mathcal{J}^{\mu\nu}\,\mathbf{\Psi}\right]. \tag{3.62}$$

For functions $\mathbf{\Psi}$ that do not transform as scalars, there might be anti-symmetric components of this tensor. If there are components of the tensor $\mathcal{T}^{\mu\nu}$ that *are* anti-symmetric under the interchange $\mu \rightleftharpoons \nu$, then that tensor will be unsuitable for use as a gravitational source in Part II. However, additional terms that are symmetric under the interchange $\mu \rightleftharpoons \nu$ can be added to the bracketed term in Eq. 3.61 without modifying the invariance of the Lagrangian, since $\delta\omega^{\mu\nu}$ is anti-symmetric under such interchanges. Therefore, terms can be added to the bracketed function in Eq. 3.62 that are anti-symmetric under the interchange $\beta \rightleftharpoons \mu$, but symmetric under the interchange $\mu \rightleftharpoons \nu$, in order to create a conserved symmetric energy-momentum tensor. Consider the combination $S^{\beta\mu\nu}$ defined by:

$$S^{\beta\mu\nu} \equiv \frac{i}{2}\left[\frac{\partial\mathcal{L}}{\partial(\partial_\beta\mathbf{\Psi})}\mathcal{J}^{\mu\nu}\,\mathbf{\Psi} - \frac{\partial\mathcal{L}}{\partial(\partial_\mu\mathbf{\Psi})}\mathcal{J}^{\beta\nu}\,\mathbf{\Psi} - \frac{\partial\mathcal{L}}{\partial(\partial_\nu\mathbf{\Psi})}\mathcal{J}^{\beta\mu}\,\mathbf{\Psi}\right]. \tag{3.63}$$

Since $S^{\beta\mu\nu} = -S^{\mu\beta\nu}$, and any integrable, analytic function must satisfy $[\partial_\beta, \partial_\mu]F(\vec{x}) = 0$, the Belinfante tensor [57] $T^{\mu\nu}$ defines a conserved symmetric tensor:

$$T^{\mu\nu} \equiv \mathcal{T}^{\mu\nu} + \partial_\beta S^{\beta\mu\nu}, \quad \partial_\mu T^{\mu\nu} = 0, \quad T^{\mu\nu} = T^{\nu\mu}. \tag{3.64}$$

Furthermore, the component $T^{0\nu}$ is directly related to the canonical four-momentum generated by the Lagrangian $\mathcal{P}^\nu c \equiv \int \mathcal{T}^{0\nu} d^3x$:

$$\int T^{0\nu} d^3x = \int \mathcal{T}^{0\nu} d^3x + \frac{i}{2} \int \partial_j S^{j0\nu} d^3x, \text{ or}$$

$$P^\nu = \mathcal{P}^\nu + \frac{i}{2c} \oint_{\Sigma=\partial\Omega_{(3)}} S^{j0\nu} \hat{n}_j d^2\Sigma, \tag{3.65}$$

where only spatial derivatives contribute due to the anti-symmetry under $\beta \rightleftharpoons \mu$. The surface term describes any topological contributions due to internal spin transformations. Such internal symmetry structures typically commute with the global space-time translations characterized by the overall energy-momentum of the system. For simple topologies, the net flux of $S^{j0\nu} \cdot \hat{n}_j$ through a closed asymptotic surface Σ vanishes for sufficiently localized fields. The conserved tensor $T^{\mu\nu}$ is an energy-momentum density tensor suitable to be a source of gravitational curvature [58] to be discussed later, since it is symmetric. The local continuity condition (3.64) implies that the four-momentum P^ν is time-independent (conserved), as long as the dynamic functions are sufficiently localized such that $T^{0\nu}$ vanishes on an asymptotic closed surface.

In addition, a conserved orbital angular momentum can be defined using the tensor $T^{\mu\nu}$. A form anti-symmetric under $\mu \rightleftharpoons \nu$ given by:

$$\mathcal{M}^{\beta\mu\nu} \equiv x^\mu T^{\beta\nu} - x^\nu T^{\beta\mu}, \quad \partial_\beta \mathcal{M}^{\beta\mu\nu} = 0, \tag{3.66}$$

satisfies this continuity equation just due to the symmetry of $T^{\beta\mu}$. Thus, $\mathcal{M}^{0\mu\nu}$ forms the density of a conserved angular momentum/Lorentz boost tensor for sufficiently localized energy-momentum densities:

$$M^{\mu\nu} \equiv \int \mathcal{M}^{0\mu\nu} d^3x, \quad J_m = \frac{1}{2}\epsilon_{mjk} M^{jk}, \quad K_j = M^{j0}. \tag{3.67}$$

These global forms will be time independent, since the local continuity equation implies that $\frac{d}{dt} M^{\mu\nu} = 0$ for sufficiently localized densities $\mathcal{M}^{\beta\mu\nu}$. This property describes angular momentum conservation.

3.4.1 Local gauge invariance

Suppose that a system is invariant under transformations of the phase of the wavefunction using a representation S of an N-parameter group of parameters $\underline{\alpha} = \{\alpha^1, \ldots, \alpha^N\}$, where $\mathbf{S}(\underline{\beta})\,\mathbf{S}(\underline{\alpha}) = \mathbf{S}(\underline{\phi}(\underline{\beta};\underline{\alpha}))$. The product $\underline{\phi}(\underline{\beta};\underline{\alpha})$ represents the general form of the group multiplication rule between elements. The set of transformations describing such an internal symmetry is referred to as an *internal symmetry group*. The symmetry group is *local* if the parameters depend on position

$\underline{\alpha} = \underline{\alpha}(\vec{x})$, and the transformation is often referred to as a *gauge transformation*. The state function $\mathbf{\Psi}(\vec{x})$ transforms under the internal symmetry group as follows:

$$\hat{\mathcal{S}}\mathbf{\Psi}(\vec{x}) \equiv \mathbf{\Psi}'_{\mathcal{S}}(\vec{x}) = \mathbf{S}(\underline{\alpha}(\vec{x}))\mathbf{\Psi}(\vec{x}), \quad \mathbf{S}^{\dagger}(\underline{\alpha}) = \mathbf{S}^{-1}(\underline{\alpha}), \tag{3.68}$$

where the inner product is preserved $||\mathbf{\Psi}'_{\mathcal{S}}||^2 = ||\mathbf{\Psi}||^2$, as long as the generators of the group are hermitian.

Since invariant parameters describing a system often involve momenta, one should note that:

$$\mathbf{S}^{-1}(\underline{\alpha}(\vec{x}))\frac{\hbar}{i}\frac{\partial}{\partial x^{\mu}}\mathbf{S}(\underline{\alpha}(\vec{x})) = \frac{\hbar}{i}\frac{\partial}{\partial x^{\mu}} + \frac{\hbar}{i}\partial_{\mu}\alpha^{r}\mathbf{S}^{-1}(\underline{\alpha})\frac{\partial}{\partial \alpha^{r}}\mathbf{S}(\underline{\alpha}). \tag{3.69}$$

Because of the second term on the right-hand side of Eq. 3.69, the momentum is not an invariant operator under local gauge transformations. To construct a gauge invariant operation, minimal coupling introduces a local gauge field whose inhomogeneous transformation properties precisely cancel those of the momentum operator. This gauge field will be introduced shortly.

The representation generators are defined by:

$$i\mathbf{G}_{r} \equiv \left.\frac{\partial \mathbf{S}(\underline{\alpha})}{\partial \alpha^{r}}\right|_{\underline{\alpha}\to\underline{I}}, \tag{3.70}$$

where $\underline{I}$ is the group identity element. The algebra of the generators satisfies:

$$[\mathbf{G}_{r}, \mathbf{G}_{s}] = -ic_{rs}^{n}\mathbf{G}_{n}, \; c_{rs}^{n} \equiv \left.\frac{\partial}{\partial \beta^{s}}\frac{\partial}{\partial \alpha^{r}}(\phi^{n}(\underline{\beta};\underline{\alpha}) - \phi^{n}(\underline{\alpha};\underline{\beta}))\right|_{\underline{\alpha}\to\underline{I},\underline{\beta}\to\underline{I}}. \tag{3.71}$$

The Lie structure constants have been shown in the Appendix to characterize the Lie algebra and define a group invariant metric, $g_{rn} \propto c_{rs}^{m}c_{nm}^{s}$. The generators span the group space (group closure property), necessitating that they transform into linear superpositions of the other generators:

$$\mathbf{S}(\underline{\alpha})\,\mathbf{G}_{r}\,\mathbf{S}^{-1}(\underline{\alpha}) = \oplus_{r}^{s}(\underline{\alpha})\mathbf{G}_{s}, \tag{3.72}$$

where the fundamental representation matrices $\oplus$ are as defined in Appendix Eq. A.26:

$$\oplus_{r}^{s}(\underline{\alpha}) \equiv \left.\frac{\partial}{\partial \beta^{r}}\phi^{s}(\underline{\alpha}^{-1};\underline{\phi}(\underline{\beta};\underline{\alpha}))\right|_{\underline{\beta}\to\underline{I}}. \tag{3.73}$$

The group multiplication rule defines useful *Lie structure matrices*:

$$\Theta_{r}^{s}(\underline{\alpha}) \equiv \left.\frac{\partial}{\partial \beta^{r}}\phi^{s}(\underline{\beta};\underline{\alpha})\right|_{\underline{\beta}\to\underline{I}}. \tag{3.74}$$

that must be non-singular due to the group properties. These structure matrices directly relate the generators to arbitrary derivatives on the representation matrices $\mathbf{S}$:

$$i\mathbf{G}_r \mathbf{S}(\underline{\alpha}) = \Theta_r^s(\underline{\alpha}) \frac{\partial}{\partial \alpha^s} \mathbf{S}(\underline{\alpha}). \tag{3.75}$$

This equation will be quite convenient in defining general gauge covariant derivatives that transform homogeneously under the local internal gauge group.

Gauge covariant derivatives

A gauge covariant derivative form is expected to transform simply under the gauge transformation:

$$\mathbf{D}_\mu \mathbf{\Psi} \Rightarrow \left[\mathbf{D}_\mu \mathbf{\Psi}\right]'_{\underline{\alpha}} = \mathbf{S}(\underline{\alpha}) \mathbf{D}_\mu \mathbf{\Psi}, \tag{3.76}$$

i.e., it transforms precisely as does the state function, $\mathbf{\Psi}$. The gauge potentials A_μ^r are real fields defined to have components in the internal group space, minimally coupled such that the covariant derivative transforms simply:

$$\mathbf{D}_\mu \equiv \mathbf{1}\frac{\partial}{\partial x^\mu} - \frac{q}{\hbar c} A_\mu^r i\, \mathbf{G}_r \Rightarrow \frac{q}{\hbar c} [A_\mu^r]'_{\underline{\alpha}} i \mathbf{G}_r = \frac{q}{\hbar c} A_\mu^r \mathbf{S}\, i \mathbf{G}_r\, \mathbf{S}^{-1} + \left(\partial_\mu \mathbf{S}\right) \mathbf{S}^{-1}, \tag{3.77}$$

which is defined in terms of generalized charges q, generating dimensionless couplings $\frac{q_1 q_2}{\hbar c}$, and gauge potentials of dimension $[\frac{q_{source}}{L}]$. The transformation is simplified by defining a component of the gauge field whose transformation satisfies the inhomogeneous term:

$$\partial_\mu \alpha^r(\vec{x}) \equiv \frac{q}{\hbar c} \mathrm{a}_\mu^s(\vec{x})\, \Theta_s^r(\underline{\alpha}(\vec{x})). \tag{3.78}$$

Combining this equation with Eq. 3.75, space-time derivatives of the gauge transformation can be expressed:

$$\partial_\mu \mathbf{S}(\underline{\alpha}) = \frac{q}{\hbar c} \mathrm{a}_\mu^s(\vec{x})\, i \mathbf{G}_r\, \mathbf{S}(\underline{\alpha}). \tag{3.79}$$

Therefore, the gauge covariant derivative defined in Eq. 3.77 transforms homogeneously under the local gauge transformation as long as the gauge potentials transform according to:

$$[A_\mu^r(\vec{x})]'_{\underline{\alpha}} = A_\mu^s(\vec{x}) \oplus_s^r(\underline{\alpha}(\vec{x})) + \mathrm{a}_\mu^r(\vec{x}). \tag{3.80}$$

The field strengths $\mathbf{F}_{\mu\nu}$ are $N \times N$ matrices that transform homogeneously under the internal group transformation $[\mathbf{F}_{\mu\nu}]'_{\underline{\alpha}} = \mathbf{S}\mathbf{F}_{\mu\nu}\mathbf{S}^{-1}$, where:

$$\frac{q}{\hbar c} i \mathbf{F}_{\mu\nu} \equiv -\frac{q}{\hbar c} [\mathbf{D}_\mu, \mathbf{D}_\nu] = \frac{q}{\hbar c} \left(\partial_\mu A_\nu^r - \partial_\nu A_\mu^r - \frac{q}{\hbar c} c_{mn}^r A_\mu^m A_\nu^n \right) i \mathbf{G}_r. \tag{3.81}$$

The field strengths are anti-symmetric under exchange of the space-time indices $\mu \rightleftharpoons \nu$.

In any Lagrangian or Hamiltonian model, the momentum terms should contain the minimally coupled form that transforms homogeneously under the gauge group:

$$\mathbf{S}(\underline{\alpha}(\vec{x}))\left(\mathbf{1}\hat{p}_{\mu}-\frac{q}{c}A^{r}_{\mu}(\vec{x})\mathbf{G}_{r}\right)=\left(\mathbf{1}\hat{p}_{\mu}-\frac{q}{c}A^{r}_{\mu}(\vec{x})\mathbf{G}_{r}\right)\mathbf{S}(\underline{\alpha}(\vec{x})), \tag{3.82}$$

where $(\mathbf{1}\hat{p}_{\mu}-\frac{q}{c}A^{r}_{\mu}(\vec{x})\mathbf{G}_{r})=\frac{\hbar}{i}\,\mathbf{D}_{\mu}$ is a hermitian operator. Such forms are consistent with the canonical proper-time formulations previously discussed.

The Jacobi identity algebraically relates commutators of operators and matrices $[\mathbf{M},[\mathbf{N},\mathbf{R}]]+[\mathbf{R},[\mathbf{M},\mathbf{N}]]+[\mathbf{N},[\mathbf{R},\mathbf{M}]]=0$. This homogeneous relationship provides equations analogous to the homogeneous Maxwell equations for electromagnetism:

$$\epsilon^{\alpha\mu\nu\beta}\left[\partial_{\mu}F^{r}_{\nu\beta}-\frac{q}{\hbar c}\,A^{m}_{\mu}c^{r}_{mn}F^{n}_{\nu\beta}\right]=0. \tag{3.83}$$

These equations are purely a consequence of the local gauge covariance, independent of the physical model.

3.4.2 Local Lagrangian dynamics

Models with local internal symmetries have been quite successful in describing a variety of physical phenomena. The most-established such model is the theory of electromagnetism, described in Section 1.3.2, which remains invariant under the gauge transformation, $A_{\mu}(\vec{x})\rightarrow A'_{\mu}(\vec{x})=A_{\mu}(\vec{x})+\partial_{\mu}\alpha(\vec{x})$, where the symmetry group representation $\mathbf{S}$ is the one-parameter abelian group U(1), the unitary group. The group parameter $\alpha(\vec{x})$ is free to take on any (differentiable) functional form without modifying any measurable field. Similarly, the gluon fields that serve as the quantum chromodynamic (QCD) potentials between strongly interacting quarks transform inhomogeneously under the group of unitary 3×3 matrices of unit determinant, SU(3). The three local (i.e., space-time-dependent) group parameters $\alpha^{r}(\vec{x})$ represent “angles” that mix the *color* states of transformed quantities (quarks, gluons). Lagrangian densities that are invariant under local internal symmetry transformations will be described in this section.

An invariant that depends on, at most, a single derivative of the functions can depend upon the generally complex spinor function $\mathbf{\Psi}(\vec{x})$, its minimally coupled first derivative $\vec{\mathbf{D}}\mathbf{\Psi}(\vec{x})$, and the hermitian field strengths $\{F^{r}_{\mu\nu}\}$. Therefore, suppose that $\mathcal{L}=\mathcal{L}(\mathbf{\Psi}(\vec{x}),\vec{\mathbf{D}}\mathbf{\Psi}(\vec{x}),\mathbf{F}_{\mu\nu})$. The Lagrangian fields that are conjugate to derivatives of the dynamic functions will be defined by:

$$(\mathbf{\Pi}^{\nu})_{b}\equiv\frac{\partial\mathcal{L}}{\partial(\mathbf{D}_{\mu}\mathbf{\Psi})_{b}},\qquad H^{\mu\nu}_{r}\equiv 2\frac{\partial\mathcal{L}}{\partial F^{r}_{\mu\nu}}, \tag{3.84}$$

where $H_r^{\mu\nu}$ maintains the same anti-symmetry under exchange of the space-time indices as that satisfied by the field strengths. The Noether currents defined by global symmetry transformations:

$$J_r^{\mu} = \frac{i}{\hbar} \mathbf{\Pi}^{\mu} \mathbf{G}_r \mathbf{\Psi} + h.c., \tag{3.85}$$

need not be conserved under the local transformations. The term $h.c.$ in these equations will represent the hermitian conjugate of the prior terms, if those terms are complex.

Since $\frac{\partial \mathcal{L}}{\partial(\partial_\mu \mathbf{\Psi})} = \frac{\partial \mathcal{L}}{\partial(\mathbf{D}_\mu \mathbf{\Psi})}$, the Euler–Lagrange equations obtained by varying the spinor functions are as follows:

$$\delta W[\mathbf{\Psi}] = 0 \Rightarrow \frac{\partial \mathcal{L}}{\partial \Psi_a} = \mathbf{\Pi}_b^{\mu} \left[\delta_{ba} \overleftarrow{\partial}_{\mu} + i \frac{q}{\hbar c} A_\mu^r (\mathbf{G}_r)_{ba} \right]. \tag{3.86}$$

Likewise, if the action is extremal under variations in the gauge potentials, one obtains:

$$\delta W[A_\mu^n] = 0 \Rightarrow \frac{q}{c} J_n^{\mu} = \left[\delta_n^m \partial_\nu + \frac{q}{\hbar c} A_\nu^r c_{rn}^m \right] H_m^{\mu\nu}. \tag{3.87}$$

Since the Lagrangian is assumed to be invariant under local gauge transformations $\frac{\delta \mathcal{L}}{\delta \alpha^r} = 0$, the gauge currents can be shown to satisfy:

$$\left(\delta_n^s \partial_\mu + \frac{q}{\hbar c} A_\mu^r c_{rn}^s \right) J_s^{\mu} = \frac{1}{2\hbar} F_{\mu\nu}^r c_{rn}^m H_m^{\mu\nu}. \tag{3.88}$$

Thus, the currents J_r^{μ} are not conserved unless the symmetry group is abelian ($c_{rm}^s = 0$ for all r, m, and s).

The conserved energy-momentum tensor form obtained directly from the Lagrangian is given by:

$$T_\nu^{\mu} = \mathbf{\Pi}^{\mu} \mathbf{D}_\nu \mathbf{\Psi} + h.c. + H_r^{\mu\beta} F_{\nu\beta}^r - \delta_\nu^{\mu} \mathcal{L}, \quad \partial_\mu T_\nu^{\mu} = 0. \tag{3.89}$$

As previously mentioned, this tensor need not be symmetric under exchange of contravariant space-time indices.

For systems that separate into material and gauge field components, internal force densities can be defined. Suppose that $\mathcal{L} = \mathcal{L}_M + \mathcal{L}_F$, where $\mathcal{L}_M$ contains gauge-covariant derivatives of the spinor functions, and $\mathcal{L}_F$ includes invariant forms involving only the field strengths. The energy-momentum forms obtained directly from the Lagrangian are then given by:

$$T_{M\ \nu}^{\mu} = \mathbf{\Pi}^{\mu} \mathbf{D}_\nu \mathbf{\Psi} + h.c. - \delta_\nu^{\mu} \mathcal{L}_M, \tag{3.90}$$

$$T_{F\ \nu}^{\mu} = H_r^{\mu\beta} F_{\nu\beta}^r - \delta_\nu^{\mu} \mathcal{L}_F. \tag{3.91}$$

Since the summed energy-momentum form $\mathcal{T}^{\mu}_{M\ \nu} + \mathcal{T}^{\mu}_{F\ \nu}$ is conserved, the individual divergences construct internal force densities exchanged between the material and gauge field components:

$$\partial_{\mu}\mathcal{T}^{\mu}_{M\ \nu} \equiv f_{\nu} = -\partial_{\mu}\mathcal{T}^{\mu}_{F\ \nu}. \tag{3.92}$$

As long as the covariant derivative $\mathbf{D}_{\nu}$ transforms under Lorentz transformations as does ∂_{ν}, a term associated with the internal spin transformations can be added to develop a symmetric energy-momentum tensor for the material component, as was done in Eq. 3.64:

$$\mathcal{S}^{\beta\mu\nu} \equiv \frac{i}{2}\left[\frac{\partial\mathcal{L}_M}{\partial(\mathbf{D}_{\beta}\mathbf{\Psi})}\mathcal{J}^{\mu\nu}\,\mathbf{\Psi} - \frac{\partial\mathcal{L}_M}{\partial(\mathbf{D}_{\mu}\mathbf{\Psi})}\mathcal{J}^{\beta\nu}\,\mathbf{\Psi} - \frac{\partial\mathcal{L}_M}{\partial(\mathbf{D}_{\nu}\mathbf{\Psi})}\mathcal{J}^{\beta\mu}\,\mathbf{\Psi}\right] + h.c.,$$

$$T^{\mu\nu}_{M} \equiv \mathcal{T}^{\mu\nu}_{M} + \partial_{\beta}\mathcal{S}^{\beta\mu\nu}, \quad \partial_{\mu}T^{\mu\nu}_{M} = 0, \quad T^{\mu\nu}_{M} = T^{\nu\mu}_{M}. \tag{3.93}$$

Since $\partial_{\mu}\partial_{\beta}\mathcal{S}^{\beta\mu\nu} = 0$, the force density generated on the symmetric tensor $T^{\mu\nu}_{M}$ is the same as that in Eq. 3.92, $f^{\nu} = \partial_{\mu}T^{\mu\nu}_{M}$. Most reasonable, Lorentz invariant forms for the field strength component of the energy-momentum form will be symmetric under exchange of the space-time indices. The force density acting upon the material component is therefore given by:

$$f_{\nu} = -\frac{q}{c}J^{\alpha}_{r}\,F^{r}_{\alpha\nu} + \frac{1}{2}\frac{q}{\hbar c}A^{r}_{\nu}\,c^{m}_{rs}\,F^{s}_{\alpha\beta}H^{\alpha\beta}_{m}. \tag{3.94}$$

This force/power density describes the interaction of the gauge field with $\mathbf{\Psi}$.

Standard electromagnetism is reproduced using the abelian U(1) algebra (i.e., vanishing structure constants) with the identification $H^{\mu\nu} = \frac{1}{4\pi}F^{\mu\nu}$, where the $4\pi = \Omega_{(2)}$ is the dimensionless surface area (solid angle) of a closed two-dimensional boundary. For electromagnetism, $q\,J^{\mu}$ represents the four-current density, Eq. 3.87 gives Gauss' Law ($\mu = 0$) and the Ampere/Maxwell Law ($\mu \neq 0$), Eq. 3.83 represents the homogeneous Faraday Law and absence of magnetic monopoles, and Eq. 3.94 is the Lorentz force equation. More complicated models (such as QCD) will involve the appropriate group structure constants, with the gauge potential A^{r}_{ν} representing the carriers of the gauge interaction between the dynamic particles represented by $\mathbf{\Psi}$.

3.4.3 Substantive quantum flows

Substantive derivatives are derivatives along the tangent vector $\vec{U} \equiv \frac{d\vec{x}}{dc\tau} = \frac{\vec{u}}{c}$ that characterizes a trajectory or streamline of laminar flow. Such derivatives are convenient for characterizing relativistic dynamics in terms of the proper times of the flowing components of a continuum system. Substantive dynamics will also be quite useful for describing co-gravitating constituents of quantum systems. A substantive Lagrangian form will depend upon space-time gradients only through the derivative $\mathbf{D\Psi}$ in a manner defined as follows:

$$\mathcal{L}_S = \mathcal{L}_S(\mathbf{\Psi}, \mathbf{D\Psi}), \quad \text{where} \quad \mathbf{D\Psi} \equiv U^\mu \mathbf{D}_\mu \mathbf{\Psi}. \tag{3.95}$$

The canonically conjugate functions $\mathbf{\Pi}^\mu$ then take on a general form proportional to the tangent vector:

$$\mathbf{\Pi}^\mu = \frac{\partial \mathcal{L}_S}{\partial \mathbf{D\Psi}} U^\mu \equiv \mathbf{\Pi}\, U^\mu. \tag{3.96}$$

All results from the previous section can then be specialized using this substitution.

In particular, the Euler–Lagrange equations for the spinor functions $\mathbf{\Psi}$,

$$(\mathbf{\Pi}\, U^\mu)\left[\overleftarrow{\partial}_\mu \mathbf{1} + i\frac{q}{\hbar c} A^r_\mu\, (\mathbf{G}_r)\right] = \left(\frac{\partial \mathcal{L}_S}{\partial \mathbf{\Psi}}\right)_{\mathbf{D\Psi}}, \tag{3.97}$$

include a term involving the divergence of the four-velocity. The gauge currents:

$$J^\mu_r = \left(\frac{i}{\hbar}\mathbf{\Pi}\mathbf{G}_r \mathbf{\Psi} + h.c.\right) U^\mu \equiv \rho^{(G)}_r\, u^\mu, \tag{3.98}$$

directly flow along this local four-velocity, with the proportionality defining a gauge density component, $\mathbf{G}_r$. It should be noted that there will be two-component energy-momentum flows, due to the gauge field in the gauge-covariant derivative:

$$\mathcal{T}^{\mu}_{S\ \nu} = U^\mu\, \mathbf{\Pi}\, \mathbf{D}_\nu \mathbf{\Psi} + h.c. - \delta^\mu_\nu \mathcal{L}_S. \tag{3.99}$$

In particular, the energy flux vector for the substantive flow:

$$\mathcal{T}^{0}_{S\ k} = U^0 \left(\mathbf{\Pi}\, \partial_k \mathbf{\Psi} + h.c. - \rho^{(G)}_r\, q A^r_k\right), \tag{3.100}$$

contains a "normal" flow component, combined with a gauge flow component. In condensed matter physics, certain macroscopic quantum fluids manifest such energy flows, usually referred to as superfluidity [59]. In gravitational physics, the classical four-velocities of freely gravitating systems can be calculated using well-defined *geodesic equations*, allowing the analysis of Lagrangian systems described in terms of such flows in a straightforward manner.

4

Group theory in quantum mechanics

Quantum states are best represented as vectors in a complex space. Matrix elements of operations that act on those states mimic relationships familiar to linear algebra, which can be completely described in terms of group theory. The generators for infinitesimal space-time translations have classical correspondence with the momentum-energy of the system. These observations demonstrate the potential usefulness of formal group theory in describing quantum dynamics. This chapter will briefly discuss useful aspects of group theory in formulating quantum mechanics.

4.1 Quantum mechanics and the Galilean group

4.1.1 Vector and ray representations

The Poincaré group structure relates the generators describing boosts, momentum, and energy by $[K_j, P_k] = i\delta_{jk}H$, whereas the corresponding commutator for the inhomogeneous Galilean group gives a c-number, where the structure constants vanish. This group center only allows one to construct non-relativistic quantum representations up to a phase, the so-called ray representations, which satisfy $U(a)\,U(b) = e^{i\zeta(a,b)}U(a \cdot b)$ with a non-vanishing phase. The phase is directly related to the mass of the system, and group constraints on the phase result in a derivation of the Schrödinger equation as the dynamical equation for a non-relativistic system. In addition, gauge invariance will be shown to be a consequence of the Galilean pre-symmetry of a non-relativistic system [60]. Since the Poincaré group structure will be shown to allow vector representations with vanishing phase, such results will not be applicable to representations of the relativistic dynamics described by the Poincaré algebra.

4.1.2 The Galilean group

The homogeneous Galilean group includes rotations and non-relativistic velocity boosts from one inertial frame to another. The inhomogeneous Galilean group additionally includes space and time translations of the coordinates:

$$\begin{aligned} \underline{x}' &= \mathbf{R}\underline{x} + \underline{v}t + \underline{a} \\ t' &= t + \tau . \end{aligned} \tag{4.1}$$

Here, $\mathbf{R}$ represents an arbitrary spatial rotation, $\underline{v}$ represents a velocity boost, $\underline{a}$ represents an arbitrary spatial translation, and τ represents an arbitrary temporal translation. For three spatial and one temporal dimensions, this group has ten parameters. If the group element is represented by $G = (\mathbf{R}, \underline{v}, \underline{a}, \tau)$, then the group operation for successive transformations satisfies:

$$(\mathbf{R}_2, \underline{v}_2, \underline{a}_2, \tau_2) \cdot (\mathbf{R}_1, \underline{v}_1, \underline{a}_, \tau_1) = (\mathbf{R}_2\mathbf{R}_1, \mathbf{R}_2\underline{v}_1 + \underline{v}_2, \mathbf{R}_2\underline{a}_1 + \underline{v}_2\tau_1 + \underline{a}_2, \tau_2 + \tau_1). \tag{4.2}$$

The identity element is $I = (\mathbf{1}, \underline{0}, \underline{0}, 0)$, and the inverse of element G is given by $G^{-1} = (\mathbf{R}^{-1}, -\mathbf{R}^{-1}\underline{v}, -\mathbf{R}^{-1}\underline{a} + \mathbf{R}^{-1}\underline{v}\tau, -\tau)$.

4.1.3 Unitary ray representation of the Galilean group

Next, a unitary ray representation $U(G')\,U(G) = e^{i\,\xi(G';G)}U(G' \cdot G)$ as described in the Appendix D.1.1 will be examined. The form of the general local factor for the Galilean group can be determined [61] to be:

$$\xi(G_2; G_1) = \frac{M}{2\hbar}\left(\underline{a}_2 \cdot \mathbf{R}_2\underline{v}_1 - \underline{v}_2 \cdot \mathbf{R}_2\underline{a}_1 + \underline{v}_2 \cdot \mathbf{R}_2\underline{v}_1\tau_1\right), \tag{4.3}$$

where M is an arbitrary real constant which must have units of mass. This non-vanishing (except for $M = 0$) local exponent determines the form of quantum mechanics on a Galilean manifold (the Schrödinger Hamiltonian). As described in the appendix, if an equivalence transformation $U'(G) = e^{i\,\zeta(G)}U(G)$ can be found such that the new local exponents vanish, then the representation is equivalent to a vector representation. Such a transformation produces new generators:

$$\left.\frac{\partial U'(G)}{\partial b^r}\right|_{G=I} = -\frac{i}{\hbar}\hat{g}_r + i\left.\frac{\partial \zeta(G)}{\partial b^r}\right|_{G=I}\hat{1}, \tag{4.4}$$

which is therefore the same as changing the generators by constant multiples of the identity operator. This means that if one is able to find constants such that $\hat{g}'_r = \hat{g}_r + \zeta_r\hat{1}$ with $[\hat{g}'_r, \hat{g}'_s] = c^n_{rs}\hat{g}'_n$, then the representation is equivalent to a vector representation. However, for the Galilean group no such transformation can be found in the relationship $[\hat{K}_j, \hat{P}_k] = i\hbar\delta_{jk}\hat{1}$ (which follows from Eq. 4.9) to eliminate the term with M, because of the abelian nature of the velocity-boost/spatial-translation transformations in the group space.

The generators for infinitesimal transformations will be defined in the conventional way:

$$\left.\frac{\partial U(G)}{\partial \tau}\right|_{G=I} \equiv -\frac{i}{\hbar}\hat{H}, \qquad \left.\frac{\partial U(G)}{\partial a^j}\right|_{G=I} \equiv \frac{i}{\hbar}\hat{P}_j,$$
$$\left.\frac{\partial U(G)}{\partial v^j}\right|_{G=I} \equiv -\frac{i}{\hbar}\hat{K}_j, \qquad \left.\frac{\partial U(G)}{\partial \theta^j}\right|_{G=I} \equiv -\frac{i}{\hbar}\hat{J}_j, \tag{4.5}$$

where the rotations are parameterized by the angle and direction $\mathbf{R} = \mathbf{R}(\theta)$. The transformation behavior of the generators can be derived using:

$$U^\dagger(G)\,U(G')\,U(G) = e^{\xi(G';G)+\xi(G^{-1};G'\cdot G)-\xi(G^{-1};G)}U(G^{-1}\cdot G'\cdot G). \tag{4.6}$$

The transformed Galilean group element is given by:

$$G^{-1}\cdot G'\cdot G = \left(\mathbf{R}^{-1}\mathbf{R}'\mathbf{R}, \mathbf{R}^{-1}(\underline{v}' + \mathbf{R}'\underline{v} - \underline{v}), \mathbf{R}^{-1}(\underline{a}' + \underline{v}'\tau - \underline{v}\tau' + \mathbf{R}'\underline{a} - \underline{a}), \tau\right), \tag{4.7}$$

where $\mathbf{R} \equiv \mathbf{R}(\underline{\theta})$ and $\mathbf{R}' \equiv \mathbf{R}(\underline{\theta}')$. Therefore, the generators transform as follows:

$$U^\dagger(G)\hat{H}U(G) = \hat{H} + v_j R_{jk}(\underline{\theta})\hat{P}_k + \frac{M}{2}\underline{v}\cdot\underline{v}\hat{1}, \tag{4.8}$$

$$U^\dagger(G)\hat{P}_jU(G) = R_{jk}(\underline{\theta})\hat{P}_k + Mv_j\hat{1}, \tag{4.9}$$

$$U^\dagger(G)\hat{K}_jU(G) = R_{jk}(\underline{\theta})\hat{K}_k - R_{jk}(\underline{\theta})\hat{P}_k + M(a_j - v_j\tau)\hat{1}, \tag{4.10}$$

$$\begin{aligned} U^\dagger(G)\hat{J}_jU(G) = {} & R_{jk}(\underline{\theta})\hat{J}_k - v_m(\mathcal{J}_j)_{mk}\,R_{ks}(\underline{\theta})\hat{K}_s + a_m(\mathcal{J}_j)_{mk}\,R_{ks}(\underline{\theta})\hat{P}_s \\ & + \frac{M}{2}(\mathcal{J}_j)_{mk}[a_m v_k - a_k v_m]\hat{1}, \end{aligned} \tag{4.11}$$

where the matrix $\mathcal{J}_j \equiv \left.\frac{\partial}{\partial\theta^j}\mathbf{R}(\underline{\theta})\right|_{\underline{\theta}=0}$, and repeated indices are summed. The terms involving M result directly from the non-vanishing local exponent.

4.1.4 Quantum mechanics on a Galilean manifold

The form of wave mechanics for non-relativistic quantum physics is already contained in the transformations 4.8–4.11. First, notice that:

$$U^\dagger(G)\frac{\hat{P}^2}{2M}U(G) = \frac{\hat{P}^2}{2M} + v_j R_{jk}(\underline{\theta})\hat{P}_k + \frac{M}{2}\underline{v}\cdot\underline{v}\hat{1}, \tag{4.12}$$

follows directly from Eq. 4.9. In addition, the Hamiltonian transforms according to Eq. 4.8. This means that:

$$U^{\dagger}(G)\left[\hat{H}-\frac{\hat{P}^2}{2M}\right]U(G)=\hat{H}-\frac{\hat{P}^2}{2M}, \tag{4.13}$$

i.e., this operator commutes with all operations in the group. By Shur's lemma, one can conclude that $\hat{H}-\frac{\hat{P}^2}{2M}=\hat{H}_{internal}$, where the operator $\hat{H}_{internal}$ must act only on an internal space that commutes with all generators of the Galilean group. Thus, the Hamiltonian satisfies $\hat{H}=\frac{\hat{P}^2}{2M}+\hat{H}_{internal}$, where M is a real parameter with units of mass.

One can examine possible forms of the operator $\hat{H}_{internal}$. If the internal system has no structure, then $\hat{H}_{internal}$ is just a constant multiple of the identity operator for the full group space. However, if there is structure, then the space of constituents no longer possesses the full symmetry of the Galilean group. However, there is Galilean *pre-symmetry* in the sense that as "interactions" weaken, the full Galilean invariance becomes restored.

Consider, for instance, an internal Hamiltonian that is a function of only the internal coordinates $\underline{x}$ and $\underline{p}$. Other internal coordinates (like spin, etc.) can be directly included if desired. The operators $\underline{x}$ and $\underline{p}$ commute with all generators of the full group. Suppose that $u(G)$ is a unitary ray representation of the Galilean group that operates in the internal parameter space. The mass parameter in the local exponent for the internal group will be specified by μ. The internal group will be assumed *not* to have internal translational invariance, and to evolve in time, but to perhaps asymptotically regain the full Galilean symmetry. Thus, there is only a need to consider transformations of the system under the homogeneous Galilean group $G_H=(\mathbf{R},\underline{v},\underline{0},0)$. This system with inhomogeneous Galilean pre-symmetry has generators that satisfy:

$$u^{\dagger}(G_H)\hat{H}_{int}u(G_H)=\hat{H}_{int}+v_jR_{jk}(\underline{\theta})\hat{p}_k+\frac{\mu}{2}\underline{v}\cdot\underline{v}\hat{1}, \tag{4.14}$$

$$u^{\dagger}(G_H)\hat{p}_ju(G_H)=R_{jk}(\underline{\theta})\hat{p}_k+\mu v_j\hat{1}, \tag{4.15}$$

$$u^{\dagger}(G_H)\hat{x}_ju(G_H)=R_{jk}(\underline{\theta})\hat{x}_k, \tag{4.16}$$

where again μ is a real parameter labeling the ray representation of the internal Galilean space.

The Galilean group and local gauge transformations

The time evolution of the position operator satisfies:

$$\hat{x}_j(\tau)\equiv e^{i\tau\hat{H}_{int}/\hbar}\hat{x}_je^{-i\tau\hat{H}_{int}/\hbar}. \tag{4.17}$$

If one defines a velocity $\dot{x}_j \equiv \left.\frac{d\hat{x}_j}{d\tau}\right|_{\tau=0}$, then a Heisenberg equation of motion for the operator $\dot{x}_j$ is directly obtained:

$$\dot{x}_j = \frac{i}{\hbar}[\hat{H}_{int}, \hat{x}_j]. \tag{4.18}$$

This commutator can be directly tranformed using the internal ray representation:

$$u^\dagger(G_H)\dot{x}_j u(G_H) = \frac{i}{\hbar}[\hat{H}_{int} + v_m\, R_{mn}\, \hat{p}_n,\, R_{jk}\, \hat{x}_k] = R_{jk}\dot{x}_k + v_j\hat{1}. \tag{4.19}$$

A direct comparison of Eqs. 4.15 and 4.19 demonstrates that:

$$u^\dagger(G_H)[\hat{p}_j - \mu\dot{x}_j]u(G_H) = R_{jk}[\hat{p}_j - \mu\dot{x}_j]. \tag{4.20}$$

Examining Eq. 4.16, this implies that the form $\hat{p}_j - \mu\dot{x}_j \equiv A_j(|\hat{\underline{x}}|)$ must be a function of a rotational and Galilean boost invariant parameter. The only possibility is the length of the position operator. Therefore:

$$\dot{\underline{x}} = \frac{\hat{\underline{p}} - \underline{A}(|\hat{\underline{x}}|)}{\mu}. \tag{4.21}$$

The form of the Hamiltonian can be obtained by first noticing that $[\dot{x}_j, \hat{x}_k] = -i\hbar\delta_{jk}/\mu$, so that $[\dot{x}\cdot\dot{x}, \hat{x}_k] = -\frac{2i\hbar}{\mu}\dot{x}_k$. Comparing $\frac{i}{\hbar}\left[\frac{\mu}{2}|\dot{\underline{x}}|^2, \hat{x}_k\right] = \dot{x}_k$ and $\frac{i}{\hbar}[\hat{H}_{int}, \hat{x}_k] = \dot{x}_k$, one obtains:

$$\left[\hat{H}_{int} - \frac{\mu}{2}|\dot{\underline{x}}|^2, \hat{x}_k\right] = 0. \tag{4.22}$$

Therefore, this difference must be purely a function of $\hat{\underline{x}}$:

$$\hat{H}_{int} - \frac{\mu}{2}|\dot{\underline{x}}|^2 \equiv V(\underline{x}). \tag{4.23}$$

More so, since both terms of the left-hand side of the previous equation are rotationally invariant, the right-hand side must depend only on the length $|\underline{x}|$. Therefore, the internal Hamiltonian for this example is given by:

$$\hat{H}_{int} = \frac{|\hat{\underline{p}} - \underline{A}(|\hat{\underline{x}}|)|^2}{2\mu} + V(|\hat{\underline{x}}|). \tag{4.24}$$

Therefore, the Hamiltonian of a general quantum system with a Galilean pre-symmetry of the internal group operators under rotations and velocity boosts takes the form:

$$\hat{H} = \frac{\hat{P}^2}{2M} + \frac{|\hat{\underline{p}} - \underline{A}(|\hat{\underline{x}}|)|^2}{2\mu} + V(|\hat{\underline{x}}|). \tag{4.25}$$

A few points should be emphasized:

- Rotational invariance requires the appearance of central potentials, as expected;

- If one defines:

$$m_1 \equiv \frac{M}{2}\left(1+\sqrt{1-4\frac{\mu}{M}}\right)$$
$$m_2 \equiv \frac{M}{2}\left(1-\sqrt{1-4\frac{\mu}{M}}\right)$$
$$\underline{p}_1 \equiv \frac{m_1}{M}\underline{P}+\underline{p}$$
$$\underline{p}_2 \equiv \frac{m_1}{M}\underline{P}-\underline{p},$$

  then the two-particle Schrödinger equation follows: $\frac{\underline{P}^2}{2M}+\frac{\underline{p}^2}{2\mu}=\frac{\underline{p}_1^2}{2m_1}+\frac{\underline{p}_2{}^2}{2m_2}$.
- The inclusion of the gauge potentials was a direct consequence of the inhomogeneous Galilean pre-symmetry.

4.2 Quantization conditions on gauge fields

The topology of the mapping of the internal group transformations onto the coordinate space-time representation of a system determines some of its characteristic properties without regard to the specific dynamical model. Such properties often manifest in quantum flows, such as superfluids. Examples of quantization in quantum flows will be briefly discussed.

Liquid ^{4}He is a bose system that undergoes a phase transition into an isotropic superfluid due to zero-point motions comparable to the van der Waals bindings between the atoms. The gauge potential for the one-parameter internal symmetry group is traditionally given by the superfluid velocity $\vec{v}_{(s)}=\{v^0_{(s)},\mathbf{v}_{(s)}\}$, which is irrotational:

$$\vec{v}_{(s)}=-\frac{\hbar}{m}\vec{\nabla}\alpha, \qquad \nabla\cdot\mathbf{v}_{(s)}=0. \tag{4.26}$$

The gauge covariant derivative takes the form:

$$\mathbf{D}_\mu=\mathbf{1}\partial_\mu-\frac{q}{\hbar c}v_{(s)\mu}i\mathbf{G}\Rightarrow\mathbf{G}=\mathbf{1},\text{ where } q=mc. \tag{4.27}$$

The form of the effective "charge" follows from the requirement that, since the superflow velocity is a velocity, then $\frac{q}{\hbar}$ must have dimensions of inverse length. The parameter α changes additively under gauge transformations.

The quantization condition follows from the mapping of the gauge parameter into configuration space:

$$\oint_C dx^\mu\partial_\mu\alpha=\pm N_{(1)}\Omega_{(1)}=-\oint_C\frac{m}{\hbar}v_{(s)\mu}dx^\mu, \tag{4.28}$$

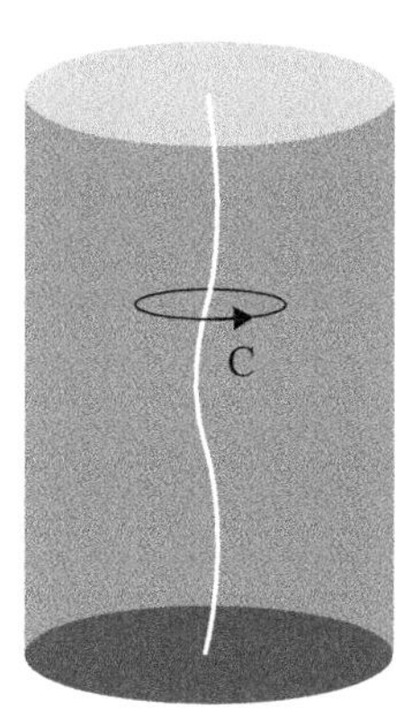

Figure 4.1 Space-like closed curve C enclosing region of broken connectivity.

where C is a space-like curve demonstrated in Figure 4.1. Here, $\Omega_{(1)}$ is the length of the compact, closed one-dimensional gauge parameter, usually 2π, and the integer $N_{(1)}$ is often referred to as the *winding number* of the curve, or the *degree* of the topological mapping [61] of the gauge parameter into configuration space. The winding number can be non-vanishing only if the topology of the space contained within the closed space-like curve does not allow that curve to be shrunk to a point. Such multiply connected topologies occur when vortices are present. Since $\Omega_{(1)}$ is an unbounded, closed curve, it could parameterize the boundary of a two-surface. Thus, the circulation of the superflow velocity is quantized in units of $\Omega_{(1)}\frac{\hbar}{m} \rightarrow \frac{h}{m}$:

$$\oint_{\text{space-like}} v_{(s)\mu} dx^{\mu} = N_{(1)} \Omega_{(1)} \frac{\hbar}{m}. \tag{4.29}$$

Quantized superfluid vortex states with non-vanishing $N_{(1)}$ can be shown to satisfy the equations of two-dimensional electrodynamics, with like circulations repelling and unlike circulations attracting.

Experiments using Bose–Einstein condensates of atoms [62] demonstrate uniform arrays of quantized vortex states in non-uniform density distributions. This implies that the quantized gauge charges $\frac{q}{c} = \frac{m}{\hbar}$ associated with the vortices are independent of material density. The quantization condition is expressed in units of $\frac{\hbar}{m}$ rather than $\hbar$, i.e., the gauge field is related to the superfluid *velocity*. This exemplifies a topological and geometric implementation of an interaction, similar to the case of gravitation. The geometry of the quantum flow thus introduces emergent interactions between topological structures.

As a second example, the three-parameter space of internal transformations of SU(2) characterized by an angle α with periodicity 4π, and a direction vector $\hat{\alpha}$ of unit norm, will be examined. The local structure constants of this group of rotations are the Levi–Civita anti-symmetric symbol components $c_{mn}^{\ \ r} = \epsilon_{mnr}$, and the group metric will be taken to satisfy $g_{mn} \equiv -\frac{1}{2} c_{ms}{}^{r} c_{nr}{}^{s} = \delta_{mn}$. The group of 2×2 transformation matrices given by $\mathbf{R}(\underline{\alpha}) = \mathbf{1} \cos\frac{\alpha}{2} + \hat{\alpha} \cdot \underline{\boldsymbol{\sigma}} \sin\frac{\alpha}{2}$, where $\hat{\alpha}^r \hat{\alpha}_r = 1$, define the internal parameter space. The group composition rule follows immediately from matrix multiplication. The Lie structure matrices resulting from this rule are given by:

$$\Theta_j{}^k = \delta_j^k + \frac{\alpha^m}{2} \epsilon_{mj}{}^k + \left(\frac{\alpha}{2} \cot\frac{\alpha}{2} - 1\right)(\delta_j^k - \hat{\alpha}_j \hat{\alpha}^k). \tag{4.30}$$

The fundamental representation matrices $\oplus_j{}^k(\underline{\beta}) \oplus_k{}^m(\underline{\alpha}) = \oplus_j{}^m(\underline{\phi}(\underline{\beta}; \underline{\alpha}))$ are:

$$\oplus_j{}^k(\underline{\alpha}) = \cos\alpha \, \delta_j^k + (1 - \cos\alpha)\hat{\alpha}_j \hat{\alpha}^k + (\sin\alpha)\hat{\alpha}^m \epsilon_{mj}{}^k, \tag{4.31}$$

where indices are raised and lowered using the metric Kronecker delta $g^{jk} = \delta^{jk}$. Notice that for SU(2), the fundamental representation matrices satisfy $\oplus_j{}^k(2\pi\hat{\alpha}) = \delta_j^k$, while $\mathbf{R}(2\pi\hat{\alpha}) = -\mathbf{1}$. This directly demonstrates the difference between the global structure of the two representations, each of which satisfy the same local algebra. For SU(2), the length of the compact closed one-dimensional gauge parameter $\Omega_{(1)}$ is 4π.

A particular type of breaking of a larger symmetry group into a smaller group that maintains less symmetry will be illustrative of a broader class of quantization conditions. The directional characteristics of certain polycrystalline materials were recognized in rocks and described as *texture* in the early nineteenth century [63]. A texture is a local distribution of orientations, which often reflect topological structure or defects. For instance, superfluid ^{3}He pairs the fermionic atoms in a spatial angular momentum $\ell = 1$ (p-wave) state. A weak spin-orbit coupling introduces an anisotropic phase breaking the full rotational symmetry but preserving a sub-symmetry about a given (local) direction in the gauge parameter space [64].

The texture will be parameterized by a local unit direction with components $\hat{z}^r$ in the gauge parameter space, and $\hat{z}^r \hat{z}_r = 1$. An example texture plotted in the group parameter space is demonstrated in Figure 4.2. Transformations about this local direction $\alpha^r = \alpha \hat{z}^r$ define the remaining symmetry group for the system. Because of the preservation of the norm of the orientation $\partial_\mu(\hat{z}^r \hat{z}_r) = 0$, the group orientations, $\hat{z}^r \perp \partial_\mu \hat{z}^r \perp \hat{z}^m \partial_\mu \hat{z}^s \epsilon_{ms}{}^r$, are mutually orthogonal in the group space. Therefore, the gauge potentials can be generally expressed in the form:

$$\vec{A}^r = \vec{A}^{(G)} \hat{z}^r + Q^{(1)} \vec{\nabla} \hat{z}^r + Q^{(2)} \hat{z}^m (\vec{\nabla} \hat{z}^s) \epsilon_{ms}{}^r \equiv \vec{A}^{(G)} \hat{z}^r + \vec{A}^{(1)r} + \vec{A}^{(2)r}, \tag{4.32}$$

where the functions $Q^{(s)}$ have dimensions of charge. Using the Lie structure matrices and fundamental representation matrices, Eq. 3.80 defines the behavior of the gauge potentials under a local gauge transformation:

$$[\vec{A}^{(G)}(\vec{x})]'_S = \vec{A}^{(G)}(\vec{x}) + \vec{\mathrm{a}}^r(\vec{x}) \hat{z}_r = \vec{A}^{(G)}(\vec{x}) + \frac{\hbar c}{q} \vec{\nabla} \alpha. \tag{4.33}$$

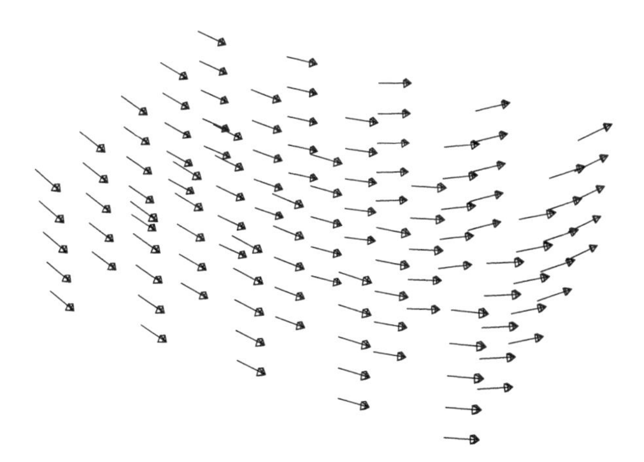

Figure 4.2 Example oriented texture $\hat{z}^r(\vec{x})$.

Thus, the gauge potentials component $\vec{A}^{(G)}$ transforms, as do those for an abelian gauge group. The transformation properties for the other components can likewise be obtained. Defining the orthonormal triad orientations from Eq. 4.32 as $\{\hat{w}_{(1)r}, \hat{w}_{(2)r}, \hat{z}_r\}$, and the components $\vec{A}^{(b)} \equiv \hat{w}_{(b)r}\vec{A}^r$, the potentials transform according to:

$$\begin{aligned}
[\vec{A}^{(1)}(\vec{x})]'_{\mathcal{S}} &= \cos\alpha\, \vec{A}^{(1)}(\vec{x}) + \sin\alpha\, \vec{A}^{(2)}(\vec{x}) + \vec{\mathrm{a}}^{(1)}(\vec{x}),\\
[\vec{A}^{(2)}(\vec{x})]'_{\mathcal{S}} &= -\sin\alpha\, \vec{A}^{(1)}(\vec{x}) + \cos\alpha\, \vec{A}^{(2)}(\vec{x}) + \vec{\mathrm{a}}^{(2)}(\vec{x}),\\
\frac{q}{\hbar c}\vec{\mathrm{a}}^{(1)}(\vec{x}) &= \sin\frac{\alpha}{2}\left[\ \cos\frac{\alpha}{2}\,\hat{w}_{(1)r} + \sin\frac{\alpha}{2}\,\hat{w}_{(2)r}\right]\vec{\nabla}\hat{z}^r,\\
\frac{q}{\hbar c}\vec{\mathrm{a}}^{(2)}(\vec{x}) &= \sin\frac{\alpha}{2}\left[-\sin\frac{\alpha}{2}\,\hat{w}_{(1)r} + \cos\frac{\alpha}{2}\,\hat{w}_{(2)r}\right]\vec{\nabla}\hat{z}^r.
\end{aligned} \tag{4.34}$$

The contribution of non-trivial Lie structure matrices to the transformation of gauge potentials is evident in Eq. 4.34. The coordinate mapping reflected in the spatial gradients transforms in a differing topology from the group components of the potentials.

Next, examine the "magnetic-like" field strength component associated with the remaining gauge symmetry. The vorticity associated with circulation of the vector potential is generally defined using [64–66].

$$\begin{aligned}
\Omega^r_m &\equiv \tfrac{1}{2}\epsilon_{jkm}F^r_{jk},\\
\mathbf{\Omega}^r &= \nabla\times\mathbf{A}^r - \frac{1}{2}\frac{q}{\hbar c}\epsilon_{wv}{}^r\mathbf{A}^w\times\mathbf{A}^v.
\end{aligned} \tag{4.35}$$

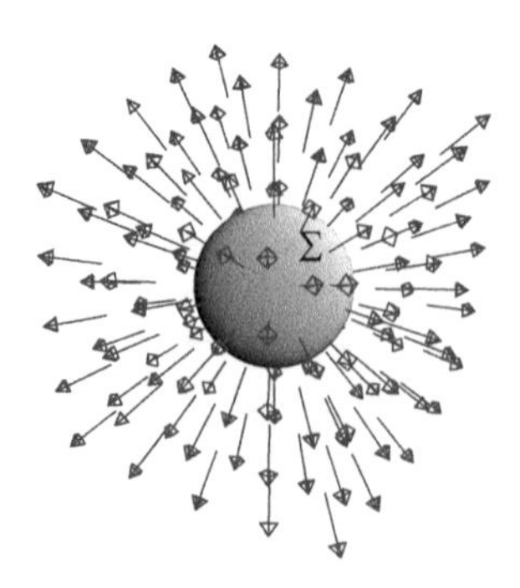

Figure 4.3 Texture with non-vanishing degree $N_{(2)}$.

Defining $\mathbf{\Omega}^{(G)} \equiv \mathbf{\Omega}^r \hat{z}_r$, and noting that $(\nabla \times \mathbf{A}^r)\hat{z}_r = \nabla \times \mathbf{A}^{(G)} + \mathbf{A}^r \times \nabla \hat{z}_r$, the gauge invariant vorticity satisfies:

$$\mathbf{\Omega}^{(G)} = \nabla \times \mathbf{A}^{(G)} + \left[Q^{(2)} - \frac{1}{2}\frac{q}{\hbar c}(Q^{(1)\,2} + Q^{(2)\,2}) \right] \epsilon_{msr} \nabla \hat{z}^m \times \nabla \hat{z}^s \, \hat{z}^r. \quad (4.36)$$

Other than the second term on the right, an integral over a closed surface of $\mathbf{\Omega}^{(G)}$ in Eq. 4.36 would yield a vanishing answer. However, consider an integral over the closed surface Σ demonstrated in Figure 4.3. For the mapping $\underline{\alpha} = \underline{\alpha}(\vec{x})$, with locally integrable coordinates and constant charges $Q_{eff} = [Q^{(2)} - \frac{1}{2}\frac{q}{\hbar c}(Q^{(1)\,2} + Q^{(2)\,2})]$, one can conclude that [64,67]:

$$\oint_\Sigma \mathbf{\Omega}^{(G)} \cdot \hat{n} d^2\Sigma = Q_{eff} \oint_\Sigma \epsilon_{msr} \nabla \hat{z}^m \times \nabla \hat{z}^s \, \hat{z}^r \cdot \hat{n} d^2\Sigma$$

$$= N_{(2)} Q_{eff} \oint \epsilon_{msr} \, d\hat{z}^m \times d\hat{z}^s \, \hat{z}^r = N_{(2)} \Omega_{(2)} Q_{eff}, \quad (4.37)$$

where the integer $N_{(2)}$ is the degree of the mapping of the group area into configuration space, and $\Omega_{(2)}$ is the solid angle in the group parameter space. The non-abelian structure of the larger broken-symmetry group, combined with the topology of the texture, are reflected in quantized monopole sources for the vorticity [67]. Closed volumes in more complicated groups could likewise generate quantized charges with degree of mapping $N_{(3)}$ and compact volume $\Omega_{(3)}$, which could parameterize a boundary of a four-parameter space.

4.3 Quantum mechanics and special relativity

4.3.1 Linear spinor field equations

The Dirac equation utilizes a matrix algebra to construct a linear operator relationship between the energy and the momentum. As demonstrated in the previous chapter, when one is constructing a scattering formalism for relativistic quantum systems, the off-diagonal (off-shell) nature of the intermediate states considerably complicates the implementation of the separation of the purely kinematic variables of the non-interacting components of a multi-particle system from the dynamical variables of the interacting components, due to the generally non-linear dispersion relationships between energies and momenta. One is able to demonstrate the general cluster decomposability of a multiparticle system if the off-shell behavior is parametric, the Lorentz frame of all complete representations of states is unaltered, and the resolvants satisfy linear dispersions [47,48,68] (as is the case using the resolvants resulting from the Dirac equation). Thus, linear dispersion equations are quite convenient for the implementation of physical cluster decomposability in general, non-perturbative, formal scattering theory.

It is therefore advantageous to extend the Lorentz group to include operators whose matrix elements reduce to the Dirac matrices [55] for spin $\frac{1}{2}$ systems but generally require that the form $\hat{\Gamma}^\mu \hat{P}_\mu$ be a scalar operation [69]. The form $\hat{\Gamma}^\mu$ will be seen to be associated with the particle type current, where $\hat{P}_\mu$ is the generator for space-time translations. The extended Lorentz group can then consistently incorporate space-time translations to form an extended Poincaré group. Finite-dimensional representations of the extended Poincaré group can then be constructed in a manner analogous to the use of the Little group of transformations on the standard state vectors in the construction of finite-dimensional representations for spin and helicity in the standard Poincaré algebra.

$$\hat{U}(\underline{b})|\psi_\lambda\, \vec{a}\rangle = \sum_{\lambda'} |\psi'_\lambda\, \vec{z}(\underline{b}; \vec{a})\rangle\, D^{\lambda'}{}_{\lambda}(\underline{b}; \vec{a}) \tag{4.38}$$

In what follows, the group structure of the appropriate extended Lorentz algebra will be developed. Finite-dimensional representations of this group will be expressed in terms of spinors, and matrix representations will be developed for two systems of interest.

4.3.2 Extended Lorentz group

The algebra of the Lorentz group was developed in Chapter 1 and Appendix A.1.3. The finite-dimensional representations of the extended group will be constructed by developing a spinor representation of the algebra. The elements $\underline{b}$ of the complete group of transformations will include three parameters representing angles, three boost parameters, and four group parameters, $\vec{\omega}$ associated with the generators $\hat{\Gamma}^\beta$ [70]. Later, it will be expanded to include the space-time translations $\vec{a}$.

Extended Lorentz group commutation relations

The extended group commutation relations will be chosen to be consistent with the Dirac matrices as follows:

$$[J_j, J_k] = i\hbar\epsilon_{jkm} J_m, \tag{4.39}$$

$$[J_j, K_k] = i\hbar\epsilon_{jkm} K_m, \tag{4.40}$$

$$[K_j, K_k] = -i\hbar\epsilon_{jkm} J_m, \tag{4.41}$$

$$[\Gamma^0, \Gamma^k] = \frac{i}{\hbar} K_k, \tag{4.42}$$

$$[\Gamma^0, J_k] = 0, \tag{4.43}$$

$$[\Gamma^0, K_k] = -i\hbar\Gamma^k, \tag{4.44}$$

$$[\Gamma^j, \Gamma^k] = -\frac{i}{\hbar}\epsilon_{jkm} J_m, \tag{4.45}$$

$$[\Gamma^j, J_k] = i\hbar\epsilon_{jkm}\Gamma^m, \tag{4.46}$$

$$[\Gamma^j, K_k] = -i\hbar\delta_{jk}\Gamma^0. \tag{4.47}$$

The operators Γ^β are dimensionless, while the angular momentum J_k and boost K_j generators have dimensions of $\hbar$.

In the following development, the reduced Planck constant $\hbar$ will be temporarily set to unity for clarity of presentation. At any time the reader can replace an angular momentum or boost generator by $\underline{J} \to \frac{\underline{J}}{\hbar}$, $\underline{K} \to \frac{\underline{K}}{\hbar}$ to re-establish units, if so desired. A Casimir operator can be constructed for this group of the form:

$$C_\Gamma = \underline{J} \cdot \underline{J} - \underline{K} \cdot \underline{K} + \Gamma^0\Gamma^0 - \underline{\Gamma} \cdot \underline{\Gamma}. \tag{4.48}$$

The underlined vectors have spatial components only. This operator can directly be verified to commute with all generators of the group.

General representations for the group algebra can be conveniently developed using ladder operators $\Delta_k^{(\pm)}$ defined by:

$$\Delta_k^{(\pm)} \equiv \Gamma^k (\pm) i K_k, \tag{4.49}$$

and the scalar form:

$$\Delta_J^{(\pm)} \equiv \underline{J} \cdot \underline{\Delta}^{(\pm)}. \tag{4.50}$$

Details of the construction are demonstrated in Appendix D.2.

Group metric

For non-abelian groups, a group metric can be developed from the adjoint representation in terms of the structure constants. For the algebra represented by:

$$[\hat{G}_r, \hat{G}_s] = -i(c_s)_r^m \hat{G}_m \tag{4.51}$$

the group metric can be defined by:

$$\eta_{ab} \equiv (c_a)_r^s (c_b)_s^r. \tag{4.52}$$

The non-vanishing components of the extended Lorentz group metric are given by:

$$\eta_{J_m J_n}^{(EL)} = -6\delta_{mn}, \quad \eta_{K_s K_n}^{(EL)} = +6\delta_{mn}, \quad \eta_{\Gamma^\mu \Gamma^\nu}^{(EL)} = +6\eta_{\mu\nu}. \tag{4.53}$$

It is interesting to note that the group structure of the extended Lorentz group generates the Minkowski metric. Neither the group structure of the usual Lorentz group nor that of the Poincaré group can generate the Minkowski metric due to the abelian nature of the generators for infinitesimal space-time translations. This property is quite useful for incorporating general metric dynamics and geometry into any microscopic dynamics associated with overall group structure and local gauge transformations.

Spinors

Most of the important particle types that make up common matter can be microscopically represented using multicomponent *spinors*. The spinor components maintain well-defined relationships under the transformations of relevance to the particle type. Generally, a spinor is a finite-dimensional vector form whose components mix under a matrix representation satisfying an algebra of transformations. For instance, a commonly utilized spinor is a three-vector, whose indices transform under rotation. The *dual* of a spinor can be defined as a conjugate form to a given spinor whose inner product with itself or another spinor will yield an invariant in the given group of transformations. Since the group operators can generally be written in terms of spinors, the spinor formalism is very convenient for constructing arbitrary representations of the group of transformations.

Supplement: SU(2) spinors

A general eigenstate for the SU(2) generators $\hat{J}^2$ and $\hat{J}_z$ can be expressed as a monomial product of two spinors χ_+ and χ_-. A representation of the angular momentum operators that satisfies the algebra is given by:

$$\hat{J}_z = \frac{\hbar}{2}\left[\chi_+ \partial_+ - \chi_- \partial_-\right], \quad \hat{J}_\pm \equiv \hat{J}_x \pm i\hat{J}_y = \hbar\chi_\pm \partial_\mp, \tag{4.54}$$

where $\partial_\pm \equiv \frac{\partial}{\partial \chi_\pm}$. An eigenstate satisfying $\hat{J}^2 \psi_M^J = J(J+1)\hbar^2 \psi_M^J$ and $\hat{J}_z \psi_M^J = M\hbar \psi_M^J$ is given by:

$$\psi_M^J = N^J \frac{\chi_-^{J-M} \chi_+^{J+M}}{\sqrt{(J-M)!(J+M)!}}. \tag{4.55}$$

This demonstrates a straightforward method using spinors to construct general states of a given algebra.

The extended Lorentz group has both angular momentum raising/lowering operators $\hat{J}_\pm$, as well as particle-type raising/lowering operators $\Delta_J^{(\pm)}$. Thus, there will be four spinor types, designated by $\chi_\pm^{(\pm)}$. Indices within the parenthesis will refer to changes in the eigenvalues of the operator $\hat{\Gamma}^0$, while the lack of such parentheses refer to eigenvalues of $\hat{J}_z$.

Conjugate spinor forms

Conjugate spinor forms for the operators that satisfy Eqs. D.3–D.18 can be constructed using spinor operations:

$$J_z = \frac{1}{2}\left[\chi_+^{(+)}\partial_+^{(+)} + \chi_+^{(-)}\partial_+^{(-)} - \chi_-^{(+)}\partial_-^{(+)} - \chi_-^{(-)}\partial_-^{(-)}\right] \tag{4.56}$$

$$J_\pm = \chi_\pm^{(+)}\partial_\mp^{(+)} + \chi_\pm^{(-)}\partial_\mp^{(-)} \tag{4.57}$$

$$\Gamma^0 = \frac{1}{2}\left[\chi_+^{(+)}\partial_+^{(+)} - \chi_+^{(-)}\partial_+^{(-)} + \chi_-^{(+)}\partial_-^{(+)} - \chi_-^{(-)}\partial_-^{(-)}\right] \tag{4.58}$$

$$\Delta_{Jz}^{(\pm)} = (\pm)\left[\chi_+^{(\pm)}\partial_+^{(\mp)} - \chi_-^{(\pm)}\partial_-^{(\mp)}\right] \tag{4.59}$$

$$\Delta_{J\pm}^{(\pm)} = (\pm)2\chi_\pm^{(\pm)}\partial_\mp^{(\mp)}, \tag{4.60}$$

where $\partial_a^{(b)} \equiv \frac{\partial}{\partial \chi_a^{(b)}}$. This representation provides a convenient mechanism to construct spinor and matrix representations.

Symmetry behavior of spinor forms

Once spinor operator forms have been constructed for the group generators, various symmetry properties can be directly explored. For instance, examine the behavior of the operators under the transformation of a spinor into its dual given by:

$$\chi_\pm^{(\pm)} \leftrightarrow \bar{\chi}_\pm^{(\mp)}. \tag{4.61}$$

Under this replacement, the angular momentum and Γ^0 operators can be seen to transform as:

$$\underline{J} \leftrightarrow \underline{\bar{J}} \tag{4.62}$$

$$\Gamma^0 \leftrightarrow -\bar{\Gamma}^0. \tag{4.63}$$

The commutation relations are preserved for the various generators in the bar representation. For the Dirac case, this will be seen to represent a "particle-anti-particle" symmetry of the system, and it will represent a general symmetry under negation of the eigenvalues of the operator Γ^0.

4.3.3 General construction of states

General spinor states can be constructed using the operation of the raising operators $\Delta_J^{(+)}$ and J_+ on the minimal state $\psi^{\Gamma,J}_{\gamma_{min},-J}$, where Γ is the parameter associated with the Casimir operator eigenvalue, J parameterizes the overall angular momentum with z-component $M\hbar$, and γ is the eigenvalue of Γ^0. A general form for $\psi^{\Gamma,J}_{\gamma,M}$ can be constructed using methods familiar from SU(2), resulting in a form:

$$\psi^{\Gamma,J}_{\gamma,M} = A^{\Gamma J}\sqrt{\frac{(J-M)!}{(J+M)!(2J)!}}[x-y]^{\Gamma-J}\chi_+^{(+)M+\gamma}\chi_+^{(-)M-\gamma}$$
$$\times\left[\frac{\partial}{\partial x}+\frac{\partial}{\partial y}\right]^{J+M} x^{J-\gamma}\,y^{J+\gamma}\Bigg|_{\substack{x=\chi_+^{(+)}\chi_-^{(-)}\\ y=\chi_-^{(+)}\chi_+^{(-)}.}} \tag{4.64}$$

The action of the spinor forms of the operators given in Eqs. 4.56–4.60 results in the following set of equations:

$$\hat{C}_\Gamma\psi^{\Gamma,J}_{\gamma,M} = 2\Gamma(\Gamma+2)\psi^{\Gamma,J}_{\gamma,M}, \tag{4.65}$$

$$\hat{J}^2\psi^{\Gamma,J}_{\gamma,M} = J(J+1)\psi^{\Gamma,J}_{\gamma,M}, \tag{4.66}$$

$$\hat{J}_z\psi^{\Gamma,J}_{\gamma,M} = M\psi^{\Gamma,J}_{\gamma,M}, \tag{4.67}$$

$$\hat{\Gamma}^0\psi^{\Gamma,J}_{\gamma,M} = \gamma\psi^{\Gamma,J}_{\gamma,M}, \tag{4.68}$$

$$\hat{J}_\pm\psi^{\Gamma,J}_{\gamma,M} = \sqrt{(J\pm M+1)(J\mp M)}\psi^{\Gamma,J}_{\gamma,M\pm 1}, \tag{4.69}$$

$$\hat{\Delta}_J^{(\pm)}\psi^{\Gamma,J}_{\gamma,M} = (\pm)(\Gamma+1)[J(\mp)\gamma]\psi^{\Gamma,J}_{\gamma\pm 1,M}. \tag{4.70}$$

Equation 4.70 clearly allows the construction of finite-dimensional representations if J is included in the eigenvalue spectrum of $\hat{\Gamma}^0$.

Number of states

The order of the spinor polynomial of the finite-dimensional state with the Casimir value $\Gamma = J_{max}$ can be determined by examining the minimal state from which other states can be constructed using the raising operators and orthonormality:

$$\psi^{\Gamma,\Gamma}_{-\Gamma,-\Gamma} = A^{\Gamma J} \chi_{-}^{(-)2\Gamma}. \tag{4.71}$$

The general state involves spinor products of the order:

$$\chi_{+}^{(+)a} \chi_{-}^{(+)b} \chi_{+}^{(-)c} \chi_{-}^{(-)d}. \tag{4.72}$$

A complete basis of states requires then that $a + b + c + d = 2\Gamma$. By direct counting, this yields the number of states for a complete basis:

$$N_\Gamma = \frac{1}{3}(\Gamma + 1)(2\Gamma + 1)(2\Gamma + 3). \tag{4.73}$$

For instance, $N_0 = 1$, $N_{\frac{1}{2}} = 4$, $N_1 = 10$, $N_{\frac{3}{2}} = 20$, and so on.

A single J basis with $(2J + 1)^2$ states does not cover this space of spinors. However, one can directly verify that:

$$N_{J_{max}} = \sum_{J=J_{min}}^{J_{max}} (2J + 1)^2, \tag{4.74}$$

where J_{min} is zero for integral systems and $\frac{1}{2}$ for half-integral systems. Thus, the parameter Γ represents the maximal angular momentum state of the system:

$$J \leq \Gamma = J_{max}, \tag{4.75}$$

which contains all angular momenta in integral steps up to the maximum J_{max}.

4.3.4 Spinor metrics

As is the case with Dirac spinors, invariant amplitudes can be defined using dual spinors, so that under transformations the inner product is a scalar:

$$\begin{aligned} < \bar{\psi}|\phi > &= < \bar{\psi}'|\phi' > \\ \psi_a^\dagger g_{ab} \phi_b &= (D_{ca}\psi_a)^\dagger \, g_{cd} \, (D_{db}\psi_b) , \end{aligned} \tag{4.76}$$

where the Dirac conjugate spinor is defined from the transposed hermitian conjugate spinor and the spinor metric g_{ab} using:

$$\bar{\psi}_b \equiv \psi_a^\dagger g_{ab}. \tag{4.77}$$

This means that the spinor metric $(\mathbf{g})_{ab} \equiv g_{ab}$ should satisfy:

$$\mathbf{g} = \mathbf{D}^\dagger \mathbf{g} \mathbf{D}. \tag{4.78}$$

Notice also that for any spinor, two subsequent Dirac conjugates yield the original spinor, $\bar{\bar{\mathbf{u}}} = \mathbf{u}$, as long as $\mathbf{g}^\dagger = \mathbf{g}$ and $\mathbf{g}^2 = \mathbf{1}$. In addition, the eigenvalues of the hermitian angular momentum and Γ^0 will be given by real numbers:

$$\underline{J}^\dagger = \underline{J}, \quad \Gamma^{0\dagger} = \Gamma^0. \tag{4.79}$$

Since the spinor metric is likewise hermitian, it satisfies:

$$\mathbf{g}\mathbf{\Gamma^0} = \mathbf{\Gamma^0}\mathbf{g}, \quad \mathbf{g}\underline{\mathbf{J}} = \underline{\mathbf{J}}\mathbf{g}. \tag{4.80}$$

Using these conditions, matrix representations of the spinor metric can be constructed.

From Eq. 4.70 it can be seen that the representation is finite-dimensional if $\Gamma = J$. Within the $J = \Gamma$ subspace, the metric can be assumed to be proportional to the identity matrix for angular momentum, since it commutes with all of the generators of angular momentum. One can then generally show that:

$$\begin{aligned} \mathbf{g}^{\Gamma:J';J}_{M'\gamma';M\gamma} &= g^{\Gamma J}_{\gamma} \delta_{J'J}\delta_{M'M}\delta_{\gamma'\gamma}, \\ \mathbf{\Delta}^{(\pm)\,\Gamma:J';J}_{M'\gamma';M\gamma} &= \Delta^{(\pm)\,\Gamma,\gamma}_{J'M';JM}\delta_{\gamma',\gamma(\pm)1}. \end{aligned} \tag{4.81}$$

If the condition $\mathbf{g}\mathbf{\Delta}^{(\pm)}_{\mathbf{J}} = -\mathbf{\Delta}^{(\pm)}_{\mathbf{J}}\mathbf{g}$ is satisfied, then the metric can be chosen such that:

$$g^{\Gamma J}_{\gamma} = (-)^{\Gamma-\gamma}, \tag{4.82}$$

which implies that $\mathbf{g}^2 = \mathbf{1}$. This form allows a definition of projectors:

$$\mathcal{P}^{(J)}_{\gamma} = \tfrac{1}{2}(\mathbf{1} + (-)^{\Gamma-\gamma}\mathbf{g})\mathcal{P}^{(J)}, \tag{4.83}$$

where $\mathcal{P}^{(J)}$ projects into the angular momentum J subspace. The projector $(\mathcal{P}^{(J)}_{\gamma})^2 = \mathcal{P}^{(J)}_{\gamma}$ projects out the particular component of a spinor with index (J, γ).

By expanding the anti-commutator $\{\mathbf{g}, \mathbf{\Delta}^{\mp}_{\mathbf{J}}\}$, this then implies that the vector components of the boost generator and Γ matrices must satisfy:

$$\mathbf{g}\underline{\mathbf{\Gamma}} = -\underline{\mathbf{\Gamma}}\mathbf{g}, \quad \mathbf{g}\underline{\mathbf{K}} = -\underline{\mathbf{K}}\mathbf{g}. \tag{4.84}$$

The Eqs. 4.80 and 4.84 can be used to identify relationships between some matrix representations of these generators and their hermitian conjugates. From Eq. A.46 in Appendix A, the form of the boost generator K in the adjoint representation is anti-hermitian. This implies that a representation of the spatial components $\mathbf{\Gamma}^k$ should likewise be anti-hermitian in order to satisfy commutation relations like Eq. 4.44. Thus, for such a representation, all of the group generators $\mathbf{X} = \{\mathbf{J}^k, \mathbf{K}^k, \mathbf{\Gamma}^\mu\}$ satisfy:

$$\mathbf{X}^\dagger = \mathbf{g}\mathbf{X}\mathbf{g}. \tag{4.85}$$

An example representation will next be demonstrated.

Construction of $\Gamma = \frac{1}{2}$ *systems*

The forms of the matrices corresponding to $\Gamma = \frac{1}{2}$ are expected to have dimensionality $N_{\frac{1}{2}} = 4$ and can be expressed in terms of the Pauli spin matrices as shown below:

$$\Gamma^0 = \frac{1}{2}\begin{pmatrix} \mathbf{1} & \mathbf{0} \\ \mathbf{0} & -\mathbf{1} \end{pmatrix} = \frac{1}{2}\mathbf{g} \qquad \underline{\mathbf{J}} = \frac{1}{2}\begin{pmatrix} \underline{\sigma} & \mathbf{0} \\ \mathbf{0} & \underline{\sigma} \end{pmatrix}$$
$$\underline{\Gamma} = \frac{1}{2}\begin{pmatrix} \mathbf{0} & \underline{\sigma} \\ -\underline{\sigma} & \mathbf{0} \end{pmatrix} \qquad \underline{\mathbf{K}} = -\frac{i}{2}\begin{pmatrix} \mathbf{0} & \underline{\sigma} \\ \underline{\sigma} & \mathbf{0} \end{pmatrix}. \tag{4.86}$$

The Γ^μ matrices can directly be seen to be proportional to a representation of the Dirac matrices [55,57]. Examining this lowest-dimensional representation of the extended Lorentz group, the particular representation developed as the covering group is seen to be a subgroup of SL(4,C).

Higher-dimensional representations can be constructed directly. The generalizations of the Dirac matrices for $\Gamma = 1$ are demonstrated in Appendix D.2.1.

4.3.5 Conserved current

A single-particle wave equation can be developed for configuration space eigenstates of the operator $\hat{\Gamma}^\mu\, \hat{P}_\mu$,

$$\mathbf{\Gamma}^\mu\, i\partial_\mu \mathbf{\Psi}(x) = \lambda\, \mathbf{\Psi}(x), \tag{4.87}$$

which implies that:

$$\partial_\mu[\overline{\mathbf{\Psi}(x)}\, \mathbf{\Gamma}^\mu\, \mathbf{\Psi}(x)] = 0. \tag{4.88}$$

The conserved current defined in Eq. 4.88 is hermitian, even though the spinor metric need not be related to the Γ^0 matrix in general. This is true because $\mathbf{g}\mathbf{\Gamma}^\beta = \mathbf{\Gamma}^{\beta *}\mathbf{g}$. For Dirac particles, this current represents the probability current. Scattering equations can be developed to express the evolution of the general physical parameter (particle-type density) represented by this operator.

4.3.6 An extended Poincaré group

Extended Poincaré group closure

The equations presented thus far are valid for internal transformations on such systems. However, once space-time translations are included, any additional commutation relations must result in a self-consistent set of generators [71]. In this section, the group structure will be minimally expanded to include space-time translations, in order to develop a group that can be used for global transformations.

Any set of operators $\hat{A}$, $\hat{B}$, $\hat{C}$ is expected to satisfy the Jacobi identity,

$$[\hat{A}, [\hat{B}, \hat{C}]] + [\hat{C}, [\hat{A}, \hat{B}]] + [\hat{B}, [\hat{C}, \hat{A}]] = 0, \tag{4.89}$$

which is a purely algebraic constraint upon such operators. An attempt to only include the four-momentum operators in addition to the extended Lorentz group operators does not produced a closed-group structure, due to Jacobi relations of the sort $[P_j, [\Gamma^0, \Gamma^k]]$. The non-vanishing of this commutator in the Jacobi identity implies a non-vanishing commutator between Γ^μ and P_ν, and this commutator must connect to an operator, which then commutes with Γ^μ to yield a four-momentum operator P_ν. Since the four-momentum operators self-commute, this therefore requires the introduction of at least one additional operator that will be referred to as $\mathcal{M}_T$. Physical properties of observables connected to this additional operator will be examined.

A closed set of extended Poincaré operators

The non-vanishing commutators involving the operators $\hat{P}_\mu$ and $\hat{\mathcal{M}}_T$ that satisfy the Jacobi identities are given by:

$$[J_j, P_k] = i\hbar\epsilon_{jkm} P_m, \tag{4.90}$$

$$[K_j, P_0] = -i\hbar P_j, \tag{4.91}$$

$$[K_j, P_k] = -i\hbar\delta_{jk} P_0, \tag{4.92}$$

$$[\Gamma^\mu, P_\nu] = i\delta^\mu_\nu \mathcal{M}_T c, \tag{4.93}$$

$$[\Gamma^\mu, \mathcal{M}_T] = \frac{i}{c}\eta^{\mu\nu} P_\nu, \tag{4.94}$$

where units have been temporarily restored. An extended Poincaré group Casimir operator can be constructed using the Lorentz invariants:

$$\mathcal{C}_m \equiv \mathcal{M}_T^2 c^2 - \eta^{\beta\nu} P_\beta P_\nu. \tag{4.95}$$

The label m in the Casimir $\mathcal{C}_m$ will parameterize the eigenstates that will be developed to construct a finite dimensional representation. Due to the form of the group Casimir, eigenvalues of the hermitian operator $\mathcal{M}_T$ will be referred to as the *transverse mass* of the state.

The extended Lorentz subgroup Casimir operator defined in Eq. 4.48 has the following non-trivial commutation relations with the additional operators:

$$[\mathcal{M}_T, C_\Gamma] = i(\Gamma^\mu P_\mu + P_\mu \Gamma^\mu), \tag{4.96}$$

$$[P_0, C_\Gamma] = -i(\Gamma^0 \mathcal{M}_T + \mathcal{M}_T \Gamma^0 - (K_j P_j + P_j K_j)), \tag{4.97}$$

$$[P_j, C_\Gamma] = i(\Gamma^k \mathcal{M}_T + \mathcal{M}_T \Gamma^k - K_j P_0 - P_0 K_j + \epsilon_{jkm}(J_k P_m + P_m J_k)), \tag{4.98}$$

where any repeated indices, Greek or Latin, are presumed to be summed over all respective values. Quantum state vectors representing mutually commuting operators satisfy:

$$\begin{aligned}
\hat{\mathcal{C}}_m |m, \Gamma, \gamma, J, s_z\rangle &= m^2 c^2 |m, \Gamma, \gamma, J, s_z\rangle, \\
\hat{C}_\Gamma |m, \Gamma, \gamma, J, s_z\rangle &= 2\Gamma(\Gamma+2) |m, \Gamma, \gamma, J, s_z\rangle, \\
\hat{\Gamma}^0 |m, \Gamma, \gamma, J, s_z\rangle &= \gamma |m, \Gamma, \gamma, J, s_z\rangle, \\
\hat{J}^2 |m, \Gamma, \gamma, J, s_z\rangle &= J(J+1)\hbar^2 |m, \Gamma, \gamma, J, s_z\rangle, \\
\hat{J}_z |m, \Gamma, \gamma, J, s_z\rangle &= s_z \hbar |m, \Gamma, \gamma, J, s_z\rangle,
\end{aligned} \tag{4.99}$$

where m^2 is generally a continuous real parameter, and all other parameters are discrete. An analysis for local factors done in Appendix D.3.1 demonstrates that a vector representation can be developed for the extended Poincaré group.

Extended Poincaré group metrics

As discussed in Appendix A.1.1, a group metric can generally be developed from the adjoint representation in terms of the structure constants. The non-vanishing group metric elements generated by the structure constants of this extended Poincaré group are given by:

$$\eta^{(EP)}_{J_m\, J_n} = -8\delta_{m,n} \eta^{(EP)}_{K_m\, K_n} = +8\delta_{m,n} \tag{4.100}$$

$$\eta^{(EP)}_{\Gamma^\mu\, \Gamma^\nu} = 8\eta^{\mu\,\nu} \tag{4.101}$$

where $\eta^{\mu\,\nu}$ is the usual Minkowski metric of the Lorentz group. The Minkowski metric is again seen to be non-trivially generated by the extended Γ algebra.

Unitary representations of the extended Poincaré group

As previously demonstrated, a natural set of basis states for the extended Poincare algebra can be specified in terms of the group Casimir $\mathcal{C}_m$, the extended Lorentz subgroup Casimir C_Γ, and mutually commuting eigenvalues of Γ^0, J^2, and J_z, given by $|m, \Gamma, \gamma, J, s_z\rangle$. Here, m^2 represents the eigenvalue of the extended Poincaré group Casimir operator, and the eigenvalue of the extended Lorentz subgroup Casimir, $\Gamma = J_{max}$. One should note, however, that not all generators of either the Lorentz group or the extended Lorentz group are hermitian. Thus, the finite-dimensional representations are not generally unitary (since $e^{i\zeta\hat{\mathcal{O}}}$ is not unitary if $\hat{\mathcal{O}}$ is not hermitian). To construct states that have correspondence with the usual particle states, eigenstates of $\hat{P}_\mu$ should be developed, along with appropriate spinor metrics. In quantum field theory, finite-dimensional representations of particle states are constructed by developing *standard state* vectors of the particles. Lorentz transformations upon a standard state vector give all general states of the particle.

The standard state is an eigenstate of the mutually commuting additional operators, P_β and $\mathcal{M}_T$, of the extended Poincaré group, and the eigenvalues of these operators will be denoted $\underline{w}_{(s)} \equiv (p_{(s)0}, p_{(s)x}, p_{(s)y}, p_{(s)z}, q_{(s)})$, where the $p_{(s)\beta}$ are eigenvalues of the four-momentum operator, and $q_{(s)}$ is the eigenvalue of $\mathcal{M}_T$. Under a general extended Lorentz transformation $\Lambda_f^{(E)r}$:

$$w'_f = \Lambda_f^{(E)r} w_r. \tag{4.102}$$

There will be generally some subgroup of transformations, which will be denoted $\mathcal{L}^{(s)}$, that will leave the standard vector invariant:

$$w_{(s)f} = \mathcal{L}_f^{(s)r} w_{(s)r}. \tag{4.103}$$

The set of such transformations $\mathcal{L}^{(s)}$ form a group that is known as the *little group* of transformations, associated with the vector $\underline{w}_{(s)}$. For example, if the standard state is that of a massive particle at rest $\underline{w}_{(s)} = ((-mc, 0, 0, 0), 0)$, any extended Lorentz transformation that does not change the time component of the momentum or $q_{(s)}$, i.e., any pure rotation, will leave this standard state invariant. Since rotations are expressed using finite-dimensional representations (rotation matrices), an angular momentum eigenstate with this momentum has well-defined unitary and finite-dimensional properties under this transformation.

There are many other extended Lorentz transformations that will not leave the standard state vector invariant. Out of this set of transformations, one can pick a particular Lorentz transformation that will result in any given momentum p_β:

$$p_\beta = \mathbf{C}_\beta^{(s)\mu}(\vec{p}) p_{(s)\mu}. \tag{4.104}$$

Since Lorentz transformations leave $\mathcal{M}_T$ invariant, the eigenvalue of this remains $q_{(s)}$. The chosen set of matrices $\mathbf{C}_\beta^{(s)\mu}$ are known as the *complementary set* of extended Lorentz transformations to the little group of $\underline{w}_{(s)}$. Since the complementary set are Lorentz matrices, the resultant four-momentum will have the same invariant rest mass as the standard-state system.

The eigenvalues used to label the standard state vector must commute with $\hat{P}_\beta$ and $\hat{\mathcal{M}}_T$. The group Casimir $\hat{\mathcal{C}}_m$ and $\hat{J}^2$ commute with an arbitrary standard state vector. An additional invariant can be constructed from the pseudo-vector:

$$\begin{aligned}
\hat{W}_\alpha &\equiv i\epsilon_{\alpha\beta\mu\nu}\, \hat{\Gamma}^\beta\, \hat{\Gamma}^\mu \eta^{\nu\lambda} \hat{P}_\lambda, \\
\hat{W}_0 &= \underline{\hat{J}} \cdot \underline{\hat{P}}, \\
\underline{\hat{W}} &= \underline{\hat{K}} \times \underline{\hat{P}} + \underline{\hat{J}} \hat{P}_0.
\end{aligned} \tag{4.105}$$

where the anti-symmetric tensor $\epsilon_{\alpha\beta\mu\nu}$ is defined by:

$$\epsilon_{\alpha\beta\mu\nu} \equiv \begin{cases} +1 & \text{for } (\alpha\beta\mu\nu) \text{ an even permutation of } (0,1,2,3), \\ -1 & \text{for } (\alpha\beta\mu\nu) \text{ an odd permutation of } (0,1,2,3), \\ 0 & \text{for any two indices equal.} \end{cases} \tag{4.106}$$

The Lorentz invariant $\hat{W}^2 \equiv \hat{W}_\alpha \eta^{\alpha\beta} \hat{W}_\beta$ also commutes with $\hat{P}_\beta$ and $\hat{\mathcal{M}}_T$, since $[\hat{W}_\beta, \hat{P}_\mu] = 0 = [\hat{W}_\beta, \hat{\mathcal{M}}_T]$. One should also notice that this covariant four-vector is orthogonal to the four-momentum operator, $\hat{P}_\mu \eta^{\mu\nu} \hat{W}_\nu = 0$, due to the antisymmetric form defining $\hat{W}_\alpha$.

General quantum states will be constructed from the standard state vector, which will be labeled $|\vec{p}_{(s)}, q_{(s)}, m, J, \kappa\rangle$, where κ will label the discrete indexes of the litte group of $\vec{p}_{(s)}, q_{(s)}$. A quantum state with arbitrary momentum is defined using the complementary transformation:

$$|\vec{p}, q_{(s)}, m, J, \kappa\rangle \equiv U(\mathbf{C}^{(s)}(\vec{p}))|\vec{p}_{(s)}, q_{(s)}, m, J, \kappa\rangle. \tag{4.107}$$

For a general Lorentz transformation $\mathbf{\Lambda}^{(L)}$, the four-momentum operator satisfies:

$$U^{-1}(\mathbf{\Lambda}^{(L)}) \hat{P}_\beta U(\mathbf{\Lambda}^{(L)}) = \Lambda_\beta^{(L)\mu} \hat{P}_\mu. \tag{4.108}$$

Thus, the four-momentum of the transformed state satisfies:

$$\begin{aligned} \hat{P}_\beta |\vec{p}, q_{(s)}, m, J, \kappa\rangle &= U(\mathbf{C}^{(s)}) U^{-1}(\mathbf{C}^{(s)}) \hat{P}_\beta U(\mathbf{C}^{(s)}) |\vec{p}_{(s)}, q_{(s)}, m, J, \kappa\rangle \\ &= U(\mathbf{C}^{(s)}) \mathbf{C}_\beta^{(s)\mu}(\vec{p}) \hat{P}_\mu |\vec{p}_{(s)}, q_{(s)}, m, J, \kappa\rangle = p_\beta |\vec{p}, q_{(s)}, m, J, \kappa\rangle, \end{aligned} \tag{4.109}$$

by the definition of the complementary set of transformations. The transformed state is thus a four-momentum eigenstate with the expected eigenvalue.

Next, consider a general Lorentz transformation on the four-momentum eigenstate. Notice that:

$$\begin{aligned} \hat{P}_\beta \left(U(\mathbf{\Lambda}^{(L)}) |\vec{p}, q_{(s)}, m, J, \kappa\rangle \right) &= U(\mathbf{\Lambda}^{(L)}) U^{-1}(\mathbf{\Lambda}^{(L)}) \hat{P}_\beta U(\mathbf{\Lambda}^{(L)}) |\vec{p}, q_{(s)}, m, J, \kappa\rangle \\ &= \mathbf{\Lambda}_\beta^{(L)\mu} p_\mu \left(U(\mathbf{\Lambda}^{(L)}) |\vec{p}, q_{(s)}, m, J, \kappa\rangle \right). \end{aligned} \tag{4.110}$$

This means that $\left(U(\mathbf{\Lambda}^{(L)})|\vec{p}, q_{(s)}, m, J, \kappa\rangle\right)$ is an eigenstate of the four-momentum operator with eigenvalue $\mathbf{\Lambda}^{(L)} \vec{p}$. One can generally conclude that:

$$U(\mathbf{\Lambda}^{(L)}) |\vec{p}, q_{(s)}, m, J, \kappa\rangle = \sum_{\kappa'} |(\mathbf{\Lambda}^{(L)} \vec{p}, q_{(s)}, m, J, \kappa'\rangle Q_{\kappa'\kappa}^{(s)}(\mathbf{\Lambda}^{(L)}, \vec{p}) \tag{4.111}$$

since the set of discrete indices completely describe the momentum eigenstate.

To determine the linear coefficients $Q^{(s)}_{\kappa'\kappa}(\mathbf{\Lambda}^{(L)}, \vec{p})$, the state $|\mathbf{\Lambda}^{(L)}\vec{p}, q_{(s)}, m, J, \kappa\rangle$ can be directly related to the standard state via the definition:

$$|\mathbf{\Lambda}^{(L)}\vec{p}, q_{(s)}, m, J, \kappa\rangle \equiv U(\mathbf{C}^{(s)}(\mathbf{\Lambda}^{(L)}\vec{p}))|\vec{p}_{(s)}, q_{(s)}, m, J, \kappa\rangle. \quad (4.112)$$

Combining Eqs. 4.111, 4.107, and 4.112, and using the group property $U(\mathbf{\Lambda}^{(L_2)})U(\mathbf{\Lambda}^{(L_1)}) = U(\mathbf{\Lambda}^{(L_2)} \cdot \mathbf{\Lambda}^{(L_1)})$, one can ascertain that:

$$U\left(\mathbf{C}^{(s)^{-1}}(\mathbf{\Lambda}^{(L)}\vec{p}) \cdot \mathbf{\Lambda}^{(L)} \cdot \mathbf{C}^{(s)}(\vec{p})\right)|\vec{p}_{(s)}, q_{(s)}, m, J, \kappa\rangle$$
$$= \sum_{\kappa'} |\vec{p}_{(s)}, q_{(s)}, m, J, \kappa'\rangle Q^{(s)}_{\kappa'\kappa}(\mathbf{\Lambda}^{(L)}, \vec{p}). \quad (4.113)$$

One should note that the group element within the operator U leaves the standard momentum $\vec{p}_{(s)}$ invariant. This means that this element must be within the little group of this standard vector. Since the little group has been chosen to have a finite-dimensional and unitary representation, the momentum states are unitary and transform under this representation. The little group element associated with a general Lorentz transformation on a general momentum eigenstate is thus given by:

$$\mathcal{L}(\mathbf{\Lambda}^{(L)}, \vec{p}) \equiv \mathbf{C}^{(s)^{-1}}(\mathbf{\Lambda}^{(L)}\vec{p}) \cdot \mathbf{\Lambda}^{(L)} \cdot \mathbf{C}^{(s)}(\vec{p}), \quad (4.114)$$

where the coefficients $Q^{(s)}_{\kappa'\kappa}(\mathbf{\Lambda}^{(L)}, \vec{p})$ are defined in terms of the matrix representation of the little group of the standard state vector $\vec{p}_{(s)}$.

Massive particle states

Standard massive particle states will transform under the Poincaré subgroup of the extended group. The mass is a Lorentz invariant satisfying $p_\mu \eta^{\mu\nu} p_\nu = -(mc)^2$, requiring that the eigenvalue $q_{(s)}$ of the operator $\mathcal{M}_T$ must vanish if the particle mass is to correspond to the Casimir of the extended group. Massive particles at rest are invariant under arbitrary rotations, which are the set of transformations that make up the little group for this standard state. The states will thus be of the form:

$$|(-mc, 0, 0, 0), q_{(s)} = 0, m, J, s_z\rangle,$$

which have the discrete label κ given by the z-component of spin s_z of the standard state, and vanishing value when operated on by $\mathcal{M}_T$. The matrices $Q^{(s)}_{\kappa'\kappa}$ are just the rotation group matrices $D^{(J)}_{s_z', s_z}$. These will represent the standard particle states from which unitary representations can be constructed. States of finite momenta are constructed using pure Lorentz boosts upon the standard state vector:

$$|\underline{u}, m, J, s_z\rangle \equiv L(\underline{u})|(-mc, 0, 0, 0), q_{(s)} = 0, m, J, s_z\rangle \quad (4.115)$$

where the four-velocities $\vec{u} \cdot \vec{u} = -1$ for massive states. Finite-dimensional, unitary transformations involving general Lorentz transformations, $\mathbf{\Lambda}$, give:

$$U(\mathbf{\Lambda})|\underline{u}, m, J, s_z\rangle = \sum_{s_z'=-J}^{J} |\underline{\mathbf{\Lambda} u}, m, J, s_z'\rangle D^{(J)}_{s_z', s_z}(R_W(\mathbf{\Lambda}, \underline{u})), \tag{4.116}$$

where $R_W(\mathbf{\Lambda}, \underline{u}) = L^{-1}(\underline{\mathbf{\Lambda} u})\mathbf{\Lambda} L(\underline{u})$ is called the *Wigner rotation* [72]. This rotation represents the little group element for the subgroup of transformations that leave the contravariant standard state vector $\vec{p}_{(s)} = (mc, 0, 0, 0)$ invariant. Thus, such states satisfy:

$$\begin{aligned}
\hat{C}_m|\underline{u}, m, J, s_z\rangle &= (mc)^2|\underline{u}, m, J, s_z\rangle, \\
\hat{P}_\beta|\underline{u}, m, J, s_z\rangle &= m u_\beta|\underline{u}, m, J, s_z\rangle, \\
\eta^{\alpha\beta}\hat{P}_\alpha\hat{P}_\beta|\underline{u}, m, J, s_z\rangle &= -(mc)^2|\underline{u}, m, J, s_z\rangle, \\
\hat{\mathcal{M}}_T|\underline{u}, m, J, s_z\rangle &= 0, \\
\hat{J}^2|\underline{u}, m, J, s_z\rangle &= J(J+1)\hbar^2|\underline{u}, m, J, s_z\rangle, \\
\hat{W}^2|\underline{u}, m, J, s_z\rangle &= (mc)^2 J(J+1)\hbar^2|\underline{u}, m, J, s_z\rangle.
\end{aligned} \tag{4.117}$$

For massive particles, the four-momentum is a time-like four-vector, and the operator $\hat{W}_\alpha$ is a space-like four-vector directly related to the spin. The invariant normalization will be chosen as discussed in the previous chapter:

$$\langle \underline{u}', m, J', s_z'|\underline{u}, m, J, s_z\rangle = \frac{u^0\delta^3(\underline{u}' - \underline{u})}{(mc)^2}\delta_{J'J}\delta_{s_z's_z} = \frac{p^0}{mc}\delta^3(\underline{p}' - \underline{p})\delta_{J'J}\delta_{s_z's_z}. \tag{4.118}$$

Massless particle states

For massless particles, the standard state vector is light-like, $\vec{p}_{(s)} \cdot \vec{p}_{(s)} = 0$, and the particles have no rest frame. If the overall group Casimir is given by $(m_T c)^2$, with a generally non-vanishing transverse mass $m_T c$, the standard state should be of the form:

$$|(-1, 0, 0, 1), q_{(s)} = m_T c, m_T, J, \lambda\rangle,$$

where the discrete label λ is to be determined.

Again, pure Lorentz transformations leave the eigenvalue of the transverse mass operator $\hat{\mathcal{M}}_T$ invariant. The set of Lorentz transformations that leave the contravariant vector $(1, 0, 0, 1)$ invariant will be directly calculated. It is quite apparent that the one-parameter group of rotations about the z-axis will leave this vector invariant:

$$\mathbf{\Lambda}_z \equiv \begin{pmatrix} 1 & 0 & 0 & 0 \\ 0 & \cos\phi & -\sin\phi & 0 \\ 0 & \sin\phi & \cos\phi & 0 \\ 0 & 0 & 0 & 1 \end{pmatrix}. \tag{4.119}$$

There are two additional types of Lorentz transformations that leave this vector invariant:

$$\mathbf{\Lambda}_x \equiv \begin{pmatrix} 1+\frac{v_x^2}{2} & 0 & -v_x & -\frac{v_x^2}{2} \\ 0 & 1 & 0 & 0 \\ -v_x & 0 & 1 & v_x \\ \frac{v_x^2}{2} & 0 & -v_x & 1-\frac{v_x^2}{2} \end{pmatrix}, \quad \mathbf{\Lambda}_y \equiv \begin{pmatrix} 1+\frac{v_y^2}{2} & -v_y & 0 & -\frac{v_y^2}{2} \\ -v_y & 1 & 0 & v_y \\ 0 & 0 & 1 & 0 \\ \frac{v_y^2}{2} & -v_y & 0 & 1-\frac{v_y^2}{2} \end{pmatrix}. \tag{4.120}$$

The transformations $\mathbf{\Lambda}_x$, $\mathbf{\Lambda}_y$, and $\mathbf{\Lambda}_z$ are non-commuting, and only the rotation along the z-axis has a finite-dimensional unitary representation. Therefore, the standard state will be an eigenstate of the infinitesimal generator of $\mathbf{\Lambda}_z$. Appendix D.4 will examine the Hilbert space for the massless particles in more detail.

The standard state for massless particles have a discrete value for the invariant component of angular momentum along the direction of motion, defined as the *helicity*. The helicities λ must be integers or half-integers, and for massless particles take on either of the two extreme values. Various group generators satisfy the following actions upon the standard state:

$$\begin{aligned}
\hat{C}_m|(-1,0,0,1), q_{(s)} = m_T c, J, \lambda\rangle &= (m_T c)^2 |(-1,0,0,1), q_{(s)} = m_T c, J, \lambda\rangle, \\
\hat{P}_\alpha \eta^{\alpha\beta} \hat{P}_\beta |(-1,0,0,1), q_{(s)} = m_T c, J, \lambda\rangle &= 0, \\
\hat{\mathcal{M}}_T|(-1,0,0,1), q_{(s)} = m_T c, J, \lambda\rangle &= m_T c|(-1,0,0,1), q_{(s)} = m_T c, J, \lambda\rangle, \\
\hat{J}^2|(-1,0,0,1), q_{(s)} = m_T c, J, \lambda\rangle &= J(J+1)\hbar^2|(-1,0,0,1), q_{(s)} \\
&= m_T c, J, \lambda\rangle, \\
\hat{W}_0|(-1,0,0,1), q_{(s)} = m_T c, J, \lambda\rangle &= \lambda\hbar|(-1,0,0,1), q_{(s)} = m_T c, J, \lambda\rangle, \\
\hat{W}_z|(-1,0,0,1), q_{(s)} = m_T c, J, \lambda\rangle &= -\lambda\hbar|(-1,0,0,1), q_{(s)} = m_T c, J, \lambda\rangle.
\end{aligned} \tag{4.121}$$

The complementary set of Lorentz transformations that takes a massless particle in the standard state to an arbitrary momentum $\mathbf{C}^{(so)}(\vec{p})$ will be chosen to be a Lorentz boost along the standard momentum (z-axis) to the final momentum, followed by a pure rotation from the standard direction into the direction of the final momentum:

$$\mathbf{C}^{(so)}(\vec{p}) \equiv \mathbf{R}(\underline{u}_{\vec{p}})\mathbf{L}(|\underline{p}|\underline{u}_z), \tag{4.122}$$

where $\underline{u}_{\vec{p}}$ is a unit vector in the direction of the final momentum. A massless state with arbitrary momentum is defined using this complementary transformation:

$$|\vec{p}, q_{(s)} = mc, J, \lambda\rangle \equiv U(\mathbf{C}^{(so)}(\vec{p}))|(-1, 0, 0, 1), q_{(s)} = mc, J, \lambda\rangle. \tag{4.123}$$

The form of the resultant four-momentum $p_\beta = -\mathbf{C}_\beta^{(so)0}(\vec{p}) + \mathbf{C}_\beta^{(so)z}(\vec{p})$ can be used to determine the form of the complementary transformation. In particular, the Lorentz boost along the z-axis satisfies:

$$\begin{aligned}
\mathbf{L}_0^0(|\underline{p}|) &= \frac{1+|\underline{p}|^2}{2|\underline{p}|} = \mathbf{L}_3^3(|\underline{p}|), \\
\mathbf{L}_0^3(|\underline{p}|) &= \frac{1-|\underline{p}|^2}{2|\underline{p}|} = \mathbf{L}_3^0(|\underline{p}|), \\
\mathbf{L}_1^1(|\underline{p}|) &= 1 = \mathbf{L}_2^2(|\underline{p}|),
\end{aligned} \tag{4.124}$$

with all other components vanishing.

The little group element associated with a general Lorentz transformation on a general massless eigenstate is thus given by:

$$\mathcal{L}^{(so)}(\mathbf{\Lambda}^{(L)}, \vec{p}) \equiv \mathbf{C}^{(so)^{-1}}(\mathbf{\Lambda}^{(L)}\vec{p}) \cdot \mathbf{\Lambda}^{(L)} \cdot \mathbf{C}^{(so)}(\vec{p}). \tag{4.125}$$

The transformation of the massless standard state by this element results in a rotation about the standard direction (see Appendix D.4):

$$U(\mathcal{L}^{(so)}(\mathbf{\Lambda}^{(L)}, \vec{p}))|\vec{p}_{(s)}, q_{(s)} = m_T c, J, \lambda\rangle \equiv e^{-i\lambda\phi(\mathbf{\Lambda}^{(L)},\vec{p})}|\vec{p}_{(s)}, q_{(s)} = m_T c, J, \lambda\rangle. \tag{4.126}$$

Therefore, a general Lorentz transformation on an arbitrary massless state satisfies:

$$U(\mathbf{\Lambda}^{(L)})|\vec{p}, q_{(s)} = m_T c, J, \lambda\rangle \equiv e^{-i\lambda\phi(\mathbf{\Lambda}^{(L)},\vec{p})}|\mathbf{\Lambda}^{(L)}\vec{p}, q_{(s)} = m_T c, J, \lambda\rangle. \tag{4.127}$$

Group operations upon the massless states are given below:

$$
\begin{aligned}
\hat{\mathcal{C}}_m|\vec{p}, q_{(s)} = m_T c, J, \lambda\rangle &= (m_T c)^2|\vec{p}, q_{(s)} = m_T c, J, \lambda\rangle, \\
\hat{P}_\beta|\vec{p}, q_{(s)} = m_T c, J, \lambda\rangle &= p_\beta|\vec{p}, q_{(s)} = m_T c, J, \lambda\rangle, \\
\eta^{\alpha\beta}\hat{P}_\alpha\hat{P}_\beta|\vec{p}, q_{(s)} = m_T c, J, \lambda\rangle &= 0, \\
\hat{\mathcal{M}}_T|\vec{p}, q_{(s)} = m_T c, J, \lambda\rangle &= m_T c|\vec{p}, q_{(s)} = m_T c, J, \lambda\rangle, \\
\hat{J}^2|\vec{p}, q_{(s)} = m_T c, J, \lambda\rangle &= J(J+1)\hbar^2|\vec{p}, q_{(s)} = m_T c, J, \lambda\rangle, \\
\hat{W}_0|\vec{p}, q_{(s)} = m_T c, J, \lambda\rangle &= \lambda\hbar|\underline{p}||\vec{p}, q_{(s)} = m_T c, J, \lambda\rangle, \\
\eta^{\alpha\beta}\hat{W}_\alpha\hat{W}_\beta|\vec{p}, q_{(s)} = m_T c, J, \lambda\rangle &= 0.
\end{aligned}
\tag{4.128}
$$

For massless particles, the four-momentum and the operator $\hat{W}_\alpha$ are distinct null vectors. The invariant normalization will be chosen as discussed in the previous chapter:

$$
\langle\vec{p}\,', q_{(s)} = m_T c, J', \lambda'|\vec{p}, q_{(s)} = m_T c, J, \lambda\rangle = \frac{p^0}{p^0_{(s)}}\delta^3(\underline{p}' - \underline{p})\delta_{J'J}\delta_{\lambda'\lambda}. \tag{4.129}
$$

Unitary extended group transformations

For more general transformations by group elements $\underline{g}$ conjugate to the standard state group element $\underline{g}_{(s)} = \underline{w}_{(s)}$, the little group of transformations $\mathcal{R}$ will leave the standard state group element invariant, $\mathcal{R}\underline{g}_{(s)} = \underline{g}_{(s)}$. Since angular momenta are hermitian operators, operations involving this subgroup of transformations will be generalized rotations on the standard state vector that will leave the standard state group element invariant. Using the complementary group of transformations, boosted states can be defined:

$$
|\underline{g}, m, J, \kappa\rangle \equiv U(\mathbf{C}(\underline{g}))|\underline{g}_{(s)}, m, J, \kappa\rangle, \tag{4.130}
$$

where $\mathbf{C}(\underline{g})\underline{g}_{(s)} = \underline{g}$. A unitary representation can then be constructed using:

$$
U(\mathcal{M})|\underline{g}, m, J, \kappa\rangle = \sum_{\kappa'}|\mathcal{M}\underline{g}, m, J, \kappa'\rangle D^{(J)}_{\kappa';\kappa}(\mathcal{R}(\mathcal{M}, \underline{g})), \tag{4.131}
$$

where the little group element satisfies:

$$
\mathcal{R}(\mathcal{M}, \underline{g}) \equiv \mathbf{C}^{-1}(\mathcal{M}\underline{g})\mathcal{M}\mathbf{C}(\underline{g}). \tag{4.132}
$$

The group structure of this extended Lorentz and Poincaré group has been explored elsewhere [70,71].

4.3.7 Linear wave equation for single particle states

The previous results allow general finite-dimensional expressions of the operator $\Gamma^\mu P_\mu$ to be developed. Eigenstates of this operator should give linear operator dispersion for energy and momenta in a wave equation. It is straightforward to calculate the commutators of the various group generators with this Lorentz invariant operator:

$$[J_k, \Gamma^\mu P_\mu] = 0 \tag{4.133}$$

$$[K_k, \Gamma^\mu P_\mu] = 0 \tag{4.134}$$

$$[\Gamma^k, \Gamma^\mu P_\mu] = i\Gamma^k \mathcal{M}_T + i\sum_{mn} \epsilon_{kmn} J_m P_n - iK_k P_0 \tag{4.135}$$

$$[\Gamma^0, \Gamma^\mu P_\mu] = i\Gamma^0 \mathcal{M}_T + i\sum_j K_j P_j \tag{4.136}$$

$$[P_\beta, \Gamma^\mu P_\mu] = -i\mathcal{M}_T P_\beta \tag{4.137}$$

$$[\mathcal{M}_T, \Gamma^\mu P_\mu] = -i\eta^{\beta\nu} P_\beta P_\nu \tag{4.138}$$

One should note that, from Eq. 4.138, the transverse mass only commutes with $\Gamma^\mu P_\mu$ for massless particles. Similarly, from 4.137 the four-momentum operator only commutes with $\Gamma^\mu P_\mu$ if the transverse mass vanishes. Finally, from Eq. 4.136, the particle type operator Γ^0 only commutes with $\Gamma^\mu P_\mu$ acting upon the standard state of massive particles, but *not* the general momentum state of massive particles.

Massive particles

Since pure Lorentz transformations commute with the operator $\Gamma^\mu P_\mu$ from Eqs. 4.133 and 4.134, mixed wavefunctions involving the standard state vectors $|\vec{p}_{(s)}, q_{(s)}, m, J, \kappa\rangle$ and finite-dimensional spinors $|\xi_a^{(\Gamma)}\rangle$ from the extended Lorentz group can be defined that satisfy:

$$\left\langle \xi_a^{(\Gamma)} \middle| \hat{\Gamma}^\mu \hat{P}_\mu \middle| \vec{p}_{(s)}, q_{(s)}, m, J, \kappa \right\rangle = \sum_b \left\langle \xi_a^{(\Gamma)} \middle| \hat{\Gamma}^\mu \middle| \xi_b^{(\Gamma)} \right\rangle p_{(s)\mu} \left\langle \xi_b^{(\Gamma)} \middle| \vec{p}_{(s)}, q_{(s)}, m, J, \kappa \right\rangle. \tag{4.139}$$

For massive particles, eigen-spinors of the particle type operator Γ^0 as previously developed have eigenvalues γ, resulting in a linear wave equation for the standard wavefunction given by:

$$\boldsymbol{\Gamma}^\mu \hat{P}_\mu \boldsymbol{\Psi}_{(\gamma)}^{(\Gamma)}(m\vec{u}_{(s)}, J, s_z) = -(\gamma)mc\boldsymbol{\Psi}_{(\gamma)}^{(\Gamma)}(m\vec{u}_{(s)}, J, s_z) \tag{4.140}$$

where the standard massive spinor is defined by:

$$\boldsymbol{\Psi}_{(\gamma)}^{(\Gamma)}(m\vec{u}_{(s)}, J, s_z) \equiv \left\langle \xi_{(\gamma)}^{(\Gamma)} \middle| \underline{u}_{(s)}, m, J, s_z \right\rangle, \tag{4.141}$$

and bold-faced quantities are matrix-/spinor-valued. The particle type eigenvalue satisfies $-\Gamma \leq \gamma \leq \Gamma = J_{max}$. For Dirac spinors, the upper components correspond to $\gamma = +\frac{1}{2}$, while the lower components correspond to $\gamma = -\frac{1}{2}$. One of the most pleasing aspects of these linear spinor forms is that there are no negative energy or negative mass solutions, only conjugate particle types labeled γ with positive semi-definite mass.

The relation can then be Lorentz transformed to a general spinor equation using the complementary set of transformations for massive particle states $U(\mathbf{C}^{(s)}(\vec{p}))$:

$$\frac{1}{\Gamma}\mathbf{\Gamma}^{\mu}\hat{P}_{\mu}\mathbf{\Psi}_{(\gamma)}^{(\Gamma)}(m\vec{u}, J, s_z) = -\frac{\gamma}{\Gamma}mc\mathbf{\Psi}_{(\gamma)}^{(\Gamma)}(m\vec{u}, J, s_z), \tag{4.142}$$

where the transformed spinor is defined by:

$$\mathbf{\Psi}_{(\gamma)}^{(\Gamma)}(m\vec{u}, J, s_z) \equiv \langle \xi_{(\gamma)}^{(\Gamma)} | \underline{u}, m, J, s_z \rangle, \tag{4.143}$$

γ is the particle type in the standard state, and the mass m is a positive definite parameter. The spinors $\mathbf{\Psi}_{(\gamma)}^{(\Gamma)}$ can be chosen to satisfy the orthonormality condition:

$$\begin{aligned}\mathbf{\Psi}_{(\gamma)}^{(\Gamma)\dagger}(\vec{p}, J, s_z) \cdot \mathbf{g}^{(\Gamma)} \cdot \mathbf{\Psi}_{(\gamma')}^{(\Gamma)}(\vec{p}, J', s_z') &\equiv \bar{\mathbf{\Psi}}_{(\gamma)}^{(\Gamma)}(\vec{p}, J, s_z) \cdot \mathbf{\Psi}_{(\gamma')}^{(\Gamma)}(\vec{p}, J', s_z') \\ &= (-)^{\Gamma-\gamma}\delta_{\gamma,\gamma'}\delta_{J,J'}\delta_{s_z,s_z'}.\end{aligned} \tag{4.144}$$

In a configuration space basis, this gives an eigenvalue differential equation for the spinor field. It is clear that since for two non-interacting subsystems the components of the energy-momentum of clusters satisfying this wave equation are additive, such linear dispersions make explicit clustering properties more apparent. It should then be straightforward to include the kinematic variables of a non-interacting cluster in a purely parametric way when calculating the dynamics of an off-shell, off-diagonal subsystem.

For example, for a free Dirac particle ($\Gamma = \frac{1}{2}$), the momentum space four-spinors that satisfy Eq. 4.142 take the forms:

$$\begin{aligned}\mathbf{\Psi}_{(+\frac{1}{2})}^{(\frac{1}{2})}\left(\vec{p}, \frac{1}{2}, s_z\right) &\equiv \mathbf{u}(\vec{p}, s_z) = \frac{\sqrt{E+mc^2}}{2mc^2}\begin{pmatrix}\chi(s_z) \\ \frac{\underline{\sigma}\cdot pc}{E+mc^2}\chi(s_z)\end{pmatrix} \\ \mathbf{\Psi}_{(-\frac{1}{2})}^{(\frac{1}{2})}\left(\vec{p}, \frac{1}{2}, s_z\right) &\equiv \mathbf{v}(\vec{p}, -s_z) = \frac{\sqrt{E+mc^2}}{2mc^2}\begin{pmatrix}\frac{\underline{\sigma}\cdot pc}{E+mc^2}\chi(s_z) \\ \chi(s_z)\end{pmatrix}\end{aligned} \tag{4.145}$$

where the two-spinors χ from SU(2) are given by:

$$\chi\left(+\frac{1}{2}\right) \equiv \begin{pmatrix}1\\0\end{pmatrix}, \quad \chi\left(-\frac{1}{2}\right) \equiv \begin{pmatrix}0\\1\end{pmatrix} \tag{4.146}$$

The four-spinors are normalized using the spinor metric $\mathbf{g}$ from Eq. 4.86:

$$\mathbf{u}^{\dagger}(\vec{p}, s_z)\, \mathbf{g}^{(\frac{1}{2})}\, \mathbf{u}(\vec{p}, s_z) = 1 = -\mathbf{v}^{\dagger}(\vec{p}, s_z)\, \mathbf{g}^{(\frac{1}{2})}\, \mathbf{v}(\vec{p}, s_z), \tag{4.147}$$

which defines the conjugate Dirac spinors $\bar{\mathbf{u}}$ and $\bar{\mathbf{v}}$:

$$\bar{\mathbf{u}}(\vec{p}, s_z) \equiv \mathbf{u}^{\dagger}(\vec{p}, s_z)\, \mathbf{g}^{(\frac{1}{2})}, \;\; \bar{\mathbf{v}}(\vec{p}, s_z) \equiv \mathbf{v}^{\dagger}(\vec{p}, s_z)\, \mathbf{g}^{(\frac{1}{2})}, \tag{4.148}$$

As previously stated, one should note that for Dirac spinors the spinor metric is directly proportional to the matrix representation $\mathbf{\Gamma}^0$.

The four-spinors are normalized such that:

$$\bar{\mathbf{u}}(\vec{p}, s_z')\, \mathbf{u}(\vec{p}, s_z) = \delta_{s_z' s_z} = -\bar{\mathbf{v}}(\vec{p}, s_z')\, \mathbf{v}(\vec{p}, s_z), \tag{4.149}$$

$$\bar{\mathbf{u}}(\vec{p}, s_z')\, \mathbf{v}(\vec{p}, s_z) = 0 = \bar{\mathbf{v}}(\vec{p}, s_z')\, \mathbf{u}(\vec{p}, s_z), \tag{4.150}$$

$$\sum_{s_z} \mathbf{u}(\vec{p}, s_z)\, \bar{\mathbf{u}}(\vec{p}, s_z) = -\frac{\mathbf{\Gamma}^{\beta} p_{\beta} - mc\Gamma\mathbf{1}}{mc}, \tag{4.151}$$

$$\sum_{s_z} \mathbf{v}(\vec{p}, s_z)\, \bar{\mathbf{v}}(\vec{p}, s_z) = -\frac{\mathbf{\Gamma}^{\beta} p_{\beta} + mc\Gamma\mathbf{1}}{mc}. \tag{4.152}$$

Supplement: Positive energies versus the Dirac sea

The negative energy solutions of the Klein–Gordon and Dirac equations pose a dilemma once the charged particle is radiatively coupled to the electromagnetic field. If the charge can radiatively transition to a lower energy state, it would cascade down continuously if there is no lowest energy.[1] Dirac's attempt to address this problem involved filling all negative-energy Fermi levels with electrons consistent with the Pauli exclusion principle.[2] The so-called vacuum state then consists of a negative-energy sea of electrons. A positron with positive energy is interpreted as a *hole* in the sea, i.e., the absence of an electron of negative energy.

However, such an interpretation eliminates one's ability to use the Dirac equation to describe a single particle, rather it must be considered a many-particle theory. The linear spinor formulation does not have negative energy solutions; instead the sign is associated with the particle-type eigenvalue. There is no need to introduce additional degrees of freedom once radiative coupling is included.

[1] J.D. Bjorken and S.D. Drell, *Relativistic Quantum Mechanics*, McGraw-Hill, New York (1964), Section 5.1.
[2] P.A.M. Dirac, *Proc. Roy. Soc. (London)*, **A126**, 360 (1930).

Massless particles

Massless particles satisfy $\vec{p} \cdot \vec{p} = 0$, but need not have vanishing transverse mass, $\mathcal{M}_T$. Massless spinors:

$$\mathbf{\Psi}^{(\Gamma)}(\vec{p}, m_T, J, \lambda) \equiv \langle \xi^{(\Gamma),\gamma} | \vec{p}, q_{(s)} = m_T, J, \lambda \rangle, \tag{4.153}$$

satisfy the equation:

$$\mathbf{\Gamma}^\mu \hat{P}_\mu \mathbf{\Psi}^{(\Gamma)}(\vec{p}, m_T, J, \lambda) = 0. \tag{4.154}$$

Equation 4.138 demonstrates that m_T remains a good quantum number for massless states.

For example, for a free Dirac neutrino ($\Gamma = \frac{1}{2}$), the momentum space four-spinors that satisfy Eq. 4.154, with standard momentum $\{-1, 0, 0, 1\}$ take the forms:

$$\mathbf{\Psi}_+^{(\frac{1}{2})}\left(\vec{p}, m_T, \frac{1}{2}, +\frac{1}{2}\right) \equiv \mathbf{u}(\vec{p}, m_T, +) = \frac{1}{\sqrt{2}} \begin{pmatrix} \frac{p_z}{|p|} \\ \frac{p_x + i p_y}{|p|} \\ 1 \\ 0 \end{pmatrix} \overset{\vec{p} \to \vec{p}_{(s)}}{\Rightarrow} \frac{1}{\sqrt{2}} \begin{pmatrix} 1 \\ 0 \\ 1 \\ 0 \end{pmatrix},$$

$$\mathbf{\Psi}_+^{(\frac{1}{2})}\left(\vec{p}, m_T, \frac{1}{2}, -\frac{1}{2}\right) \equiv \mathbf{u}(\vec{p}, m_T, -) = \frac{1}{\sqrt{2}} \begin{pmatrix} -\frac{p_x - i p_y}{|p|} \\ \frac{p_z}{|p|} \\ 0 \\ -1 \end{pmatrix} \overset{\vec{p} \to \vec{p}_{(s)}}{\Rightarrow} \frac{1}{\sqrt{2}} \begin{pmatrix} 0 \\ 1 \\ 0 \\ -1 \end{pmatrix}, \tag{4.155}$$

while those with standard momentum $\{-1, 0, 0, -1\}$ or $\{1, 0, 0, 1\}$ take the forms:

$$\mathbf{\Psi}_-^{(\frac{1}{2})}\left(\vec{p}, m_T, \frac{1}{2}, +\frac{1}{2}\right) \equiv \mathbf{v}(\vec{p}, m_T, -) = \frac{1}{\sqrt{2}} \begin{pmatrix} -\frac{p_z}{|p|} \\ -\frac{p_x + i p_y}{|p|} \\ 1 \\ 0 \end{pmatrix} \overset{\vec{p} \to \vec{p}_{(s)}}{\Rightarrow} \frac{1}{\sqrt{2}} \begin{pmatrix} -1 \\ 0 \\ 1 \\ 0 \end{pmatrix},$$

$$\mathbf{\Psi}_-^{(\frac{1}{2})}\left(\vec{p}, m_T, \frac{1}{2}, -\frac{1}{2}\right) \equiv \mathbf{v}(\vec{p}, m_T, +) = \frac{1}{\sqrt{2}} \begin{pmatrix} \frac{p_x - i p_y}{|p|} \\ -\frac{p_z}{|p|} \\ 0 \\ -1 \end{pmatrix} \overset{\vec{p} \to \vec{p}_{(s)}}{\Rightarrow} \frac{1}{\sqrt{2}} \begin{pmatrix} 0 \\ -1 \\ 0 \\ -1 \end{pmatrix}. \tag{4.156}$$

The standard state limits on the right of these equations are clearly not eigenstates of $\hat{\Gamma}^0$, differing from those of massive particles. The four-spinors are normalized such that:

$$\mathbf{u}^\dagger(\vec{p}, m_T, \lambda')\, \mathbf{u}(\vec{p}, m_T, \lambda) = \delta_{\lambda'\lambda} = \mathbf{v}^\dagger(\vec{p}, m_T, \lambda')\, \mathbf{v}(\vec{p}, m_T, \lambda), \tag{4.157}$$

$$\mathbf{u}^\dagger(\vec{p}, m_T, \lambda')\, \mathbf{v}(\vec{p}, m_T, \lambda) = 0 = \mathbf{v}^\dagger(\vec{p}, m_T, \lambda')\, \mathbf{u}(\vec{p}, m_T, \lambda), \tag{4.158}$$

$$\bar{\mathbf{u}}(\vec{p}, m_T, \lambda')\, \mathbf{u}(\vec{p}, m_T, \lambda) = 0 = \bar{\mathbf{v}}(\vec{p}, m_T, \lambda')\, \mathbf{v}(\vec{p}, m_T, \lambda), \tag{4.159}$$

$$\bar{\mathbf{u}}(\vec{p}, m_T, \lambda')\, \mathbf{v}(\vec{p}, m_T, \lambda) = -\delta_{\lambda'\lambda} = \bar{\mathbf{v}}(\vec{p}, m_T, \lambda')\, \mathbf{u}(\vec{p}, m_T, \lambda), \tag{4.160}$$

$$\sum_\lambda \mathbf{u}(\vec{p}, m_T, \lambda)\, \bar{\mathbf{u}}(\vec{p}, m_T, \lambda) = \frac{\mathbf{\Gamma}^\beta p_\beta}{-|\underline{p}|} = \frac{\mathbf{\Gamma}^\beta p_\beta}{p_0}, \tag{4.161}$$

$$\sum_\lambda \mathbf{v}(\vec{p}, m_T, \lambda)\, \bar{\mathbf{v}}(\vec{p}, m_T, \lambda) = \frac{\mathbf{\Gamma}^\beta p_\beta}{|\underline{p}|} = \frac{\mathbf{\Gamma}^\beta p_\beta}{p_0}. \tag{4.162}$$

One should note that the normalization for the massless particle spinors differ somewhat from the normalization of massive particle spinors.

A particular set of projection operations are particularly useful for building physical models. Consider the matrices:

$$\mathbf{\Pi}_{RL} \equiv \frac{1}{2}\begin{pmatrix} +1 & 0 & +1 & 0 \\ 0 & +1 & 0 & +1 \\ +1 & 0 & +1 & 0 \\ 0 & +1 & 0 & +1 \end{pmatrix} \tag{4.163}$$

$$\mathbf{\Pi}_{LR} \equiv \frac{1}{2}\begin{pmatrix} +1 & 0 & -1 & 0 \\ 0 & +1 & 0 & -1 \\ -1 & 0 & +1 & 0 \\ 0 & -1 & 0 & +1 \end{pmatrix} \tag{4.164}$$

The matrices are projections $\mathbf{\Pi}_{RL}^2 = \mathbf{\Pi}_{RL}$, $\mathbf{\Pi}_{LR}^2 = \mathbf{\Pi}_{LR}$, $\mathbf{\Pi}_{RL}\mathbf{\Pi}_{LR} = 0$. In the usual Dirac nomenclature, these projections are given by $\mathbf{\Pi}_{RL} = \frac{1}{2}(1 + \gamma_5)$ and $\mathbf{\Pi}_{LR} = \frac{1}{2}(1 - \gamma_5)$, where $\gamma_5 \equiv i\gamma^0\gamma^1\gamma^2\gamma^3$ in terms of the Dirac gamma matrices. These projections have the following actions upon the standard state spinors for massless particles:

$$\begin{aligned}
&\mathbf{\Pi}_{RL}\mathbf{u}(\vec{p}_{(s)}, m_T, +) = +1\mathbf{u}(\vec{p}_{(s)}, m_T, +), \quad \mathbf{\Pi}_{LR}\mathbf{u}(\vec{p}_{(s)}, m_T, +) = 0, \\
&\mathbf{\Pi}_{RL}\mathbf{u}(\vec{p}_{(s)}, m_T, -) = 0, \quad \mathbf{\Pi}_{LR}\mathbf{u}(\vec{p}_{(s)}, m_T, -) = +1\mathbf{u}(\vec{p}_{(s)}, m_T, -), \\
&\mathbf{\Pi}_{RL}\mathbf{v}(\vec{p}_{(s)}, m_T, +) = 0, \quad \mathbf{\Pi}_{LR}\mathbf{v}(\vec{p}_{(s)}, m_T, +) = +1\mathbf{v}(\vec{p}_{(s)}, m_T, +), \\
&\mathbf{\Pi}_{RL}\mathbf{v}(\vec{p}_{(s)}, m_T, -) = +1\mathbf{v}(\vec{p}_{(s)}, m_T, -), \quad \mathbf{\Pi}_{LR}\mathbf{v}(\vec{p}_{(s)}, m_T, -) = 0.
\end{aligned} \tag{4.165}$$

Thus, the projector $\mathbf{\Pi}_{RL}$ is seen to project into standard states with right-handed ($\lambda = +\frac{1}{2}$) particles with positive standard-state energies, and left-handed ($\lambda = -\frac{1}{2}$) particles with negative standard-state energies. Likewise, $\mathbf{\Pi}_{LR}$ is seen to project into standard states with left-handed particles with positive standard-state energies, and right-handed particles with negative standard-state energies. Neutrinos seem to phenomenologically fall within the latter category with regards to weak interaction dynamics.

For massless particles, since the helicity is Lorentz-invariant, projections $\mathbf{\Pi}_{RL}$, $\mathbf{\Pi}_{LR}$ on the spinor generate fixed helicity states. However, for massive particles, a velocity boost can change one helicity state into the opposite helicity. These projections instead produce states of well-defined *chirality*. States of differing chirality mix only through the mass term in the equations of motion.

4.3.8 Finite translations

One can include translations by the mutually commuting operators $\hat{\mathcal{M}}_T$ and $\hat{P}_\beta$ to construct general unitary transformations on the extended group with finite dimensional representations generally expressed in Eq. 4.131. Such unitary operations will be parameterized as $U(\mathcal{M}, \vec{a}, \alpha)$. Pure translations can be directly represented by:

$$U(\mathcal{I}, \vec{a}, \alpha) = e^{-ia^\beta \hat{P}_\beta} e^{-i\alpha \mathcal{M}_T}. \tag{4.166}$$

A unitary representation can be constructed using:

$$U(\mathcal{M}, \vec{a}, \alpha)|\underline{g}, m, J, \kappa\rangle = e^{-ia^\beta p_\beta} e^{-i\alpha m_T} \sum_{\kappa'} |\mathcal{M}\underline{g}, m, J, \kappa'\rangle D^{(J)}_{\kappa';\kappa}(\mathcal{R}(\mathcal{M}, \underline{g})), \tag{4.167}$$

where the little group element $\mathcal{R}(\mathcal{M}, \underline{g})$ is given in Eq. 4.132, and the parameter $\underline{g} = \{\vec{p}, m_T\}$ specifies the momentum and eigenvalue of the transverse mass operator. The group algebra requires $m^2 = m_T^2 - \eta^{\alpha\beta} p_\alpha p_\beta$, indicating that massive particles have vanishing transverse mass, and that the group Casimir for massless particles is the transverse mass.

The operators $\hat{\Gamma}^\mu$ transform under translations as follows:

$$e^{ia^\beta \hat{P}_\beta} \hat{\Gamma}^\mu e^{-ia^\beta \hat{P}_\beta} = \hat{\Gamma}^\mu + a^\mu \hat{\mathcal{M}}_T \tag{4.168}$$

$$e^{i\alpha \hat{\mathcal{M}}_T} \hat{\Gamma}^\mu e^{-i\alpha \hat{\mathcal{M}}_T} = \hat{\Gamma}^\mu + \alpha \eta^{\mu\nu} \hat{P}_\nu \tag{4.169}$$

Since the transverse mass and momenta mutually commute, one can directly examine translations of the operation $\hat{\Gamma}^\mu \hat{P}_\mu$:

$$e^{ia^\beta \hat{P}_\beta} \hat{\Gamma}^\mu \hat{P}_\mu e^{-ia^\beta \hat{P}_\beta} = \hat{\Gamma}^\mu \hat{P}_\mu + a^\mu \hat{P}_\mu \hat{\mathcal{M}}_T \tag{4.170}$$

$$e^{i\alpha \hat{\mathcal{M}}_T} \hat{\Gamma}^\mu \hat{P}_\mu e^{-i\alpha \hat{\mathcal{M}}_T} = \hat{\Gamma}^\mu \hat{P}_\mu + \alpha \hat{P}_\mu \eta^{\mu\nu} \hat{P}_\nu. \tag{4.171}$$

Equation 4.170 indicates that the operation $\hat{\Gamma}^\mu \hat{P}_\mu$ is invariant under space-time translations only if:

- the transverse mass vanishes (as is true for massive states), or
- the translation is the null translation associated with a massless state.

This operation is invariant under translations conjugate to the transverse mass operator only if the particle is massless. Therefore, massive particle states cannot involve a translation with non-vanishing α. Likewise, only α can parameterize those null translations associated with a massless particle.

The implications of these constraints can be examined relative to the proper frame parameterization of particle trajectories. For a massive free particle with trajectory $\vec{x}(t)$, the proper time given by:

$$c\tau = -\frac{x^\mu(t) p_\mu}{\sqrt{-\vec{p}\cdot\vec{p}}} = \frac{Et - \underline{p}\cdot\underline{x}(t)}{mc} \tag{4.172}$$

serves as the affine parameter $\xi = c\tau$, to be utilized in equations such as the geodesic equations, $\frac{du^\beta}{d\xi} + \Gamma^\beta_{\mu\nu} u^\mu u^\nu = 0$. However, this equation clearly cannot be used for massless trajectories. For massless particles, the Lorentz invariant length α conjugate to the transverse mass can be used as the affine parameter $\xi = \alpha$ describing the trajectories. For this reason, one might refer to m_T as the *affine mass*, since it generates translations along light-like trajectories. Massless particles with non-vanishing transverse mass provide a straightforward mechanism for the mixing of massless states along null trajectories (see Appendix D.6). Such particles represent fixed helicity states (consistent with massless particles) decoupled from the considerations of the states of opposite type necessary in consistent discussions of massive particles.

The extended Poincaré group $\mathcal{P}_X = (\vec{\omega}, \underline{u}, \underline{\theta}, \vec{a}, \alpha)$ is a fifteen-parameter group of transformations that includes four Dirac boosts $\vec{\omega}$, three Lorentz velocity boosts $\underline{u}$, three rotation angles $\underline{\theta}$, four space-time translations $\vec{a}$, and an affine null translation α. The usual Lorentz group is just the subgroup of these transformations given by $(\vec{0}, \underline{u}, \underline{\theta}, \vec{0}, 0)$. Likewise, the Poincaré group is the subgroup extending the Lorentz group to include translations $(\vec{0}, \underline{u}, \underline{\theta}, \vec{a}, 0)$. In the interest of brevity and relevance to the present manuscript, results directly exhibited in the Poincaré subgroup will be emphasized in the remaining sections of this chapter.

4.3.9 Discrete particle transformations

Much of particle theory (in flat space-time) is founded on the assumption of the local invariance of the laws of physics under Poincaré transformations, which generate a vector representation of the group algebra $\hat{U}(\mathbf{\Lambda}_{(2)}, \vec{a}_{(2)})\hat{U}(\mathbf{\Lambda}_{(1)}, \vec{a}_{(1)}) = \hat{U}(\mathbf{\Lambda}_{(2)} \cdot \mathbf{\Lambda}_{(1)}, \vec{a}_{(2)} + \mathbf{\Lambda}_{(2)}\vec{a}_{(1)})$. This implies that the rules governing a quantum cluster should be the same irrespective of the space-time position, speed, or orientation of an observer's inertial Lorentz frame. This assumption leads to the conservation of energy, momentum, and angular momentum for any physical process within the cluster.

Besides those proper Lorentz transformations continuously connected to the identity, discrete space-time Lorentz transformations such as parity $\mathcal{P}$ (demonstrated in Eq. 1.14) and time reversal $\mathcal{T}$ (demonstrated in Eq. 1.15) can have dynamical symmetry properties. In addition, charge conjugation $\hat{\mathcal{C}}$ has been seen to describe a discrete transformation. The effect of these discrete transformations on quantum fields will next be developed.

Parity

The state transformation operator representing parity will be denoted $\hat{\mathcal{P}} \equiv \hat{U}(\mathcal{P}, \vec{0})$. Developing a general ray representation for the discrete transformation, note that:

$$\hat{\mathcal{P}}\hat{\mathcal{P}} = \hat{U}(\mathcal{P}, \vec{0})\hat{U}(\mathcal{P}, \vec{0}) = e^{2i\delta_{\mathcal{P}}}\hat{U}(\mathcal{P} \cdot \mathcal{P}, \vec{0}) = e^{2i\delta_{\mathcal{P}}}\hat{1}, \tag{4.173}$$

since $\mathcal{P} \cdot \mathcal{P} = \mathbf{1}$. A general composition rule defines the action of a parity operation upon the group generators:

$$\hat{\mathcal{P}}\hat{U}(\mathbf{\Lambda}, \vec{a})\hat{\mathcal{P}}^{-1} = \hat{U}(\mathcal{P} \cdot \mathbf{\Lambda} \cdot \mathcal{P}^{-1}, \mathcal{P}\vec{a}). \tag{4.174}$$

By examining infinitesimal space-time translations, the four-momentum must satisfy:

$$\begin{aligned} \hat{\mathcal{P}} i \hat{H} \hat{\mathcal{P}}^{-1} &= i\hat{H}, \\ \hat{\mathcal{P}} i \hat{P}_k \hat{\mathcal{P}}^{-1} &= -i\hat{P}_k. \end{aligned} \tag{4.175}$$

Since parity conventionally leaves the energy invariant, this means that it must be a linear operation $\hat{\mathcal{P}} i \hat{\mathcal{O}} \hat{\mathcal{P}}^{-1} = i\hat{\mathcal{P}}\hat{\mathcal{O}}\hat{\mathcal{P}}^{-1}$. Also, since the transverse mass is Lorentz-invariant, it remains unchanged under the parity operation. For infinitesimal Lorentz transformations, direct substitution of the Lorentz matrices $\mathcal{P}$ and

J, **K** from Eq. A.46 in Appendix A yields:

$$\begin{aligned}\hat{\mathcal{P}}\hat{\mathcal{M}}_T\hat{\mathcal{P}}^{-1} &= \hat{\mathcal{M}}_T,\\ \hat{\mathcal{P}}\hat{H}\hat{\mathcal{P}}^{-1} &= \hat{H},\\ \hat{\mathcal{P}}\hat{P}_k\hat{\mathcal{P}}^{-1} &= -\hat{P}_k,\\ \hat{\mathcal{P}}\hat{J}_k\hat{\mathcal{P}}^{-1} &= \hat{J}_k,\\ \hat{\mathcal{P}}\hat{K}_k\hat{\mathcal{P}}^{-1} &= -\hat{K}_k.\end{aligned} \tag{4.176}$$

In addition, the commutators $[\hat{\Gamma}^\mu, \hat{P}_\nu] = i\delta^\mu_\nu \hat{\mathcal{M}}_T$ and $[\hat{\Gamma}^0, \hat{\Gamma}^k] = i\hat{K}_k$ imply:

$$\begin{aligned}\hat{\mathcal{P}}\hat{\Gamma}^0\hat{\mathcal{P}}^{-1} &= \hat{\Gamma}^0,\\ \hat{\mathcal{P}}\hat{\Gamma}^k\hat{\mathcal{P}}^{-1} &= -\hat{\Gamma}^k.\end{aligned} \tag{4.177}$$

Thus, $\hat{\mathcal{M}}_T$, $\hat{H}$, $\hat{\Gamma}^0$, $\hat{P}^\mu\eta_{\mu\nu}\hat{P}^\nu$ and $\hat{\Gamma}^\beta\hat{P}_\beta$ are scalar operators, $\hat{P}_j$, $\hat{K}_j$, and $\hat{\Gamma}^j$ are vector operators, $\hat{W}_0$ is a pseudo-scalar operator, and $\hat{J}_k$ and $\hat{W}_k$ are pseudo-vector operators, as defined by the behavior of these operators under parity transformations.

Expressed as an action upon a massive quantum state describing particle ψ:

$$\hat{\mathcal{P}}|\psi : \vec{p}, m, J, s_z\rangle \equiv e^{i\delta_{\mathcal{P}\psi}}|\psi : \mathcal{P}\vec{p}, m, J, s_z\rangle. \tag{4.178}$$

The angular momentum parameters are invariant under the parity transformation. Notice that, as defined $\hat{\mathcal{P}}^2 = \hat{1}$, which implies that the parity factor for particle ψ must satisfy $e^{i\delta_{\mathcal{P}\psi}} \equiv \mathcal{P}_\psi = \pm 1$, or:

$$\hat{\mathcal{P}}|\psi : \vec{p}, m, J, s_z\rangle \equiv \eta^P_{(\gamma)}|\psi : \mathcal{P}\vec{p}, m, J, s_z\rangle = \pm|\psi : \mathcal{P}\vec{p}, m, J, s_z\rangle. \tag{4.179}$$

Here, $\mathcal{P}\cdot\{p^0, \underline{p}\} = \{p^0, -\underline{p}\}$. The $\pm$ sign defines the *parity* of the massive particle.

The action of a parity transformation upon a massless state can be determined by first examining the standard state:

$$\begin{aligned}&\hat{J}_z|\psi : \{1,0,0,1\}, m_T, J, \lambda\rangle = \lambda|\psi : \{1,0,0,1\}, m_T, J, \lambda\rangle,\\ &\hat{\mathcal{P}}|\psi : \{1,0,0,1\}, m_T, J, \lambda\rangle \equiv e^{i\delta_{\mathcal{P}}(\lambda)}|\psi : \{1,0,0,-1\}, m_T, J, \lambda\rangle.\end{aligned} \tag{4.180}$$

A rotation by π about the y-axis $\mathbf{R}_y(\pi)$ generates the same change in the standard momentum as does the parity transformation. Thus:

$$\hat{\mathcal{P}}|\psi : \vec{p}_{(so)}, m_T, J, \lambda\rangle = e^{i\tilde{\delta}_{\mathcal{P}}(\lambda)}\hat{U}(\mathbf{R}_y(\pi))|\psi : \vec{p}_{(so)}, m_T, J, -\lambda\rangle. \tag{4.181}$$

The complementary transformation $\mathbf{C}^{(so)}(\underline{p}) = \mathbf{R}(\underline{u}_p)\mathbf{L}_z(|\underline{p}|)$ can then be used to determine the action of a parity transformation upon the general momentum state:

$$\begin{aligned}
\hat{\mathcal{P}}|\psi : \vec{p}, m_T, J, \lambda\rangle &= \hat{\mathcal{P}}\hat{U}(\mathbf{C}^{(so)}(\underline{p}))\hat{\mathcal{P}}^{-1}\hat{\mathcal{P}}|\psi : \vec{p}_{(so)}, m_T, J, \lambda\rangle \\
&= e^{i\tilde{\delta}_{\mathcal{P}}(\lambda)}\hat{U}(\mathbf{R}(\underline{u}_p))\hat{U}(\mathbf{L}_{-z}(|\underline{p}|))\hat{U}(\mathbf{R}_y(\pi))|\psi : \vec{p}_{(so)}, m_T, J, -\lambda\rangle \\
&= e^{i\tilde{\delta}_{\mathcal{P}}(\lambda)}\hat{U}(\mathbf{R}(\underline{u}_p)\mathbf{R}_y(\pi)\mathbf{R}^{-1}(-\underline{u}_p))\hat{U}(\mathbf{R}(-\underline{u}_p)) \\
&\quad \times \hat{U}(\mathbf{L}_z(|\underline{p}|))|\psi : \vec{p}_{(so)}, m_T, J, -\lambda\rangle.
\end{aligned} \tag{4.182}$$

One can directly determine that the transformation $\mathbf{R}(\underline{u}_p)\mathbf{R}_y(\pi)\mathbf{R}^{-1}(-\underline{u}_p)$ transforms the z-axis into itself, and thus is just a rotation about the z-axis. Therefore, a parity transformation acts upon a massless state as follows:

$$\begin{aligned}
\hat{\mathcal{P}}|\psi : \vec{p}, m_T, J, \lambda\rangle &= e^{i\tilde{\delta}_{\mathcal{P}}(\lambda)}\, e^{-i\theta(\underline{u}_p)\lambda}|\psi : \mathcal{P}\vec{p}, m_T, J, -\lambda\rangle, \\
\hat{U}(\mathbf{R}(\underline{u}_p)\mathbf{R}_y(\pi)\mathbf{R}^{-1}(-\underline{u}_p)) &\equiv e^{\frac{i}{\hbar}\theta(\underline{u}_p)\hat{J}_z},
\end{aligned} \tag{4.183}$$

where again $\mathcal{P} \cdot \{p^0, \underline{p}\} = \{p^0, -\underline{p}\}$. Notice that for massless states the condition $\mathcal{P} \cdot \mathcal{P} = \mathbf{1}$ implies only that $\tilde{\delta}_{\mathcal{P}}(\lambda) + \tilde{\delta}_{\mathcal{P}}(-\lambda) = 2\pi n$, where n is an integer or zero. Little more can be said about the parity of massless states, due to the multiple connectedness of the rotation group.

Time reversal

The state transformation operator representing the improper Lorentz operation of time reversal will be denoted $\hat{\mathcal{T}} \equiv \hat{U}(\mathcal{T}, \vec{0})$. As was done with parity, a general ray representation for the discrete transformation satisfies:

$$\hat{\mathcal{T}}\hat{\mathcal{T}} = \hat{U}(\mathcal{T}, \vec{0})\hat{U}(\mathcal{T}, \vec{0}) = e^{2i\delta_{\mathcal{T}}}\hat{U}(\mathcal{T} \cdot \mathcal{T}, \vec{0}) = e^{2i\delta_{\mathcal{T}}}\hat{1}, \tag{4.184}$$

since $\mathcal{T} \cdot \mathcal{T} = \mathbf{1}$, and $\det \mathcal{T} = -1$. A general composition rule defines the action of a time-reversal operation upon the group generators:

$$\hat{\mathcal{T}}\hat{U}(\mathbf{\Lambda}, \vec{a})\hat{\mathcal{T}}^{-1} = \hat{U}(\mathcal{T} \cdot \mathbf{\Lambda} \cdot \mathcal{T}^{-1}, \mathcal{T}\vec{a}). \tag{4.185}$$

Again examining infinitesimal space-time translations, the four-momentum must satisfy:

$$\begin{aligned}
\hat{\mathcal{T}}i\hat{H}\hat{\mathcal{T}}^{-1} &= -i\hat{H}, \\
\hat{\mathcal{T}}i\hat{P}_k\hat{\mathcal{T}}^{-1} &= i\hat{P}_k.
\end{aligned} \tag{4.186}$$

As was the case for parity, the time reversal transformation conventionally leaves the energy invariant. Were this not the case, time reversal would generate a state with negative energy (energy less than that of the vacuum). This means that time-reversal must be an anti-linear operation $\hat{\mathcal{T}}i\hat{\mathcal{O}}\hat{\mathcal{T}}^{-1} = -i\hat{\mathcal{T}}\hat{\mathcal{O}}\hat{\mathcal{T}}^{-1}$. Using this convention, the action of a time-reversal transformation on the operator $\hat{\Gamma}^0 \hat{P}_0$ acting on the standard state of a massive particle should also not result in a sign change in the mass or type of the particle.

Since transverse mass translations are Lorentz-invariant, they should remain unchanged under time reversal. The commutator $[\hat{\Gamma}^0, \hat{P}_0] = i\hat{\mathcal{M}}_T$ defines the behavior of the transverse mass generator under time reversal. For infinitesimal Lorentz transformations, direct substitution of the Lorentz matrices $\mathcal{T}$ and $\mathbf{J}, \mathbf{K}$ from Eq. A.46 in Appendix A yields:

$$\begin{aligned} \hat{\mathcal{T}} \hat{\mathcal{M}}_T \hat{\mathcal{T}}^{-1} &= -\hat{\mathcal{M}}_T, \\ \hat{\mathcal{T}} \hat{H} \hat{\mathcal{T}}^{-1} &= \hat{H}, \\ \hat{\mathcal{T}} \hat{P}_k \hat{\mathcal{T}}^{-1} &= -\hat{P}_k, \\ \hat{\mathcal{T}} \hat{J}_k \hat{\mathcal{T}}^{-1} &= -\hat{J}_k, \\ \hat{\mathcal{T}} \hat{K}_k \hat{\mathcal{T}}^{-1} &= \hat{K}_k. \end{aligned} \tag{4.187}$$

This means that affine translations for massless particles do not change sign under a time-reversal transformation. In addition, the commutators $[\hat{\Gamma}^\mu, \hat{P}_\nu] = i\delta^\mu_\nu \hat{\mathcal{M}}_T$ and $[\hat{\Gamma}^0, \hat{\Gamma}^k] = i\hat{K}_k$ imply:

$$\begin{aligned} \hat{\mathcal{T}} \hat{\Gamma}^0 \hat{\mathcal{T}}^{-1} &= \hat{\Gamma}^0, \\ \hat{\mathcal{T}} \hat{\Gamma}^k \hat{\mathcal{T}}^{-1} &= -\hat{\Gamma}^k. \end{aligned} \tag{4.188}$$

Thus, $\hat{\Gamma}^\beta \hat{P}_\beta$ is an invariant operator under time-reversal transformations.

The action of a time-reversal transformation upon a massive state can be determined by first examining its action upon the standard state $\vec{p}_{(s)}$=$\{$mc,$\underline{0}\}$:

$$\begin{aligned} \hat{J}_z |\psi : \vec{p}_{(s)}, m, J, s_z\rangle &= s_z |\psi : \vec{p}_{(s)}, m, J, s_z\rangle, \\ \hat{\mathcal{T}} |\psi : \vec{p}_{(s)}, m, J, s_z\rangle &\equiv e^{i\delta_{\mathcal{T}}(s_z)} |\psi : \vec{p}_{(s)}, m, J, -s_z\rangle. \end{aligned} \tag{4.189}$$

The angular momentum raising/lowering operators $\hat{J}_\pm = \hat{J}_x \pm i\hat{J}_y$ transform under time reversal as follows:

$$\hat{\mathcal{T}} \hat{J}_\pm \hat{\mathcal{T}}^{-1} = -\hat{J}_\mp. \tag{4.190}$$

Since $\hat{J}_\pm |J, s_z\rangle = \sqrt{J(J+1) - s_z(s_z \pm 1)}\hbar |J, s_z \pm 1\rangle \equiv h_\pm(s_z)\hbar |J, s_z \pm 1\rangle$, action of these operators upon the standard state yields:

$$\begin{aligned} \hat{\mathcal{T}} \hat{J}_\pm |\psi : \vec{p}_{(s)}, m, J, s_z\rangle &= \hat{\mathcal{T}} \hat{J}_\pm \hat{\mathcal{T}}^{-1} \hat{\mathcal{T}} |\psi : \vec{p}_{(s)}, m, J, s_z\rangle \\ e^{i\delta_{\mathcal{T}}(s_z \pm 1)} h_\pm(s_z) |J, -(s_z \pm 1)\rangle &= -e^{i\delta_{\mathcal{T}}(s_z)} h_\mp(-s_z) |J, -s_z \mp 1\rangle. \end{aligned} \tag{4.191}$$

Since $h_\pm(s_z) = h_\mp(-s_z)$, this requires that $\delta_\mathcal{T}(s_z \pm 1) = \delta_\mathcal{T}(s_z) + \pi$. Solving, the phase is given by $\delta_\mathcal{T}(s_z) = \delta_\mathcal{T}(J) + (J - s_z)\pi$. Thus:

$$\hat{\mathcal{T}}|\psi : \vec{p}_{(s)}, m, J, s_z\rangle = (-)^{J-s_z} e^{i\delta_\mathcal{T}(J)} |\psi : \vec{p}_{(s)}, m, J, -s_z\rangle. \tag{4.192}$$

This demonstrates the action of time reversal upon a standard massive state. Since the complementary set of transformations bringing the standard state to states with finite momenta are pure Lorentz boost, and $\hat{\mathcal{T}} i \hat{K}_k \hat{\mathcal{T}}^{-1} = -i\hat{K}_k$, then $\hat{\mathcal{T}}\hat{U}(\mathbf{L}(\underline{p}/mc))\hat{\mathcal{T}}^{-1} = \hat{U}(\mathbf{L}(-\underline{p}/mc))$.

Therefore, the action of a time-reversal transformation upon a general massive state is given by:

$$\begin{aligned}\hat{\mathcal{T}}|\psi : \vec{p}, m, J, s_z\rangle &\equiv (-)^{J-s_z}\, \eta^T_{(\gamma)}\, |\psi : \mathcal{P}\vec{p}, m, J, -s_z\rangle \\ &= (-)^{J-s_z} e^{i\delta_\mathcal{T}(J)} |\psi : \mathcal{P}\vec{p}, m, J, -s_z\rangle,\end{aligned} \tag{4.193}$$

where again, the parity operation satisfies $\mathcal{P} \cdot \{p^0, \underline{p}\} = \{p^0, -\underline{p}\}$. The action of two subsequent time-reversal operations is given by:

$$\begin{aligned}\hat{\mathcal{T}}^2|\psi : \vec{p}, m, J, s_z\rangle &= (-)^{J+s_z} e^{-i\delta_\mathcal{T}(-J)} \hat{\mathcal{T}}|\psi : \mathcal{P}\vec{p}, m, J, -s_z\rangle \\ &= (-)^{2J} |\psi : \vec{p}, m, J, s_z\rangle,\end{aligned} \tag{4.194}$$

where the anti-linearity of time reversal cancels the overall phase factor $\delta_\mathcal{T}(-J)$. It should be noted that the overall action of two time-reversal operations will produce an overall sign change for half-integral spin systems.

To determine the action of a time-reversal transformation upon a massless state, consider the standard state:

$$\begin{aligned}\hat{J}_z|\psi : \{1,0,0,1\}, m_T, J, \lambda\rangle &= \lambda|\psi : \{1,0,0,1\}, m_T, J, \lambda\rangle, \\ \hat{P}_z|\psi : \{1,0,0,1\}, m_T, J, \lambda\rangle &= |\psi : \{1,0,0,1\}, m_T, J, \lambda\rangle, \\ \hat{P}_0|\psi : \{1,0,0,1\}, m_T, J, \lambda\rangle &= -|\psi : \{1,0,0,1\}, m_T, J, \lambda\rangle.\end{aligned} \tag{4.195}$$

This means that:

$$\hat{\mathcal{T}}|\psi : \{1,0,0,1\}, m_T, J, \lambda\rangle \equiv e^{i\tilde{\delta}_\mathcal{T}(\lambda)} |\psi : \{1,0,0,-1\}, m_T, J, -\lambda\rangle. \tag{4.196}$$

This means that time reversal does not change the helicity of the standard state $\hat{W}_0 = \underline{\hat{P}} \cdot \underline{\hat{J}}$. As was done in determining the parity of a massless state, one observes that a rotation by π about the y-axis $\mathbf{R}_y(\pi)$ generates the same change in the standard state parameters as does the time-reversal transformation. Thus:

$$\hat{\mathcal{T}}|\psi : \vec{p}_{(so)}, m_T, J, \lambda\rangle = e^{i\tilde{\delta}_\mathcal{T}(\lambda)} \hat{U}(\mathbf{R}_y(\pi))|\psi : \vec{p}_{(so)}, m_T, J, \lambda\rangle. \tag{4.197}$$

Again using the complementary transformation $\mathbf{C}^{(so)}(\underline{p}) = \mathbf{R}(\underline{u}_p)\mathbf{L}_z(|\underline{p}|)$ to generate the general momentum state, one concludes that:

$$\hat{\mathcal{T}}|\psi : \vec{p}, m_T, J, \lambda\rangle = e^{i\tilde{\delta}_T(\lambda)}\, e^{-i\theta(\underline{u}_p)\lambda}|\psi : \mathcal{P}\vec{p}, m_T, J, \lambda\rangle, \tag{4.198}$$

where again $\mathcal{P} \cdot \{p^0, \underline{p}\} = \{p^0, -\underline{p}\}$, and $\theta(\underline{u}_p)$ is as defined in Eq. 4.183. The action of two subsequent time-reversal operations on massless states is given by:

$$\begin{aligned}\hat{\mathcal{T}}^2|\psi : \vec{p}, m_T, J, \lambda\rangle &= e^{-i\tilde{\delta}_T(\lambda)}\, e^{i\theta(\underline{u}_p)\lambda}\hat{\mathcal{T}}|\psi : \mathcal{P}\vec{p}, m_T, J, \lambda\rangle \\ &= e^{2i\theta(\underline{u}_p)\lambda}|\psi : \vec{p}, m_T, J, \lambda\rangle,\end{aligned} \tag{4.199}$$

since $\theta(-\underline{u}_p) = -\theta(\underline{u}_p)$.

Charge conjugation

Charge conjugation is defined as the operation that changes a particle into its corresponding anti-particle, without any additional change. Expressed as an action upon a quantum state describing particle ψ:

$$\hat{\mathcal{C}}|\psi : \vec{p}, m, J, \kappa\rangle \equiv C_\psi|\bar{\psi} : \vec{p}, m, J, \kappa\rangle. \tag{4.200}$$

Notice that, as defined $\hat{\mathcal{C}}^2 = \hat{1}$, which implies that the charge conjugation factor for particle ψ must satisfy $C_\psi = \pm 1$.

4.4 Particles and the Poincaré group

This section will explore the foundations of relativistic quantum field theory. Particle states will be developed as unitary vector representations of the (extended) Poincaré group using standard developments [73]. The operator product representation of quantum states will introduce the creation and annihilation operators that are so useful for keeping track of the statistics of various particles.

4.4.1 Construction of quantum fields

There are various mechanisms available for keeping track of the quantum statistics of multi-particle systems. In solid state and nuclear physics, Slater determinants have been useful for incorporating the quantum statistics of fermions. However, for relativistic systems, the possibilities of particle creations and annihilations are properties that should also be described. Quantum fields constructed using the operator product representation of quantum states are quite convenient for incorporating both the quantum statistics and relativistic transformation of particles. The quantum statistics of a particle will be described by the commutation or anti-commutation properties of the quantum field that represents that particle.

The particle eigenstates are described in terms of discrete parameters J, κ describing the internal spin/helicity of the particle, and continuous parameters like energy-momentum that describe the unique kinematics of the particle. These particle states are assumed to be eigenstates of the linear operator $\hat{\Gamma}^\mu \hat{P}_\mu$. For brevity in the development, the quantum numbers defining the particle will be described by the single parameter q whenever convenient. A relativistic quantum field in configuration space-time can be built from the creation and annihilation operators associated with that particle. The construction and properties of such quantum fields is the focus of this section.

Multiparticle states

In the operator product representation, quantum states are constructed by action of operators upon an invariant, normalized vacuum, which is devoid of net quantum numbers or independent structure. Particle states are eigenstates of the hermitian number operator represented by $\hat{N}(q) \equiv \hat{a}^\dagger(q)\hat{a}(q)$. The state of a single fermion with quantum number q is represented $|q\rangle = \hat{a}_F^\dagger(q)|vac\rangle$, where the vacuum is annihilated by $\hat{a}_F|vac\rangle = 0$. The incorporation of the Pauli exclusion principle requires that the statistics of fermions be anti-symmetric under pairwise exchange, which is incorporated by requiring the fermionic operators anticommute, $\{\hat{a}_F(q'), \hat{a}_F(q)\} = 0$. The normalization of the states:

$$\langle q'|q\rangle = \delta(q', q) = \langle vac|\hat{a}_F(q')\hat{a}_F^\dagger(q)|vac\rangle = \langle vac|\{\hat{a}_F(q'), \hat{a}_F^\dagger(q)\}|vac\rangle \tag{4.201}$$

then requires that the anti-commutator satisfies:

$$\{\hat{a}_F(q'), \hat{a}_F^\dagger(q)\} = \delta(q', q)\hat{1} = \frac{p^0}{p^0_{(s)}}\delta(\underline{p}' - \underline{p})\delta_{m'_T, m_T}\delta_{\kappa', \kappa}\hat{1}. \tag{4.202}$$

Similarly, the state of a single boson with quantum number q is represented $|q\rangle = \hat{a}_B^\dagger(q)|vac\rangle$. Bose statistics requires that two bosons be symmetric under pairwise exchange, which is incorporated by requiring the bosonic operators commute, $[\hat{a}_B(q'), \hat{a}_B(q)] = 0$. The normalization of the states:

$$\langle q'|q\rangle = \delta(q', q) = \langle vac|\hat{a}_B(q')\hat{a}_B^\dagger(q)|vac\rangle = \langle vac|[\hat{a}_B(q'), \hat{a}_B^\dagger(q)]|vac\rangle \tag{4.203}$$

then requires that the commutator satisfies:

$$\left[\hat{a}_B(q'), \hat{a}_B^\dagger(q)\right] = \delta(q', q)\hat{1}, \tag{4.204}$$

for bosonic operators.

Multiparticle states are constructed by the action of several creation operators upon the vacuum state. For fermions, the anti-commutation relation requires that $\hat{a}_F^\dagger(q)\hat{a}_F^\dagger(q) = 0$, which prevents any two fermions from having an identical set of quantum numbers q. However, this is not the case for bosons. A bosonic system with n_q particles labeled by quantum numbers q is given by:

$$|n_q\rangle = \frac{(\hat{a}_B^\dagger(q))^{n_q}}{\sqrt{n_q!}}|vac\rangle. \tag{4.205}$$

Since for fermions $\hat{a}_F^\dagger(q)\hat{a}_F^\dagger(q) \doteq 0$, a general N-parameter multiparticle state is expressed in operator product notation:

$$|n_{q_1}, \ldots, n_{q_N}\rangle = \prod_{s=1}^{N} \frac{(\hat{a}^\dagger(q_s))^{n_{q_s}}}{\sqrt{n_{q_s}!}}|vac\rangle. \tag{4.206}$$

The state is normalized appropriate to the particles' quantum statistics:

$$\langle n_{q'_1}, \ldots, n_{q'_N}|n_{q_1}, \ldots, n_{q_N}\rangle = \sum_{\mathcal{P}_a} (\mp)^{\mathcal{P}_a} \delta_{n_{q_{a1}} n_{q_1}} \delta(q_{a1}, q_1) \ldots \delta_{n_{q_{aN}} n_{q_N}} \delta(q_{aN}, q_N), \tag{4.207}$$

where the $\mp$ refers to fermions/bosons, and $\mathcal{P}_a$ is the number of permutations that brings the labels $\{a1, a2, \ldots, aN\}$ to standard order $\{1, 2, \ldots, N\}$.

Action by a creation operator with a new set of quantum numbers upon a multiparticle state defines a state with the new quantum numbers appended to the far-right of the prior parameters. Thus, creation and annihilation operators modify the state as follows:

$$\hat{a}^\dagger(q_{N+1})|n_{q_1}, \ldots, n_{q_N}\rangle = |n_{q_1}, \ldots, n_{q_N}, 1_{q_{N+1}}\rangle, \tag{4.208}$$

$$\hat{a}^\dagger(q_s)|n_{q_1}, \ldots, n_{q_N}\rangle = \left(\frac{1 \mp 1}{2}\right) \sqrt{n_{q_s}+1}|n_{q_1}, \ldots, n_{q_{s-1}}, n_{q_s}+1, n_{q_{s+1}}, \ldots, n_{q_N}\rangle, \tag{4.209}$$

$$\hat{a}(q_s)|n_{q_1}, \ldots, n_{q_N}\rangle = (\mp)^{N-s} \sqrt{n_{q_s}}|n_{q_1}, \ldots, n_{q_{s-1}}, n_{q_s}-1, n_{q_{s+1}}, \ldots, n_{q_N}\rangle. \tag{4.210}$$

In what follows, the properties of the single particle operators will be developed.

Transformation properties

The transformation properties of the creation and annihilation operators are directly obtained from the action of the unitary form of the Poincaré operator upon the quantum state vectors. First, the vacuum state is presumed to be invariant under Poincaré transformations $\hat{U}(\mathbf{\Lambda}, \vec{a})|vac\rangle = |vac\rangle$. From Eq. 4.111 the creation/annihilation operators satisfy:

$$\hat{U}(\mathbf{\Lambda}, \vec{x})\hat{a}^\dagger(p, \kappa)\hat{U}^\dagger(\mathbf{\Lambda}, \vec{x}) = e^{-i\vec{x}\cdot\mathbf{\Lambda}\vec{p}} \textstyle\sum_{\kappa'} \hat{a}^\dagger(\mathbf{\Lambda}p, \kappa')Q^{(s)}_{\kappa'\kappa}(\mathbf{\Lambda}, p), \quad (4.211)$$

$$\hat{U}(\mathbf{\Lambda}, \vec{x})\hat{a}(p, \kappa)\hat{U}^\dagger(\mathbf{\Lambda}, \vec{x}) = e^{+i\vec{x}\cdot\mathbf{\Lambda}\vec{p}} \textstyle\sum_{\kappa'} \hat{a}(\mathbf{\Lambda}p, \kappa')Q^{(s)*}_{\kappa'\kappa}(\mathbf{\Lambda}, p), \quad (4.212)$$

where $Q^{(s)}_{\kappa'\kappa}(\mathbf{\Lambda}, p) \Rightarrow D^{(J)}_{s_z' s_z}(R_W(\mathbf{L}, \vec{p}))$ for massive particles, and $Q^{(s)}_{\kappa'\kappa}(\mathbf{\Lambda}, p) \Rightarrow e^{-\lambda\phi(\mathbf{L},\vec{p})}\delta_{\lambda'\lambda}$ for massless particles. These relationships define the transformation properties of the momentum space creation and annihilation operators defining particle states.

Quantum fields

Next, configuration space representations of the prior particle operators, defined as *quantum fields*, will be developed. The form of the asymptotic photon field was developed in Eq. 2.48. Motivated by this prior discussion, a general spinor field $\hat{\phi}_b(\vec{x})$ will be defined to transform according to:

$$\hat{U}(\mathbf{\Lambda}, \vec{a})\hat{\phi}_b(\vec{x})\hat{U}^\dagger(\mathbf{\Lambda}, \vec{a}) = \sum_{b'} \mathcal{D}_{bb'}(\mathbf{\Lambda}^{-1})\hat{\phi}_{b'}(\mathbf{\Lambda}\vec{x} + \vec{a}). \quad (4.213)$$

This formula is analogous to the transformation of standard three-vectors under rotations and translations of the axes. The matrices, $\mathcal{D}$, need not be unitary, but they must form a finite-dimensional representation of the homogeneous Lorentz group, $\mathcal{D}(\mathbf{\Lambda}_2^{-1})\mathcal{D}(\mathbf{\Lambda}_1^{-1}) = \mathcal{D}(\mathbf{\Lambda}_2^{-1}\mathbf{\Lambda}_1^{-1})$. Since the creation and annihilation operators completely span the Hilbert space of the particle, the quantum field can be expanded in terms of these operators. In order to account for the correct behavior of the fields under translations, the expansion must be of the form:

$$\hat{\phi}_b(\vec{x}) = \sum_{\kappa} \int \frac{p^0_{(s)}}{p^0} d^3p\, \eta(p)\, e^{i\vec{x}\cdot\vec{p}}\mathbf{u}_b(\vec{p}, \kappa)\hat{a}(\vec{p}, \kappa), \quad (4.214)$$

since automatically $\hat{U}(\mathbf{1}, \vec{a})\hat{\phi}_b(\vec{x})\hat{U}^\dagger(\mathbf{1}, \vec{a}) = \hat{\phi}_b(\vec{x} + \vec{a})$, and the normalization factor $\eta(p)$ must be consistent with Eq. 4.213.

Directly substituting Eq. 4.214 into Eq. 4.213, gives the transformation properties of the spinors:

$$\mathbf{u}_b(\mathbf{\Lambda}\vec{p}, \kappa) = \left(\frac{\eta(\vec{p})}{\eta(\mathbf{\Lambda}\vec{p})}\right) \sum_{b'} \sum_{\kappa'} \mathcal{D}_{bb'}(\mathbf{\Lambda})\mathbf{u}_b(\vec{p}, \kappa')Q^{(s)-1}_{\kappa'\kappa}(\mathbf{\Lambda}, \vec{p}), \quad (4.215)$$

where unitarity of the little group representations $Q^{(s)*}_{\kappa'\kappa} = Q^{(s)-1}_{\kappa\kappa'}$ has been utilized. In particular, the standard state spinor defines the general spinor form from the representation of the complementary transformation giving the momentum:

$$\mathbf{u}_b(\vec{p}, \kappa) = \sum_{b'} \mathcal{D}_{bb'}(\mathbf{C}^{(s)}(\vec{p}))\mathbf{u}_b(\vec{p}_{(s)}, \kappa), \quad (4.216)$$

where the normalization factor cannot depend upon momentum if the spinor normalization is to be preserved. Also, the standard state spinor must properly transform under any element of the little group of transformations $\mathcal{L}$ that leaves the standard momentum invariant, giving:

$$\mathbf{u}_b(\vec{p}_{(s)}, \kappa) = \sum_{b'} \sum_{\kappa'} \mathcal{D}_{bb'}(\mathcal{L}) \mathbf{u}_{b'}(\vec{p}_{(s)}, \kappa') Q^{-1}_{\kappa'\kappa}(\mathcal{L}, \vec{p}_{(s)}). \tag{4.217}$$

4.4.2 Non-relativistic correspondence

As previously mentioned, the normalizations of momentum and configuration quantum states will be chosen to have close correspondence with non-relativistic quantum states, given in Eqs. 2.18 and 2.19. For non-relativistic systems, the momentum and configuration space-state vectors are related via:

$$\langle \mathbf{x} | \mathbf{p}; m \rangle = \frac{e^{\frac{i}{\hbar}\mathbf{p}\cdot\mathbf{x}}}{(2\pi\hbar)^{3/2}}. \tag{4.218}$$

This means that a configuration space-state vector is related to the momentum eigenstate via $|\mathbf{x}\rangle = \int d^3p \, \frac{e^{-\frac{i}{\hbar}\mathbf{p}\cdot\mathbf{x}}}{(2\pi\hbar)^{3/2}} |\mathbf{p}\rangle$. Generalizing to the relativistic normalization, this becomes:

$$|\mathbf{x}\rangle = \int \frac{p^0_{(s)} d^3p}{\epsilon(\mathbf{p})} \frac{e^{-\frac{i}{\hbar}\mathbf{p}\cdot\mathbf{x}}}{(2\pi\hbar)^{3/2}} |\vec{p}, m\rangle, \tag{4.219}$$

where the standard energy is mc^2 for massive particles, $1 \cdot c$ for massless particles, and $\epsilon(\mathbf{p}) = c\sqrt{|\mathbf{p}|^2 + (mc)^2}$.

One can therefore construct a configuration space-time quantum state of the form:

$$|\vec{x}\rangle = \left[\int \frac{p^0_{(s)} c d^3p}{\epsilon(\mathbf{p})} \frac{e^{-\frac{i}{\hbar}(\mathbf{p}\cdot\mathbf{x} - \epsilon(\mathbf{p})t)}}{(2\pi\hbar)^{3/2}} \hat{a}^\dagger(\vec{p}) \right] |vac\rangle \equiv \hat{\phi}^\dagger(\vec{x}) |vac\rangle. \tag{4.220}$$

The operator $\hat{\phi}(\vec{x})$ is a solution to the relativistic wave equation for a scalar system of mass m, $\hbar^2(\nabla^2 - \frac{1}{c^2}\frac{\partial^2}{\partial t^2})\hat{\phi}(\vec{x}) = (mc)^2 \hat{\phi}(\vec{x})$, and is defined to be the *quantum field* describing a single non-interacting scalar particle. General quantum fields will be such configuration space operators that are solutions to the equations of motion describing specific particle types.

The quantum field $\hat{\phi}(\vec{x})$ will satisfy the same quantum statistics that defines the system as incorporated in the particles creation and annihilation operators $\hat{a}(\vec{p}), \hat{a}^\dagger(\vec{p})$. The overlap amplitude between the configuration states at differing space-time points is referred to as the *propagator* Δ_m, given by:

$$\Delta_m(\vec{x}-\vec{y}) \equiv \langle \vec{x}|\vec{y}\rangle = \langle vac|\hat{\phi}(\vec{x})\hat{\phi}^\dagger(\vec{y})|vac\rangle$$
$$= \int \frac{p^0_{(s)}c\, d^3p}{\epsilon(\mathbf{p})} \int \frac{q^0_{(s)}c\, d^3q}{\epsilon(\mathbf{q})} \frac{e^{\frac{i}{\hbar}(\mathbf{p}\cdot\mathbf{x}-\epsilon(\mathbf{p})t)}}{(2\pi\hbar)^{3/2}} \frac{e^{-\frac{i}{\hbar}(\mathbf{q}\cdot\mathbf{y}-\epsilon(\mathbf{q})t)}}{(2\pi\hbar)^{3/2}} \langle vac|\hat{a}(\vec{p})\hat{a}^\dagger(\vec{q})|vac\rangle. \tag{4.221}$$

The operator product representation quite conveniently evaluates vacuum expectation values by commuting/anti-commuting annihilation operators to the right until they act upon the vacuum and vanish:

$$\langle vac|\hat{a}(\vec{p})\hat{a}^\dagger(\vec{q})|vac\rangle = \langle vac|[\hat{a}(\vec{p}),\hat{a}^\dagger(\vec{q})]_\pm \mp \hat{a}^\dagger(\vec{q})\hat{a}(\vec{p})|vac\rangle = \frac{\epsilon(\mathbf{p})}{p^0_{(s)}c}\delta^3(\mathbf{p}-\mathbf{q}). \tag{4.222}$$

Therefore, the propagator for the scalar field $\hat{\phi}(\vec{x})$ takes the form:

$$\Delta_m(\vec{x}-\vec{y}) = \frac{1}{(2\pi\hbar)^3}\int \frac{p^0_{(s)}c\, d^3p}{\epsilon(\mathbf{p})} e^{\frac{i}{\hbar}\vec{p}\cdot(\vec{x}-\vec{y})}, \tag{4.223}$$

where in the exponential $p^0 = \epsilon(\mathbf{p})/c$. Since the integrand contains the Lorentz invariant measure, the propagator $\Delta_m(\vec{x}-\vec{y})$ is a Lorentz invariant, depending only upon the invariant interval between the space-time points. It should be noted that in the large mass limit,

$$\Delta_m(\vec{x}-\vec{y})|_{x^0=y^0} \overset{m\to\infty}{\Rightarrow} \delta^3(\mathbf{x}-\mathbf{y}). \tag{4.224}$$

Thus, the equal-time scalar propagator satisfies the expected non-relativistic normalization.

For a general (normalized) spinor, a spinor-valued quantum field is given by:

$$\hat{\mathbf{\Phi}}^{(m,\Gamma,J)}_{(\gamma)}(\vec{x}) \equiv \sum_\kappa \int \frac{p^0_{(s)}c\, d^3p}{\epsilon(\mathbf{p})} \frac{e^{\frac{i}{\hbar}(\mathbf{p}\cdot\mathbf{x}-\epsilon(\mathbf{p})t)}}{(2\pi\hbar)^{3/2}} \mathbf{u}^{(\Gamma)}_{(\gamma)}(\vec{p},m,J,\kappa)\hat{a}^{(\Gamma)}_{(\gamma)}(\vec{p},m,J,\kappa), \tag{4.225}$$

where γ represents the eigenstate of $\hat{\Gamma}^0$ in the standard state, J represents the Casimir eigenvalue for the spin, Γ represents the Casimir eigenvalue for the extended Lorentz group, and m represents the Casimir eigenvalue for the extended Poincaré group. Massless states will be specified by vanishing mass, but general transverse mass m_T. For a scalar field, the spinor form is given by $\mathbf{u}^{(0)}_{(0)}(\vec{p},m,0,0)=1$. Explicitly, the spinor-valued component of the field $\hat{\Phi}^{(\Gamma)}_{(\gamma)a}$ involves the corresponding spinor component $u^{(\Gamma)}_{(\gamma)a}$. Defining the matrix-valued function:

$$\left(\mathbf{\Pi}^{(\Gamma)}_{(\gamma)}(\vec{p},m,J)\right)_{ab} \equiv \sum_\kappa u^{(\Gamma)}_{(\gamma)a}(\vec{p},m,J,\kappa)\bar{u}^{(\Gamma)}_{(\gamma)b}(\vec{p},m,J,\kappa), \tag{4.226}$$

the spinor anti-commutator/commutator satisfies:

$$\left[\hat{\Phi}^{(\Gamma)}_{(\gamma)a}(\vec{x}), \hat{\bar{\Phi}}^{(\Gamma)}_{(\gamma)b}(\vec{y})\right]_{\pm} = \left(\mathbf{\Pi}^{(\Gamma)}_{(\gamma)}(\frac{\hbar}{i}\frac{\partial}{\partial(\vec{x}-\vec{y})}, m, J)\right)_{ab} \Delta_m(\vec{x}-\vec{y}). \qquad (4.227)$$

One should note that for $\gamma \neq 0$, the normalization requires multiplication of the hermitian adjoint field by the spinor metric **g**. The causal properties of the commutation relations will be explored next.

4.4.3 Causality and quantum statistics

Communications generally involve parameters propagating conserved physical quantities. Since observables should be constructed out of bi-linear pairs of the field operators (such as the number operator $\hat{N} = \hat{a}^\dagger \hat{a}$ or $\hat{\psi}^\dagger(x)\hat{\psi}(x)$), the commutators describing the group dynamics of a quantum system can ultimately be related to commutation or anti-commutation relationships between the field operators.

$$[\hat{\mathcal{O}}, \hat{\psi}^\dagger \hat{\psi}] = [\hat{\mathcal{O}}, \hat{\psi}^\dagger]_\pm \, \hat{\psi} \mp \hat{\psi}^\dagger \, [\hat{\mathcal{O}}, \hat{\psi}]_\pm, \qquad (4.228)$$

where the $[A, B]_\pm$ refers to an anti-commutator/commutator between the operators. The concept of microscopic causality as requiring the vanishing of the commutator or anti-commutator of the field associated with related indistinguishable particles outside of the light cone (i.e., for space-like separations) will be developed. Such a restriction disallows any communication between observables constructed from those fields. The group structure of the Poincaré algebra will be shown (in both a pedestrian and formal manner) to connect bosonic behaviors to integral spin systems, and fermionic behaviors to half-integral spin systems. The group structure and symmetry properties of the field representations will likewise be shown to determine the form of the field equation satisfied by a particular representation, as well as higher spin product representations.

Supplement: Spin and statistics, in brief

One can gain insight into the connection of spin with statistics under interchange by straightforward physical demonstrations. Examine the following figure representing the two indistinguishable systems as arrows. In the figure, two vertically oriented systems are connected via a strap with fixed attachments

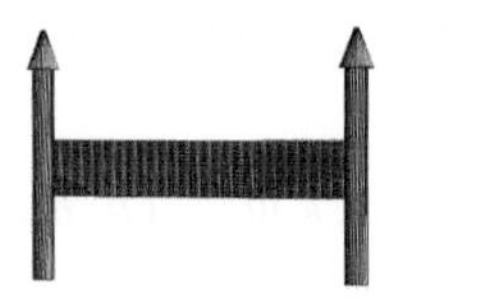 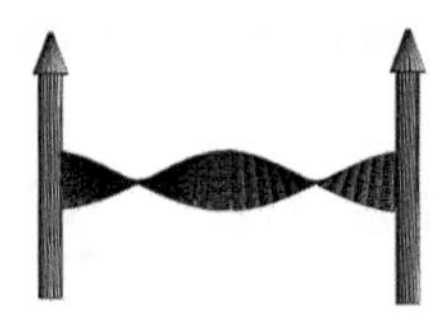

on the sides of each system. The systems are exchanged in spatial location without undergoing a rotation themselves. Such an exchange (which can be demonstrated with a belt) necessarily places a twist of 2π on the strap (the twist can be avoided only if each system undergoes a rotation of π during the exchange, which will yield the same result as that discussed here). The attached system returns to its initial un-twisted state only if one of the components undergoes a rotation of 2π to remove the twist.

As demonstrated in Appendix A.1.3 rotations for spin-$\frac{1}{2}$ systems satisfy the transformations for the covering group SU(2) given by the 2×2 matrix:

$$\mathbf{S}(\underline{\theta}) = \mathbf{1}\cos\left(\frac{\theta}{2}\right) + i\hat{\theta}\cdot\underline{\sigma}\sin\left(\frac{\theta}{2}\right),$$

while the corresponding rotations for spin-1 systems are parameterized by rotation matrices of the orthogonal group O(3) given by:

$$R_{jk} = \frac{1}{2}Tr(\underline{\sigma}_j\mathbf{S}^{-1}\underline{\sigma}_k\mathbf{S}), \tag{4.229}$$

where $\mathbf{S}$ is the corresponding matrix transformation for SU(2) given above. For a rotation of 2π, a spin-$\frac{1}{2}$ system is seen to undergo a sign change, $\mathbf{S}(2\pi\hat{\theta}) = -\mathbf{1}$, while a spin-1 system does not, $R_{jk} = \delta_{jk}$ (because the sign gets squared in the transformation). Thus, spin-$\frac{1}{2}$ systems are anti-symmetric under pair-wise exchange, while spin-1 systems are symmetric under pair-wise exchanges.

Construction of a causal scalar field

The form of the propagator in Eq. 4.223 can be directly evaluated for space-like separations. Since $\Delta_m(\vec{X})$ is a Lorentz invariant, it can be evaluated in the frame of reference where the spatial components of the separation represents the proper distance between the events, $\vec{X} = \{0, \mathbf{X}^*\}$, where $|\mathbf{X}^*| \equiv \sqrt{\vec{X}\cdot\vec{X}}$. Expressing the momentum in terms of the dimensionless four-velocity $\mathbf{p} = mc\mathbf{u}$, the propagator for massive particles takes the form:

$$\begin{aligned}\Delta_m(\{0, \mathbf{X}^*\}) &= \frac{\hbar(mc)^2}{2\pi^2|\mathbf{X}^*|}\int_0^\infty \frac{u\,du}{\sqrt{1+u^2}}\sin\left(\frac{mc|\mathbf{X}^*|}{\hbar}u\right)\\ &= \frac{(mc)^2}{2\pi^2\hbar^2|\mathbf{X}^*|}K_1\left(\frac{mc|\mathbf{X}^*|}{\hbar}\right),\end{aligned} \tag{4.230}$$

where $K_1(\zeta)$ is the first-order modified Bessel function of the second kind. Since the propagator is clearly even in $|\mathbf{X}^*|$, it *does not* vanish at space-like separations, as required for microscopic causality. At distances much less than the Compton

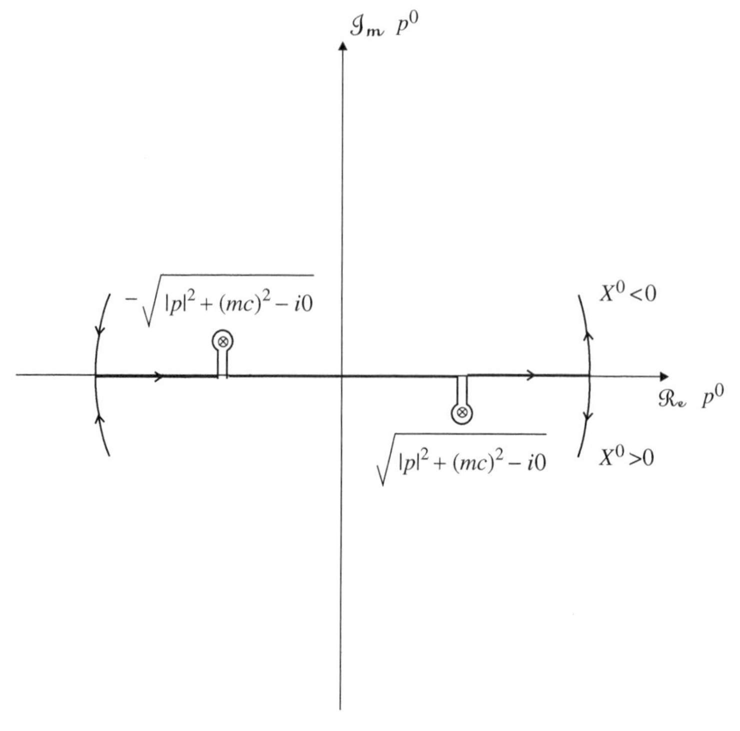

Figure 4.4 Contour integrals for causal propagator.

wavelength of the particle, $\Delta_m(\{0, \mathbf{X}^*\})$ takes the form:

$$\Delta_m(\{0, \mathbf{X}^*\}) \overset{\frac{mc}{\hbar} \ll \frac{1}{|\mathbf{X}^*|}}{\Rightarrow} \frac{1}{2\pi^2|\mathbf{X}^*|^2}. \tag{4.231}$$

This clearly demonstrates that coherent quantum systems manifest non-vanishing correlations over space-like separations.

Prior to constructing a causal scalar field, it is useful to develop a causal propagator that serves as a Green's function for the scalar wave equation. The integrand in Eq. 4.223 can be expressed as an integral over all values of energy p^0c, as long as the time signature is taken to appropriately project positive energies. Consider the following integral:

$$\mathcal{I}_m \equiv \int_{-\infty}^{\infty} dp^0 \frac{e^{-\frac{i}{\hbar}p^0 X^0}}{(p^0)^2 - |\mathbf{p}|^2 - (mc)^2 + i0^+} \tag{4.232}$$

evaluated using the contour displayed in Figure 4.4. If the time interval X^0 is negative, a contour (at infinity) that is closed upwards in the complex plane (Im $p^0 > 0$) will converge to the desired integral $\mathcal{I}_m$, while if the time inteval X^0 is positive, the contour must be closed in the lower half plane (Im $p^0 < 0$) to converge to the answer. The resultant residues at the respective poles give:

$$\mathcal{I}_m = 2\pi i \left[\frac{e^{-\frac{i}{\hbar}\sqrt{|\mathbf{p}|^2+(mc)^2}\, X^0}}{2\sqrt{|\mathbf{p}|^2+(mc)^2}} \Theta(X^0) + \frac{e^{+\frac{i}{\hbar}\sqrt{|\mathbf{p}|^2+(mc)^2}\, X^0}}{2\sqrt{|\mathbf{p}|^2+(mc)^2}} \Theta(-X^0) \right]. \tag{4.233}$$

The factors in the denominators are just what is needed for the Lorentz-invariant measure in Eq. 4.233. Therefore, the causal scalar propagator can be expressed in a manifestly covariant form given by:

$$\Delta_m(\vec{X}) = -\frac{1}{\pi i} \frac{p^0_{(s)}}{(2\pi\hbar)^3} \int d^4 p \, \frac{e^{\frac{i}{\hbar}\vec{p}\cdot\vec{X}}}{\vec{p}\cdot\vec{p} + (mc)^2 - i0^+}, \tag{4.234}$$

where the 0^+ implies a vanishing limiting value for a positive real parameter.

One can examine the action of the Klein–Gordon wave operator upon Δ_m. Using the configuration space representation of the momentum operators $\hat{p}_\beta = \frac{\hbar}{i}\frac{\partial}{\partial x^\beta}$, the Lorentz invariant operator satisfies:

$$\begin{aligned}
[\hat{\vec{p}} \cdot \hat{\vec{p}} + (mc)^2]\Delta_m(\vec{X}) &= [-\hbar^2(\nabla^2 - \tfrac{\partial^2}{\partial X^{0^2}}) + (mc)^2]\Delta_m(\vec{X}) \\
&= -\tfrac{1}{\pi i}\tfrac{p^0_{(s)}}{(2\pi\hbar)^3} \int d^4 p \, e^{\frac{i}{\hbar}\vec{p}\cdot\vec{X}} = -\tfrac{\hbar}{i}\, 2p^0_{(s)}\delta^4(\vec{X}).
\end{aligned} \tag{4.235}$$

Therefore, up to dimensional and normalization factors, this propagator is just the Green's function for the Klein–Gordon equation.

As previously noted, the form of the propagator for space-like separations in Eq. 4.230 is even in the proper separation between the points $\sqrt{(\vec{x}-\vec{y})\cdot(\vec{x}-\vec{y})}$. This means that it might be possible to construct a causal quantum field consisting of an annihilation component for the particle, and a canceling creation component for the associated anti-particle, as long as the coefficients are properly chosen. Consider the following field:

$$\hat{\boldsymbol{\Psi}}(\vec{x}) \equiv \int \frac{mc^2\, d^3p}{\epsilon(\mathbf{p})} \left[\zeta_+ \frac{e^{\frac{i}{\hbar}(\mathbf{p}\cdot\mathbf{x} - \epsilon(\mathbf{p})t)}}{(2\pi\hbar)^{3/2}} \hat{a}(\vec{p}, m) + \zeta_- \frac{e^{-\frac{i}{\hbar}(\mathbf{p}\cdot\mathbf{x} - \epsilon(\mathbf{p})t)}}{(2\pi\hbar)^{3/2}} \hat{a}^\dagger(\vec{p}, m) \right]. \tag{4.236}$$

The anti-commutator/commutator of this field with its adjoint satisfies:

$$[\hat{\boldsymbol{\Psi}}(\vec{x}), \hat{\boldsymbol{\Psi}}^\dagger(\vec{y})]_\pm = |\zeta_+|^2 \Delta_m(\vec{x}-\vec{y}) \pm |\zeta_-|^2 \Delta_m(\vec{y}-\vec{x}), \tag{4.237}$$

while the anti-commutator/commutator of this field with itself satisfies:

$$[\hat{\boldsymbol{\Psi}}(\vec{x}), \hat{\boldsymbol{\Psi}}(\vec{y})]_\pm = \zeta_+\zeta_-(1 \pm 1)\Delta_m(\vec{x}-\vec{y}). \tag{4.238}$$

It is clear from Eq. 4.238 that such a field only self-commutes at all points for a field that satisfies Bose statistics, corresponding to the lower signs (commutators, rather than anti-commutators). Using commutation relations in Eq. 4.237, the scalar field

($J = 0$) indeed commutes outside of the light cone (since $\Delta_m(\vec{x} - \vec{y}) = \Delta_m(\vec{y} - \vec{x})$), satisfying expected microscopic causality, as long as $|\zeta_+| = |\zeta_-|$. Therefore, the spin-zero causal field must satisfy Bose–Einstein statistics. Non-relativistic correspondence of normalization is then maintained as long as $|\zeta_+| = |\zeta_-| = \frac{1}{\sqrt{2}}$.

Construction of general, massive causal spinor fields

The previous result constructing a scalar causal field motivates an examination of the form:

$$\hat{\mathbf{\Psi}}^{(\Gamma)}_{(\gamma)}(\vec{x}) \equiv \sum_{s_z} \int \frac{mc^2\, d^3p}{\epsilon(\mathbf{p})} \left[\zeta^{(\Gamma)}_{(\gamma)}(J) \frac{e^{\frac{i}{\hbar}(\mathbf{p}\cdot\mathbf{x} - \epsilon(\mathbf{p})t)}}{(2\pi\hbar)^{3/2}} \mathbf{u}^{(\Gamma)}_{(\gamma)}(\vec{p}, m, J, s_z) \hat{a}^{(\Gamma)}_{(\gamma)}(\vec{p}, m, J, s_z) \right.$$

$$\left. + \zeta^{(\Gamma)}_{(-\gamma)}(J) \frac{e^{-\frac{i}{\hbar}(\mathbf{p}\cdot\mathbf{x} - \epsilon(\mathbf{p})t)}}{(2\pi\hbar)^{3/2}} \mathbf{v}^{(\Gamma)}_{(\gamma)}(\vec{p}, m, J, s_z) \hat{a}^{(\Gamma)\dagger}_{(-\gamma)}(\vec{p}, m, J, s_z) \right]. \quad (4.239)$$

for a spinor field. The field should be causal, satisfy expected Lorentz transformation properties, satisfy the equations of motion for linear spinor fields, and have well-defined behaviors under parity, time reversal, and charge conjugation.

Under Lorentz transformations, the spinor field is expected to satisfy:

$$\hat{U}(\mathbf{\Lambda}, \vec{b}) \hat{\mathbf{\Psi}}^{(\Gamma)}_{(\gamma)}(\vec{x}) \hat{U}^{-1}(\mathbf{\Lambda}, \vec{b}) = \mathbf{D}^{(\Gamma)}(\mathbf{\Lambda}^{-1}) \cdot \hat{\mathbf{\Psi}}^{(\Gamma)}_{(\gamma)}(\mathbf{\Lambda}\vec{x} + \vec{b}), \quad (4.240)$$

where the spinor components are understood for the bold-faced vector/matrix forms $(\mathbf{D} \cdot \mathbf{\Psi})_b \equiv \sum_d \mathbf{D}_{bd} \Psi_d$. The transformation of creation and annihilation operators are given in Eqs. 4.211 and 4.212. Substitution of these forms into Eq. 4.240, and noting that $\vec{p} \cdot \vec{x} + \mathbf{\Lambda}\vec{p} \cdot \vec{b} = \mathbf{\Lambda}\vec{p} \cdot (\mathbf{\Lambda}\vec{x} + \vec{b})$, the tranformation properties relating the spinor coefficients are given by:

$$\sum_{s'_z} D^{(J)*}_{s_z s'_z}(R_W(\mathbf{\Lambda}, \vec{p})) u^{(\Gamma)}_{(\gamma)b}(\vec{p}, J, s'_z) = \sum_d \mathbf{D}^{(\Gamma)}_{bd}(\mathbf{\Lambda}^{-1}) u^{(\Gamma)}_{(\gamma)d}(\mathbf{\Lambda}\vec{p}, J, s_z), \quad (4.241)$$

$$\sum_{s'_z} D^{(J)}_{s_z s'_z}(R_W(\mathbf{\Lambda}, \vec{p})) v^{(\Gamma)}_{(\gamma)b}(\vec{p}, J, s'_z) = \sum_d \mathbf{D}^{(\Gamma)}_{bd}(\mathbf{\Lambda}^{-1}) v^{(\Gamma)}_{(\gamma)d}(\mathbf{\Lambda}\vec{p}, J, s_z). \quad (4.242)$$

These relationships constrain the form of the spinor coefficients $\mathbf{v}^{(\Gamma)}_{(\gamma)}$. Define the little group matrix representations of the angular momentum operators as having matrix elements $(\mathcal{J}^{(J)}_k)_{s_z s'_z}$. Then, for pure rotations on the standard massive state, Eq. 4.242 implies that:

$$\sum_{s'_z} (\mathcal{J}^{(J)*}_k)_{s_z s'_z} \mathbf{u}^{(\Gamma)}_{(\gamma)}(\vec{p}_{(s)}, J, s'_z) = \mathbf{J}^{(\Gamma)}_k \cdot \mathbf{u}^{(\Gamma)}_{(\gamma)}(\vec{p}_{(s)}, J, s_z), \quad (4.243)$$

$$-\sum_{s'_z} (\mathcal{J}^{(J)}_k)_{s_z s'_z} \mathbf{v}^{(\Gamma)}_{(\gamma)}(\vec{p}_{(s)}, J, s'_z) = \mathbf{J}^{(\Gamma)}_k \cdot \mathbf{v}^{(\Gamma)}_{(\gamma)}(\vec{p}_{(s)}, J, s_z). \quad (4.244)$$

The form of the angular momentum components for the little group of rotations is well-known:

$$(\mathcal{J}_z^{(J)})_{s_z s_z'} = s_z' \delta_{s_z, s_z'}, \; (\mathcal{J}_x^{(J)} \pm i \mathcal{J}_y^{(J)})_{s_z s_z'} = \sqrt{(J \mp s_z')(J \pm s_z' + 1)} \delta_{s_z, s_z' \pm 1}. \quad (4.245)$$

One can directly verify these relations imply that the angular momentum matrices satisfy:

$$-(\mathcal{J}_k^{(J)*})_{s_z, s_z'} = (-)^{s_z - s_z'} (\mathcal{J}_k^{(J)})_{-s_z, -s_z'} \quad (4.246)$$

Combining Eqs. 4.244 and 4.246, the spinor $\mathbf{v}_{(\gamma)}^{(\Gamma)}(\vec{p}_{(s)}, J, s_z)$ transforms under rotations like $\mathbf{v}_{(\gamma)}^{(\Gamma)}(\vec{p}_{(s)}, J, s_z) \propto (-)^{J+s_z} \mathbf{u}_{(\pm\gamma)}^{(\Gamma)}(\vec{p}_{(s)}, J, -s_z)$. Substitution of this relationship into Eq. 4.242 gives the general form of the spinor $\mathbf{v}_{(\gamma)}^{(\Gamma)}$ consistent with Lorentz transformation properties:

$$\mathbf{v}_{(\gamma)}^{(\Gamma)}(\vec{p}, J, s_z) \propto (-)^{J+s_z} \mathbf{u}_{(\pm\gamma)}^{(\Gamma)}(\vec{p}, J, -s_z). \quad (4.247)$$

Next, consider whether the spinor field satisfies the needed linear spinor field equation:

$$\mathbf{\Gamma}^{\beta} \cdot \frac{\hbar}{i} \frac{\partial}{\partial x^{\beta}} \hat{\mathbf{\Psi}}_{(\gamma)}^{(\Gamma)}(\vec{x}) = -(\gamma) mc \hat{\mathbf{\Psi}}_{(\gamma)}^{(\Gamma)}(\vec{x}). \quad (4.248)$$

(Note, the γ label for the eigenvalue of $\hat{\Gamma}^0$ should not be confused with the Lorentz factor $\frac{1}{\sqrt{1-\beta^2}}$ describing relativistic kinematics). From the spinor relations $\mathbf{\Gamma}^{\beta} p_{\beta} \mathbf{u}_{(\gamma)}^{(\Gamma)}(\vec{p}_{(s)}, J, s_z) = -(\gamma) mc \mathbf{u}_{(\gamma)}^{(\Gamma)}(\vec{p}_{(s)}, J, s_z)$, one concludes from Eq. 4.247 that the spinor $\mathbf{v}_{(\gamma)}^{(\Gamma)}$ takes the form:

$$\mathbf{v}_{(\gamma)}^{(\Gamma)}(\vec{p}, J, s_z) = (-)^{J+s_z} \mathbf{u}_{(-\gamma)}^{(\Gamma)}(\vec{p}, J, -s_z). \quad (4.249)$$

With this substitution, the general spinor field (4.239) satisfies both the spinor field equation and the expected Lorentz transformation properties. As previously mentioned, Eq. 4.248 has only positive mass solutions. There is no need to introduce any *Dirac sea* of filled negative energy states. The signature of the eigenvalue of the linear field operator $\mathbf{\Gamma}^{\beta} \cdot \frac{\hbar}{i} \frac{\partial}{\partial x^{\beta}}$ is given by the sign of the eigenvalue of $\hat{\Gamma}^0$ on the standard state.

Thus, linear spinor fields satisfy linear dispersion relations, allowing direct construction of unitary, cluster decomposable scattering amplitudes for arbitrary spins, without a need for perturbative renormalizability. The linear differential operator takes the mass off-shell in an elegant manner, maintaining the Lorentz frame of reference, as well as energy-momentum relationships. This property, along with the generation of the Minkowski metric form directly from the group structure, makes these fields quite convenient in geometric approaches to quantum scattering.

The causal properties of the spinor field will be explored next. The spinor propagator of the form $[\hat{\mathbf{\Psi}}^{(\Gamma)}_{(\gamma)}(\vec{x}), \hat{\bar{\mathbf{\Psi}}}^{(\Gamma)}_{(\gamma')}(\vec{y})]_\pm$ is expected to have well-defined Lorentz transformation properties, where the bar includes the spinor metric $\mathbf{g}^{(\Gamma)}$ as defined in Eq. 4.144. The anti-commutator/commutator will involve outer products of the spinors as demonstrated in Eq. 4.227. These outer products are related to projection operators:

$$\mathbf{\Pi}^{(\Gamma)}_{(\gamma)}(\vec{p}, m, J) \equiv \sum_{s_z} \mathbf{u}^{(\Gamma)}_{(\gamma)}(\vec{p}, m, J, s_z)\bar{\mathbf{u}}^{(\Gamma)}_{(\gamma)}(\vec{p}, m, J, s_z) = (-)^{\Gamma-\gamma}\tilde{\mathbf{\Pi}}^{(\Gamma)}_{(\gamma)}(\vec{p}, m, J), \tag{4.250}$$

where $(\tilde{\mathbf{\Pi}}^{(\Gamma)}_{(\gamma)}(\vec{p}, m, J))^2 = \tilde{\mathbf{\Pi}}^{(\Gamma)}_{(\gamma)}(\vec{p}, m, J)$. The sign factor $(-)^{\Gamma-\gamma}$ comes from the spinor metric $\mathbf{g}^{(\Gamma)}$, canceling the signature from the invariant spinor inner product. Thus the spinor field anti-commutator/commutator satisfies:

$$\begin{aligned}[\hat{\mathbf{\Psi}}^{(\Gamma)}_{(\gamma)}(\vec{x}), \hat{\bar{\mathbf{\Psi}}}^{(\Gamma)}_{(\gamma')}(\vec{y})]_\pm = \delta_{\gamma,\gamma'} \int \frac{mc^2 d^3p}{\epsilon(|\mathbf{p}|)} \Big[&|\zeta^{(\Gamma)}_{(\gamma)}(J)|^2 (-)^{\Gamma-\gamma}\tilde{\mathbf{\Pi}}^{(\Gamma)}_{(\gamma)}(\vec{p}, m, J) e^{\frac{i}{\hbar}\vec{p}\cdot(\vec{x}-\vec{y})} \\ &+ \pm |\zeta^{(\Gamma)}_{(-\gamma)}(J)|^2 (-)^{\Gamma+\gamma}\tilde{\mathbf{\Pi}}^{(\Gamma)}_{(-\gamma)}(\vec{p}, m, J) e^{-\frac{i}{\hbar}\vec{p}\cdot(\vec{x}-\vec{y})}\Big].\end{aligned} \tag{4.251}$$

Thus, the causal properties of the field are directly related to those of the projectors $\tilde{\mathbf{\Pi}}^{(\Gamma)}_{(\gamma)}$.

One should note that the analytic continuation to negative four-momentum of a spinor of particle type γ has the same eigenvalue as a spinor of particle type $-\gamma$:

$$\begin{aligned}\mathbf{\Gamma}^\beta p_\beta \mathbf{u}^{(\Gamma)}_{(\gamma)}(-\vec{p}, m, J, s_z) &= (\gamma) mc\,\mathbf{u}^{(\Gamma)}_{(\gamma)}(-\vec{p}, m, J, s_z), \\ \mathbf{\Gamma}^\beta p_\beta \mathbf{u}^{(\Gamma)}_{(-\gamma)}(\vec{p}, m, J, s_z) &= (\gamma) mc\,\mathbf{u}^{(\Gamma)}_{(-\gamma)}(\vec{p}, m, J, s_z).\end{aligned} \tag{4.252}$$

Therefore, the spinors are linearly related:

$$\mathbf{u}^{(\Gamma)}_{(-\gamma)}(\vec{p}, m, J, s_z) = \sum_{s'_z} \mathbf{u}^{(\Gamma)}_{(\gamma)}(-\vec{p}, m, J, s'_z) B_{s'_z s_z}(\vec{p}, J). \tag{4.253}$$

The normalization condition for the spinors implies that the matrices $\mathbf{B}$ are either anti-unitary or unitary:

$$(-)^{\Gamma+\gamma}\delta_{s'_z, s_z} = (-)^{\Gamma-\gamma} \sum_{s''_z} B^*_{s''_z s'_z}(\vec{p}, J) B_{s''_z s_z}(\vec{p}, J), \tag{4.254}$$

which means that the projectors satisfy:

$$\tilde{\mathbf{\Pi}}^{(\Gamma)}_{(-\gamma)}(\vec{p}, m, J) = \frac{\sum_{s_z} \mathbf{u}^{(\Gamma)}_{(-\gamma)}(\vec{p}, m, J, s_z)\bar{\mathbf{u}}^{(\Gamma)}_{(-\gamma)}(\vec{p}, m, J, s_z)}{(-)^{\Gamma+\gamma}} = \tilde{\mathbf{\Pi}}^{(\Gamma)}_{(\gamma)}(-\vec{p}, m, J). \tag{4.255}$$

In these expressions, the general off-mass-shell eigenstates satisfy:

$$\mathbf{\Gamma}^\beta p_\beta \mathbf{u}^{(\Gamma)}_{(\gamma)}(\vec{p}, \sqrt{-\vec{p}\cdot\vec{p}}/c, J, s_z) = -(\gamma)\sqrt{-\vec{p}\cdot\vec{p}}\ \mathbf{u}^{(\Gamma)}_{(\gamma)}(\vec{p}, \sqrt{-\vec{p}\cdot\vec{p}}/c, J, s_z), \tag{4.256}$$

where the positive value of the square root is explicit. The forms of the relevant relationships between spinor forms will be examined in order to ascertain their behaviors under discrete transformations shortly.

Equation 4.255 allows the anti-commutator/commutator to be conveniently expressed in the following form:

$$[\hat{\mathbf{\Psi}}^{(\Gamma)}_{(\gamma)}(\vec{x}), \hat{\bar{\mathbf{\Psi}}}^{(\Gamma)}_{(\gamma)}(\vec{y})]_\pm = \tilde{\mathbf{\Pi}}^{(\Gamma)}_{(\gamma)}\left(\frac{\hbar}{i}\frac{\partial}{\partial\vec{X}}, m, J\right) \times \left[|\zeta^{(\Gamma)}_{(\gamma)}(J)|^2(-)^{\Gamma-\gamma}\Delta_m(\vec{X}) \pm |\zeta^{(\Gamma)}_{(-\gamma)}(J)|^2(-)^{\Gamma+\gamma}\Delta_m(-\vec{X})\right]_{\vec{X}=\vec{x}-\vec{y}}. \tag{4.257}$$

This general form is seen to depend only upon the coordinate difference $\vec{x}-\vec{y}$. The spinor field is causal only if Eq. (4.257) vanishes for space-like separations (i.e., outside of the light cone). Again, for space-like intervals the form of the scalar propagator is even in the coordinate difference, $\Delta_m(\vec{X}) = \Delta_m(-\vec{X})$, as was seen in Eq. 4.230. To satisfy microscopic causality, the field parameters must satisfy:

$$|\zeta^{(\Gamma)}_{(\gamma)}(J)|^2 = \mp(-)^{2\gamma}|\zeta^{(\Gamma)}_{(-\gamma)}(J)|^2. \tag{4.258}$$

This implies that $|\zeta^{(\Gamma)}_{(\gamma)}(J)| = |\zeta^{(\Gamma)}_{(-\gamma)}(J)|$ and $1 = \mp(-1)^{2\gamma}$. Since $\Gamma = J_{max}$, the signature of γ is the same as the signature of J. Therefore, the causal spinor field must associate the quantum statistics with the spin of the field as follows:

$$\begin{array}{l}\text{Fermi statistics, anti-commuting fields (upper sign)} \rightarrow 2J, 2\Gamma \text{ are odd integers,}\\ \text{Bose statistics, commuting fields (lower sign)} \rightarrow J, \Gamma \text{ are integers.}\end{array} \tag{4.259}$$

This is the *spin-statistics theorem* for general linear spinor fields.

4.4.4 Parity transformations of fields

The action of the improper Lorentz transformation of spatial reflection upon a quantum state vector given in Eq. 4.179 implies that the creation operators satisfy:

$$\hat{\mathcal{P}}\hat{a}^{(\Gamma)\dagger}_{(\gamma)}(\vec{p}, m, J, s_z)\hat{\mathcal{P}}^{-1} = \eta^P_{(\gamma)}\,\hat{a}^{(\Gamma)\dagger}_{(\gamma)}(\mathcal{P}\vec{p}, m, J, s_z), \tag{4.260}$$

where the phase, which can be chosen to satisfy $\eta^P_{(\gamma)} = \pm 1$ for any particle type, gives the intrinsic parity of particle type γ. Thus, the causal field transforms according to:

$$\hat{\mathcal{P}}\,\hat{\mathbf{\Psi}}^{(\Gamma)}_{(\gamma)}(\vec{x})\,\hat{\mathcal{P}}^{-1} \equiv \sum_{s_z}\int \frac{mc^2\,d^3p}{\epsilon(\mathbf{p})}$$
$$\times\left[\eta^{P*}_{(\gamma)}\zeta^{(\Gamma,J)}_{(\gamma)}\frac{e^{\frac{i}{\hbar}\vec{p}\cdot\vec{x}}}{(2\pi\hbar)^{3/2}}\mathbf{u}^{(\Gamma)}_{(\gamma)}(\vec{p},m,J,s_z)\hat{a}^{(\Gamma)}_{(\gamma)}(\mathcal{P}\vec{p},m,J,s_z)\right.$$
$$\left.+(-)^{J+s_z}\eta^{P}_{(-\gamma)}\zeta^{(\Gamma,J)}_{(-\gamma)}\frac{e^{-\frac{i}{\hbar}\vec{p}\cdot\vec{x}}}{(2\pi\hbar)^{3/2}}\mathbf{u}^{(\Gamma)}_{(-\gamma)}(\vec{p},m,J,-s_z)\hat{a}^{(\Gamma)\dagger}_{(-\gamma)}(\mathcal{P}\vec{p},m,J,s_z)\right].\tag{4.261}$$

Therefore, a relationship between a spinor with argument $\vec{p}$ and one with argument $\mathcal{P}\vec{p}$ should be found if the field is to transform in an appropriate manner.

The spinor metric is quite convenient for relating parity transformed spinors due to the specific properties from Section 4.3.4, in particular $\mathbf{g}^{(\Gamma)}\mathbf{\Gamma}^0 = \mathbf{\Gamma}^0\mathbf{g}^{(\Gamma)}$, $\mathbf{g}^{(\Gamma)}\mathbf{J} = \mathbf{J}\mathbf{g}^{(\Gamma)}$, $\mathbf{g}^{(\Gamma)}\mathbf{\Gamma}^k = -\mathbf{\Gamma}^k\mathbf{g}^{(\Gamma)}$, and $\mathbf{g}^{(\Gamma)}\mathbf{g}^{(\Gamma)} = \mathbf{1}^{(\Gamma)}$. These relationships imply that:

$$\mathbf{\Gamma}^{\beta}(\mathcal{P}p)_{\beta}\left(\mathbf{g}^{(\Gamma)}\mathbf{u}^{(\Gamma)}_{(\gamma)}(\vec{p},m,J,s_z)\right) = -(\gamma)mc\left(\mathbf{g}^{(\Gamma)}\mathbf{u}^{(\Gamma)}_{(\gamma)}(\vec{p},m,J,s_z)\right),$$
$$\text{i.e. } \mathbf{u}^{(\Gamma)}_{(\gamma)}(\vec{p},m,J,s_z) \propto \left(\mathbf{g}^{(\Gamma)}\mathbf{u}^{(\Gamma)}_{(\gamma)}(\mathcal{P}\vec{p},m,J,s_z)\right).\tag{4.262}$$

From the form of the spinor metric generated from Eq. 4.82, and the convention for the standard state vectors, the proportionality constant must satisfy:

$$\mathbf{u}^{(\Gamma)}_{(\gamma)}(\vec{p},m,J,s_z) = (-)^{\gamma-\Gamma}\left(\mathbf{g}^{(\Gamma)}\mathbf{u}^{(\Gamma)}_{(\gamma)}(\mathcal{P}\vec{p},m,J,s_z)\right).\tag{4.263}$$

Substitution of the form Eq. 4.263 into Eq. 4.261 has an intriguing consequence. If the parity-transformed spinor field is to be appropriately related to the spinor field itself, the intrinsic parities of the particle and antiparticle must be related by:

$$(-)^{\gamma-\Gamma}\,\eta^{P*}_{(\gamma)} = (-)^{-\gamma-\Gamma}\,\eta^{P}_{(-\gamma)}\,,\text{ or}$$
$$\eta^{P}_{(-\gamma)} = (-)^{2\gamma}\,\eta^{P*}_{(\gamma)}.\tag{4.264}$$

Since $\Gamma = J_{max}$, the signature $(-)^{2\gamma} = (-)^{2J}$. This means that the intrinsic parity of a fermion-anti-fermion state *must be odd*! The experimental observations of the ground state of positronium (e^+, e^-) and the pseudoscalar π^0 meson $(q, \bar{q})$ as composite parity -1 systems are consistent with spatially symmetric s-wave $(\ell = 0)$ states, as predicted using Eq. 4.264. The resultant action of a parity transformation on a spinor field is given by:

$$\hat{\mathcal{P}}\,\hat{\mathbf{\Psi}}^{(\Gamma)}_{(\gamma)}(\vec{x})\,\hat{\mathcal{P}}^{-1} = (-)^{\gamma-\Gamma}\,\eta^{P*}_{(\gamma)}\mathbf{g}^{(\Gamma)}\,\hat{\mathbf{\Psi}}^{(\Gamma)}_{(\gamma)}(\mathcal{P}\vec{x}).\tag{4.265}$$

It should be noted that although electromagnetic and strong interactions seem to be invariant under parity transformations, parity conservation is maximally violated by weak interactions. Only left-handed neutrinos or right-handed anti-neutrinos couple to other particles during weak interactions. As previously noted in Eq. 4.183, a parity transformation on a massless particle reverses its helicity, generating a state that violates this phenomenology.

4.4.5 Time-reversal transformations of fields

The action of the improper Lorentz transformation of time reversal upon a quantum state vector given in Eq. 4.193 implies that the creation operators satisfy:

$$\hat{\mathcal{T}}\,\hat{a}^{(\Gamma)\dagger}_{(\gamma)}(\vec{p},m,J,s_z)\,\hat{\mathcal{T}}^{-1}=(-)^{J-s_z}\,\eta^{T}_{(\gamma)}\,\hat{a}^{(\Gamma)\dagger}_{(\gamma)}(\mathcal{P}\vec{p},m,J,-s_z), \tag{4.266}$$

where the phase factor $\eta^{T}_{(\gamma)}$ defines the intrinsic time reversal phase of particle type γ. Thus, recalling that the time-reversal operation is anti-linear, the causal field transforms according to:

$$\begin{aligned}\hat{\mathcal{T}}\,\hat{\boldsymbol{\Psi}}^{(\Gamma)}_{(\gamma)}(\vec{x})\,\hat{\mathcal{T}}^{-1}&\equiv\sum_{s_z}\int\frac{mc^2\,d^3p}{\epsilon(\mathbf{p})}\\&\times\Bigg[(-)^{J-s_z}\eta^{T*}_{(\gamma)}\zeta^{(\Gamma,J)*}_{(\gamma)}\frac{e^{-\frac{i}{\hbar}\vec{p}\cdot\vec{x}}}{(2\pi\hbar)^{3/2}}\mathbf{u}^{(\Gamma)*}_{(\gamma)}(\vec{p},m,J,s_z)\hat{a}^{(\Gamma)}_{(\gamma)}(\mathcal{P}\vec{p},m,J,-s_z)\\&+(-)^{2J}\eta^{T}_{(-\gamma)}\zeta^{(\Gamma,J)*}_{(-\gamma)}\frac{e^{+\frac{i}{\hbar}\vec{p}\cdot\vec{x}}}{(2\pi\hbar)^{3/2}}\mathbf{u}^{(\Gamma)*}_{(-\gamma)}(\vec{p},m,J,-s_z)\hat{a}^{(\Gamma)\dagger}_{(-\gamma)}(\mathcal{P}\vec{p},m,J,-s_z)\Bigg].\end{aligned} \tag{4.267}$$

Therefore, a relationship between a spinor with argument $\vec{p}$ and its complex conjugate with argument $\mathcal{P}\vec{p}$ should be found if the field is to transform in an appropriate manner.

As previously noted in Eq. 4.246, a $2J+1$ dimensional representation of angular momentum satisfies $-(\mathbf{J}^{(J)*}_k)_{\sigma,\bar{\sigma}}=(-)^{\sigma,-\bar{\sigma}}(\mathbf{J}^{(J)}_k)_{-\sigma-\bar{\sigma}}$. This transformation transposes the indices, up to a sign. It is convenient to define the conjugation matrix $\mathbf{C}^{(J)}$ given by:

$$\mathbf{C}^{(J)}_{\sigma\bar{\sigma}}\equiv(-)^{J+\sigma}\,\delta_{\sigma,-\bar{\sigma}},\quad\left(\mathbf{C}^{(J)}\right)^2=(-)^{2J}\mathbf{1}^{(J)}. \tag{4.268}$$

Equation 4.246 can then be rewritten as:

$$\mathbf{C}^{(J)}\,\mathbf{J}^{(J)*}_k=-\mathbf{J}^{(J)}_k\,\mathbf{C}^{(J)}. \tag{4.269}$$

Γ representations for these matrices can be constructed from these $2J+1$ dimensional forms.

In the representation for which $\mathbf{\Gamma}^0$ is diagonal, a spinor representation of this conjugation matrix can be directly constructed:

$$\mathbf{C}^{(\Gamma)} = \begin{pmatrix} \mathbf{C}^{(J_{min})} & \dots & \mathbf{0}^{(J_{min})} & \dots & \mathbf{0} & \dots & \mathbf{0} \\ \mathbf{0}^{(J_{min})} & \dots & \mathbf{0}^{(J_{min})} & \dots & \mathbf{0} & \dots & \mathbf{0} \\ \mathbf{0}^{(J_{min})} & \dots & \mathbf{C}^{(J_{min})} & \dots & \mathbf{0} & \dots & \mathbf{0} \\ \dots & \dots & \dots & \dots & \dots & \dots & \dots \\ \mathbf{0} & \dots & \mathbf{0} & \dots & \mathbf{C}^{(J_{max})} & \dots & \mathbf{0}^{(J_{max})} \\ \mathbf{0} & \dots & \mathbf{0} & \dots & \mathbf{0}^{(J_{max})} & \dots & \mathbf{0}^{(J_{max})} \\ \mathbf{0} & \dots & \mathbf{0} & \dots & \mathbf{0}^{(J_{max})} & \dots & \mathbf{C}^{(J_{max})} \end{pmatrix}, \tag{4.270}$$

where there are $2J+1$ copies of each $\mathbf{C}^{(J)}$ on the diagonal corresponding to the γ values for that J, and $J_{min}=0$ for bosons, $J_{min}=\frac{1}{2}$ for fermions. The angular momenta J take all values from $J_{min} \leq J \leq \Gamma = J_{max}$ in integer increments. For the $\Gamma=\frac{1}{2}$ Dirac spinors in the given representation, the conjugation matrix is directly related to a product of Dirac matrices:

$$\mathbf{C}^{(1/2)} = \begin{pmatrix} 0 & -1 & 0 & 0 \\ 1 & 0 & 0 & 0 \\ 0 & 0 & 0 & -1 \\ 0 & 0 & 1 & 0 \end{pmatrix} = -\gamma^x \gamma^z = -4\mathbf{\Gamma}^x \mathbf{\Gamma}^z. \tag{4.271}$$

It is clear that Eq. 4.269 implies that:

$$\mathbf{C}^{(\Gamma)}\, \mathbf{J}_k^* = -\mathbf{J}_k\, \mathbf{C}^{(\Gamma)}. \tag{4.272}$$

Since $[\mathbf{\Gamma}^0, \mathbf{J}_k]=0$, and the standard state vectors are eigenstates of $\mathbf{\Gamma}^0$, one can likewise determine that:

$$\mathbf{C}^{(\Gamma)}\, \mathbf{\Gamma}^{0*} = +\mathbf{\Gamma}^0\, \mathbf{C}^{(\Gamma)}. \tag{4.273}$$

Commutators like $[\mathbf{\Gamma}^0, \mathbf{K}_k] = -i\mathbf{\Gamma}^k$ demonstrate that the commutation signatures of $\mathbf{K}_k$ and $\mathbf{\Gamma}^k$ with $\mathbf{C}^{(\Gamma)}$ have the opposite signs. In the present representation, $\mathbf{J}_k^\dagger = \mathbf{J}_k$ and $\mathbf{K}_k^\dagger = -\mathbf{K}_k$. Therefore:

$$\mathbf{C}^{(\Gamma)}\, \mathbf{K}_k^* = +\mathbf{K}_k\, \mathbf{C}^{(\Gamma)}, \tag{4.274}$$

$$\mathbf{C}^{(\Gamma)}\, \mathbf{\Gamma}^{k*} = -\mathbf{\Gamma}^k\, \mathbf{C}^{(\Gamma)}. \tag{4.275}$$

These relations provide the needed direct connection between the complex conjugate spinors. The complex conjugate spinor with component s_z is directly related through the operation of $\mathbf{C}^{(\Gamma)}$ to the spinor with component $-s_z$. The proportionality constant can be determined by application of $\mathbf{J}_z$ on the standard state. Thus:

$$\mathbf{u}_{(\gamma)}^{(\Gamma)*}(\vec{p}, m, J, s_z) = (-)^{J-s_z}\, \mathbf{C}^{(\Gamma)}\mathbf{u}_{(\gamma)}^{(\Gamma)}(\mathcal{P}\vec{p}, m, J, -s_z), \tag{4.276}$$

describes the action of the conjugation matrix upon the spinors.

It should be noted that in Eq. 4.267 the translation factor satisfies $e^{\pm\frac{i}{\hbar}\mathcal{P}\vec{p}\cdot x} = e^{\mp\frac{i}{\hbar}\vec{p}\cdot\mathcal{T}x}$. If a straightforward relationship for the time-reversed field is to be found, after substitution of Eq. 4.276 into Eq. 4.267, the phase factors must satisfy:

$$(-)^{2J-2s_z}\eta_{(\gamma)}^{T*}\zeta_{(\gamma)}^{(\Gamma)*} = \kappa^T \zeta_{(\gamma)}^{(\Gamma)}$$
$$(-)^{2J}(-)^{J+s_z}\eta_{(-\gamma)}^{T}\zeta_{(-\gamma)}^{(\Gamma)*} = \kappa^T (-)^{J+s_z}\zeta_{(-\gamma)}^{(\Gamma)}, \tag{4.277}$$

which implies that the intrinsic time-reversal properties of the particle/anti-particle must satisfy:

$$(-)^{2J}\,\frac{\eta_{(\gamma)}^{T*}}{\eta_{(-\gamma)}^{T}} = \left(\frac{\zeta_{(\gamma)}^{(\Gamma)}\zeta_{(-\gamma)}^{(\Gamma)*}}{\zeta_{(\gamma)}^{(\Gamma)*}\zeta_{(-\gamma)}^{(\Gamma)}}\right). \tag{4.278}$$

The resultant action of a time-reversal transformation on a spinor field is given by:

$$\hat{\mathcal{T}}\,\hat{\mathbf{\Psi}}_{(\gamma)}^{(\Gamma)}(\vec{x})\,\hat{\mathcal{T}}^{-1} = \left(\frac{\zeta_{(\gamma)}^{(\Gamma)*}}{\zeta_{(\gamma)}^{(\Gamma)}}\right)\eta_{(\gamma)}^{T*}\,\mathbf{C}^{(\Gamma)}\,\hat{\mathbf{\Psi}}_{(\gamma)}^{(\Gamma)}(\mathcal{T}\vec{x}). \tag{4.279}$$

This transformation has its most direct interpretation if the field coefficients $\zeta_{(\gamma)}^{(\Gamma)}$ are real, in which case time reversal just reverses coordinate time and conjugates spinor components. To date, no experiments have been done that involve direct measurements of macroscopic time reversal (a reversal of clock motions). There have been experiments involving an asymmetry in the rate of oscillation of the meson K^0 into its antiparticle $\bar{K}^0$, versus the rate of oscillation of $\bar{K}^0$ into K^0 that suggest a direct measurement of microscopic time-reversal non-invariance [74]. Many of the fundamental equations describing microscopic physical models display time-reversal invariance. One set of relevant exceptions are those equations that describe the evolution of the cosmos as a whole. Equations describing the Big Bang of standard cosmology exhibit a direction of temporal progression.

4.4.6 Charge conjugation transformations of fields

The action of charge conjugation upon a quantum state vector given in Eq. 4.200 implies that the creation operators satisfy:

$$\hat{\mathcal{C}}\,\hat{a}_{(\gamma)}^{(\Gamma)\dagger}(\vec{p}, m, J, s_z)\,\hat{\mathcal{C}}^{-1} = \eta_{(\gamma)}^{C}\,\hat{a}_{(-\gamma)}^{(\Gamma)\dagger}(\vec{p}, m, J, s_z), \tag{4.280}$$

where the phase factor $\eta^C_{(\gamma)}$ defines the intrinsic charge conjugation phase of particle type γ. Thus, the causal field transforms according to:

$$\hat{\mathcal{C}}\,\hat{\boldsymbol{\Psi}}^{(\Gamma)}_{(\gamma)}(\vec{x})\,\hat{\mathcal{C}}^{-1} \equiv \sum_{s_z}\int \frac{mc^2\,d^3p}{\epsilon(\mathbf{p})}$$

$$\times\left[\eta^{C*}_{(\gamma)}\zeta^{(\Gamma,J)}_{(\gamma)}\frac{e^{\frac{i}{\hbar}\vec{p}\cdot\vec{x}}}{(2\pi\hbar)^{3/2}}\mathbf{u}^{(\Gamma)}_{(\gamma)}(\vec{p},m,J,s_z)\hat{a}^{(\Gamma)}_{(-\gamma)}(\vec{p},m,J,s_z)\right.$$

$$\left.+(-)^{J+s_z}\eta^{C}_{(-\gamma)}\zeta^{(\Gamma,J)}_{(-\gamma)}\frac{e^{-\frac{i}{\hbar}\vec{p}\cdot\vec{x}}}{(2\pi\hbar)^{3/2}}\mathbf{u}^{(\Gamma)}_{(-\gamma)}(\vec{p},m,J,-s_z)\hat{a}^{(\Gamma)\dagger}_{(\gamma)}(\vec{p},m,J,s_z)\right]. \quad (4.281)$$

Therefore, a relationship between a spinor with arguments $\vec{p}, s_z$ and the complex conjugate of its anti-spinor with arguments $\vec{p}, -s_z$ should be found if the field is to transform in an appropriate manner.

A transposition matrix in the index γ analogous to the conjugation matrix in Eq. 4.268 is needed to connect a particle of type γ to one of type $-\gamma$. It is convenient to define the type-transposition matrix $\mathbf{\Gamma}^{(J)}_E$ given by:

$$\left(\mathbf{\Gamma}^{(J)}_E\right)_{\gamma\bar{\gamma}} \equiv (-)^{J-\gamma}\,\delta_{\gamma,-\bar{\gamma}}\,\mathbf{1}^{(J)}, \quad \left(\mathbf{\Gamma}^{(\Gamma)}_E\right)^2 = (-)^{2J}\mathbf{1}. \quad (4.282)$$

The convention has been chosen so states with $\gamma = +J$ will transpose with positive signature. For the $\Gamma = \frac{1}{2}$ Dirac spinors in the given representation, the type transposition matrix is directly related to a product of Dirac matrices:

$$\mathbf{\Gamma}^{(\frac{1}{2})}_E = \begin{pmatrix} 0 & 0 & 1 & 0 \\ 0 & 0 & 0 & 1 \\ -1 & 0 & 0 & 0 \\ 0 & -1 & 0 & 0 \end{pmatrix} = \gamma^0\gamma_5,\ \mathbf{\Gamma}^{(1)}_E = \begin{pmatrix} \mathbf{1} & \mathbf{0} & \mathbf{0} & \mathbf{0} \\ \mathbf{0} & \mathbf{0} & \mathbf{0} & \mathbf{1} \\ \mathbf{0} & \mathbf{0} & -\mathbf{1} & \mathbf{0} \\ \mathbf{0} & \mathbf{1} & \mathbf{0} & \mathbf{0} \end{pmatrix}. \quad (4.283)$$

The most crucial property of this matrix is that it transposes eigenvalues of $\mathbf{\Gamma}^0$, while commuting with angular momentum:

$$\mathbf{\Gamma}_E\mathbf{\Gamma}^0 = -\mathbf{\Gamma}^0\mathbf{\Gamma}_E, \quad \mathbf{\Gamma}_E\mathbf{J}_k = +\mathbf{J}_k\mathbf{\Gamma}_E. \quad (4.284)$$

Commutation properties with the other generators consistent with the Jacobi identities can be determined in a straightforward manner:

$$\mathbf{\Gamma}_E\mathbf{\Gamma}^k = +\mathbf{\Gamma}^k\mathbf{\Gamma}_E, \quad \mathbf{\Gamma}_E\mathbf{K}_k = -\mathbf{K}_k\mathbf{\Gamma}_E, \quad \mathbf{\Gamma}_E\mathbf{C}^{(\Gamma)} = +\mathbf{C}^{(\Gamma)}\mathbf{\Gamma}_E. \quad (4.285)$$

The type transposition matrix $\mathbf{\Gamma}_E$ is shown to be related to generalized chirality projections in Appendix D.8. When combined with the $\mathbf{C}^{(\Gamma)}$ commutation relations given in Eqs. 4.272, 4.273, and 4.275, the $\mathbf{\Gamma}_E$ relations in Eq. 4.284 and Eq. 4.285

provide the needed spinor transposition for any of the group generators **G**:

$$\mathbf{\Gamma}_E \mathbf{C}^{(\Gamma)} \mathbf{G}^* = -\mathbf{G}\mathbf{C}^{(\Gamma)} \mathbf{\Gamma}_E, \quad \left(\mathbf{\Gamma}_E \mathbf{C}^{(\Gamma)}\right)^2 = \mathbf{1}. \tag{4.286}$$

This implies that a representation of spinors can be found that satisfy:

$$\mathbf{u}^{(\Gamma)*}_{(-\gamma)}(\vec{p}, m, J, -s_z) = (-)^{J+s_z}\, \mathbf{\Gamma}_E \mathbf{C}^{(\Gamma)} \mathbf{u}^{(\Gamma)}_{(\gamma)}(\vec{p}, m, J, s_z). \tag{4.287}$$

For the $\Gamma = \frac{1}{2}$ Dirac spinors in the given representation, the charge conjugation matrix $\mathbf{\Gamma}_E \mathbf{C}^{(\Gamma)}$ is directly related to a Dirac matrix:

$$\mathbf{\Gamma}_E \mathbf{C}^{(\Gamma)} = \begin{pmatrix} 0 & 0 & 0 & -1 \\ 0 & 0 & 1 & 0 \\ 0 & 1 & 0 & 0 \\ -1 & 0 & 0 & 0 \end{pmatrix} = -i\gamma^y. \tag{4.288}$$

The general charge-conjugation matrix is seen to involve the matrix product of particle-type transposition with complex conjugation.

If a straightforward relationship for the charge-conjugated field is to be found, after substitution of Eq. 4.287 into Eq. 4.281 the phase factors must satisfy:

$$\begin{aligned} (-)^{J+s_z} \eta^{C*}_{(\gamma)} \zeta^{(\Gamma)}_{(\gamma)} &= \kappa^C (-)^{J+s_z} \zeta^{(\Gamma)*}_{(-\gamma)}, \\ \eta^C_{(-\gamma)} \zeta^{(\Gamma)}_{(-\gamma)} &= \kappa^C \zeta^{(\Gamma)*}_{(+\gamma)}, \end{aligned} \tag{4.289}$$

which implies the intrinsic charge conjugation properties of the particle/anti-particle must satisfy:

$$\frac{\eta^{C*}_{(\gamma)}}{\eta^C_{(-\gamma)}} = \frac{\left|\zeta^{(\Gamma)}_{(-\gamma)}\right|^2}{\left|\zeta^{(\Gamma)}_{(+\gamma)}\right|^2} = 1. \tag{4.290}$$

The resultant action of charge conjugation on a spinor field is given by:

$$\hat{\mathcal{C}}\, \hat{\mathbf{\Psi}}^{(\Gamma)}_{(\gamma)}(\vec{x})\, \hat{\mathcal{C}}^{-1} = \left(\frac{\zeta^{(\Gamma)}_{(\gamma)}}{\zeta^{(\Gamma)*}_{(-\gamma)}}\right) \eta^{C*}_{(\gamma)}\, \mathbf{\Gamma}_E \mathbf{C}^{(\Gamma)} \left(\hat{\mathbf{\Psi}}^{(\Gamma)\dagger}_{(\gamma)}(\vec{x})\right)^T. \tag{4.291}$$

This transformation has its most direct interpretation if the field coefficients for the particle and anti-particle components satisfy $\zeta^{(\Gamma)}_{(\gamma)} = \zeta^{(\Gamma)*}_{(-\gamma)}$, in which case charge-conjugation takes the adjoint of the field and transposes spinor components. It should be noted that although electromagnetic and strong interactions seem to be invariant under charge conjugation, charge-conjugation conservation is violated by weak interactions.

For completeness, a justification of the operation of particle-type transposition being referred to as charge conjugation will be given. The linear spinor field equation for a system with a local gauge symmetry takes the generic form:

$$\mathbf{\Gamma}^\beta \left(\frac{\hbar}{i} \frac{\partial}{\partial x^\beta} - \frac{q}{c} A_\beta(\vec{x}) \right) \hat{\mathbf{\Psi}}^{(\Gamma)}_{(\gamma)}(\vec{x}) = -(\gamma) mc \hat{\mathbf{\Psi}}^{(\Gamma)}_{(\gamma)}(\vec{x}), \tag{4.292}$$

where $A_\beta(\vec{x})$ is a real gauge field with charge coupling q. The adjoint spinor field satisfies:

$$\mathbf{\Gamma}^{\beta *} \left(-\frac{\hbar}{i} \frac{\partial}{\partial x^\beta} - \frac{q}{c} A_\beta(\vec{x}) \right) \left(\hat{\mathbf{\Psi}}^{(\Gamma)\dagger}_{(\gamma)}(\vec{x}) \right)^T = -(\gamma) mc \left(\hat{\mathbf{\Psi}}^{(\Gamma)\dagger}_{(\gamma)}(\vec{x}) \right)^T. \tag{4.293}$$

Motivated by the relation from Eq. 4.286 demonstrating that:

$$\mathbf{\Gamma}_E \mathbf{C}^{(\Gamma)} \, \mathbf{\Gamma}^{\beta *} = -\mathbf{\Gamma}^\beta \, \mathbf{\Gamma}_E \mathbf{C}^{(\Gamma)}$$

a multiplication of Eq. 4.293 by the charge conjugation matrix from the left results in the equation:

$$\begin{aligned} \mathbf{\Gamma}^\beta \left(\frac{\hbar}{i} \frac{\partial}{\partial x^\beta} + \frac{q}{c} A_\beta(\vec{x}) \right) & \left[\mathbf{\Gamma}_E \mathbf{C}^{(\Gamma)} \left(\hat{\mathbf{\Psi}}^{(\Gamma)\dagger}_{(\gamma)}(\vec{x}) \right)^T \right] \\ &= -(\gamma) mc \left[\mathbf{\Gamma}_E \mathbf{C}^{(\Gamma)} \left(\hat{\mathbf{\Psi}}^{(\Gamma)\dagger}_{(\gamma)}(\vec{x}) \right)^T \right]. \end{aligned} \tag{4.294}$$

Therefore, defining the charge conjugated field $\hat{\mathbf{\Psi}}^{(\Gamma)CC}_{(\gamma)}(\vec{x}) \equiv \hat{\mathcal{C}} \, \hat{\mathbf{\Psi}}^{(\Gamma)}_{(\gamma)}(\vec{x}) \, \hat{\mathcal{C}}^{-1}$, the equation of motion satisfied by this field is given by:

$$\mathbf{\Gamma}^\beta \left(\frac{\hbar}{i} \frac{\partial}{\partial x^\beta} + \frac{q}{c} A_\beta(\vec{x}) \right) \hat{\mathbf{\Psi}}^{(\Gamma)CC}_{(\gamma)}(\vec{x}) = -(\gamma) mc \, \hat{\mathbf{\Psi}}^{(\Gamma)CC}_{(\gamma)}(\vec{x}). \tag{4.295}$$

A comparison of Eq. 4.295 to Eq. 4.292 demonstrates that the charge conjugated field has a charge opposite that of the field.

4.4.7 Summary of discrete transformations on causal spinor fields

In summary of the previous discussions, the form of a causal spinor field that has the expected properties under parity, time reversal, and charge conjugation is given by:

$$\begin{aligned} \hat{\mathbf{\Psi}}^{(\Gamma)}_{(\gamma)}(\vec{x}) \equiv \frac{1}{\sqrt{2}} \sum_{s_z} \int \frac{mc^2 \, d^3p}{\epsilon(\mathbf{p})} & \left[\frac{e^{\frac{i}{\hbar}(\mathbf{p}\cdot\mathbf{x} - \epsilon(\mathbf{p})t)}}{(2\pi\hbar)^{3/2}} \mathbf{u}^{(\Gamma)}_{(\gamma)}(\vec{p}, m, J, s_z) \hat{a}^{(\Gamma)}_{(\gamma)}(\vec{p}, m, J, s_z) \right. \\ & \left. + (-)^{J+s_z} \frac{e^{-\frac{i}{\hbar}(\mathbf{p}\cdot\mathbf{x} - \epsilon(\mathbf{p})t)}}{(2\pi\hbar)^{3/2}} \mathbf{u}^{(\Gamma)}_{(-\gamma)}(\vec{p}, m, J, -s_z) \hat{a}^{(\Gamma)\dagger}_{(-\gamma)}(\vec{p}, m, J, s_z) \right]. \end{aligned} \tag{4.296}$$

The normalization has been chosen to have non-relativistic correspondence with the usual form $\langle t, \mathbf{x} | t, \mathbf{y} \rangle$. The field is causal in that it anticommutes/commutes outside of the light cone (i.e., where $(\vec{x} - \vec{y}) \cdot (\vec{x} - \vec{y}) > 0$ is a space-like separation), explicitly exhibited by $[\hat{\boldsymbol{\Psi}}^{(\Gamma)}_{(\gamma)}(\vec{x}), \hat{\boldsymbol{\Psi}}^{(\Gamma)}_{(\gamma)}(\vec{y})]_{(-)^{2\Gamma+1}} = 0$ and $[\hat{\boldsymbol{\Psi}}^{(\Gamma)}_{(\gamma)}(\vec{x}), \hat{\boldsymbol{\Psi}}^{(\Gamma)\dagger}_{(\gamma)}(\vec{y})]_{(-)^{2\Gamma+1}} = 0$, according to the spin $J_{max} = \Gamma$. The fields transform under Poincaré transformations according to:

$$\hat{U}(\boldsymbol{\Lambda}, \vec{a}) \left[\hat{\boldsymbol{\Psi}}^{(\Gamma)}_{(\gamma)}(\vec{x}) \right]_b \hat{U}^\dagger(\boldsymbol{\Lambda}, \vec{a}) = \sum_{b'} \mathcal{D}^{(\Gamma)}_{bb'}(\boldsymbol{\Lambda}^{-1}) \left[\hat{\boldsymbol{\Psi}}^{(\Gamma)}_{(\gamma)}(\boldsymbol{\Lambda}\vec{x} + \vec{a}) \right]_{b'}. \tag{4.297}$$

One might note that the field does not satisfy vanishing anticommutation or commutation relations with its adjoint for time-like separations $(\vec{x} - \vec{y}) \cdot (\vec{x} - \vec{y}) < 0$.

The free causal spinor field satisfies the equation of motion:

$$\boldsymbol{\Gamma}^\beta \frac{\hbar}{i} \frac{\partial}{\partial x^\beta} \hat{\boldsymbol{\Psi}}^{(\Gamma)}_{(\gamma)}(\vec{x}) = -(\gamma) mc \, \hat{\boldsymbol{\Psi}}^{(\Gamma)}_{(\gamma)}(\vec{x}). \tag{4.298}$$

Therefore, linear spinor fields satisfy an equation of motion that is *linear* in space-time derivatives. As previously mentioned, the conserved particle type current $\mathcal{J}^\beta_{(\gamma)}(\vec{x})$ given by:

$$\mathcal{J}^\beta_{(\gamma)}(\vec{x}) \equiv \hat{\bar{\boldsymbol{\Psi}}}^{(\Gamma)}_{(\gamma)}(\vec{x}) \boldsymbol{\Gamma}^\beta \hat{\boldsymbol{\Psi}}^{(\Gamma)}_{(\gamma)}(\vec{x}), \qquad \frac{\partial}{\partial x^\beta} \mathcal{J}^\beta_{(\gamma)}(\vec{x}) = 0, \tag{4.299}$$

defines a contravariant vector field, whose contravariant index is derived from the generators $\hat{\Gamma}^\beta$. The adjoint representation of these group generators define a group metric $\eta_{\mu\nu}$, which has well-defined group properties on the tangent space of group transformations. Under arbitrary coordinate transformations from flat space coordinates $\vec{x}$ to curvilinear coordinates $\vec{X}$, the group metric on this index will undergo the usual transformation $\eta_{\mu\nu} \to g_{\alpha\beta}(\vec{X}) = \frac{\partial x^\mu}{\partial X^\alpha} \eta_{\mu\nu} \frac{\partial x^\nu}{\partial X^\beta}$. The momentum operator $\hat{p}_\beta = \frac{\hbar}{i} \frac{\partial}{\partial x^\beta}$ transforms as a covariant vector under this transformation. The linear spinor form $\hat{\Gamma}^\beta \hat{p}_\beta$ remains a Lorentz invariant suitable for incoporation into gravitation using the principle of equivalence, which will be discussed in Part II.

The spinor field transforms under parity as given by:

$$\hat{\mathcal{P}} \, \hat{\boldsymbol{\Psi}}^{(\Gamma)}_{(\gamma)}(\vec{x}) \, \hat{\mathcal{P}}^{-1} = (-)^{\gamma - \Gamma} \, \eta^{P*}_{(\gamma)} \, \mathbf{g}^{(\Gamma)} \, \hat{\boldsymbol{\Psi}}^{(\Gamma)}_{(\gamma)}(\mathcal{P}\vec{x}), \tag{4.300}$$

where the intrinsic parities of the particle or anti-particle must satisfy:

$$\eta^P_{(-\gamma)} = (-)^{2\gamma} \, \eta^{P*}_{(\gamma)} = (-)^{2J} \, \eta^{P*}_{(\gamma)}, \tag{4.301}$$

as is experimentally observed. The phase can always be chosen such that the parity signature of any particle satisfies $\eta^P_{(-\gamma)} = \pm 1$. The spinor metric $\mathbf{g}^{(\Gamma)}$ is as defined in Eq. 4.82.

The spinor field transforms under time reversal as given by:

$$\hat{\mathcal{T}}\,\hat{\boldsymbol{\Psi}}^{(\Gamma)}_{(\gamma)}(\vec{x})\,\hat{\mathcal{T}}^{-1} = \eta^{T*}_{(\gamma)}\,\mathbf{C}^{(\Gamma)}\,\hat{\boldsymbol{\Psi}}^{(\Gamma)}_{(\gamma)}(\mathcal{T}\vec{x}), \tag{4.302}$$

where the intrinsic time-reversal phases of the particle or anti-particle must satisfy:

$$\eta^{T}_{(-\gamma)} = (-)^{2J}\,\eta^{T*}_{(\gamma)}. \tag{4.303}$$

Using the anti-linear nature of the time reversal operator, since $\eta^{T*}_{(\gamma)}\eta^{T}_{(\gamma)} = 1$, the field satisfies the expected behavior under two time reversals:

$$\left(\hat{\mathcal{T}}\right)^2\,\hat{\boldsymbol{\Psi}}^{(\Gamma)}_{(\gamma)}(\vec{x})\,\left(\hat{\mathcal{T}}^{-1}\right)^2 = (-)^{2J}\,\hat{\boldsymbol{\Psi}}^{(\Gamma)}_{(\gamma)}(\vec{x}). \tag{4.304}$$

The conjugation matrix $\mathbf{C}^{(\Gamma)}$ is defined in Eq. 4.270.

The spinor field transforms under charge conjugation as given by:

$$\hat{\mathcal{C}}\,\hat{\boldsymbol{\Psi}}^{(\Gamma)}_{(\gamma)}(\vec{x})\,\hat{\mathcal{C}}^{-1} = \eta^{C*}_{(\gamma)}\,\boldsymbol{\Gamma}_E\mathbf{C}^{(\Gamma)}\left(\hat{\boldsymbol{\Psi}}^{(\Gamma)\dagger}_{(\gamma)}(\vec{x})\right)^T, \tag{4.305}$$

where the intrinsic charge-conjugation phases of the particle or anti-particle must satisfy:

$$\eta^{C}_{(-\gamma)} = \eta^{C*}_{(\gamma)} \tag{4.306}$$

The particle-type transposition matrix $\boldsymbol{\Gamma}_E$ is defined in Eq. 4.282.

Multiple discrete transformations TP, CP, and CTP

There are a few multiple discrete transformation of physical systems that are of interest. A parity transformation, followed by time reversal, yields the result:

$$\hat{\mathcal{T}}\hat{\mathcal{P}}\,\hat{\boldsymbol{\Psi}}^{(\Gamma)}_{(\gamma)}(\vec{x})\,\hat{\mathcal{P}}^{-1}\hat{\mathcal{T}}^{-1} = (-)^{\gamma-\Gamma}\,\eta^{P}_{(\gamma)}\,\eta^{T*}_{(\gamma)}\,\mathbf{C}^{(\Gamma)}\,\mathbf{g}^{(\Gamma)}\,\hat{\boldsymbol{\Psi}}^{(\Gamma)}_{(\gamma)}(-\vec{x}). \tag{4.307}$$

This transformation reflects space-time through the origin. Many symmetric invariant field equations, like the Klein–Gordon equation, are invariant under this combination.

A parity transformation, followed by charge conjugation, yields the result:

$$\hat{\mathcal{C}}\hat{\mathcal{P}}\,\hat{\boldsymbol{\Psi}}^{(\Gamma)}_{(\gamma)}(\vec{x})\,\hat{\mathcal{P}}^{-1}\hat{\mathcal{C}}^{-1} = (-)^{\gamma-\Gamma}\,\eta^{P*}_{(\gamma)}\,\eta^{C*}_{(\gamma)}\,\boldsymbol{\Gamma}_E\,\mathbf{C}^{(\Gamma)}\left[\hat{\bar{\boldsymbol{\Psi}}}^{(\Gamma)}_{(\gamma)}(\mathcal{P}\vec{x})\right]^T. \tag{4.308}$$

This symmetry, though conserved by electromagnetic and strong interactions, is weakly violated in weak interactions such as K^0_L, K^0_S meson decays, such as the asymmetry in the decays $K^0_L \rightarrow \pi^- e^+ \nu_e$ vs $K^0_L \rightarrow \pi^+ e^- \bar{\nu}_e$. Flavor mixing among three or more generations provides a microscopic mechanism for incorporating CP symmetry violation.

A parity transformation, followed by time reversal, then charge conjugation, yields the result:

$$\hat{\mathcal{C}}\hat{\mathcal{T}}\hat{\mathcal{P}}\,\hat{\mathbf{\Psi}}^{(\Gamma)}_{(\gamma)}(\vec{x})\,\hat{\mathcal{P}}^{-1}\hat{\mathcal{T}}^{-1}\hat{\mathcal{C}}^{-1} = (-)^{\gamma+\Gamma}\,\eta^{P}_{(\gamma)}\,\eta^{T*}_{(\gamma)}\,\eta^{C*}_{(\gamma)}\,\mathbf{\Gamma}_E\mathbf{g}^{(\Gamma)}\left[\hat{\mathbf{\Psi}}^{(\Gamma)\dagger}_{(\gamma)}(-\vec{x})\right]^T. \tag{4.309}$$

The standard model for microscopic physics is invariant under CTP transformations. The factors to the left of $\mathbf{\Gamma}_E$ are phases, so that any bilinear form involving the field with its adjoint will not display these phases. A statement of the well-known *CTP theorem* is that any causal, local model that generates a hermitian Hamiltonian density, and is invariant under proper Lorentz transformations, will be invariant under the combined transformations of parity, time reversal, and charge conjugation. The reason for this invariance is that there must be an even set of space-time indices if the Lagrangian form is to be a Lorentz invariant. This must be reflected in an even number of generators with space-time indices. As shown in Appendix D.8, exchanges of the operator $\mathbf{\Gamma}_E\mathbf{g}^{(\Gamma)}$ with any of the generators involves only a change in sign. Thus, bilinear products will remain unchanged.

Part II

General relativity

5

Fundamentals of general relativity

5.1 From special to general relativity

In general relativity, the "force of gravity", which directly couples to a gravitating body through its mass, is replaced by relationships between geometric coordinates. This can be done because the *inertial mass* that relates the acceleration of a body from rest (a purely geometrical aspect) to the force through Newton's second law $\mathbf{F} = m\mathbf{a}$, is the same as the *gravitational mass* that couples the gravitational acceleration $\mathbf{g}$ to that body $\mathbf{F}_{gravitation} = m\mathbf{g}$. Newton tested this equivalence using various pendulums, and Eotvos [75] in 1889 verified the equivalence of inertial and gravitational mass to better than one part in 10^9. Gravity attracts different masses in a way that results in the differing masses having the same accelerations. This tenet embodies the equivalence principle, which will be discussed next. Since bodies of vastly differing constitutions and masses gravitate equivalently, one can then construct the trajectories of general gravitating masses in terms of geometric *geodesics* (special curves in the space-time), independent of the mass, charge, or internal structure of the gravitating body.

5.1.1 The principle of equivalence

The principle of equivalence forms the conceptual foundation of early formulations of the theory of general relativity. For present purposes, the *principle of equivalence* will be stated as follows: *At every space-time point in an arbitrary gravitational field, it is possible to choose a locally inertial coordinate system such that (within a sufficiently small region of that point) the laws of nature take the same form as in an unaccelerated Minkowski coordinate system in the absence of gravity.* Such an assertion inherently relates the inertial mass to the gravitational mass.

In Part I, the characteristics of the geometry of space-time were described in terms of the (invariant) distance between nearby points in the geometry, as parameterized using various coordinate systems. For special relativity, this relationship was established in terms of the Minkowski metric by the invariant interval $ds^2 = d\xi^\mu \, \eta_{\mu\nu} \, d\xi^\nu$, where $\eta = \begin{pmatrix} -1 & 0 & 0 & 0 \\ 0 & 1 & 0 & 0 \\ 0 & 0 & 1 & 0 \\ 0 & 0 & 0 & 1 \end{pmatrix}$ and $\vec{d\xi} = \begin{pmatrix} d\xi^t \\ d\xi^x \\ d\xi^y \\ d\xi^z \end{pmatrix}$. The renamed special coordinates ξ^μ will represent contravariant components of inertial locally freely falling coordinates from Chapter 1. In general relativity, one assumes that arbitrary curvilinear coordinates x^β used by an observer can be expressed in terms of locally freely falling coordinates ξ^μ describing gravitationally free motion for the system under consideration. It should be noted that coordinate systems are chosen for the convenience of observers, and have no independent compulsion.

Generally, the relationship between coordinate systems allows one to use the chain rule for differentials:

$$d\xi^\mu = \frac{\partial \xi^\mu}{\partial x^\alpha} dx^\alpha \tag{5.1}$$

to determine a space-time-dependent form for the components of the metric:

$$ds^2 = d\xi^\mu \eta_{\mu\nu} d\xi^\nu = dx^\alpha \left(\frac{\partial \xi^\mu}{\partial x^\alpha} \, \eta_{\mu\nu} \, \frac{\partial \xi^\nu}{\partial x^\beta} \right) dx^\beta \equiv \sum_{\alpha\beta} dx^\alpha g_{\alpha\beta}(\vec{x}) \, dx^\beta . \tag{5.2}$$

The curvilinear metric $g_{\alpha\beta}(\vec{x}) = \frac{\partial \xi^\mu}{\partial x^\alpha} \, \eta_{\mu\nu} \, \frac{\partial \xi^\nu}{\partial x^\beta}$ is generally coordinate-dependent. The most general background upon which derivatives are well-defined is referred to as a *manifold*. Metric space(-time) manifolds are referred to as *Riemannian* manifolds. The behaviors of multi-component forms and vectors on these geometries will be examined next.

5.1.2 Vectors and tensors

As is familiar to any student of introductory physics, dimensional analysis begins one's attempts to model physical parameters with mathematical structures. Dimensioned parameters (such as length, mass, and time) cannot be arguments of arbitrary functional forms but must maintain multi-linear, rational relationships of units amongst both sides of a given equation. General structures that maintain multi-linear relationships under a specified set of transformations have defined mathematical properties.

More complex physical observables (such as four-momenta and field strengths) often take the form of multicomponent parameters with well-defined properties under coordinate transformations. Such scalars, vectors, and arbitrary rank multilinear forms are generally described as *tensors*. As was the case with special relativity described in Part I, vectors will be of two types: contravariant vectors, whose components will be labeled with raised indices, and covariant vectors that transform under the inverse of the transformation for contravariant vectors, whose components will be labeled using lowered indices.

Contravariant vectors

Tangent vectors on the space-time are multi-component objects whose components depend upon the orientation of the basis vectors in a well-defined manner. A contravariant vector will be distinguished using an arrow $\vec{A}$, and the basis vectors for the space-time will be denoted using $\vec{e}_\beta$, where $\beta = 0, 1, 2, 3$. The basis vectors satisfy the transformation equation:

$$\vec{e}'_\mu = \frac{\partial x^\beta}{\partial x'^\mu}\, \vec{e}_\beta. \tag{5.3}$$

One should note that in a *coordinate basis* for which any point on the manifold has a unique coordinate, basis vectors transform like partial derivatives using the chain rule:

$$\vec{e}_\beta = \frac{\partial}{\partial x^\beta}, \qquad \vec{e}^{\,(o)}_\mu = \frac{\partial}{\partial \xi^\mu}. \tag{5.4}$$

This means that for a coordinate basis, the basis vectors commute $[\vec{e}_\alpha, \vec{e}_\beta] = 0$ (integrability). For a general non-coordinate basis, the commutation of the basis vectors must result in a linear combination of the basis vectors $[\vec{e}_\alpha, \vec{e}_\beta] = c^\lambda_{\alpha\beta}\vec{e}_\lambda$. For instance, on a group manifold, the structure constants $c^\lambda_{\alpha\beta}$ are independent of the point on the manifold, and the basis vectors are the generators for infinitesimal group parameter transformations. The contravariant vector $\vec{A}$ can be expanded using components relative to any complete set of basis vectors:

$$\vec{A} = A^\beta\, \vec{e}_\beta = A'^\mu\, \vec{e}'_\mu \quad \Rightarrow \quad A'^\mu = \frac{\partial x'^\mu}{\partial x^\beta}\, A^\beta \tag{5.5}$$

The transformation of the components of the contravariant vector motivates the traditional use of raised indices associated with the upper parameter in the partial derivative in the chain rule.

Covariant vectors and one-forms

The basis one-forms dual to the coordinate basis tangent vectors $\vec{e}_\mu$ are given by the differentials, $\mathtt{d}x^\mu$, where the operation $\mathtt{d}$ gives the local variation of a function under an arbitrary infinitesimal displacement according to the (coordinate independent) chain rule $\mathtt{d}f = \mathtt{d}x^\mu \frac{\partial}{\partial x^\mu} f$. A bi-linear inner product can be defined on the basis one-forms and basis vectors within a general affine space (that need not be a metric space) specified by $\langle \mathtt{d}x^\mu, \vec{e}_\nu\rangle = \delta^\mu_\nu$.

The basis forms transform using the inverse transformation of the basis vectors:

$$\mathrm{d}x'^{\alpha} = \frac{\partial x'^{\alpha}}{\partial x^{\mu}} \, \mathrm{d}x^{\mu}. \tag{5.6}$$

A general one-form has components relative to the particular basis given by $\mathtt{B} = B_{\mu} \, \mathrm{d}x^{\mu}$. The components of $\mathtt{B}$ transform, as do the components of a covariant vector under coordinate transformations:

$$B'_{\beta} = \frac{\partial x^{\mu}}{\partial x'^{\beta}} \, B_{\mu}. \tag{5.7}$$

In a metric space the invariance of the inner product $\langle \mathtt{B}, \vec{A} \rangle$ defines the covariant vector as linearly related to its contravariant dual through the metric:

$$B_{\mu} \equiv g_{\mu\alpha} \, B^{\alpha}. \tag{5.8}$$

General *p-forms* are convenient for defining oriented surfaces for integration. An anti-symmetric wedge product between forms satisfying $\mathrm{d}x^{\alpha} \wedge \mathrm{d}x^{\beta} = -\mathrm{d}x^{\beta} \wedge \mathrm{d}x^{\alpha}$ is defined in order to construct higher-order exterior forms. A general p-form $\mathtt{f}$ has components defined by $\mathtt{f} = f_{\mu_1 \cdots \mu_k} \mathrm{d}x^{\mu_1} \wedge \cdots \wedge \mathrm{d}x^{\mu_k}$. Multi-variable integrands over an oriented surface $\Omega_{(s)}$ then follow the usual Jacobian transformation under a change in variables:

$$\int_{\Omega_{(s)}} F(\vec{x}) \, \mathrm{d}x^1 \wedge \cdots \wedge \mathrm{d}x^s = \int_{\Omega'_{(s)}} F(\vec{x}(\vec{x}')) \begin{vmatrix} \frac{\partial x^1}{\partial x'^1} & \cdots & \frac{\partial x^1}{\partial x'^s} \\ \cdots & \cdots & \cdots \\ \frac{\partial x^s}{\partial x'^1} & \cdots & \frac{\partial x^s}{\partial x'^s} \end{vmatrix} \mathrm{d}x'^1 \wedge \cdots \wedge \mathrm{d}x'^s \tag{5.9}$$

Therefore, integration is an operation upon the space of forms, while differentiation is an operation in the space of tangent vectors.

General tensors of rank (k,s)

General multi-linear tensor forms can be defined by their transformation properties under coordinate changes. A tensor with contravariant rank k and covariant rank s has components that are defined to transform according to the rule:

$$T'^{\alpha_1 \cdots \alpha_k}_{\mu_1 \cdots \mu_s} = \frac{\partial x^{\nu_1}}{\partial x'^{\mu_1}} \cdots \frac{\partial x^{\nu_s}}{\partial x'^{\mu_s}} T^{\beta_1 \cdots \beta_k}_{\nu_1 \cdots \nu_s} \frac{\partial x'^{\alpha_1}}{\partial x^{\beta_1}} \cdots \frac{\partial x'^{\alpha_k}}{\partial x^{\beta_k}}. \tag{5.10}$$

Using this general definition, contravariant vectors are rank (1,0) tensors, covariant vectors are rank (0,1) tensors, and k-forms are fully anti-symmetric tensors of rank (0,k).

5.1.3 Geodesics

Since the motions of gravitating bodies can be characterized in terms of certain geometric trajectories, those trajectories on the geometry satisfy special uniqueness conditions. The construction and properties of these unique curves, which are referred to as *geodesics*, will be the topic of this section.

Relativity of geodesics

As previously stated, the relating of geometry to the physics of gravitation comes through the postulate of general relativity called the *strong equivalence principle*, which states that it is always possible in an arbitrary gravitational field to choose a local set of freely falling coordinates such that the laws of physics are the same as those without gravity. In this manner, gravity is eliminated as a dynamical force, because it (locally) acts the same on all objects. Besides the equivalence of inertial mass and gravitational mass, a direct consequence of this principle is that light must be bent by gravity, and that in being bent there is a gravitational redshift (or blueshift, a change in color/energy due to gravity). In a local freely falling frame, a free particle should have no acceleration due to gravity, and thus it should satisfy Newton's first law of motion in a frame of reference parameterized in terms of the locally inertial time and position coordinates $(\xi^t, \xi^x, \xi^y, \xi^z)$:

$$\frac{d^2\xi^\beta}{d\tau^2} = 0, \tag{5.11}$$

where τ represents the particle's proper time, or an appropriate proper affine parameter. If the coordinates are the proper-frame coordinates of the particle, this quantity gives the proper acceleration of the particle.

However, convenient coordinates are often *not* freely falling, and therefore bodies fixed in those coordinates experience an apparent gravitational force, since the coordinates are non-inertial. For instance, an observer at rest on the surface of the Earth maintains a fixed radial location relative to its center. One can rewrite Eq. 5.11 for inertial motion in terms of the (non-freely falling, non-inertial) coordinates x^μ of a *fiducial* observer using the following manipulations involving the chain rule for derivatives:

$$\frac{d}{d\tau}\left(\frac{d\xi^\beta}{d\tau}\right) = \frac{d}{d\tau}\sum_\mu\left(\frac{\partial\xi^\beta}{\partial x^\mu}\frac{dx^\mu}{d\tau}\right) = \sum_\mu \frac{\partial\xi^\beta}{\partial x^\mu}\left(\frac{d^2x^\mu}{d\tau^2} + \sum_{\lambda\nu}\Gamma^\mu_{\lambda\nu}\frac{dx^\lambda}{d\tau}\frac{dx^\nu}{d\tau}\right) = 0$$

or

$$\left(\frac{d^2x^\mu}{d\tau^2} + \sum_{\lambda\nu}\Gamma^\mu_{\lambda\nu}\frac{dr^\lambda}{d\tau}\frac{dr^\nu}{d\tau}\right) = 0. \tag{5.12}$$

The last equation is known as the *geodesic equation*. It describes the trajectory of freely falling systems in terms of the local coordinates of the observers in a "gravitational field". One sees that the effect that appears to be gravitational using coordinates x^μ is contained in the term with the *affine connections* Γ, where the connections satisfy:

$$\Gamma^{\mu}_{\lambda\nu} = \frac{\partial x^{\mu}}{\partial \xi^{\beta}} \frac{\partial^2 \xi^{\beta}}{\partial x^{\lambda} \partial x^{\nu}}. \tag{5.13}$$

For coordinates satisfying the usual integrability conditions, the connections are symmetric under exchange of the lower two indices $\Gamma^{\mu}_{\lambda\nu} = \Gamma^{\mu}_{\nu\lambda}$.

One should note that despite the convenience of using the vector symbol $\vec{x}$ to specify the four coordinates, the position coordinate *does not* generally transform as a vector. The components are merely a set of coordinates. Likewise, the connections will not be found to transform under coordinate transformations as a tensor.

Uniqueness of geodesics

The geodesic equation can be obtained by minimizing the space-time distance between two points (as measured by the invariant interval), which gives Newton's first law of motion in terms of particles moving along the "straight lines" (or geodesics) when there are no external forces:

$$\delta \int ds = \delta \int \sqrt{\sum_{\mu\nu} dx^{\mu} g_{\mu\nu} dx^{\nu}} = \delta \int \sqrt{\dot{x}^{\mu} g_{\mu\nu}(\vec{x}) \dot{x}^{\nu}}\, d\tau = 0,$$

the variation set to zero being equivalent to finding the path of minimum distance. A geodesic is thus a unique curve on the geometry, in the same sense that a straight line is a unique curve connecting two points in a flat geometry. This means that a "gravitating" particle picks out the unique path in the geometry that will correspond to its trajectory.

The principle of least action provides a formalism for determining the functional form of the trajectory $\vec{x}(\tau)$ that minimizes the interval path length of the motion. If the interval is minimized, so is its square, so the Lagrangian $L(\vec{x}, \dot{\vec{x}})$ can be taken to be the square of the interval. The Euler–Lagrange equations, $\frac{d}{d\tau}\frac{\partial L}{\partial \dot{x}^{\mu}} - \frac{\partial L}{\partial x^{\mu}} = 0$, result in a form for the connections:

$$L(\vec{x}, \dot{\vec{x}}) \equiv g_{\mu\nu}(\vec{x}) \dot{x}^{\mu} \dot{x}^{\nu}, \quad \frac{d}{d\tau}(2 g_{\beta\nu}(\vec{x}) \dot{x}^{\nu}) - (\partial_{\beta} g_{\mu\nu}(\vec{x})) \dot{x}^{\mu} \dot{x}^{\nu} = 0,$$

$$\frac{d^2 x^{\nu}}{d\tau^2} + \frac{1}{2} g^{\nu\mu} \left[\frac{\partial g_{\alpha\mu}}{\partial x^{\beta}} + \frac{\partial g_{\beta\mu}}{\partial x^{\alpha}} - \frac{\partial g_{\alpha\beta}}{\partial x^{\mu}} \right] \frac{dx^{\alpha}}{d\tau} \frac{dx^{\beta}}{d\tau} = 0,$$

$$\Rightarrow \Gamma^{\nu}_{\alpha\beta} = \frac{1}{2} g^{\nu\mu} \left[\frac{\partial g_{\alpha\mu}}{\partial x^{\beta}} + \frac{\partial g_{\beta\mu}}{\partial x^{\alpha}} - \frac{\partial g_{\alpha\beta}}{\partial x^{\mu}} \right], \tag{5.14}$$

where $g^{\nu\mu}$ is the inverse to the metric, $g^{\nu\mu} g_{\mu\beta} = \delta^{\nu}_{\beta}$.

Connections and geodesics

In a general curvilinear coordinate system, the basis vectors $\vec{e}_\beta$ change their orientations based upon location. The connections on a general *affine* [76] space are defined as the coefficients that relate the local change in the basis vectors, which, of course, can be expressed as a linear combination of that complete set of basis vectors:

$$\frac{\partial}{\partial x^\alpha}\,\vec{e}_\beta = \sum_\nu \Gamma^\nu_{\alpha\beta}(\vec{x})\,\vec{e}_\nu. \tag{5.15}$$

Since the Minkowski basis vectors are spatially constant $\frac{\partial}{\partial x^\alpha}\vec{e}^{(0)}_\mu = 0$, then:

$$\frac{\partial}{\partial x^\alpha}\vec{e}_\beta = \frac{\partial}{\partial x^\alpha}\left(\frac{\partial \xi^\mu}{\partial x^\beta}\vec{e}^{(0)}_\mu\right) = \frac{\partial^2 \xi^\mu}{\partial x^\alpha \partial x^\beta}\frac{\partial x^\nu}{\partial \xi^\mu}\vec{e}_\nu = \Gamma^\nu_{\alpha\beta}(\vec{x})\,\vec{e}_\nu, \tag{5.16}$$

which corresponds exactly with the definition from Eq. 5.13.

The existence of an invariant inner product or distance measure on the geometry provides a direct relationship between the metric tensor g and the connections, Γ. Arbitrary tangent vectors can be expanded in terms of the basis vectors, which define the metric components in terms of the inner product of basis vectors,

$$\vec{A}\cdot\vec{B} = A^\mu B^\nu g_{\mu\nu} = A^\mu B^\nu \vec{e}_\mu \cdot \vec{e}_\nu \quad \Rightarrow \quad g_{\mu\nu} = \vec{e}_\mu \cdot \vec{e}_\nu. \tag{5.17}$$

Since the derivative is a linear operation, and the inner product is a bi-linear operation, this means derivatives of the scalar metric components give:

$$\frac{\partial}{\partial x^\alpha} g_{\mu\nu} = \Gamma^\beta_{\alpha\mu}\, g_{\beta\nu} + g_{\mu\beta}\,\Gamma^\beta_{\alpha\nu}. \tag{5.18}$$

Therefore, changes in the coordinate basis vectors at different positions in the metric space results in connections given by:

$$\sum_\beta g_{\lambda\beta}\,\Gamma^\beta_{\mu\nu} = \frac{1}{2}\left(\frac{\partial}{\partial x^\mu} g_{\lambda\nu} + \frac{\partial}{\partial x^\nu} g_{\mu\lambda} - \frac{\partial}{\partial x^\lambda} g_{\mu\nu}\right). \tag{5.19}$$

This precisely corresponds to the geodesic form derived in Eq. 5.14. Thus, the two expressions 5.13 and 5.14 are equivalent relationships for the affine connections relating local variation of the basis vectors. The most general form for connections in an affine space is given by $\Gamma^\beta_{\mu\nu} \equiv \langle \mathbf{e}^\beta, \partial_\mu \vec{e}_\nu \rangle$, where $\mathbf{e}^\beta$ is dual to the basis tangent vector $\vec{e}_\beta$.

Supplement: Torsion on manifolds

For coordinate geometries, locations on the manifold can be uniquely determined independent of the path taken from the reference origin. However, this need not be the case in general (see e.g., Figure 5.1), and is not true for non-abelian group manifolds. Such manifolds have non-vanishing *torsion*[1] [76]. Gravitation based upon the principle of equivalence must be torsion-free [77].

In order to examine any contribution of torsion to connections, one can rewrite the equation for the affine connections Eq. 5.15 in the form $\vec{e}_\mu \vec{e}_\nu = \Gamma^\lambda_{\mu\nu}\, \vec{e}_\lambda$. As previously mentioned, following Eq. 5.4, $[\vec{e}_\mu, \vec{e}_\nu] = c^\lambda_{\mu\nu} \vec{e}_\lambda \Rightarrow \Gamma^\lambda_{\mu\nu} - \Gamma^\lambda_{\nu\mu} = c^\lambda_{\mu\nu}$. Therefore, any asymmetry in the lower components of the affine connections implies non-vanishing torsion. Combining this equation with Eq. 5.18 results in general connections satisfying:

$$\Gamma^\lambda_{\mu\nu} = \frac{1}{2} g^{\lambda\alpha} \left(\left(g_{\nu\alpha,\mu} + g_{\mu\alpha,\nu} - g_{\mu\nu,\alpha} \right) - \left(g_{\mu\beta}\, c^\beta_{\nu\alpha} + g_{\beta\nu}\, c^\beta_{\mu\alpha} - g_{\alpha\beta}\, c^\beta_{\mu\nu} \right) \right).$$

For coordinate geometries like general relativity, the tensors $c^\beta_{\mu\alpha}$ all vanish, yielding the standard result for the connections.

It should be emphasized that the connections *do not* transform as components of a tensor. Under coordinate transformations, the connections satisfy:

$$\Gamma'^\beta_{\mu\nu} = \frac{\partial x'^\beta}{\partial x^\alpha} \Gamma^\alpha_{\lambda\kappa} \frac{\partial x^\lambda}{\partial x'^\mu} \frac{\partial x^\kappa}{\partial x'^\nu} + \frac{\partial x'^\beta}{\partial x^\alpha} \frac{\partial^2 x^\alpha}{\partial x'^\mu \partial x'^\nu}. \tag{5.20}$$

The inhomogeneous term in the transformation of the connections has similarities to the inhomogeneous term in the gauge transformation of a gauge potential. However, a gauge field such as $A^r_\mu \mathbf{G}_r$ has a single index associated with space-time which does not transform, and two indices on the generator matrix associated with the gauge transformation matrices, while all indices of the connections transform.

5.1.4 Covariant derivatives

The *covariant derivative* of a contravariant component of a vector field is defined by:

$$\frac{\partial}{\partial x^\alpha} \vec{A} = \left(\frac{\partial A^\nu}{\partial x^\alpha} + \Gamma^\nu_{\alpha\beta} A^\beta \right) \vec{e}_\nu \equiv A^\nu_{;\alpha}\, \vec{e}_\nu. \tag{5.21}$$

[1] T.J. Willmore, *An Introduction to Differential Geometry*. Oxford University Press, Oxford (1959).

Notice that $\frac{\partial}{\partial x'^{\mu}}\vec{A} = \frac{\partial x^{\alpha}}{\partial x'^{\mu}}\frac{\partial}{\partial x^{\alpha}}\vec{A}$. This implies that $A'^{\kappa}_{;\beta} = \frac{\partial x^{\alpha}}{\partial x'^{\mu}} A^{\nu}_{;\alpha}\frac{\partial x'^{\beta}}{\partial x^{\nu}}$, i.e., $A^{\nu}_{;\alpha}$ transforms as components of a mixed contravariant-covariant tensor of rank (1, 1).

One can similarly define the covariant derivative in terms of the covariant components of a tangent vector, and components of a 1-form. One requires that the function $\vec{A}\cdot\vec{B} = A^{\beta} g_{\beta\alpha} B^{\alpha} = A^{\beta} B_{\beta}$ be an invariant. Since the covariant derivative should satisfy the Liebnitz rule,

$$\frac{\partial}{\partial x^{\alpha}}\left(A^{\beta}B_{\beta}\right) = A^{\beta}_{;\alpha}B_{\beta} + A^{\beta}B_{\beta;\alpha}.$$

Expanding both sides of this expression yields the form of the covariant derivative:

$$B_{\beta;\alpha} \equiv \frac{\partial B_{\beta}}{\partial x^{\alpha}} - \Gamma^{\lambda}_{\alpha\beta}B_{\lambda}, \tag{5.22}$$

for the covariant components of $\vec{B}$. The covariant derivative of a general rank (R,S) tensor likewise introduces factors involving derivatives of the basis vectors and duals when expressed in terms of its components:

$$\left(T^{\mu_1\cdots\mu_R}_{\beta_1\cdots\beta_S}\right)_{;\alpha} \equiv \frac{\partial}{\partial x^{\alpha}}T^{\mu_1\cdots\mu_R}_{\beta_1\cdots\beta_S} + \sum_{r=1}^{R}\Gamma^{\mu_r}_{\alpha\nu_r}T^{\mu_1\cdots\nu_r\cdots\mu_R}_{\beta_1\cdots\beta_S} - \sum_{s=1}^{S}\Gamma^{\lambda_s}_{\alpha\beta_s}T^{\mu_1\cdots\mu_R}_{\beta_1\cdots\lambda_s\cdots\beta_S}, \tag{5.23}$$

where the corresponding contravariant tensor component μ_r is replaced by ν_r and summed, and likewise the corresponding covariant tensor component β_s is replaced by λ_s and summed.

A fairly convenient form for the sum over indices of connections can be obtained by recalling the formula $\log(\det \mathbf{M}) = Tr(\log \mathbf{M})$ from linear algebra. Applying this to relation $g \equiv \det \mathbf{g} = \det((g_{\mu\nu}))$, one obtains

$$\Gamma^{\nu}_{\lambda\nu} = \frac{1}{\sqrt{-g}}\frac{\partial}{\partial x^{\lambda}}\sqrt{-g}. \tag{5.24}$$

This allows the covariant divergence of contravariant vectors to be expressed in the elegant form:

$$V^{\beta}_{;\beta} = \frac{1}{\sqrt{-g}}\frac{\partial}{\partial x^{\beta}}\left(\sqrt{-g}\,V^{\beta}\right). \tag{5.25}$$

One must take care to include additional connection factors for the covariant divergence of tensors beyond rank one.

5.1.5 Newtonian correspondence

Given the geodesic equation, one can find a direct correspondence to the equations of Newtonian mechanics. For Newtonian gravitation, the gravitational field is "weak" in the sense that the space-time coordinates of the non-inertial observers are very close to those of the freely falling system ($t \simeq \tau$). Also, Newtonian mechanics is concerned with slowly moving systems compared to the motion of light. One can directly relate Newton's gravitational force equation with the geodesic equation as follows:

$$m_{inertial}\,\mathbf{a} = \mathbf{F}_{gravity} \Rightarrow m_{inertial}\frac{d^2\mathbf{r}}{dt^2} = -m_{grav}\,\nabla\Phi,$$

$$\frac{d^2 r^\lambda}{d\tau^2} + \Gamma^\lambda_{\mu\nu}\frac{dr^\mu}{d\tau}\frac{dr^\nu}{d\tau} = 0 \Rightarrow \frac{d^2 r^j}{dt^2} \approx -\Gamma^j_{00}c^2, \tag{5.26}$$

where the terms involving $\frac{dr^j}{d\tau} \approx \frac{dr^j}{dt} \ll c$ are neglected in the geodesic equation.

Because of the equivalence of inertial mass and gravitational mass $m_{inertial} = m_{grav} \equiv m$, the gravitational potential can directly be related to the connection:

$$\Gamma^j_{00} \simeq \frac{1}{c^2}\frac{\partial}{\partial r^j}\Phi, \quad \text{where} \quad \Phi = -\left(\frac{G_N M}{r}\right). \tag{5.27}$$

In this equation M is the mass of the *source* of the Newtonian gravitation field. For slow speeds and weak gravitational fields, one can directly find an approximate form for the metric g using the equation for the connection Γ given in Eq. 5.14. For static fields, $\frac{\partial g_{\mu\nu}}{\partial r^0} = \frac{\partial g_{\mu\nu}}{\partial ct} = 0$, so that the metric satisfies:

$$\Gamma^j_{00} \simeq -\frac{1}{2}\eta^{jk}\frac{\partial}{\partial r^k}g_{00} \simeq \frac{\partial}{\partial r^j}\left(\frac{G_N M}{r}\right).$$

Since the metric must be the same as that of special relativity when one is far away from the source of the gravitational field, $g_{00} \Rightarrow -1$ for $r \to \infty$. This means that the metric can be very closely determined by the form:

$$g_{00} \simeq -1 + \frac{2\,G_N\,M}{c^2\,r}. \tag{5.28}$$

This result embodies much of the philosophy of general relativity. The metric g is a purely geometric quantity that determines how one defines "distance" in the space-time. By choosing this form for the metric, one is able to obtain the same results as a gravitational force field that satisfies Newton's law of gravity using only a curved space-time. It is particularly interesting to note that a modification of the part of the space-time distance metric that deals with the time-time displacement (time- rather than space-metric modification) is dominant in order to emulate Newtonian gravitation.

Geodesics revisited

The geodesic equation has been shown to describe the trajectory of a freely falling object. It also represents the evolution of the tangent vector of that object defined by its four-velocity $u^\mu = \frac{dx^\mu}{d\tau}$ (for time-like trajectories). More generally, the geodesic motions of both massive and massless particles can be described in terms of the evolution of the tangent vectors with respect to an affine parameter λ that quantifies proper progression along the trajectory. In terms of this tangent vector $\vec{U}$, the general motions described in Eq. 5.12 are given by:

$$\left(\frac{dU^\mu}{d\lambda} + \Gamma^\mu_{\beta\nu} U^\beta U^\nu \right) = 0, \qquad \text{where} \quad U^\mu = \frac{dx^\mu}{d\lambda}. \tag{5.29}$$

For massive particles following time-like trajectories, $\lambda = c\tau$ and $\vec{U} = \frac{\vec{u}}{c}$. Since massless particles follow null trajectories $ds^2 = 0 = -(dc\tau)^2$, the proper time is not useful for parameterizing such trajectories, and an appropriate parameterization consistent with Eq. 5.29 should be used.

Supplement: Why massless particles must gravitate

Consider a massive particle that falls from rest through a gravitational potential $\delta\Phi$. After this fall, the energy is given by $mc^2 + m\delta\Phi = mc^2(1 + \frac{\delta\Phi}{c^2})$. The kinetic energy is then extracted for other uses, and the mass is brought back to rest. Suppose that the rest energy is then converted into photons, all of which are elastically reflected upwards without loss of energy. If the photons are collected at the original starting point of the mass without loss of gravitational energy, they could in principle be reassembled into the original mass at rest, resulting in a perpetual energy source.

Therefore, energy conservation requires that the photons are *redshifted* as they climb through the gravitational potential:

$$E_{bottom} = E_{top}\left(1 + \frac{\delta\Phi}{c^2}\right), \qquad \frac{h\nu_B}{h\nu_T} = \frac{\lambda_T}{\lambda_B} = 1 + \frac{\delta\Phi}{c^2},$$

which defines the redshift parameter $\frac{\delta\lambda}{\lambda}$ in terms of the gravitational potential.

If the tangent vector can be parameterized by the coordinates x^μ, then the geodesic equation takes the form:

$$U^\beta \left(\frac{\partial U^\mu}{\partial x^\beta} + \Gamma^\mu_{\beta\nu} U^\nu \right) = U^\beta U^\mu_{;\beta} = 0. \tag{5.30}$$

This implies that in some sense, components of the four-velocity are conserved along the geodesic motion. This is referred to as *parallel transport* of this vector along the trajectory.

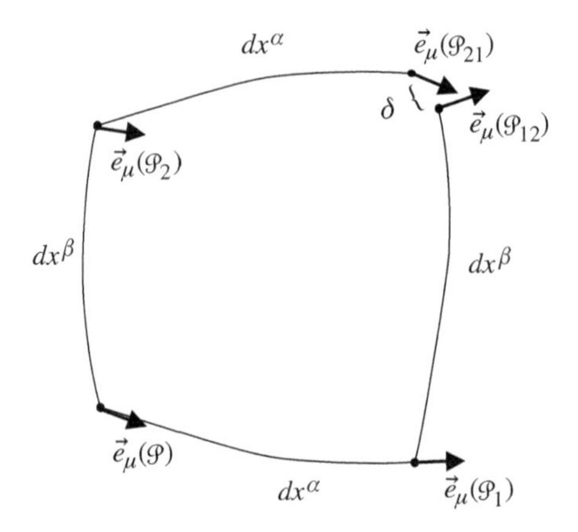

Figure 5.1 Construction of a curvature measure.

5.1.6 Curvature

The *curvature* of a surface can be parameterized by how vectors maintain their direction as they are parallel-transported on the surface. For instance, a vector oriented along the prime meridian at the North Pole can be transported along that meridian to the Equator while maintaining its parallel orientation with that curve of longitude on the surface of the Earth. The vector is perpendicular to the equator, and it can be transported along the equator to the 90° longitude maintaining its orientation, then parallel transported along that meridian back to the North Pole, where it will be perpendicular to its original orientation due to the curvature of the Earth. A formal mechanism to ensure that the components of a vector $\vec{A}$ are maintained relative to a curve $x^\mu(\zeta)$ is given by the equation of parallel transport:

$$\frac{dA^\mu}{d\zeta} + \frac{dx^\beta}{d\zeta}\Gamma^\mu_{\beta\nu}A^\nu = 0 = U^\beta_{(\zeta)}A^\mu_{;\beta}. \tag{5.31}$$

The concept of parallel transport of vectors allows one to quantify the curvature of the surface. The second form directly shows that derivatives of components along the tangent vector of transport vanish.

A local measure of the curvature can be constructed by examining the difference in a vector parallel transported via differing infinitesimal paths to a nearby point as illustrated in Figure 5.1. For a torsion-free coordinate geometry, the points $\mathcal{P}_{12} = \mathcal{P}_{21}$ are coincident, i.e., the parameter δ in Figure 5.1 vanishes. This is a necessary condition for points in the geometry to be represented by path-independent coordinates. As previously mentioned, if the principle of equivalence is satisfied, the geometry is torsion-free [77]. For instance, a freely falling observer (using local Minkowski coordinates) that parallel-transports three vectors forming a triangle satisfying $\vec{A} + \vec{B} + \vec{C} = \vec{0}$, should not have the legs of that triangle break apart during the geodesic motion. A Riemann geometry is a general coordinate geometry characterized by a metric.

If there is non-vanishing curvature, then $\vec{e}_\mu(\mathcal{P}_{12}) \neq \vec{e}_\mu(\mathcal{P}_{21})$ when examined along differing paths characterized by infinitesimal displacements (dx^α, dx^β). Since the basis set is complete, the difference can be expressed as coefficients of those basis vectors given by:

$$\left[\frac{\partial}{\partial x^\alpha}, \frac{\partial}{\partial x^\beta}\right] \vec{e}_\mu \equiv -\mathcal{R}^\nu_{\ \mu\alpha\beta}\, \vec{e}_\nu, \tag{5.32}$$

where the sign corresponds to the traditional definition of the *Riemann curvature tensor*. Whereas this commutation operating on integrable scalar functions must vanish, the basis vectors mix under the operation. The commutation of covariant differentiation on vector components can be seen to satisfy $A^\nu_{\ ;\alpha\,;\beta} - A^\nu_{\ ;\beta\,;\alpha} = \mathcal{R}^\nu_{\ \mu\alpha\beta} A^\mu$ and $B_{\mu\,;\alpha\,;\beta} - B_{\mu\,;\beta\,;\alpha} = -\mathcal{R}^\nu_{\ \mu\alpha\beta} B_\nu$.

Using the relation defined in Eq. 5.15, the curvature tensor can be expressed in terms of the connections via:

$$-\mathcal{R}^\nu_{\ \mu\alpha\beta} = \frac{\partial}{\partial x^\alpha} \Gamma^\nu_{\ \beta\mu} - \frac{\partial}{\partial x^\beta} \Gamma^\nu_{\ \alpha\mu} + \Gamma^\lambda_{\ \beta\mu} \Gamma^\nu_{\ \alpha\lambda} - \Gamma^\lambda_{\ \alpha\mu} \Gamma^\nu_{\ \beta\lambda}. \tag{5.33}$$

This tensor can alternatively be expressed directly in terms of second derivatives of the metric. After considerable algebra,

$$\begin{aligned}\mathcal{R}^\nu_{\ \mu\alpha\beta} = g^{\nu\kappa} \Bigg[& \frac{1}{2} \left(\frac{\partial^2 g_{\kappa\alpha}}{\partial x^\beta \partial x^\mu} - \frac{\partial^2 g_{\kappa\beta}}{\partial x^\alpha \partial x^\mu} + \frac{\partial^2 g_{\mu\beta}}{\partial x^\alpha \partial x^\kappa} - \frac{\partial^2 g_{\kappa\alpha}}{\partial x^\beta \partial x^\mu} \right) \\ & + g_{\lambda\zeta} \left(\Gamma^\lambda_{\ \alpha\kappa} \Gamma^\zeta_{\ \mu\beta} - \Gamma^\lambda_{\ \beta\kappa} \Gamma^\zeta_{\ \mu\alpha} \right) \Bigg].\end{aligned} \tag{5.34}$$

This form will be convenient for examining some of the symmetries of the Riemann tensor.

5.1.7 Symmetries of the Riemann tensor

A general space-time tensor of rank (1,3) has $4^4 = 256$ independent components. However, the various symmetries of the Riemann tensor considerably reduces its number of independent components. The symmetries of this tensor are itemized below:

- Anti-symmetry due to commutation of covariant derivatives:

$$\mathcal{R}^\nu_{\ \mu\alpha\beta} = -\mathcal{R}^\nu_{\ \mu\beta\alpha}; \tag{5.35}$$

- Cyclicity on covariant components:

$$\mathcal{R}^{\nu}_{\ \mu\alpha\beta}+\mathcal{R}^{\nu}_{\ \alpha\beta\mu}+\mathcal{R}^{\nu}_{\ \beta\mu\alpha}=0; \tag{5.36}$$

- Symmetry under exchange of covariant pairs:

$$g_{\lambda\nu}\,\mathcal{R}^{\nu}_{\ \mu\alpha\beta}=g_{\alpha\lambda}\,\mathcal{R}^{\lambda}_{\ \beta\nu\mu},\ \text{or}\ \mathcal{R}_{\nu\mu\alpha\beta}=\mathcal{R}_{\alpha\beta\nu\mu}. \tag{5.37}$$

These symmetries reduce the number of independent components to 20.

Two additional tensors constructed from the Riemann tensor will be important in this discussion. The *Ricci tensor* is defined by:

$$\mathcal{R}_{\mu\beta}\equiv\mathcal{R}^{\lambda}_{\ \mu\lambda\beta}=\mathcal{R}_{\beta\mu}. \tag{5.38}$$

The symmetry of this tensor follows directly from Eq. 5.37. A further contraction of the Ricci tensor produces the *Ricci scalar* (sometimes called the Ricci curvature scalar):

$$\mathcal{R}\equiv g^{\mu\beta}\mathcal{R}_{\mu\beta}. \tag{5.39}$$

These tensors will be crucial in constructing the Einstein tensor shortly.

The Bianchi identities

Covariant derivatives of the Riemann tensor satisfy a general algebraic symmetry due to the Jacobi identity. If one denotes the covariant derivative operation on a tensor component by ∇_α, then algebraically $[\nabla_\kappa,[\nabla_\beta,\nabla_\alpha]]+[\nabla_\beta,[\nabla_\alpha,\nabla_\kappa]]+[\nabla_\alpha,[\nabla_\kappa,\nabla_\beta]]=0$. This relationship then implies the *Bianchi identities*:

$$R^{\nu}_{\ \mu\alpha\beta\,;\kappa}+R^{\nu}_{\ \mu\kappa\alpha\,;\beta}+R^{\nu}_{\ \mu\beta\kappa\,;\alpha}=0. \tag{5.40}$$

These identities are analogous to the inhomogeneous Maxwell equations resulting from gauge covariant derivatives.

Kinematic invariants and Killing vectors

One of the most convenient characteristics of conservative force fields is that kinematic parameters can be defined that allow some problems to be addressed using conservation principles, rather than having to solve complicated equations of dynamics. It is hopeful that one can similarly find kinematic parameters on a geometry that satisfy certain types of conservation properties. One type of conserved parameters can be found when the functional form of the metric g is invariant under coordinate transformations proportional to changes along a local tangent vector $\vec{\Upsilon}$. Formally, this means that under an infinitesimal local coordinate change along the tangent vector $x^\mu\to x'^\mu=x^\mu+\delta a\,\Upsilon^\mu$, the form of all of the metric components remain invariant, i.e., $g_{\mu\nu}(\vec{x})\to g'_{\mu\nu}(\vec{x}\,')=g_{\mu\nu}(\vec{x}\,')$. The tangent vector $\vec{\Upsilon}$ is referred to as the generator of an *isometry* on the manifold.

In order to determine the equation satisfied by such vectors on the geometry, examine the transformation of the metric:

$$g_{\mu\nu}(\vec{x}) = \frac{\partial x'^{\alpha}}{\partial x^{\mu}} g'_{\alpha\beta}(\vec{x}\,') \frac{\partial x'^{\beta}}{\partial x^{\nu}} = \frac{\partial x'^{\alpha}}{\partial x^{\mu}} g_{\alpha\beta}(\vec{x}\,') \frac{\partial x'^{\beta}}{\partial x^{\nu}} . \tag{5.41}$$

Substitution of the coordinate change x'^{μ} for infinitesimal δa into Eq. 5.41 yields the equation:

$$g_{\alpha\nu}(\partial_{\mu}\Upsilon^{\alpha}) + g_{\mu\beta}(\partial_{\nu}\Upsilon^{\beta}) + (\partial_{\lambda}g_{\mu\nu})\Upsilon^{\lambda} = 0 ,$$

which can be simplified to give:

$$\Upsilon_{\mu;\nu} + \Upsilon_{\nu;\mu} = 0, \quad \Upsilon^{\mu}_{;\mu} = 0. \tag{5.42}$$

The first equation is known as *Killing's* equation. The second conservation equation of 5.42 follows directly from the first by multiplying by the inverse metric.

If $\vec{U}$ represents the tangent vector of a geodesic, one can demonstrate that the component of $\vec{U}$ along the Killing vector is conserved along that geodesic. This can be shown by examining the form:

$$U^{\alpha}(U^{\beta}\Upsilon_{\beta})_{;\alpha} = U^{\alpha}U^{\beta}_{;\alpha}\Upsilon_{\beta} + U^{\alpha}U^{\beta}\Upsilon_{\beta;\alpha} = U^{\alpha}U^{\beta}_{;\alpha}\Upsilon_{\beta} + \frac{1}{2}U^{\alpha}U^{\beta}(\Upsilon_{\beta;\alpha} + \Upsilon_{\alpha;\beta}).$$

The first term on the right vanishes due to the geodesic equation $U^{\alpha}U^{\beta}_{;\alpha} = 0$, while the second term vanishes due to Killing's equation, 5.42. Thus:

$$U^{\alpha}(U^{\beta}\Upsilon_{\beta})_{;\alpha} = 0 \quad \Rightarrow \quad \vec{U} \cdot \vec{\Upsilon} = \text{constant along a geodesic.} \tag{5.43}$$

Therefore, the value of this inner product is maintained during geodesic motion.

Certain types of Killing vectors can be determined in a straightforward manner. If the metric is independent of a particular coordinate x^{a}, then one can directly construct a Killing vector associated with that symmetry. Specifically, suppose that $\frac{\partial}{\partial x^{a}} g_{\mu\nu}(\vec{x}) = 0$. Then the components of a vector field defined by $\Upsilon^{\beta} \equiv \delta^{\beta}_{a}$ will be shown to automatically satisfy Killing's equation. Note that the covariant form of this vector is given by $\Upsilon_{\mu} = g_{\mu\beta}\Upsilon^{\beta} = g_{\mu a}$. Examining the form of Eq. 5.42,

$$\Upsilon_{\mu;\nu} + \Upsilon_{\nu;\mu} = \partial_{\nu}g_{\mu a} + \partial_{\mu}g_{\nu a} - g^{\lambda\beta}(\partial_{\nu}g_{\mu\beta} + \partial_{\mu}g_{\nu\beta} - \partial_{\beta}g_{\mu\nu})g_{\lambda a} = \frac{\partial}{\partial x^{a}} g_{\mu\nu} = 0. \tag{5.44}$$

Therefore, $\Upsilon_{\mu} = g_{\mu a}$ indeed represents covariant components of a Killing vector on the geometry.

This result is useful for finding temporally conserved forms in time-independent geometries. Such geometries have a time-like Killing vector with components given by $g_{\beta 0}$. Recall that in flat space-time, the combination $p^{0}_{(s)}\vec{U}$ represents the four-momentum of a massive or massless particle, where $p^{0}_{(s)}$ represents the standard state energy, and $\vec{U}$ is the dimensionless four-velocity of the particle. Using Eq. 5.43 to define an energy $E \equiv -p^0_{(s)} U^\beta g_{\beta 0} = -p^0_{(s)} U_0$, this kinematic form will be conserved by a particle undergoing geodesic motion. Similar conserved quantities can be constructed in geometries that are independent of angular translations (angular momentum), transverse motions, etc.

5.2 Einstein's equations

In Section 5.1.5 a modification of the space-time metric was demonstrated to give Newton's gravitation law in the non-relativistic, weak field limit. It now remains to construct a general form connecting the sources of arbitrarily strong gravitational fields to curvatures in the space-time geometry. One generally characterizes empty space as being devoid of energy-momentum density but not necessarily devoid of curvature. However, that curvature is expected to somehow be coupled to mass or energy-momentum sources *somewhere*. The Einstein field equations will demonstrate a general covariant form coupling a geometrically conserved curvature to dynamically conserved energies and momenta.

5.2.1 Correspondence with classical gravitation

The quantities derived from the curvature tensors discussed in Section 5.1.6 are generally tensor forms that transform covariantly under arbitrary coordinate transformations. One therefore needs to connect a covariant curvature form to a covariant energy-momentum form that can source gravity as given by Newton's laws in the appropriate limits. This covariant energy-momentum function T should also vanish in empty space $T \overset{\text{empty}}{\Rightarrow} 0$. An appropriate form will be provided by the conserved energy-momentum tensor.

Conserved gravitational sources

In order to incorporate more general kinematic situations and motions into the geometrodynamics of "gravitating" systems, it is useful to discuss one of the most fundamental phenomenological characteristic properties of the local physical universe. The energy and momentum can flow only from one region of the universe into another, and there seems to be no evidence allowing "local" energy or momentum to enter or leave the universe. Energy conservation can be mathematically expressed in terms of the energy density u and the energy flow (flux) vector Q^j using the equation:

$$\frac{\partial}{\partial t} u + \vec{\nabla} \cdot \vec{Q} = 0.$$

One obtains the energy within a region of space-time by appropriately integrating the density u over that volume. Similarly the conservation of momentum can be expressed in terms of the momentum density π^j and the momentum flux tensor T^{jk} using the equation:

$$\frac{\partial}{\partial t}\pi^j + \frac{\partial}{\partial \xi^k} T^{jk} = 0.$$

From special relativity, the space-time indices can be combined into a general energy-momentum tensor that is conserved:

$$\mathbf{T} = \begin{pmatrix} \mathbf{T}^{00} = u & \mathbf{T}^{0k} = Q^k/c \\ \mathbf{T}^{j0} = \pi^j c & \mathbf{T}^{jk} = T^{jk} \end{pmatrix} \text{ with } \sum_{\beta=0}^{3} \frac{\partial}{\partial \xi^\beta} \mathbf{T}^{\mu\beta} = 0.$$

Once curvilinear basis vectors are introduced for general relativity, derivatives of the basis vectors modify this conservation condition into the form:

$$T^{\mu\beta}_{;\beta} = 0. \tag{5.45}$$

This energy-momentum tensor contains the mass-energy of the classical "gravitational" sources and can be used to couple the energies and momenta of the sources to a correspondingly conserved geometrical object in order to construct the general equation satisfied by the geometry of general gravitating systems.

Geometrically conserved curvatures

The geometric object that can be constructed that satisfies the proper conservation properties involves the Ricci tensor $\mathcal{R}_{\mu\beta}$ and its contraction the Ricci scalar $\mathcal{R}$. If one examines $g^{\mu\beta}\times$ Eq. 5.40 from the Bianchi identity, and contract the indices ν and α, the following condition is obtained:

$$\mathcal{R}_{;\kappa} - \mathcal{R}^{\mu}_{\kappa;\mu} - \mathcal{R}^{\nu}_{\kappa;\nu} = 0. \tag{5.46}$$

The special tensor that Einstein generated $\mathcal{G}$ takes advantage of this geometric conservation property:

$$\mathcal{G}_{\mu\beta} \equiv \mathcal{R}_{\mu\beta} - \frac{1}{2} g_{\mu\beta}\mathcal{R}, \quad \mathcal{G}^{\mu}_{\beta;\mu} = 0. \tag{5.47}$$

Besides being covariantly conserved, this tensor is symmetric under exchange of the indices μ and β. The metric tensor $g_{\mu\beta}$ also satisfies these properties, as will any constant multiple of this tensor. The *Einstein field equations*:

$$\mathcal{G}_{\mu\beta} - \Lambda\, g_{\mu\beta} = -\frac{8\pi G_N}{c^4}\mathbf{T}_{\mu\beta}, \tag{5.48}$$

couple the geometry to a symmetric energy-momentum tensor, as long as the *cosmological constant*, Λ, is a true constant. The symmetry of the energy-momentum tensor is physically meaningful, since T^{0k} = energy density flux vector = momentum density = T^{k0}, and non-symmetric stresses T^{jk} would generate non-equilibrium local torques on the energy distribution. Einstein's field equations directly couple gravity to the "stuff" that gravitates.

An alternative form of the Einstein field equations is obtained by contracting Eq. 5.48 with the inverse metric $g^{\mu\beta}$. Recognizing that $g^{\mu\beta}g_{\mu\beta} = \delta^{\mu}_{\mu} = 4$, this contraction yields $\mathcal{R} - 2\mathcal{R} - 4\Lambda = -\frac{8\pi G_N}{c^4}\mathbf{T}^{\lambda}{}_{\lambda}$. This results in the form:

$$\mathcal{R}_{\mu\beta} = -\frac{8\pi G_N}{c^4}\left(\mathbf{T}_{\mu\beta} - \frac{1}{2}g_{\mu\beta}\mathbf{T}^{\lambda}{}_{\lambda} + g_{\mu\beta}\frac{\Lambda c^4}{8\pi G_N}\right). \tag{5.49}$$

A positive cosmological constant can be interpreted as contributing some type of zero-point density as a gravitational source.

The factor $-\frac{8\pi G_N}{c^4}$ in the field equations is needed in order to generate the equations of local Newtonian gravitation in the weak-field, slow-speed limit. This can be seen by recalling the Newtonian form of the connection Eq. 5.27 and metric Eq. 5.28 obtained from the geodesic equation. From Eq. 5.33, the Ricci tensor then satisfies $\mathcal{R}_{00} \simeq -\frac{1}{c^2}\nabla^2\Phi(r)$. A static, symmetric form for the energy-momentum tensor is given by $\mathbf{T}_{00} \simeq \rho$, with $\mathbf{T}^{\lambda}{}_{\lambda} \simeq -\rho$ in terms of energy density (*not* mass density). Therefore, the Poisson equation for gravitation $-\nabla^2\Phi(r) = -4\pi G_N \rho(r)/c^2$ (neglecting the cosmological factor) is appropriately reproduced by Eq. 5.49.

It should be noted that Einstein field equations are *local*. This means that nearby sources dominate the local geometry, irrespective of overall cosmological behaviors. In this sense, a small cosmological constant will not dominate the behaviors within gravitationally bound systems but might dominate behaviors between unbound co-moving regions.

Supplement: Relativity, absolute space, and absolute motion

The epistemology of absolute motions has been a long-standing problem in natural philosophy.[2,3,4] The classical ideas of Newton established absolute space and time, with accelerations defined relative to this absolute. A common analysis of these ideas involves examining a bucket of water. If the bucket of water is set rotating relative to the distant stars, the meniscus of the fluid rises against the bucket due to its centripetal motion, as the water comes to rest relative to the surface of the bucket.

[2] H.G. Alexander, *The Leibniz-Clarke Correspondence*. Manchester University Press, Barnes & Noble, New York (1970).

[3] E. Mach, *The Science of Mechanics: A Critical & Historical Account of Its Development*. The Open Court Publishing Co., Chicago (1960).

[4] L. Sklar, *Space, Time, and Spacetime*. The University of California Press, Berkeley (1974).

Mach argued that this behavior would likewise arise with a stationary bucket of water if all distant stars were rotated with the same angular velocity, i.e., that *all* motions and accelerations are relative to the predominant matter background. It was the hope of many that general relativity inherently incorporated *Mach's principle* into a formal structure. However, the Einstein field equations are differential equations, and as such, say little about any boundary conditions. Complete solutions to Einstein field equations require the imposition of boundary conditions. Even if the equations of motion are completely relative, there can still be motions relative to the boundary conditions. Thus, the field equations themselves do not completely describe the cosmology with regards to Mach's principle.

In particular, Big Bang cosmology (which relies substantially upon general relativity) indeed infers the preferred frame of the thermal fluid background of the expanding universe in describing motions through (for instance) the CMB. The motion of the Earth has been detected through asymmetries in the otherwise uniform background radiation based upon the direction of observation. Cosmology will be further discussed in Chapter 8.

Some recent concerns involve whether space-time itself has a structure, an arena, or a set of relationships established only through events. Such examinations have been motivated by the discrete nature of quantum detections and descriptions of dynamics using general relativity. For instance, if the fundamental structure of space-time is discrete, only a random lattice would be consistent with observed isotropy. Otherwise, a regular lattice would manifest physical effects analogous to those exhibited by crystals. Local connections should be defined in terms of some probability distribution that is a function of invariant intervals. However, normalization of the probability distribution is problematic,[5] complicating efforts to eliminate ultraviolet divergences using such methods. Curvature effects from an inhomogeneous rigid lattice also considerably complicate the microscopic gravitational dynamics.

Beyond the previous concerns, microscopic space-time definitions of connectivity and differentiability on lattices are not unique and elegant. Since there is no evidence for inherently discrete space-time, only physics on a differentiable manifold of convenient coordinates will be considered in what follows.

5.2.2 Action principle

As was discussed in Chapter 3, the principle of least action is quite convenient for deriving and describing the equations of motion in microscopic physics. It is

[5] C. Moore, Comment on "Space-Time as a Causal Set", *PRL* **60**, 655 (1988).

convenient to develop an action that generates Einstein's equations. The gravitational action should be an invariant that will be extremized to give a unique solution characterized by differing coordinates. Thus, one might consider the Ricci scalar as a contributor to the gravitational action in the form:

$$W_{Grav} \propto \int d^4x \sqrt{-g}\, \mathcal{R} \quad \rightarrow \quad \mathcal{L} \propto \sqrt{-g}\, \mathcal{R}. \tag{5.50}$$

Variations of this Lagrangian density are given by:

$$\begin{aligned}\delta\left(\sqrt{-g}\,\mathcal{R}\right) &= \delta\left(\sqrt{-g}\, g^{\alpha\beta}\mathcal{R}_{\alpha\beta}\right) \\ &= (\delta\sqrt{-g})\mathcal{R} - \sqrt{-g}\, g^{\alpha\mu}(\delta g_{\mu\nu})g^{\nu\beta}\mathcal{R}_{\mu\nu} + \sqrt{-g}\, g^{\alpha\beta}(\delta\mathcal{R}_{\alpha\beta}).\end{aligned} \tag{5.51}$$

The various terms need to be factored into a common arbitrary variant to satisfy the principle of least action.

The Riemann tensor has variations that can be expressed in terms of the connections:

$$\delta\mathcal{R}^{\lambda}_{\alpha\mu\beta} = \delta\Gamma^{\lambda}_{\alpha\beta;\mu} - \delta\Gamma^{\lambda}_{\alpha\mu;\beta}, \tag{5.52}$$

where the anti-symmetric covariant derivative difference indeed acts upon the *variation* of the connections $\delta\mathbf{\Gamma}$ as components of a tensor. Contracting on the indices λ and μ in 5.52 gives an expression for variations of the Ricci tensor:

$$\delta\mathcal{R}_{\alpha\beta} = \delta\Gamma^{\lambda}_{\alpha\beta;\lambda} - \delta\Gamma^{\lambda}_{\alpha\lambda;\beta}. \tag{5.53}$$

This is the needed expression for variations of the Ricci tensor.

A particular property of the metric and Jacobian factors is convenient in order to proceed. Generally, from Eq. 5.18 one concludes that covariant derivatives of the metric components vanish, $g_{\mu\nu;\lambda}$. This implies that the same is true of the Jacobian factor $\left(\sqrt{-g}\right)_{;\lambda} = 0$. This allows the term in Eq. 5.51 involving the variation of the Ricci tensor to be written in the form:

$$\begin{aligned}\sqrt{-g}\, g^{\alpha\beta}(\delta\mathcal{R}_{\alpha\beta}) &= \left(\sqrt{-g}\, g^{\alpha\beta}\delta\Gamma^{\lambda}_{\alpha\beta}\right)_{;\lambda} - \left(\sqrt{-g}\, g^{\alpha\beta}\delta\Gamma^{\lambda}_{\alpha\lambda}\right)_{;\beta}, \\ &= \partial_\lambda\left(\sqrt{-g}\, g^{\alpha\beta}\delta\Gamma^{\lambda}_{\alpha\beta}\right) - \partial_\beta\left(\sqrt{-g}\, g^{\alpha\beta}\delta\Gamma^{\lambda}_{\alpha\lambda}\right).\end{aligned} \tag{5.54}$$

If the arbitrary metric variations that minimize the action are chosen to vanish on the bounding surface, the terms in Eq. 5.54 do not contribute to the variation in Eq. 5.51 (since they give surface integrals). Expanding the Jacobian determinant in co-factors, and using Cramer's rule for the inverse of the metric, the effective variation is given by:

$$\delta\left(\sqrt{-g}\,\mathcal{R}\right) \doteq \sqrt{-g}\left[\frac{1}{2}g^{\mu\nu}\mathcal{R} - \mathcal{R}^{\mu\nu}\right]\delta g_{\mu\nu} = -\sqrt{-g}\,\mathcal{G}^{\mu\nu}\,\delta g_{\mu\nu}. \tag{5.55}$$

This gives the expected form for the gravitational contribution to the overall action functional for a physical system:

$$W_{Grav} = -\frac{c^3}{8\pi G_N} \int d^4x \sqrt{-g}\, \mathcal{R}, \tag{5.56}$$

or $\mathcal{L}_{Grav} = -\frac{c^3}{8\pi G_N} \sqrt{-g}\, \mathcal{R}$.

Comparing the gravitational action with Einstein's equation allows the identification of a form for the energy-momentum tensor of the matter distribution. The energy-momentum tensor associated with sources of gravitation satisfies:

$$\mathbf{T}^{\mu\nu} = -\frac{2}{\sqrt{-g}} \frac{\delta}{\delta g_{\mu\nu}} (\sqrt{-g}\, L_{matter}), \tag{5.57}$$

where L_{matter} represents the Lagrangian form of the micro-physical model with the Minkowski metric replaced by the curvilinear metric, and any derivatives of vectors replaced by covariant derivatives.

Tetrad formulation

Some descriptions of microscopic phenomena are not directly expressed in terms of the metric, rather, they are best described by vector-valued functions combined with derivatives. For such systems, the *tetrad* formulation (sometimes referred to as *vierbien*) provides a convenient method for calculating the energy-momentum tensor.

A set of tetrads $\mathcal{V}_\alpha^{\hat{\mu}}$ transform the components of a curvilinear contravariant vector to those of the locally inertial coordinates consistent with the equivalence principle:

$$\mathcal{V}_\alpha^{\hat{\mu}} \equiv \frac{\partial \xi^{\hat{\mu}}}{\partial x^\alpha}, \quad g_{\alpha\beta} = \mathcal{V}_\alpha^{\hat{\mu}}\, \eta_{\hat{\mu}\hat{\nu}}\, \mathcal{V}_\beta^{\hat{\nu}}, \quad \mathcal{V}_{\hat{\mu}}^{-1\,\alpha} \equiv \frac{\partial x^\alpha}{\partial \xi^{\hat{\mu}}} = \eta_{\hat{\mu}\hat{\nu}}\, \mathcal{V}_\beta^{\hat{\nu}}\, g^{\alpha\beta}. \tag{5.58}$$

Tetrads are hybrid forms that have mixed components, one of which transforms under curvilinear transformations α, the other of which transforms under locally inertial transformations $\hat{\mu}$. Inertial vector field components can be obtained from the contravariant field components using $A^{\hat{\mu}} = \mathcal{V}_\alpha^{\hat{\mu}}\, A^\alpha$.

Several important geometric functions are directly related to the tetrads. The Jacobian in integrals satisfies:

$$\sqrt{-g} = \left| \frac{\partial(\vec{\xi})}{\partial(\vec{x})} \right| = \det \boldsymbol{\mathcal{V}}. \tag{5.59}$$

The affine connections can likewise be defined in terms of the tetrads and their inverse,

$$\Gamma^\lambda_{\alpha\beta} = \mathcal{V}_{\hat{\mu}}^{-1\,\lambda} \frac{\partial}{\partial x^\alpha} \mathcal{V}_\beta^{\hat{\mu}}. \tag{5.60}$$

Thus, all physical geometric functions can be expressed in terms of the tetrads.

The energy-momentum tensor is obtained using variation of geometric functions. One should note that variations of the metric can be expressed as:

$$\delta g_{\alpha\beta} = 2 g_{\alpha\lambda} \mathcal{V}^{-1\,\lambda}_{\hat{\mu}} \, \delta\mathcal{V}^{\hat{\mu}}_{\beta} = g_{\alpha\lambda} \mathcal{V}^{-1\,\lambda}_{\hat{\mu}} \, \delta\mathcal{V}^{\hat{\mu}}_{\beta} + g_{\beta\lambda} \mathcal{V}^{-1\,\lambda}_{\hat{\mu}} \, \delta\mathcal{V}^{\hat{\mu}}_{\alpha}. \tag{5.61}$$

When curvilinear coordinate forms are expressed as integrals over the tetrads, variations in the space-time coordinates involve functional variations in the tetrads. The energy-momentum tensor from Eq. 5.57 can be alternatively expressed as:

$$\mathbf{T}^{\alpha\beta} g_{\alpha\lambda} = \mathbf{T}^{\beta}_{\lambda} = -\frac{1}{\det \mathcal{V}} \mathcal{V}^{\hat{\mu}}_{\lambda} \frac{\delta}{\delta\mathcal{V}^{\hat{\mu}}_{\beta}} ((\det \mathcal{V})\, L_{matter}). \tag{5.62}$$

This allows the mixed energy-momentum tensor to be written in a form with no explicit dependency on the metric.

5.2.3 Energy conditions

Classical gravitating systems are expected to be sourced from local energy densities that everywhere satisfy causal properties, referred to as *energy conditions*. These conditions assert that, in some manner, observers should locally measure gravitational fields generated by time-like or light-like sources, regardless of their motions. This is consistent with an expectation that no energy source can propagate at a speed greater than that of light, thereby generating space-like effects. Unlike classical systems, quantum systems *do* exhibit space-like coherent behaviors that are consistent with all communications being causal. As will be discussed in Chapter 7, the evaporation of black holes infers such space-like coherence, thereby locally violating energy conditions. Also, as discussed in Section 2.3.3, quantum systems with significant binding might violate these conditions if the energy component of the energy-momentum tensor is sufficiently reduced relative to other components. The general forms of the classical energy conditions will be discussed below.

Using a convention requiring time-like invariants to have negative signs, the *null* and *weak* energy conditions assert that the form $\mathcal{I}_{null/weak}$ defined by:

$$\mathcal{I}_{null/weak} \equiv -u^{\mu}_{null/weak} \mathbf{T}_{\mu\beta} \, u^{\beta}_{null/weak} \leq 0,$$

should be non-positive for light-like (null) or time-like (weak) observer four-velocities $\vec{u}_{null/weak}$, where the components of the energy-momentum tensor sourcing the gravitational field in Einstein's equation are represented by $\mathbf{T}_{\mu\beta}$. These conditions constrain the sign of the particular observed component of the energy-momentum tensor associated with energy density.

The *dominant* energy condition develops the form of the four-momentum density of the gravitational source more directly, as seen by an observer with four-velocity $\vec{u}_{observer}$. This source density, which is given by $\Pi^{\mu}_{source} \equiv -T^{\mu}{}_{\beta}\, u^{\beta}_{observer}$, is expected to be time-like or light-like irrespective of the motion of the observer,

$$\mathcal{I}^{DE}_{observer} \equiv \vec{\Pi}_{source} \cdot \vec{\Pi}_{source} \leq 0,$$

where the dot product is defined by the metric of the geometry. This condition directly asserts that the local-source density can nowhere be space-like.

Supplement: Tests of general relativity

There continues to be extensive work done testing the viability of general relativity as a theory of gravitation. General relativity is a metric theory of gravity that incorporates the Einstein equivalence principle, that the laws governing experiments in a local freely falling frame are compatible with special relativity (locally Lorentz-invariant). Expert updates of the confrontation between theory and experiment continue to be given in reviews by Will.[6]

Throughout the past century, a series of experiments have verified the accuracy of general relativity in modeling the dynamics of gravitating systems. The first such experiments involved measuring the deflection of light from a distant star around the Sun during an eclipse, and explaining the observed perihelion advance in the elliptical orbit of Mercury. Later experiments directly measured the gravitational redshift of light near the Earth's surface, as well as verifying the non-dispersive nature of the gravitational lensing of light as expected from geodesic motion. The observation of energy losses in orbital motion as a decrease in the period of a binary pulsar is likewise consistent with the existence of *gravity waves* from general relativity.

However, more familiar tests of the space-time curvature consistent with general relativity involve the use of global positioning satellites (GPS) to accurately determine location. If relativistic effects were not included in the positioning calculations, there would be a deviation of accurate timing on the satellites relative to Earth-based clocks of about 39 microseconds/day. About 46 microseconds/day of this is due to gravity, with −7 microseconds/day due to the time dilation effect of special relativity. These numbers are well in excess of the needed accuracy beyond 50 *nano*seconds in the clocks for accurate position calculations. Thus, GPS systems provide daily useful tests of the accuracy of general relativity.

[6] C.M. Will, *The Confrontation between General Relativity and Experiment*, arXiv:gr-qc/0510072v2 (2006) 89 pages.

5.3 Time-independent spherically symmetric solutions

The solutions to Einstein's field equations most familiar to common experience involve spherically symmetric geometries that do not change in time, since such geometries are quite accurate in describing the gravitational fields of planets and stars. The most general form of the metric describing static spherically symmetric systems preserves the angular arc-lengths $r d\vartheta$ and $r \sin\vartheta \, d\varphi$ associated with transverse distances in spherical polar coordinates, but recognizes possible modifications in proper radial lengths and times generating gravity. A general form with orthogonal coordinates is given by:

$$ds^2 = -A(r)(cdt_D)^2 + B(r)dr^2 + r^2\left(d\vartheta^2 + \sin^2\vartheta \, d\varphi^2\right). \tag{5.63}$$

This form should be able to describe all regions of the space-time.

5.3.1 *Schwarzschild geometry*

Exterior to the matter generating the gravitation where the energy-momentum tensor vanishes $\mathbf{T}_{\mu\nu} = 0$, the field equations 5.48 take the form $\mathcal{G}_{\mu\nu} = 0 = \mathcal{R}_{\mu\nu}$. Substitution of the metric form Eq. 5.63 into this curvature equation yields equations for the metric components $A(r)$ and $B(r)$. The result is that these components satisfy $\frac{d}{dr}(A\,B) = 0$ and $\frac{d}{dr}(rA) = 1$ in the spherically symmetric static vacuum exterior to the source density.

The solution that demonstrates appropriate correspondence with Newtonian gravitation ($A \to 1$, $B \to 1$ as $r \to \infty$) was first developed by Schwarzschild:

$$ds^2 = -(cd\tau)^2 = -\left[1 - \frac{2G_N M}{c^2 r}\right](cdt_S)^2 + \frac{dr^2}{\left[1 - \frac{2G_N M}{c^2 r}\right]} + r^2\left(d\vartheta^2 + \sin^2\vartheta \, d\varphi^2\right). \tag{5.64}$$

In this formula, r is the radial parameter from the center of a planet or star of mass M. However, r is *not* a proper distance from the center; rather it is a measure of the surface area of a sphere centered at $r = 0$, $\text{Area}_r = 4\pi r^2$, or the circumference of a great circle through that coordinate $\text{Circumference}_r = 2\pi r$. The radial coordinate r is not the distance to the center of the planet in the usual sense, because radial displacements are rescaled in the metric: $dL^2_{proper} = \frac{dr^2}{[1 - \frac{2G_N M}{c^2 r}]}$. This non-equivalence of flat-space and curved-space geometrical relationships between radius and circumference, area, etc., is crucial to the interpretation of coordinatization. The invariant interval Eq. 5.64 relates the local *proper length*, ds or local *proper time* interval, $d\tau$ to the space-time coordinates t, r, ϑ and φ. This coordinatization presumes that the surface area of a sphere is related to radius, as it would be if there were no curved space-time.

5.3.2 General radially stationary forms

The more general form 5.63 can be rewritten using $B(r) = \frac{1}{1-\beta^2(r)}$ as:

$$ds^2 = -A(r)(dct_D)^2 + \frac{dr^2}{1-\beta^2(r)} + r^2 d\varpi^2, \tag{5.65}$$

where $d\varpi^2 \equiv d\vartheta^2 + \sin^2\vartheta \, d\varphi^2$ will be a shorthand notation for representing angular displacements. If the form of the geometry is asymptotically Minkowski, then the metric components satisfy:

$$\lim_{r\to\infty} A(r) = 1, \quad \lim_{r\to\infty} \beta(r) = 0.$$

A *fiducial* observer in such a geometry is defined to be an observer with fixed spatial coordinates (r, θ, ϕ). Because of the gravitational attraction, such observers must be radially accelerating. From Section 1.2.1, a convenient form for parameterizing the coordinates of an accelerating system has radially stationary solutions parameterized using mixed temporal-spatial coordinates. A uniform proper acceleration means a radially dynamic, off-diagonal metric form:

$$dct_D = \alpha \, dct + \kappa \, dr, \quad \frac{\partial \kappa}{\partial ct} = \frac{\partial \alpha}{\partial r}, \tag{5.66}$$

where the second relationship follows from integrability of the coordinates. By requiring that the metric be spatially flat for a fixed temporal coordinate, the factor $\kappa(r)$ must algebraically satisfy:

$$\kappa(r) = \pm\sqrt{\frac{\beta^2(r)}{A(r)[1-\beta^2(r)]}}. \tag{5.67}$$

If the temporal scales are chosen to match asymptotically ($\alpha \to 1$), then the radially stationary metric takes the form:

$$ds^2 = -A(r)(dct)^2 \pm 2\sqrt{\frac{\beta^2(r)}{A(r)[1-\beta^2(r)]}} dct\, dr + dr^2 + r^2 d\vartheta^2 + r^2 \sin^2\vartheta \, d\varphi^2. \tag{5.68}$$

The positive sign on the mixed term is appropriate for stars and planets. This radially stationary metric form using variables $(ct, r, \vartheta, \varphi)$ will be quite convenient for developing geometries describing general radially dynamic geometries.

A particular type of observer is special in this stationary geometry. The set of observers whose four velocities $\vec{U} = \vec{u}/c$ satisfy:

$$U^{ct}_{GS} = \sqrt{\frac{1-\beta^2(r)}{-g_{00}(r)}} \overset{\text{exterior}}{\Rightarrow} 1, \quad U^r_{GS} = -\beta, \tag{5.69}$$

all share the time of the distant observer in the exterior where $g_{00} = -(1 - \beta^2(r))$. These observers in a sense "flow" with the geometry, and are referred to as *geometrically stationary* observers. Such observers at differing radial coordinates are *co-moving* relative to each other, yet they all share the same proper time, t (in the exterior), since $U_{GS}^{ct} = \frac{dct}{dc\tau} = 1$. Just as useful, a displacement dr in the radial coordinate in Eq. 5.68 represents a proper radial distance between geometrically stationary observers (with $dct = 0 = d\vartheta = d\varphi$). This means that the radial coordinate in this metric not only measures transverse distances but is also the proper distance between a geometrically stationary observer at (ct, r) and that geometrically stationary observer at the center at that same time $(ct, 0)$. Once again, however, r is *not* a direct measure of the proper distance between a fiducial observer and the center, since the proper measurement must be simultaneous in the fiducial observer's time.

It will be illustrative to examine general solutions for time-independent spherically symmetric systems. The standard correspondence with Newtonian dynamics motivates writing the general form of the metric in terms of gravitational potentials $V(r)$ and $V_o(r)$ which vanish asymptotically. The static form with diagonal coordinates will be taken to be:

$$ds^2 = -\left(1 + \frac{2[V(r) + V_o(r)]}{c^2}\right)(dct_D)^2 + \frac{dr^2}{1 + \frac{2[V(r)]}{c^2}} + r^2 d\varpi^2. \tag{5.70}$$

The previous parameter β is given by $\beta^2(r) \equiv -\frac{2}{c^2}[V(r)]$, and $A(r) = -(1 - \beta_o^2(r))$, where $\beta_o^2(r) \equiv -\frac{2}{c^2}[V(r) + V_o(r)]$. The radially stationary form is given by:

$$ds^2 = -\left(1 - \beta_o^2(r)\right)(dct)^2 + 2\sqrt{\frac{(1 - \beta_o^2(r))}{(1 - \beta^2(r))}\beta^2(r)}\, dct\, dr + dr^2 + r^2 d\varpi^2. \tag{5.71}$$

An example geometry will be developed in the next section as an exemplar of the relationships of physical to geometric parameters.

5.3.3 Stellar hydrostatics

One of the most relevant of astrophysical phenomena is the dynamics of stars. A straightforward model that will give some insight into stellar geometrodynamics is to assume that a star behaves as a relativistic fluid, and to use the energy-momentum tensor of a fluid to drive Einstein's equations.

Fluid densities

An ideal fluid in flat space-time is a homogeneous, isotropic system that has no energy transfers due to viscosity. The energy-momentum tensor using curvilinear coordinates is obtained from the flat-space tensor by replacing the Minkowski metric η with the general space-time metric g:

$$\mathbf{T}^{\mu\nu} = P g^{\mu\nu} + (\rho + P) U^{\mu}_{(F)} U^{\nu}_{(F)} \quad \text{where} \quad U^{\mu}_{(F)}\, g_{\mu\nu}\, U^{\nu}_{(F)} = -1. \tag{5.72}$$

The functions $\vec{U}_{(F)}$ represent the dimensionless four-velocities of local fluid elements.

The tensor (5.72) is a covariant form that represents the density and pressure of an ideal fluid:

$$\rho = U^{\mu}_{(F)}\, \mathbf{T}_{\mu\nu}\, U^{\nu}_{(F)}, \tag{5.73}$$

$$P = \frac{1}{3}\left(\rho + \mathbf{T}^{\beta}_{\beta}\right). \tag{5.74}$$

The local four-velocities of the fluid are as measured by an arbitrary observer.

Hydrostatic equilibrium

The rest-frame of the fluid provides a unique frame of reference for describing its dynamics. The fluid four-velocities $\vec{U}_{(F)}$ for temporal stationary, radially symmetric (isotropic) systems are given by:

$$U^{r}_{(F)} = 0,\ U^{0}_{(F)} = \frac{1}{\sqrt{-g_{00}}},\quad U_{(F)r} = \frac{g_{r0}}{\sqrt{-g_{00}}},\quad U_{(F)0} = -\sqrt{-g_{00}}, \tag{5.75}$$

where the label (F) will be dropped subsequently. Substitution into Eq. 5.72 gives non-vanishing components for the mixed energy-momentum tensor as follows:

$$((T^{\beta}{}_{\mu})) = \begin{pmatrix} -\rho & -(\rho + P)\left(\frac{g_{r0}}{g_{00}}\right) & 0 & 0 \\ 0 & P & 0 & 0 \\ 0 & 0 & P & 0 \\ 0 & 0 & 0 & P \end{pmatrix}. \tag{5.76}$$

This is the general mixed-form energy-momentum for a radially stationary ideal fluid.

One set of equations of dynamics for the fluid can be obtained directly from the local conservation condition:

$$\mathbf{T}^{\mu\nu}_{;\nu} = 0 = \partial_{\nu}\left[P\, g^{\mu\nu} + (\rho + P) U^{\mu} U^{\nu}\right] + \Gamma^{\mu}_{\lambda\nu}\, \mathbf{T}^{\lambda\nu} + \Gamma^{\nu}_{\lambda\nu}\, \mathbf{T}^{\mu\lambda}. \tag{5.77}$$

Notice that for any static parameter A, one can replace $\partial_{\nu}(A\, U^{\nu}) = 0$, since $\frac{\partial A}{\partial t} = 0 = \frac{\partial U^0}{\partial t}$. Therefore, the term with $(\rho + P)$ in Eq. 5.77 vanishes. The metric itself also has vanishing covariant divergence $g^{\mu\nu}_{;\nu} = 0$, yielding a form of this equation given by:

$$g^{\mu\nu}\,\partial_\nu P + \Gamma^\mu_{\lambda\nu}\,(\rho + P)\,U^\lambda\,U^\nu = 0. \tag{5.78}$$

Therefore, using Eq. 5.75 and expressing the connections for a stationary system in terms of the metric, the fluid continuity equation is given by:

$$\partial_\nu P + \frac{1}{2}\left[\partial_\nu \log(-g_{00})\right](\rho + P) = 0. \tag{5.79}$$

This equation results from local conservation principles.

An additional equation can be developed from the Einstein equation:

$$T^r_r = \frac{c^4}{8\pi G_N} G^r_r = \frac{c^4}{8\pi G_N}\left[1 + \frac{g_{00} + r g'_{00}}{(g_{0r})^2 - g_{00} g_{rr}}\right] = P, \tag{5.80}$$

where the quantity in the square brackets results from a direct calculation of $G^r{}_r$. Substitution of the derivative of g_{00} from Eq. 5.79 into 5.80 allows one to formulate an equation independent of V_o in the general metric forms Eqs. 5.70 and 5.71:

$$\frac{2\,\frac{dP}{dr}}{\rho + P} = \frac{1}{r} - \frac{1}{r}\left(g_{rr} - \frac{(g_{0r})^2}{g_{00}}\right)\left(1 + \frac{8\pi G_N}{c^4} r^2\, P\right), \tag{5.81}$$

or finally:

$$\frac{2\,\frac{dP}{dr}}{\rho + P} = -\frac{1}{r}\left(\frac{\frac{8\pi G_N}{c^4} r^2\, P - \frac{2V}{c^2}}{1 + \frac{2V}{c^2}}\right). \tag{5.82}$$

This form is quite useful, since only the single potential $V(r)$ determines the form of the pressure gradient, if the equation of state is known.

Calculation of hydrostatic parameters

An example system will demonstrate the relationships between fluid parameters and geometry. The example stellar object has been chosen to have an equation of state that generates the density and pressure profiles in Figure 5.2. The units have been scaled relative to the surface radial coordinate R_o, beyond which the exterior vacuum solution is valid.

The interior energy density modifies the space-time geometry of the system from that of Minkowski. Figure 5.3 constructs the conformal diagram for this system. The diagram on the left superimposes the energy density sourcing the geometry within the boundaries of the conformal diagram. The diagram on the right illustrates radial coordinates (vertical) originally scaled in tenths, then units of the surface scale R_o. Fixed temporal curves are scaled in units of R_o. The effect of the energy density is apparent when this diagram is compared to the conformal diagram in

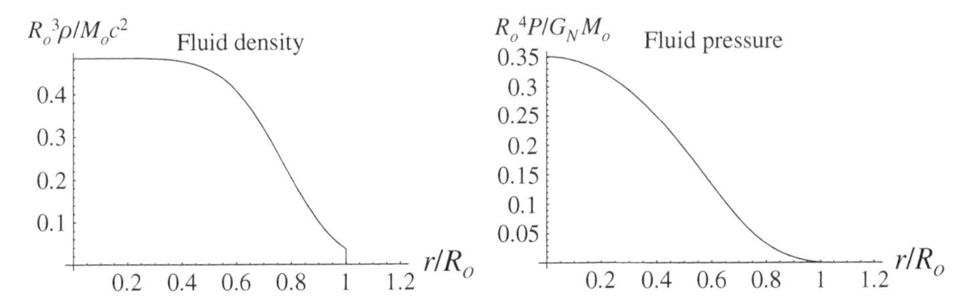

Figure 5.2 Fluid density ρ and pressure P of an example stationary spherically symmetric object.

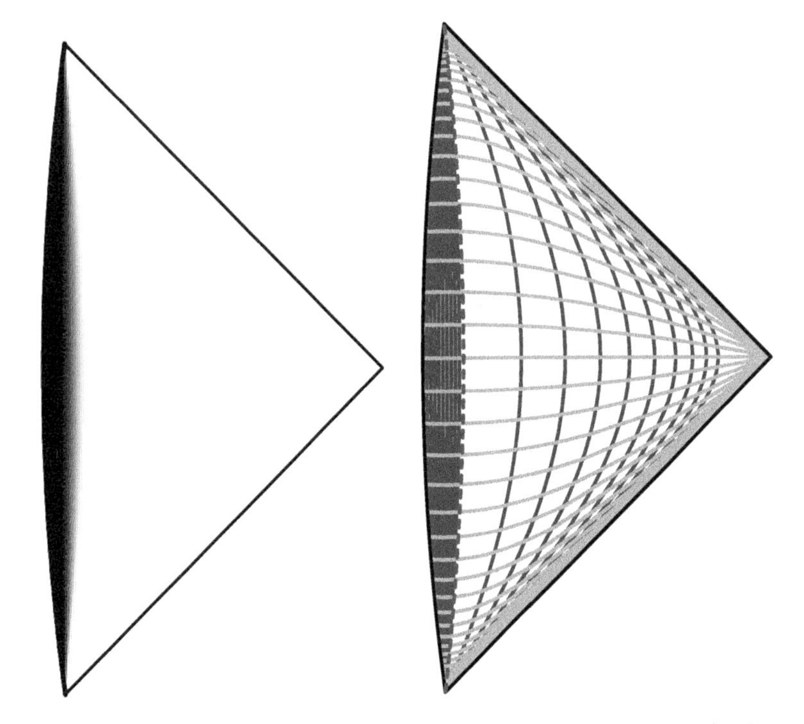

Figure 5.3 Conformal diagrams of fluid. *Left*: Energy density. *Right*: Fixed (r, ct) coordinates. The surface scale $r = R_o$ is given by the vertical dashed curve.

Figure 1.14 for Minkowski space-time, where the surface $r = 0$ is a straight vertical line.

Gravitational energy density

Densities directly associated with the gravitational field must vanish as the gravitational coupling vanishes, $G_N \to 0$. Those densities associated with microscopic fields should remain unaltered in this limit if alternative forces are arranged to maintain the microscopic configuration in the absence of gravity. Generally, these microscopic configurations are *not* the same as the equilibrium configuration in the absence of gravity.

To parameterize the gravitational dynamics, the metric potentials will be separated into microscopic and gravitational components:

$$V(r) \equiv v(r) + w_r(r), \qquad V_o(r) \equiv v_o(r) + w_o(r) - w_r(r). \tag{5.83}$$

The potentials $v(r)$ and $v_o(r)$ will be presumed to be due to microscopic interactions and generate non-vanishing values for the energy-momentum tensor in the absence of gravity $G_N \to 0$, while the potentials $w_r(r)$ and $w_o(r)$ will be presumed to generate energy-momentum contributions that vanish in the absence of gravity:

$$\lim_{G_N \to 0} \frac{w_r(r)}{G_N} = 0, \qquad \lim_{G_N \to 0} \frac{w_o(r)}{G_N} = 0. \tag{5.84}$$

From Einstein's equations, this is guaranteed if these potentials are related by the formula:

$$w_r(r) = -r \frac{d}{dr} w_o(r). \tag{5.85}$$

Given these potential functions, a form for gravitational binding energy can be developed.

A mass form associated with the geometry can be developed for the geometry. From Eq. 5.76 and Einstein's equation (shown in Appendix E, Eq. E.18), one can write:

$$V(r) = \frac{2G_N M(r)}{r} \quad \Rightarrow \quad M(r)\, c^2 = \int_0^r \rho(r') 4\pi r'^2 dr'. \tag{5.86}$$

Here, $M(r)$ has the usual interpretation of the mass of gravitational source within a sphere of radial coordinate, r. This is especially true for the metric form Eq. 5.71, for which dr' is a proper radial length between co-moving observers. On dimensional grounds, all of the potentials can also be expressed in terms of characteristic mass distributions:

$$V_o(r) \equiv \frac{2G_N M_{V_o}(r)}{r}, \quad v(r) \equiv \frac{2G_N m_v(r)}{r}, \quad w_r(r) \equiv \frac{2G_N m_w(r)}{r}, \tag{5.87}$$

with the mass form in w_o being defined from Eq. 5.85 as $m_w(r) = m_{wo}(r) - r\, m'_{wo}(r)$. The example mass distributions generating Figure 5.2 are shown in Figure 5.4. These masses include contributions from both microscopic and gravitational energies. However, using Eq. 5.83, one can decompose the fluid density as

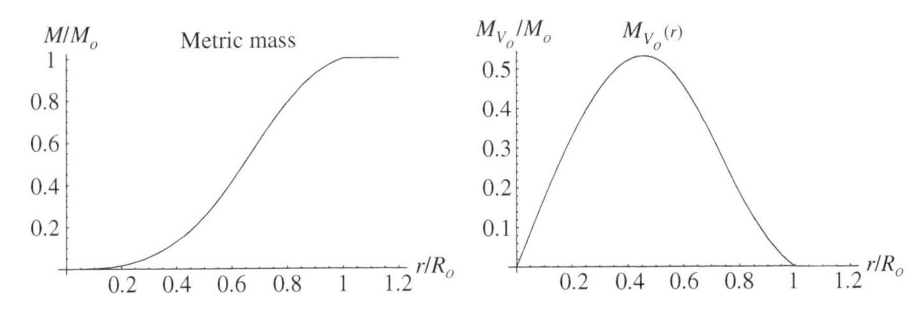

Figure 5.4 Metric mass $M(r)$ and temporal potential mass $M_{V_o}(r)$, in units of the total mass $M_o \equiv M(R_o)$.

$\rho = \rho_v + \rho_w$, where:

$$\rho_v(r) \equiv \frac{m'_v(r)c^2}{4\pi r^2} \text{ and } \rho_w(r) \equiv \frac{m'_w(r)c^2}{4\pi r^2} = -\frac{m''_{wo}(r)c^2}{4\pi r}. \tag{5.88}$$

The gravitational energy density will be given by $\rho_w(r)$.

In order to calculate the gravitational contribution to the densities, one can construct the mass distribution by calculating the work done in assembling infinitesimal spherical shells brought to the surface of an intermediate assembly from infinity. Consider the proper acceleration of a general observer with four-velocity U:

$$a^r_{proper} = \frac{d^2\xi^\lambda}{d\tau^2} \equiv \frac{\partial \xi^\lambda}{\partial x^\mu}\left[\frac{dU^\mu}{dc\tau} + \Gamma^\mu_{\alpha\beta}\, U^\alpha\, U^\beta\right] c^2. \tag{5.89}$$

Since it will be assumed that no mass distribution is ever within its Schwarzschild radius, a proper radial displacement (squared) is given from the metric by $ds^2 = (d\xi^r)^2 = g_{rr}\,(dr)^2$. For radial displacements, $\frac{d\xi^r}{dr} = \sqrt{g_{rr}}$. Also, for fiducial observers associated with a stationary fluid, all spatial components of the four-velocity vanish $U^r = 0 = U^\vartheta = U^\varphi$. This means that the proper acceleration for fiducial systems can be calculated using:

$$a^r_{proper} = \sqrt{g_{rr}}\,\Gamma^r_{00}\left(U^0\right)^2 c^2 = \frac{c^2}{2\sqrt{g_{rr}}}\frac{d}{dr}\log(-g_{00}). \tag{5.90}$$

The work done taking a unit test mass from infinity to radial coordinate r through the exterior geometry is thus given by:

$$\int_\infty^r a^r_{proper}\, d\xi^r = \int_\infty^r \frac{c^2}{2}\frac{d}{dr'}\log(-g^{ext}_{00})\, dr' = c^2 \log\sqrt{-g^{ext}_{00}}. \tag{5.91}$$

This form gives the expected Newtonian limit $\Phi_{Newton}(r) \rightarrow -\frac{G_N M(r)}{r}$. Therefore, the gravitational work done in bringing an infinitesimal mass shell $\delta m_v(r)$ from ∞

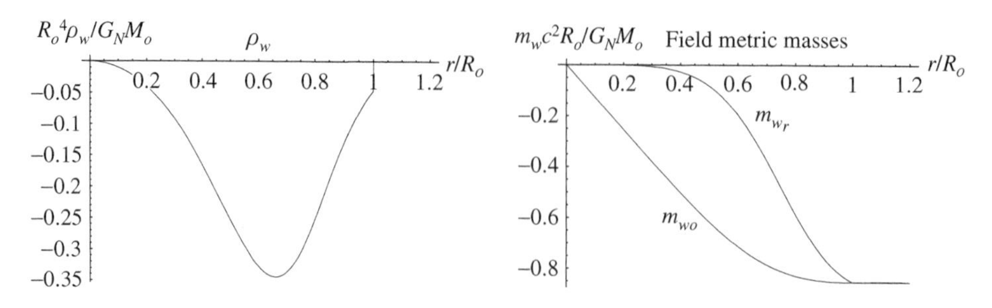

Figure 5.5 Gravitational energy density ρ_w and metric masses $m_w = m_{w_r}$ and m_{wo}.

to its final configuration is the gravitational energy:

$$\delta m_w(r)c^2 = \delta m_v(r)c^2 \log \sqrt{-g_{00}^{ext}(r)} \quad \Rightarrow \quad \rho_w(r) = \rho_v(r) \log \sqrt{-g_{00}^{ext}(r)}. \tag{5.92}$$

Recalling again that $\rho = \rho_v + \rho_w$, this gives equations that generate solutions for all densities:

$$\rho_v(r) = \frac{\rho(r)}{1 + \log \sqrt{-g_{00}^{ext}(r)}} \quad \Rightarrow \quad m_v'(r) = \frac{M'(r)}{1 + \frac{1}{2}\log\left(1 - \frac{2G_N M(r)}{c^2 r}\right)}. \tag{5.93}$$

The gravitational field energy density is then calculated using $M'(r) - m_v'(r) = m_w'(r) = -r m_{wo}''(r)$, where boundary conditions require all masses to vanish at $r = 0$. The functional forms of these parameters corresponding to the example in Figure 5.2 are displayed in Figure 5.5. The gravitational energy density vanishes at the surface and has natural units of $\frac{G_N M_o}{R_o^4}$. The mass $m_w(r)$ is the contribution of the gravitational field to the total mass $M(r)$ of the system and has natural units of $\frac{G_N M_o}{c^2 R_o}$. All of these functions vanish in the $G_N \to 0$ limit. The gravitational binding energy is given by $m_w(R_o)c^2 = m_{wo}(R_o)c^2$.

Energy conditions for stationary spherically symmetric geometries

For $V_o = 0$, writing $V(r) \equiv -\frac{G_N M(r)}{r}$, the dominant energy condition is satisfied for arbitrary radial motions. For angular motions, the condition:

$$-\left[M'(r)\right]^2 + \left[\frac{r}{2}M''(r)\right]^2 \leq 0, \tag{5.94}$$

must be valid. This condition is satisfied for any mass distribution where the interior mass grows less rapidly than it would due to a constant density $M(r) \leq \frac{4}{3}\pi r^3 \rho_o$, i.e., the interior mass must have a functional form that grows less rapidly than r^3.

For a more general metric form, the dominant energy condition is satisfied for arbitrary radial motions if:

$$\frac{V_o'}{V} - \frac{2V_o}{c^2 + 2V + 2V_o}\left(\frac{V' + V_o'}{V}\right) \leq 0 \tag{5.95}$$

Most reasonable metrics where the potential V dominates V_o will satisfy this condition.

5.3.4 Vacuum solutions

The exterior geometry satisfies pressureless vacuum solutions to Einstein's field equations. If the vacuum solution holds down to $r = 0$, the geometry has a black hole, which will be discussed in Chapter 7. Two forms of the vacuum metric will be discussed in this section.

Schwarzschild geometry

The Schwarzschild metric was derived in Eq. 5.64. It is convenient to define the Schwarzschild radius as $R_s \equiv \frac{2G_N M}{c^2}$. This mass scale has dimensions of length. The metric 5.64 can then be rewritten:

$$ds^2 = -(c\,d\tau)^2 = -\left[1 - \frac{R_S}{r}\right](c\,dt_S)^2 + \frac{dr^2}{\left[1 - \frac{R_S}{r}\right]} + r^2\left(d\vartheta^2 + \sin^2\vartheta\, d\varphi^2\right). \tag{5.96}$$

The Schwarzschild time coordinate t_S in this equation is the proper time of a very distant fiducial observer, infinitely far from the center, not a local observer. Rather, the proper time of a fiducial observer is given by $d\tau_r = \sqrt{1 - \frac{R_S}{r}}\,(dt_S)$. Similarly, the radial coordinate r is not a proper distance from the center, again it is a measure of the area of a sphere of fixed radius $A = 4\pi r^2$, or the circumference of a great circle $C = 2\pi r$. Rather, a local radially oriented measuring rod has proper length $ds_r = \frac{dr}{\sqrt{1 - \frac{R_S}{r}}}$. The metric only allows (accelerating) fiducial observers for fixed $r > R_S$.

One should note that if this vacuum solution holds for radii less than R_S, then the geometry has a coordinate singularity at $r = R_S$. However, since physical effects like tidal forces depend upon curvature components (second derivatives of the metric) which scale like $\frac{1}{R_S^2}$ at this location, all physical quantities are finite. Therefore, $r = R_S$ is *not* a physical singularity. In the interior $r < R_S$ the factors $[1 - \frac{R_S}{r}]$ in the metric change sign, reversing the roles of time-like and space-like coordinates. This metric also displays a physical singularity at $r = 0$.

An object that freely falls from rest at radial coordinate R will never be observed to reach the Schwarzschild radius in a finite time t. However, according to the

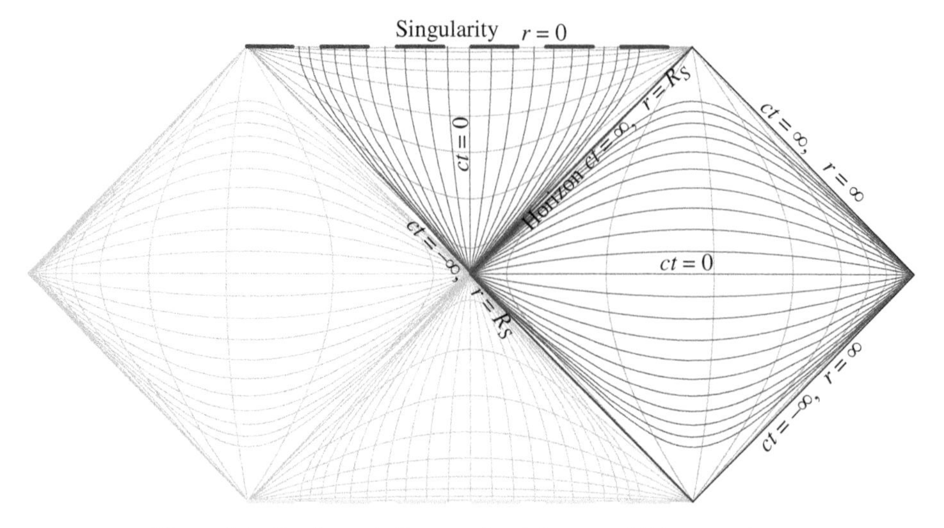

Figure 5.6 Conformal diagram for Schwarzschild space-time.

proper time of that object, it will cross the Schwarzschild radius as no particularly special location and reaches the center in a finite proper time $\tau = \frac{\pi R}{2}(\frac{R}{2G_N M})^{1/2}$.

A set of conformal coordinates (ct^*, r^*) for purely radial motions can be developed by defining "tortoise" coordinates $-[1 - \frac{R_S}{r}](c\,dt_S)^2 + \frac{dr^2}{[1-\frac{R_S}{r}]} \equiv -[1 - \frac{R_S}{r}]((c\,dt^*)^2 - (dr^*)^2)$. Closed-form solutions for these coordinates are given by:

$$ct^* = ct_S,\ r^* = \begin{cases} r + R_S \log\left(\frac{r}{R_S} - 1\right), & r \geq R_S \\ r + R_S \log\left(1 - \frac{r}{R_S}\right), & r \leq R_S \end{cases}. \tag{5.97}$$

These coordinates can be used to construct the conformal diagram in Figure 5.6. The gray portion is the *maximum analytic extension* of Schwarzschild geometry, whose bounding surfaces are either asymptotic light-like surfaces or singularities. The exterior is the right-most causal region, bounded by past light-like infinity (skri minus), future light-like infinity (skri plus), and the light-like surfaces representing the Schwarzschild radius $r = R_S$. Fixed coordinate time ct surfaces are graded in tenths about $t = 0$, then units of the Schwarzschild radius R_S. Fixed radial coordinate r surfaces are graded in tenths of the Schwarzschild radius up to $2R_S$, then in units of this scale. The center $r = 0$ is denoted by the horizontal dashed line bounding the diagram from above. The *horizon* is an outgoing light-like surface that delineates the exterior region, whose observers can potentially send signals to future light-like infinity, from the interior region, whose observers must ultimately encounter the center.

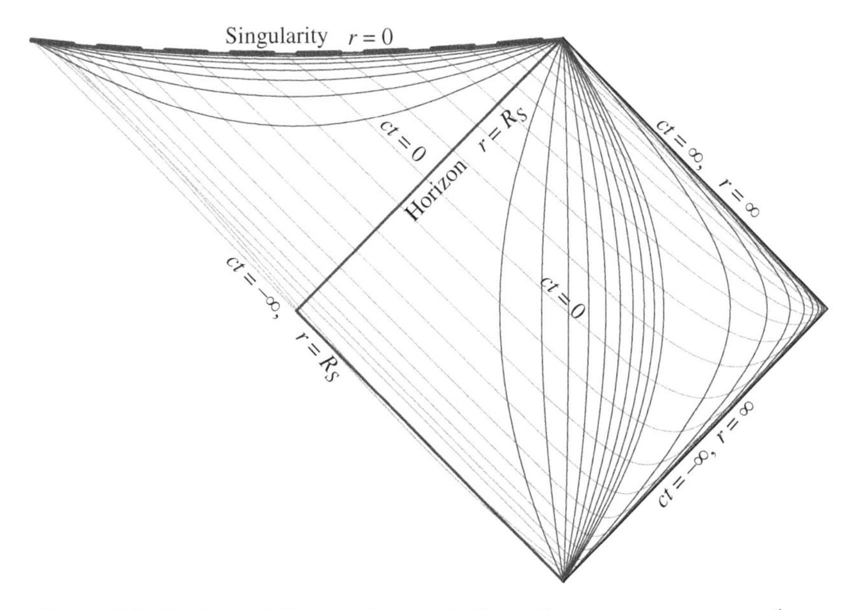

Figure 5.7 Conformal diagram for a radially stationary vacuum space-time.

In the Schwarzschild geometry, the center $r = 0$ represents a space-like surface, while the center of Minkowski space-time is time-like. This means that non-exotic (time-like or light-like) particles cannot persist indefinitely within the region $0 < r < R_s$, since they must follow trajectories that nowhere have slopes less than 45° on this diagram. They will eventually encounter the space-like center and cease their non-exotic motions.

Radially stationary geometry

A radially stationary form of the vacuum metric is sometimes referred to as the *river model* [78]. Using Eq. 5.71, the metric is given by:

$$ds^2 = -\left[1 - \frac{R_S}{r}\right](dct)^2 + 2\sqrt{\frac{R_S}{r}}dct\, dr + dr^2 + r^2 d\vartheta^2 + r^2 \sin^2\vartheta\, d\varphi^2 \,. \tag{5.98}$$

Again, one should note that fixed-time surfaces are space-like everywhere, and for the geometrically stationary observers that share a fixed time coordinate t, the displacement dr represents a proper length.

The conformal diagram of this geometry is demonstrated in Figure 5.7. The large-scale causal structure of the space-time is the same as that of the Schwarzschild metric. The diagram demonstrates the space-like nature of fixed-time surfaces, which are graded in units of R_S. Unlike the Schwarzschild time coordinate, these temporal coordinates traverse the horizon. Fixed radial coordinate r curves are time-like exterior to the horizon and space-like in the interior. In the diagram, they are graded initially in tenths of R_S, then units, and decades of this scale.

As was the case with the Schwarzschild metric, no infalling object will be observed to cross the horizon in a finite time, since the horizon is a light-like surface reaching future infinity. However, the fact that the horizon has distinct temporal coordinates means that this time cannot be directly mapped onto Schwarzschild time, for which the horizon is a $t \rightarrow \infty$ surface. This makes some direct comparisons of these metrics problematic.

Supplement: Ambiguities in Asymptopia

One of the difficulties with Schwarzschild geometry is that the temporal coordinate t_S serves as the actual proper time coordinate only for an asymptotic fiducial observer. This observer resides in a Minkowski space-time essentially bereft of gravitational influences. The region of a geometry far away from the sources of gravitation will be referred to as its *Asymptopia* [79]. The motions of a distant fiducial (fixed r) versus a freely falling Schwarzschild observer are indistinguishable, generating ambiguities and paradoxes whenever gravitational measurements which differ between the two types of observers (like Hawking radiation) are taken.

In contrast, there are a large class of geometrically stationary observers in the radially stationary (river) space-time that share the metric temporal coordinate t, which is everywhere a time-like parameter, as their local proper time. This large class of observers include asymptotic geometrically stationary observers that reside in a Minkowski space-time as a subset. Any gravitational measurement that distinguishes fiducial from geometrically stationary observers will generate unambiguous differences defining their motions.

One might question whether there is a well-defined relationship between the Asymptopias of Schwarzschild versus radially stationary geometries. There is a well-defined local expression relating the time intervals given by:

$$dct = dct_S + \frac{\sqrt{\frac{R_S}{r}}\, dr}{1 - \frac{R_S}{r}},$$

which is integrable for static R_S. The problem of relating the asymptotic observers then becomes one of finding a surface of correspondence $t_S = t$ as the boundary condition.

If one utilizes the co-moving, geometrically stationary observers whose motions follow $\frac{dr_{GS}(ct)}{dct} = -\sqrt{\frac{R_S}{r_{GS}(ct)}}$, this allows the previous relationship to be integrated into the finite expression:

$$ct_S - ct_o = \int_{ct_o}^{ct} \frac{dct'}{1 - \frac{R_S}{r_{GS}(ct')}}.$$

In this sense, Asymptopia lies in the distant still waters of the river, rather than on the rigid shoreline of an accelerating fiducial observer.

Any attempt to establish a set of fiducial observers at r_{corr} as the surface of correspondence of radially dynamic time t with Schwarzschild time t_S will encounter ambiguities in Asymptopia. An integration along the fixed ct surface will generate a local value for the Schwarzschild time given by:

$$(ct_S)|_r = ct - \int_{r_{corr}}^{r} \sqrt{\frac{R_S}{r'}} \frac{dr'}{1 - \frac{R_S}{r'}}.$$

This expression is well-defined for *any* finite surface of correspondence, even a very distant surface. However, the expression is singular as $r_{corr} \to \infty$, which disallows its direct correspondence with the Schwarzschild observer. The relationship also becomes singular as *more* distant times are calculated using *any* finite surface of correspondence. This means that the relationship between the Asymptopias of geometries that result from such Minkowski spaces that are infinitely Poincaré translated relative to each other is ambiguous.

5.4 An axially stationary rotating geometry

The vacuum solution of a rotationally dynamic space-time is given by the *Kerr* geometry [80] which can be expressed using the Boyer–Lindquist [81] metric. It is convenient to define length scales identified with the mass M and angular momentum J of the geometry given by:

$$R_M \equiv \frac{2G_N M}{c^2}, \quad R_J \equiv \frac{J}{Mc}. \tag{5.99}$$

Using these constant radial scales, the metric is given by:

$$\begin{aligned} ds^2 = &-\left(1 - \frac{R_M\, r}{r^2 + R_J^2 \cos^2\theta}\right)(dct)^2 + \frac{r^2 + R_J^2\cos^2\theta}{r(r - R_M) + R_J^2}\, dr^2 \\ &+ (r^2 + R_J^2\cos^2\theta)\, d\theta^2 - 2\,\frac{R_M\, R_J\, r\sin^2\theta}{r^2 + R_J^2\cos^2\theta}\, dct\, d\phi \\ &+ \left(r^2 + R_J^2 + \frac{R_M\, R_J^2\, r\sin^2\theta}{r^2 + R_J^2\cos^2\theta}\right)\sin^2\theta\, d\phi^2. \end{aligned} \tag{5.100}$$

This metric has a non-orthogonal temporal-angular term associated with rotational motions analogous to the temporal-radial term in Eq. 5.98. This mixed term contributes to the frame-dragging behaviors in the rotating geometry. The asymptotic form is the metric in some form of Minkowski space-time.

5.4.1 Motions on the equatorial plane

The stationary rotational metric Eq. 5.100 has somewhat complicated angular dependence. However, some of the main features of this geometry can be ascertained by examining motions on the equatorial plane $\theta = \frac{\pi}{2}$. First, if one examines purely radial motions of photons (null geodesics) on the equatorial plane, the trajectories of these massless particles must satisfy:

$$\left(\frac{dr_\gamma}{dct}\right)^2 = \frac{(r_\gamma - R_M)(r_\gamma^2 - R_M r_\gamma + R_J^2)}{r_\gamma^3}. \tag{5.101}$$

Since the vacuum solution is valid everywhere, the radial motion of photons is stationary when $\frac{dr_\gamma}{dct} = 0$ on the surfaces:

$$R_{SL} = R_M, \quad R_H^{\pm} \equiv \frac{1}{2}\left[R_M \pm \sqrt{R_M^2 - 4R_J^2}\right]. \tag{5.102}$$

The radial coordinate R_{SL} defines the *static limit*, within which all future trending trajectories must necessarily have non-vanishing angular motion, since light itself cannot have purely radial motions within this surface. The surfaces $R_H^{\pm}$ will be examined further shortly.

One can next examine the motion of a (momentarily) purely tangential photon with $\dot{r}_\gamma = 0$, described by the null geodesic equation:

$$\left[1 + \left(\frac{R_J}{r_\gamma}\right)^2 + \left(\frac{R_M}{r_\gamma}\right)\left(\frac{R_J}{r_\gamma}\right)^2\right]\left(r_\gamma \frac{d\phi_\gamma}{dct}\right)^2 - 2\left(\frac{R_M}{r_\gamma}\right)\left(\frac{R_J}{r_\gamma}\right)^2\left(r_\gamma \frac{d\phi_\gamma}{dct}\right)$$
$$-\left(1 - \frac{R_M}{r_\gamma}\right) = 0. \tag{5.103}$$

Since the equation is quadratic in $\dot{\phi}_\gamma$, there are two solutions corresponding to photons moving in opposite directions. On the surface $r_\gamma = R_M$, the solutions are:

$$R_M \frac{d\phi_\gamma}{dct} = 0 \quad \text{and} \quad R_M \frac{d\phi_\gamma}{dct} = \frac{2\frac{R_J}{R_M}}{1 + 2\left(\frac{R_J}{R_M}\right)^2}. \tag{5.104}$$

The first solution verifies that light traveling in a tangential direction opposite that of the rotation is stationary, defining a static limit for both azimuthal and radial motions. Furthermore, on the surfaces $r = R_H^\pm$, both solutions satisfy:

$$R_H^\pm \frac{d\phi_\gamma}{dct} = \frac{R_J^2}{R_M(2R_J - R_M)}, \quad \frac{dr_\gamma}{dct} = 0. \tag{5.105}$$

Thus these surfaces define *unique* light-like trajectories, which is one of the characteristics of a *horizon*, as will be discussed in Chapter 7. The surfaces $R_H^\pm$ define a region within which a fixed radial coordinate defines a space-like rather than time-like curve, since the term in the metric involving dr^2 has a negative signature within this region.

One might notice that for sufficiently large angular momenta $R_J > \frac{1}{2}R_M$, there are no real solutions for $R_H^\pm$. For instance, if one considers an electron with mass m_e and angular momentum $\frac{\hbar}{2}$, the ratio of the radial angular momentum scale to the radial mass scale is given by:

$$\frac{R_{J_e}}{R_{m_e}} = \frac{\hbar c}{4G_N m_e^2} = \left(\frac{M_P}{2m_e}\right)^2 \ggg \frac{1}{2},$$

where M_P is the Planck mass. If the vacuum solution is valid down to the origin $r = 0$, this implies that the radial coordinate is everywhere a space-like coordinate.

Several terms in the metric 5.100 become singular at $r = 0$. As will be explored in the next chapter, one expects quantum non-locality to potentially modify the behaviors of quantum sources in a manner sufficient to prevent the occurrence of singular structures.

6

Quantum mechanics in curved space-time backgrounds

The incorporation of quantum mechanics into gravitational dynamics introduces perplexing issues into modern physics. In contrast to other interactions like electromagnetism, the classical trajectory of a gravitating system is independent of the mass coupling to the gravitational field. As previously discussed, this allows the gravitation of arbitrary test particles to be described in terms of local geometry only, the basis of general relativity. Thus, the geometrodynamics of classical general relativity are most directly expressed using localized geodesics. However, quantum dynamics incorporate measurement constraints that disallow complete localization of physical systems. A coherent quantum system is not represented by a path or a classical trajectory; rather, it self-interferes throughout regions. This complicates the use of classical formulations in describing inherently quantum processes.

In addition, the equations of general relativity are complex and non-linear in the interrelations between sources and geometry, which makes solutions of even classical systems complicated. The key to describing complex systems is to determine the most useful set of parameters and coordinates that give concise predictive explanations of those systems. This chapter will develop tools for examining quantum behaviors in gravitating systems.

6.1 Quantum coherence and gravity

The behaviors of quantum objects in Minkowski space-time are well understood, despite a lack of consensus on the various interpretations (Copenhagen, many worlds, etc.) of the underlying fundamentals of the quantum world, or concerns of the completeness of quantum theory. There have also been tests of systems modeled by equations that involve both Newton's gravitational constant G_N and Planck's constant $\hbar$, as described in Section 2.3.1. These experiments indicate that gravitating quantum systems *do* maintain their coherence, demonstrating that the structure of the interaction need not break coherence in order to localize the system in the field between detections. Moreover, gravitating systems *themselves* are expected to serve as source energy densities (*co-gravitation*). Even the highly dynamic gravitating environment during the Big Bang involved redshifts of the CMB radiation without breaking the coherence of the individual quanta after decoupling. This is reasonable using the principle of equivalence, since the motions of detectors and screens should not break the coherence of any given inertial system prior to its detection.

As discussed in Chapter 2, the experimental evidence involving quantum coherence in gravity make it problematic for space-time to somehow "bubble up" to localize particles while those particles maintain their quantum coherence. As was seen with accelerating observers in special relativity, anomalous observed quanta need not be associated with the geometry itself. If an observer establishes coordinates based on micro-physical measurements, those coordinates are not expected to fluctuate due to the quantum behaviors of a gravitating system. The approach presented here presumes space-time to be an emergent construct from the measurement process which has no constraints beyond those consistent with various conservation/symmetry properties. This is perhaps as open-minded as one must be, consistent with what can be experimentally verified. Only a Planck-scale coherent object would establish gravitational decoherences of sufficient scale to examine the experimental properties of single objects beyond the co-gravitational approach taken here. The quantum behaviors of such systems would likely be described completely through the quantum properties of the sources themselves.

The formulation of space-time coordinates inherently involve the sufficient breaking of quantum coherence to establish well-defined relationships between events. More precisely stated, the construction of a viable space-time grid implies that it be disentangled from any intermediate scattering process. Generally, coordinates can be *chosen* by the observer, which is a fundamental property expressed in manifest covariance. In particular, one should always be able to choose the simplest coordinates needed to parameterize the dynamics of a system. The choice here will utilize the "mean-field" (expectation value) coordinate expressions generated by all of the co-gravitating systems. Therefore, the geometrization of gravity sourced by quantum objects is expected to take the form $\mathcal{G}_{\mu\nu} = -\frac{8\pi G_N}{c^4} \langle \mathbf{T}_{\mu\nu} \rangle + \Lambda \, g_{\mu\nu}$. The cosmological constant will be neglected in the description of the systems in this chapter. As in special relativity, the general coherence rule will be that geometric quantities associated with the space-time grid (including Lorentz frames) are expressed classically, while stationary and transitional states are expressed using quantum sources. This is consistent with the principle of equivalence.

The approach does not imply that long-range gravitational quanta are never useful constructs. As was discussed in Section 2.1.3, as well as in the supplement beginning on page 125, the quantization properties of the electric and magnetic fields follow directly from the quantum behaviors of the charged sources producing semi-classical electromagnetic fields. If gravitational quanta must involve sources and sinks, they can be described in terms of "gravitons" with as much convenience as electromagnetic radiations can be described using photons. The key point is that the exhibition of quantum coherent behavior for gravitating systems need not require the quantization of the gravitation field *itself*.

An additional comment can be made concerning the internal structure of the fundamental constituents. Because of quantum measurement constraints, there are no "point" particles in quantum mechanics, just particles with the simplest internal structure. Quantum systems cannot be completely localized upon a point, string, membrane, or any geometric structure that completely constrains any given set of non-commuting observables. For this examination of the most basic foundations, only the simplest geometric structures will be considered. More general structures are examined elsewhere in the literature [23,82].

Supplement: Divergences in perturbative gravity fields

The gravitational force is weak when compared to the microscopic forces between common particles. For instance, two protons experience an electrostatic repulsion $\frac{e^2}{G_N m_{proton}^2} = \frac{e^2}{\hbar c}\left(\frac{M_P}{m_{proton}}\right)^2 \approx 10^{36}$ times stronger than the gravitational attraction between those protons. However, unlike gravity, electromagnetism involves a dimensionless coupling $\alpha = \frac{e^2}{\hbar c} \simeq \frac{1}{137}$ that is small and can be used as the parameter in a perturbative expansion of a particular probability amplitude for a complicated process. The coupling between mass/energy sources that gravitate has dimensions $G_N = \frac{\hbar c}{M_P^2}$, and of itself cannot serve as the dimensionless parameter in a perturbative expansion without being multiplied by masses or energies involved in the calculation.

Since gravity acts between all mass/energy sources, some terms in a dimensionless perturbative amplitude of order G_N^s involve the intermediate energy E of the calculation in the form:

$$\int \frac{dE}{E}\left(\frac{G_N E^2}{\hbar c^5}\right)^s A_s(E, E_o, mc^2), \tag{6.1}$$

where the amplitude A_s is of order unity. If there is no ultraviolet cutoff at high energies, such integrals will diverge, despite the weakness of the gravitational attraction. These divergences make standard perturbative approaches for an interaction between energy sources (like gravity) problematic.[1] However,

[1] N.D. Birrell and P.C. Davies, *Quantum Fields in Curved Space*. Cambridge University Press, Cambridge (1994).

general non-perturbative approaches need not be analytic in the zero-coupling limit yet still can define weakness in the interaction, as occurs with Cooper pairing in superconductors. For this reason, the analytic properties of gravitation should be maintained whenever possible.

6.1.1 Energy conditions and quantum gravity

It is common to use energy dominance conditions to test whether arbitrary metric-based geometries satisfy the expectation from classical physics that the energy-momentum tensor used in Einstein's equation be measured by *any* observer to be either energy-like or light-like. Concisely stated, the energy conditions assert that the speed of the energy flow of known types of matter/energy should always be measured by all to be less than or equal to that of light. Any local measurement of matter/energy by any observer should measure matter to be time-like, or perhaps for quanta, light-like. Since classical gravitating systems are composed of such constituents, the overall energy-momentum tensor driving Einstein's equation is expected to likewise satisfy this condition as measured anywhere by anyone.

However, energy dominance conditions often do not hold for the expectation values of the energy-momentum tensors of quantum fields, even in Minkowski space-time, due to space-like coherence. Furthermore, if black holes evaporate in the manner described by Hawking (as will be discussed in the next chapter), the overall geometry cannot everywhere satisfy the usual energy dominance conditions, since those conditions were used to derive the classical non-decreasing area theorem. The Penrose diagram of a temporally transient black hole in Figure 6.1 will help explain the rationale. The diagram demonstrates a generic cosmology that manifests a horizon that at some future point decreases to vanishing radial scale and area, $R_H \rightarrow 0$. A *horizon* is a light-like surface that delineates a region of space-time beyond which communications can *never* be received. Recalling that on a conformal diagram all outgoing light signals have slope $+1$, any observer in the exterior can send out a signal that will reach future infinity, while all signals sent out by any observer in the interior will terminate at the space-like center $r = 0$. For such a geometry, there must be a future space-like surface terminating all outgoing light-like trajectories. In the vicinity of this space-like center, any energy-momentum tensor generating radiation in the exterior must have a space-like character. It is the excretion of this radiation that allows the area $= 4\pi \left(\frac{2G_N M}{c^2}\right)^2$ to decrease as the mass decreases. Therefore it is clear that quantum behaviors in gravitating systems motivate substantial modifications on the use of energy dominance conditions to determine the physical viability of certain geometries.

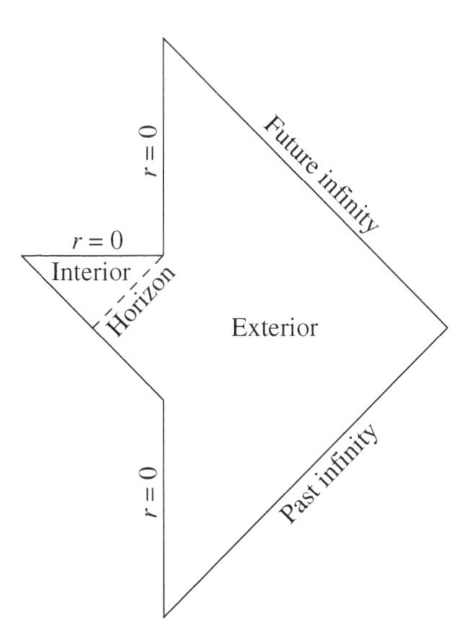

Figure 6.1 Violation of horizon area theorem by a horizon of decreasing area.

6.2 Lagrangian dynamics of quantum systems

It is relatively straightforward to examine quantum systems on curved space-time backgrounds. Generally, the flat-space form of the Lagrangian describing a given physical system can be incorporated into general relativity by just replacing the Minkowski metric form with the metric appropriate to the space-time background, and ensuring that the action has the appropriate invariance behavior under coordinate transformations (i.e., that the overall Lagrangian density has a factor of $\sqrt{-\det((g_{\mu\nu}))} = \det \mathcal{V}$). Simple gravitating quantum systems, including the scalar Klein–Gordon field, will be examined on common curved space-time backgrounds. In this section, the quantum systems will be presumed to have negligible impact upon the space-time background itself. Whereas gravitating classical objects undergo geodesic motion, quantum systems are not localized on a geodesic but rather have probability density distributions on the space-time. Local physical properties of the quantum fields on simple backgrounds will be explored.

6.2.1 The Klein–Gordon field

One of the most frequently examined of quantum systems is the Klein–Gordon field. The field equation of this scalar in Minkowski space-time results most directly by replacing the forms for the energy and momentum in the square of the non-interacting relativistic energy-momentum dispersion relation $E^2 = p^2c^2 + m^2c^4$

with their corresponding quantum operators:

$$\eta^{\mu\nu}\,\partial_\mu\partial_\nu\chi - \left(\frac{mc}{\hbar}\right)^2\chi = \left(\nabla^2 - \frac{1}{c^2}\frac{\partial^2}{\partial t^2}\right)\chi - \left(\frac{mc}{\hbar}\right)^2\chi = 0. \tag{6.2}$$

This means that in the presence of a gravitational field, the Lagrangian of a non-interacting scalar Klein–Gordon field χ takes the form:

$$\mathcal{L} = \frac{1}{2}\sqrt{-g}\,\left(\hbar^2c^2g^{\mu\nu}\,\partial_\mu\chi^*(\vec{x})\,\partial_\nu\chi(\vec{x}) + (mc^2)^2\chi^*(\vec{x})\chi(\vec{x})\right), \tag{6.3}$$

where the fields must have units of $Dim[\,|\chi(\vec{x})|^2] \sim \frac{1}{Energy\times Length^3}$.

The gravitational energy-momentum tensor for the Lagrangian 6.3 can be calculated using Eq. 5.57. Consider a general spherically symmetric geometry with a metric of the form:

$$ds^2 = g_{00}\,(dct)^2 + 2g_{0r}\,dct\,dr + g_{rr}\,dr^2 + r^2(d\vartheta^2 + \sin^2\vartheta\,d\varphi^2), \tag{6.4}$$

where the unspecified metric components are functions of time and the radial coordinate $g_{\mu\nu}(ct, r)$. The Klein–Gordon mixed energy-momentum tensor $\mathbf{T}^0{}_0$ for a field with no angular dependence ($\ell = 0$) is given by:

$$\begin{aligned}\mathbf{T}^0{}_0 = &-\frac{1}{8\pi}\left(\frac{2g_{0r}^2 - g_{00}\,g_{rr}}{g_{0r}^2 - g_{00}\,g_{rr}}\right)(mc^2)^2\chi_0^*\,\chi_0 \quad + \\ &-\frac{1}{8\pi}\left(\frac{g_{rr}(2g_{0r}^2 - g_{00}\,g_{rr})}{(g_{0r}^2 - g_{00}\,g_{rr})^2}\right)(\hbar c)^2\dot{\chi}_0^*\,\dot{\chi}_0 \quad + \\ &-\frac{1}{8\pi}\left(\frac{g_{00}^2\,g_{rr}}{(g_{0r}^2 - g_{00}\,g_{rr})^2}\right)(\hbar c)^2\chi_0^{*\prime}\,\chi_0' \\ &+\frac{1}{8\pi}\left(\frac{g_{00}\,g_{0r}\,g_{rr}}{(g_{0r}^2 - g_{00}\,g_{rr})^2}\right)(\hbar c)^2\left(\dot{\chi}_0^*\,\chi_0' + \chi_0^{*\prime}\,\dot{\chi}_0\right).\end{aligned} \tag{6.5}$$

This form (given by $-\rho$ for hydrodynamic systems) will be useful in subsequent calculations.

For systems with spherical symmetry, one can perform a partial wave expansion on the field $\chi(\vec{x})$ [23]:

$$\chi(\vec{x}) \equiv \sum_{\ell m}\chi_\ell(ct, r)\,Y_\ell^m(\theta, \phi) = \sum_{\ell m}\frac{\psi_\ell(ct, r)}{r}\,Y_\ell^m(\theta, \phi). \tag{6.6}$$

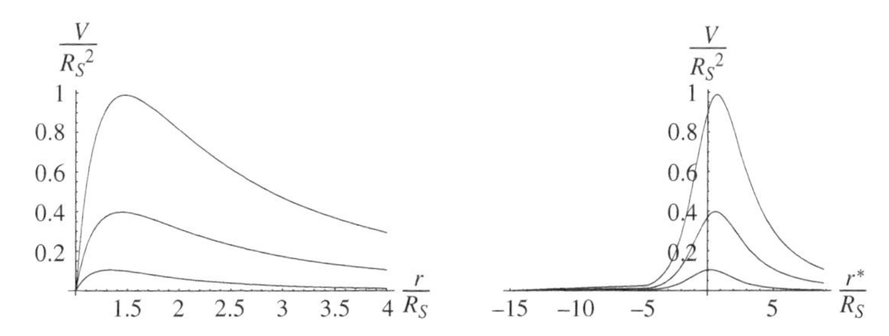

Figure 6.2 Effective potentials for motions near $r = R_S$ in Schwarzschild geometry for $\ell = 0$, $\ell = 1$, and $\ell = 2$. The potential barriers increase with increasing angular momentum.

Consider first the exterior of Schwarzschild geometry. It is convenient to examine the Euler–Lagrange equations using tortoise coordinates Eq. 5.97. The metric takes the form $g_{00} = -(1 - \frac{R_S}{r(r^*)})$, $g_{0r^*} = 0$, $g_{r^*r^*} = (1 - \frac{R_S}{r(r^*)})$. The Euler–Lagrange equations for a massless Klein–Gordon field for this geometry becomes:

$$\left(\frac{\partial^2}{\partial (ct_S)^2} - \frac{\partial^2}{\partial r^{*2}} \right) \psi_\ell(ct_S, r^*) + V_\ell^{eff}(r^*) \psi_\ell(ct_S, r^*) = 0, \tag{6.7}$$

where the effective potential for massless particles satisfies:

$$V_\ell^{eff}(r^*) \equiv \frac{r(r^*) - R_S}{r(r^*)} \left(\frac{\ell(\ell+1)}{r^2(r^*)} + \frac{R_s}{r^3(r^*)} \right). \tag{6.8}$$

Therefore, a massless quantum with asymptotic frequency ω conjugate to the Schwarzschild time satisfies the equation of motion:

$$-\frac{\partial^2}{\partial r^{*2}} \psi_\ell(r^*) + V_\ell^{eff}(r^*) \psi_\ell(r^*) = \left(\frac{\omega}{c} \right)^2 \psi_\ell(r^*). \tag{6.9}$$

This form motivates the notation of V_ℓ^{eff} as an effective potential in appropriate units.

Plots of the effective potential are given in Figure 6.2. The plot on the left expresses this potential as a function of the Schwarzschild radial coordinate r, while that on the right expresses it as a function of the conformal tortoise coordinate r^*. The tortoise coordinate r^* has a value of unity at $r = 2R_S$ and approaches $-\infty$ as $r \rightarrow R_S$. One can conclude that larger ℓ values generate a larger "centrifugal barrier" for quanta near R_S. The turning point for a given potential Eq. 6.8 can be determined by the vanishing of its derivative. For large angular momenta, this maximum occurs at $r_{TP} = 3R_S$. This represents the minimum radius for a stable circular orbit in the Schwarzschild geometry.

Classically, quanta in the vicinity of $r \approx R_S$ must have sufficient frequency to traverse the potential barrier and be asymptotically detected. The minimum frequencies classically needed to overcome the potential barrier are given by $\ell = 0$ states with $\omega \gtrsim 0.32\frac{c}{R_S}$. As might be expected, this frequency scales inversely with the mass of the geometry.

A general radially dynamic spherically symmetric geometry is obtained by replacing $V_o = 0$ in the general stationary Eq. 5.71, and giving temporal dependency in terms of the time coordinate t of the geometrically stationary observers: $g_{00} = -(1 - \frac{R_M(ct,r)}{r})$, $g_{0r} = \sqrt{\frac{R_M(ct,r)}{r}}$, $g_{rr} = 1$. The Euler–Lagrange equations for the Klein–Gordon field for this geometry becomes [83]:

$$-\frac{\partial^2 \psi_\ell}{(\partial ct)^2} + \frac{\partial}{\partial ct}\left[\sqrt{\frac{R_M}{r}}\left(\frac{\partial \psi_\ell}{\partial r}\right)\right] + \frac{\partial}{\partial r}\left[\sqrt{\frac{R_M}{r}}\left(\frac{\partial \psi_\ell}{\partial ct}\right)\right] + \frac{\partial}{\partial r}\left[\left(1 - \frac{R_M}{r}\right)\left(\frac{\partial \psi_\ell}{\partial r}\right)\right] +$$
$$-\left[\left(\frac{mc}{\hbar}\right)^2 + \frac{\ell(\ell+1) + \frac{R_M}{r}}{r^2} + \frac{1}{r}\frac{\partial}{\partial ct}\left(\sqrt{\frac{R_M}{r}}\right)\right]\psi_\ell = 0. \tag{6.10}$$

Substitution of the metric forms into Eq. 6.5 gives the energy density $\mathbf{T}_0^0$ for this field.

6.2.2 Substantive gravitating quantum flows

Despite the convenience of the coordinates of a given observer, the relevant quantum dynamics is often expressed best in terms of the affine parameter associated with the gravitating particles/fields. The proper-time derivative of local physical parameters can be expressed in terms of the substantive derivative $\frac{d}{d\lambda} = U^\beta \frac{\partial}{\partial x^\beta} \equiv U^\beta \partial_\beta$, which directly incorporates the principles of equivalence in the resultant substantive quantum flows on the geometry, as previously discussed in Section 3.4.3. In this expression, U^β represents components of the (dimensionless) four-velocity of the gravitating system in the observer frame of reference. The gravitation of such systems is generic, not requiring a specialized micro-physical model to manifest its effects. As with canonical proper-time gravitation, one expects gravitational coherence to be defined on space-like surfaces of fixed proper times of the *gravitating system*. One should recall that for quantum systems, the component of the momentum operator given by $\hat{p}_\beta = \frac{\hbar}{i}\frac{\partial}{\partial x^\beta} = \frac{\hbar}{i}\vec{e}_\beta$ is proportional to a basis tangent vector on the geometry.

In addition, kinematic cluster decomposability has been most directly realized in Chapter 3 when each coherent disentangled cluster is parameterized using its conserved proper four-velocity (Lorentz frame velocity), i.e., each cluster undergoes substantive material flow. Those solutions manifesting cluster decomposability in Minkowski space-time required the use of four-velocities $\vec{u}$ as geometric parameters (defining Lorentz frames) that do *not* go off-shell or off-diagonal for intermediate states. The geometric parameters are distinct from quantum operators like the four-momentum, whose eigenvalues *do* go off shell during quantum dynamics. Likewise, to satisfy cluster decomposability for quantum processes in curved space-time, the intermediate eigenstates defined by eigenvalues of $\hat{K} = -\vec{u} \cdot \vec{P}$ go off-diagonal, while the geometric paramter $\vec{u}$ should be preserved to provide a unique Lorentz coordinate frame for the quantum dynamics.

An example scalar field satisfying substantive flow has an action described by:

$$\begin{aligned} W = \int d^4x \sqrt{-g} \big\{ \tfrac{c}{2} U^\beta \big[\chi^* \left(\tfrac{\hbar}{i} \partial_\beta \chi - \tfrac{q}{c} A_\beta \chi \right) \\ + \left(-\tfrac{\hbar}{i} \partial_\beta \chi^* - \tfrac{q}{c} A_\beta \chi^* \right) \chi \big] + mc^2 \chi^* \chi - \tfrac{1}{16\pi} g^{\alpha\mu} g^{\beta\nu} F_{\alpha\beta} F_{\mu\nu} \big\}, \end{aligned} \tag{6.11}$$

where the homogeneous Maxwell equations relate the field strengths to the gauge potentials:

$$F_{\mu\nu} = \partial_\mu A_\nu - \partial_\nu A_\mu. \tag{6.12}$$

This form exhibits invariance of the action under a local gauge transformation of the scalar field $\chi(\vec{x}) \to e^{i\alpha(\vec{x})} \chi(\vec{x})$.

The scalar field has two components, its real and imaginary parts, or alternatively its modulus and phase. Writing $\chi = |\chi| e^{\xi_\chi}$, the action can be re-expressed as:

$$W = \int d^4x \sqrt{-g} \left\{ \left(\hbar c\, U^\beta \partial_\beta \xi_\chi + mc^2 - q\, U^\beta A_\beta \right) |\chi|^2 - \frac{1}{16\pi} g^{\alpha\mu} g^{\beta\nu} F_{\alpha\beta} F_{\mu\nu} \right\}. \tag{6.13}$$

Extrema of the action under arbitrary variations of the modulus $\delta|\chi|^2$, the phase $\delta\xi_\chi$, and the gauge potential δA_β yield three sets of Euler–Lagrange equations:

$$\hbar c\, U^\beta \partial_\beta \xi_\chi + mc^2 - q\, U^\beta A_\beta = 0, \tag{6.14}$$

$$\frac{1}{\sqrt{-g}} \partial_\beta \left[\sqrt{-g} |\chi|^2 U^\beta \right] = 0, \tag{6.15}$$

$$\frac{1}{\sqrt{-g}} \partial_\alpha \left[\sqrt{-g} \frac{g^{\alpha\mu} g^{\beta\nu} F_{\mu\nu}}{4\pi} \right] + q\, U^\beta |\chi|^2 = 0. \tag{6.16}$$

The first equation 6.14 demonstrates phase coherence in terms of the proper energies of the gravitating particle. For free particles in flat space-time, its solution takes the usual form $\xi_\chi = \vec{k} \cdot \vec{x}$. The second equation 6.15 demonstrates conservation of particle flux $\vec{J}_\chi \equiv |\chi|^2 U^\beta c$, with $(J^\beta_\chi)_{;\beta} = 0$. The third equation 6.16 takes a more familiar form if one recognizes that $\Gamma^\mu_{\nu\beta} F^{\beta\nu} = 0$ due to anti-symmetry of the field strengths. Thus, this equation yields the inhomogeneous Maxwell equations, $F^{\beta\alpha}_{;\alpha} = 4\pi q\, |\chi|^2 U^\beta = \frac{4\pi}{c} q J^\beta_\chi$.

6.2.3 Linear spinor fields

It is straightforward to extend the development of the quantum flow fields from the previous section beyond scalar fields. The linear spinor field formalism developed in Section 4.3.1 provides a convenient mechanism for incorporating gravitation into relativistic quantum mechanics. Of particular relevance, the group metric of the extended Poincaré group includes the Minkowski metric (unlike the standard Poincaré group), providing a direct connection of the group theory to general relativity. Since there are no physical theorems that exclude angular momentum contributions to the masses of fundamental particles, it is possible that particle spin contributes to the gravitational field. The Casimir invariants of the extended Poincaré group somewhat connect the mass and spin through Dirac-like operators.

The incorporation of curvilinear coordinates into the linear spinor field equations is most direct using the inverse of the tetrad $\mathcal{V}_{\beta}^{\hat{\mu}}$ from Eq. 5.58 given by $(\mathcal{V}^{-1})_{\tilde{\mu}}^{\beta} \equiv \mathcal{V}_{\tilde{\mu}}^{\beta}$. Recall also that the Jacobian of transformation from locally inertial coordinates $\xi^{\hat{\mu}}$ to curvilinear coordinates x^{β} is given by the determinant of the tetrad $\mathcal{J} \equiv \left| \frac{\partial \vec{\xi}}{\partial \vec{x}} \right| = \det \mathcal{V}$ as an alternative to $\sqrt{-g}$. The free-particle linear spinor field equation then transforms according to:

$$\boldsymbol{\Gamma}^{\tilde{\mu}} \hat{P}_{\tilde{\mu}} \, \boldsymbol{\Psi}_{(\gamma)} = \boldsymbol{\Gamma}^{\tilde{\mu}} \mathcal{V}_{\tilde{\mu}}^{\beta} \hat{P}_{\beta} \, \boldsymbol{\Psi}_{(\gamma)} = -(\gamma) mc \, \boldsymbol{\Psi}_{(\gamma)}. \tag{6.17}$$

This allows the apparatus developed for substantive flows in Section 6.2.2 to be utilized if one defines the spinor-valued four-velocity $\vec{\mathbf{U}}$ as:

$$\mathbf{U}_{\Gamma}^{\beta} \equiv \mathcal{V}_{\hat{\mu}}^{\beta} \boldsymbol{\Gamma}^{\hat{\mu}}. \tag{6.18}$$

This spinor-valued form must be appropriately sandwiched between spinor states in any (spinor) invariant expression.

The Lagrangian form $\mathcal{L}_{matter} \equiv (\det \mathcal{V}) \, L_{matter}$ for the free spinor field is thus given by:

$$L_{matter} = \frac{1}{2} \mathcal{V}_{\tilde{\mu}}^{\beta} \frac{\hbar c}{i} \left[\bar{\boldsymbol{\Psi}}_{(\gamma)} \boldsymbol{\Gamma}^{\tilde{\mu}} \left(\partial_{\beta} \boldsymbol{\Psi}_{(\gamma)} \right) - \left(\partial_{\beta} \bar{\boldsymbol{\Psi}}_{(\gamma)} \right) \boldsymbol{\Gamma}^{\tilde{\mu}} \boldsymbol{\Psi}_{(\gamma)} \right] + (\gamma) mc^2 \, \bar{\boldsymbol{\Psi}}_{(\gamma)} \boldsymbol{\Psi}_{(\gamma)}. \tag{6.19}$$

This Lagrangian yields the needed equations of motion because the extra term involving the divergence of the inverse tetrad vanishes. One demonstrates this by examining the form of the affine connections in terms of the tetrads:

$$\Gamma_{\mu\nu}^{\beta} = \frac{\partial x^{\beta}}{\partial \xi^{\tilde{\lambda}}} \frac{\partial^2 \xi^{\tilde{\lambda}}}{\partial x^{\mu} \partial x^{\nu}} = \mathcal{V}_{\tilde{\lambda}}^{\beta} \, \partial_{\mu} \mathcal{V}_{\nu}^{\tilde{\lambda}} = \mathcal{V}_{\tilde{\lambda}}^{\beta} \, \partial_{\nu} \mathcal{V}_{\mu}^{\tilde{\lambda}}. \tag{6.20}$$

The trace of the connection is related to the derivative of the Jacobian:

$$\Gamma_{\mu\lambda}^{\lambda} = \frac{1}{\sqrt{-g}} \partial_{\mu} \sqrt{-g} = \frac{1}{\det \mathcal{V}} \partial_{\mu} (\det \mathcal{V}) = \mathcal{V}_{\tilde{\alpha}}^{\lambda} \, \partial_{\lambda} \mathcal{V}_{\mu}^{\tilde{\alpha}}. \tag{6.21}$$

This term has precisely the opposite sign of that of the derivative of the inverse tetrad, yielding:

$$\partial_\beta \left[(\det \mathcal{V}) \mathcal{V}^\beta_{\tilde{\mu}} \right] = 0. \tag{6.22}$$

Thus, the Lagrangian of Eq. 6.19 yields the expected form for the free linear spinor field equation 6.17.

The conservation condition for the particle current $\frac{\partial}{\partial \xi^{\tilde{\mu}}} (\bar{\mathbf{\Psi}}_{(\gamma)} \mathbf{\Gamma}^{\tilde{\mu}} \mathbf{\Psi}_{(\gamma)}) = 0$ transforms to the curvilinear form $\frac{\partial x^\beta}{\partial \xi^{\tilde{\mu}}} \frac{\partial}{\partial x^\beta} (\bar{\mathbf{\Psi}}_{(\gamma)} \mathbf{U}^\beta_\Gamma (\vec{x}) \mathbf{\Psi}_{(\gamma)}) = 0$. Defining the local particle type current $J^\beta_\Gamma (\vec{x}) \equiv \bar{\mathbf{\Psi}}_{(\gamma)} (\vec{x}) \mathbf{U}^\beta_\Gamma (\vec{x}) \mathbf{\Psi}_{(\gamma)} (\vec{x})$, the local form of particle current conservation is as expected:

$$\partial_\beta J^\beta_\Gamma + \Gamma^\beta_{\beta\alpha} J^\alpha_\Gamma = 0 = \frac{1}{\det \mathcal{V}} \partial_\beta \left((\det \mathcal{V}) \, J^\beta_\Gamma \right). \tag{6.23}$$

Thus, the current $\mathcal{J}^\beta_\Gamma \equiv \det \mathcal{V} \, J^\beta_\Gamma$ generates a globally conserved particle-type charge $\mathcal{Q}_\Gamma$. Because the group metric $\eta_{\tilde{\mu}\tilde{\nu}}$ for the generators $\hat{\Gamma}^{\tilde{\mu}}$ defines the Minkowski metric, these local generators have invariant inner products under the general space-time metric $g_{\alpha\beta}$.

Energy-momentum tensor

The energy-momentum for the Lagrangian 6.19 can be calculated using Eq. 5.62. It is useful to note that:

$$\frac{\delta}{\delta \mathcal{V}^{\tilde{\mu}}_\beta} \det \mathcal{V} = \mathcal{V}^\beta_{\tilde{\mu}} \, \det \mathcal{V}, \tag{6.24}$$

defines the co-factor of the element of the tetrad in its determinant (by Cramer's rule). The gravitationally derived energy-momentum tensor is thus:

$$\begin{aligned} \mathcal{T}^{\alpha\beta} \, g_{\beta\lambda} &= - \left[\mathcal{V}^{\tilde{\mu}}_\lambda \, \frac{\partial}{\delta \mathcal{V}^{\tilde{\mu}}_\alpha} ((\det \mathcal{V}) \, L_{matter}) + \delta^\alpha_\lambda \, L_{matter} \right] \\ &= \left[\mathcal{V}^\alpha_{\tilde{\mu}} \, \frac{\partial}{\delta \mathcal{V}^\lambda_{\tilde{\mu}}} ((\det \mathcal{V}) \, L_{matter}) - \delta^\alpha_\lambda \, L_{matter} \right]. \end{aligned} \tag{6.25}$$

Substituting the Lagrangian form 6.19, yields an explicit energy-momentum tensor:

$$\begin{aligned} \mathcal{T}^{\alpha\beta} \, g_{\beta\lambda} &= \frac{\hbar c}{2i} \mathcal{V}^\alpha_{\tilde{\mu}} \left[\bar{\mathbf{\Psi}}_{(\gamma)} \mathbf{\Gamma}^{\tilde{\mu}} \left(\partial_\lambda \mathbf{\Psi}_{(\gamma)} \right) - \left(\partial_\lambda \bar{\mathbf{\Psi}}_{(\gamma)} \right) \mathbf{\Gamma}^{\tilde{\mu}} \mathbf{\Psi}_{(\gamma)} \right] - \delta^\alpha_\lambda L_{matter} \\ &= \frac{\hbar c}{2i} \left[\bar{\mathbf{\Psi}}_{(\gamma)} \mathbf{U}^\alpha_\Gamma \left(\partial_\lambda \mathbf{\Psi}_{(\gamma)} \right) - \left(\partial_\lambda \bar{\mathbf{\Psi}}_{(\gamma)} \right) \mathbf{U}^\alpha_\Gamma \mathbf{\Psi}_{(\gamma)} \right] - \delta^\alpha_\lambda L_{matter}. \end{aligned} \tag{6.26}$$

Since the tensor is not generated using variations of a symmetric metric, it might not itself be symmetric.

One should note that this is identical to the form of the quantum mechanically derived energy-momentum tensor using Eq. 3.51 for this Lagrangian. The Belinfante tensor defined in Eq. 3.64 represents a symmetrized tensor suitable for Einstein's equations. This tensor preserves the local conservation principle of the quantum mechanical tensor $\partial_{\tilde{\mu}}(\mathcal{T}^{\tilde{\mu}\tilde{\nu}} + \partial_{\tilde{\kappa}} S^{\tilde{\kappa}\tilde{\mu}\tilde{\nu}}) = 0$ due to the anti-symmetry properties of the tensor $S^{\tilde{\kappa}\tilde{\mu}\tilde{\nu}} = -S^{\tilde{\mu}\tilde{\kappa}\tilde{\nu}}$ from Eq. 3.63, implying that $\partial_{\tilde{\mu}}\partial_{\tilde{\kappa}} S^{\tilde{\kappa}\tilde{\mu}\tilde{\nu}} = 0$. Substituting the algebra for the generators from Eqs. 4.42 and 4.45 which imply that $\mathcal{J}^{\tilde{\mu}\tilde{\nu}} = -i[\mathbf{\Gamma}^{\tilde{\mu}}, \mathbf{\Gamma}^{\tilde{\nu}}]$, the Belinfante tensor for the linear spinor fields in flat space-time are given by:

$$S^{\tilde{\kappa}\tilde{\mu}\tilde{\nu}} = i\frac{\hbar c}{2}\,\bar{\Psi}_{(\gamma)}\left(-\mathbf{\Gamma}^{\tilde{\kappa}}[\mathbf{\Gamma}^{\tilde{\mu}}, \mathbf{\Gamma}^{\tilde{\nu}}] + \mathbf{\Gamma}^{\tilde{\mu}}[\mathbf{\Gamma}^{\tilde{\kappa}}, \mathbf{\Gamma}^{\tilde{\nu}}] + \mathbf{\Gamma}^{\tilde{\nu}}[\mathbf{\Gamma}^{\tilde{\kappa}}, \mathbf{\Gamma}^{\tilde{\mu}}]\right)\Psi_{(\gamma)}\,. \quad (6.27)$$

The curvilinear form of this tensor is obtained by the direct replacements $\mathbf{\Gamma}^{\tilde{\mu}} \to \mathbf{U}^{\alpha}(\vec{x})$. Using the approach from Eq. 3.62,

$$\mathcal{T}^{\alpha\beta} - \mathcal{T}^{\beta\alpha} = -i\left[\frac{\partial L}{\partial(\partial_\zeta \mathbf{\Psi})}\mathcal{J}^{\alpha\beta}(\vec{x})\mathbf{\Psi}\right]_{;\zeta} = -\left[S^{\zeta\alpha\beta} - S^{\zeta\beta\alpha}\right]_{;\zeta}, \quad (6.28)$$

a symmetric tensor can be constructed, if desired:

$$T^{\alpha\beta} = \mathcal{T}^{\alpha\beta} + \left[S^{\zeta\alpha\beta}\right]_{;\zeta}\,. \quad (6.29)$$

However, curvature contributions to the spin term are not excluded using this construction.

6.3 Self-gravitation

A system is considered to be *self-gravitating* if it responds exclusively to the gravitational field of its constituents. Since gravity is attractive, a coherent, self-gravitating system can consist of a single constituent, or co-gravitating constituents. *Self-generating* systems will be self-gravitating systems that have a vanishing contribution of gravitational binding energy to the overall proper energy of the system. Generic self-gravitating systems will be examined in this section.

6.3.1 Canonical proper-time dynamics in curved space-time

The canonical proper-time formulation of relativistic dynamics will presume that the gravitating quantum system maintains coherence on surfaces defined by *its* proper time. In gravitational physics this represents a non-trivial distinction. Since the various regions across which a coherent state propagates have varying gravitational potentials, space-like surfaces of simultaneity defined by fixed proper-time, τ are generally different from those defined by fixed coordinate time, t.

Canonical proper-time gravitation and geodesics

As was the case for cluster-decomposable few-particle dynamics, the problem somewhat simplifies once it is expressed using the proper parameters. Proper-time gravitation was previously examined in flat space-time in Section 2.3. The equations of motion generated for a stationary system ensure that the canonical proper energy is conserved ($\frac{dK}{d\tau}=0$). Thus, one does not expect the eigenvalues of K to change due to gravitational interactions, independent of its energetics. For a generic proper-potential form $\Phi(\mathbf{r})$, the canonical proper energy can be expressed:

$$K=\frac{\mathbf{p}\cdot\mathbf{p}}{2m}+m\,\Phi(\mathbf{r})+mc^2. \tag{6.30}$$

The potential form $\Phi(\mathbf{r})$ will be developed so that it is consistent with the geodesic motion of the expectation value of the four-velocity of the system.

The equations of motion resulting from Eq. 6.30 are given by:

$$\frac{dp_j}{d\tau}=-m\,\partial_j\Phi(\mathbf{r}),\quad \frac{dx_j}{d\tau}=\frac{p_j}{m}\quad\Rightarrow\quad \frac{d^2x^j}{d\tau^2}=-\partial_j\Phi(\mathbf{r}). \tag{6.31}$$

Since the rest mass is constant, this form is analogous to the geodesic equation,

$$\frac{d^2x^j}{d\tau^2}+\Gamma^j_{\alpha\beta}\frac{dx^\alpha}{d\tau}\frac{dx^\beta}{d\tau}=0. \tag{6.32}$$

As is the case with electronic distributions in stable atoms and previously examined in Section 2.3, the mass distribution should be stationary in a quantum self-gravitating system.

For a stationary gravitating distribution, assume that $\langle\frac{dx^\alpha}{d\tau}\rangle=\langle\frac{dx^0}{d\tau}\rangle\delta^\alpha_0$ (consistent with quantum expectation values). This defines the four-velocities as geometric parameters. Substituting the form of the connections $\Gamma^j_{\alpha\beta}$ in the geodesic equation 6.32, the proper interaction form must satisfy:

$$\frac{d^2x^j}{d\tau^2}=-\partial_j\Phi(\mathbf{r})=\frac{1}{2}\left(\frac{dx^0}{d\tau}\right)^2 g^{j\mu}\,\frac{\partial}{\partial x^\mu}g_{00}. \tag{6.33}$$

This form will be generated for various stationary energy densities.

Proper interaction for a stationary spherically symmetric metric

A stationary spherically symmetric system with a metric given by Eq. 5.70 has geometric parameters $\frac{dx^0}{d\tau}=\frac{c}{\sqrt{1+\frac{2}{c^2}(V+V_o)}}$ and $g^{rr}=1+\frac{2}{c^2}V$. This means that the potential in Eq. 6.33 satisfies the differential equation:

$$\frac{\partial}{\partial r}\Phi=\frac{1+\frac{2}{c^2}V}{1+\frac{2}{c^2}(V+V_o)}\frac{\partial}{\partial r}(V+V_o). \tag{6.34}$$

Since the potentials V and V_o can be completely solved for hydrodynamic systems, the general form of the proper interaction can be solved for a stellar hydrodynamic system from Section 5.3.3, although not generally in closed form.

This proper-potential form generally has a complicated, non-linear dependence upon the metric functions V and V_o, unless $V_o \to 0$. In that case, 6.34 considerably simplifies, giving a closed-form solution for the proper potential. This special case will be explored in some detail next.

Stationary radial gravitation

The space-time metric for a spherically symmetric geometry will be chosen as a radially stationary geometry modeled upon the uniformly accelerating systems of Section 1.2.1, with non-vanishing local densities. The non-orthogonal metric will be taken to be:

$$ds^2 = -\left(1 - \frac{R_M(r)}{r}\right)(dct)^2 + 2\sqrt{\frac{R_M(r)}{r}}\, dct\, dr + dr^2 + r^2 d\theta^2 + r^2 \sin^2\theta\, d\phi^2, \tag{6.35}$$

where the determinant of the metric $\sqrt{-\det \mathbf{g}} \equiv \sqrt{-g} = r^2 \sin\theta$ is the Jacobian of the transformation from locally inertial coordinates. The space and time in the non-orthogonal metric mix in a way that kinematically combines them in a manner analogous to accelerating systems in special relativity. This mixing allows a transparent formulation of space-like (or light-like) coherent dynamics consistent with the principles of quantum mechanics on surfaces characterized everywhere by the coordinate t.

A finite *radial mass scale* $R_M(r) \equiv 2G_N M(r)/c^2$ is the length scale of the interior mass-energy content of the system, which is directly connected to the energy density through the mixed Einstein tensor $G^0{}_0 = \frac{1}{r^2}\frac{\partial}{\partial r} R_M(r)$, i.e., $M(r) = \int_0^r \frac{c^2}{2G_N} G^0{}_0(r') r'^2\, dr' = -\int_0^r T^0{}_0(r')\, 4\pi\, r'^2\, dr'$. For a finite mass distribution, the metric takes the form of Minkowski space-time asymptotically ($r \gg R_M$). The Ricci scalar:

$$\mathcal{R} = -\frac{1}{r^3}\frac{d}{dr}\left(r^2 \frac{dR_M(r)}{dr}\right), \tag{6.36}$$

for such distributions is non-singular everywhere as long as the mass density decreases rapidly enough for small r. Since the state of the system will remain quantum coherent, any considerations of statistically derived macroscopic properties such as pressure that might modify this metric form is expected to be minimal.

Proper potential form

Substituting the metric 6.35 into Eq. 6.34 and noting that for radially stationary metrics $g^{rr} = -g_{00}$ and $\left(\frac{dx^0}{d\tau}\right)^2 = -\frac{c^2}{g_{00}}$, the proper potential satisfies:

$$\frac{\partial}{\partial r}\Phi(r) = -\frac{c^2}{2}\frac{\partial}{\partial r}\left[\frac{R_M}{r}\right] \Rightarrow \Phi(r) = -\frac{G_N M(r)}{r}. \tag{6.37}$$

This fully relativistic form is the same as the Minkowski space-time interaction examined in Section 2.3.2, and is the same form that would be generated by an orthogonal metric generalizing the geometry of Schwarzschild, with $M \to M(r)$. It is somewhat surprising that the flat-space calculation in Part I is quite relevant for the self-gravitating solutions here that will modify space-time.

6.3.2 Stationary quantum self-gravitation

The equations developed in the previous section apply for any given mass m gravitating in a field generated by a composite source mass distribution $M(r)$. A set of self-gravitating solutions for which the source mass distribution is generated by the quantum probability distribution of the mass itself will next be developed.

Probabilities, operators, and eigenvalues

Consider the stationary gravitation of a mass m due to an interior source mass distribution $M(r)$ parameterized in the preferred stationary frame of reference. The canonical proper-time formulation allows one to use the proper time of the gravitating system as the temporal coordinate, while choosing *any* convenient spatial coordinates. The spatial coordinates will be chosen to be those of the (inertial) geometrically stationary observers $(ct, r, \vartheta, \varphi)$. For those observers, dr is a proper distance, and the quantum state can be normalized using the usual prescription for stationary states:

$$1 = \int d^3r\, |\Psi(r, \vartheta, \varphi)|^2 = \int r^2 \sin\vartheta\, dr\, d\vartheta\, d\varphi\, |\Psi(r, \vartheta, \varphi)|^2. \tag{6.38}$$

Using this isotropic system, the state function that satisfies the canonical proper-energy equation for this mass should be decomposable in terms of the spherical harmonics $Y_\ell^{\ell_z}(\vartheta, \varphi)$,

$$\Psi_{n\ell\ell_z}(r, \vartheta, \varphi) = R_{n\ell}(r) Y_\ell^{\ell_z}(\vartheta, \varphi). \tag{6.39}$$

Therefore, for a coherent state with defined angular momentum, the likelihood that it will be detected within a volume specified by $\Delta\mathcal{V}^*$ is given by:

$$\mathcal{P}_{n\ell\ell_z}^{\Delta\mathcal{V}^*} = \int_{\Delta\mathcal{V}^*} d^3r\, |R_{n\ell}(r) Y_\ell^{\ell_z}(\theta, \phi)|^2, \tag{6.40}$$

where the spherical harmonics are orthonormal for integrals where the volume $\Delta\mathcal{V}^*$ ranges over the complete solid angle.

The form of the canonically conjugate momentum components in Eq. 6.30 should be consistent with the Heisenberg equations of motion:

$$\left\langle \frac{d\hat{p}_r}{d\tau} \right\rangle = \left\langle \frac{i}{\hbar}[\hat{K}, \hat{p}_r] \right\rangle = -m\, \partial_r \Phi(r), \tag{6.41}$$

where $\Phi(r)$ was obtained from the geodesic equation 6.34. For the spatially flat metric form Eq. 6.35, the square of the momentum is thus given by:

$$\hat{\mathbf{p}} \cdot \hat{\mathbf{p}} = -\hbar^2 \left\{ \frac{1}{r^2} \frac{\partial}{\partial r} \left(r^2 \frac{\partial}{\partial r} \right) - \frac{\hat{\mathbf{L}}^2}{\hbar^2 r^2} \right\}, \tag{6.42}$$

where the spherical harmonics are eigenfunctions of the orbital angular momentum operator $\hat{L}^2$.

The parameters in Eq. 6.30 analogous to those of Bohr for hydrogenic systems were developed previously in the supplement beginning on page 90. The radial scale of the solutions given in Eq. 2.86 will be reproduced here: $a \equiv \frac{\hbar^2}{G_N m^3} = \frac{\lambda_m^3}{L_P^2}$, where the reduced Compton wavelength is given by $\lambda_m \equiv \frac{\hbar}{mc}$, and the Planck length is labeled L_P.

States of vanishing angular momentum are of two types. Those satisfying the general solution in the limit that $\ell \to 0$ have vanishing density at the center and will be discussed later. A central mass solution with non-vanishing central density can also be developed. Considering only s-wave ($\ell = 0$) states, it is convenient to introduce a central reduced radial wavefunction $u_C(r/a) \propto r R_{C,\ell=0}(r)$ parameterized by dimensionless variable $\zeta \equiv r/a$. As previously done with the dynamic parameters in Section 2.3.2, the geometric parameters can also be scaled using the parameter a:

$$\begin{aligned}
&\mathcal{P}_C(r/a) \equiv \int_0^{r/a} u_C^2(\zeta')\,d\zeta', \mathcal{P}_C(\infty) = 1, \\
&\tfrac{R_M(r)}{r} = 2 \left(\tfrac{m}{M_P}\right)^4 \tfrac{a}{r} \mathcal{P}_C(r/a), \\
&-\tfrac{2ma^2}{\hbar^2} V(r) = 2 \tfrac{a}{r} \tfrac{M(r/a)}{m}, \\
&-\tfrac{2ma^2}{\hbar^2}(K - mc^2) = -2 \left(\tfrac{M_P}{m}\right)^4 \tfrac{K - mc^2}{mc^2}.
\end{aligned} \tag{6.43}$$

Using these identifications, Eq. 6.39 can then be re-written using the $\ell = 0$ version of Eq. 2.87,

$$\epsilon_C\, u_C(\zeta) = \frac{d^2 u_C(\zeta)}{d\zeta^2} + \left(\frac{2}{\zeta}\right) \left(\frac{M(\zeta)}{m}\right) u_C(\zeta), \tag{6.44}$$

where the dimensionless parameter $\epsilon_C \equiv -2\left(\frac{M_P}{m}\right)^4 \frac{K-mc^2}{m\,c^2}$ quantifies the gravitational binding energy of the mass.

For self-gravitating systems, the source mass $M(\zeta)$ will be presumed to be generated by the interior self-sourcing of the probability density. Generally, $M(\zeta)$ consists of an incoherent sum of coherent components, formally described using a mass density matrix. From Eq. 2.88, a single mass source gives:

$$M(\zeta) = \int_0^\zeta \rho_{mass}(\zeta')\,d\zeta' = \int_0^\zeta m\,u_C^2(\zeta')\,d\zeta' = m\mathcal{P}_C(\zeta). \tag{6.45}$$

The distribution indicates that the differential equation 6.44 is non-linear. It should be noted that the particle mass scale m appears nowhere in 6.44. This is completely consistent with the principle of equivalence, despite the non-linear nature of the equations. The solutions required no specific form for micro-physical interactions, consistent with the universal nature of gravitation.

Since the form of Eq. 6.44 is the same as that of Eq. 2.89, the functional forms of the solutions using general relativity are the same as those given in Section 2.3.2, as long as one uses the coordinates of the geometrically stationary observers. However, space-time curvature is included by calculating the radial mass scale used to generate the Einstein tensor for a specific mass m using $R_M(r) = 2\left(\frac{m}{M_P}\right)^4 a\,\mathcal{P}(r/a)$.

6.3.3 Einstein's equations for self-generating masses

For illustrative purposes, solutions of Einstein's equations for a mass rendering small, noticeable modifications of space-time from flatness will be demonstrated. The mass of a self-generating system will be chosen to satisfy $\left(\frac{m}{M_P}\right)^4 = 0.01$, or $m \simeq 0.316\,M_P$. The metric factor $-g_{00} = 1 - \frac{R_M(r)}{r}$ measuring temporal modifications for fiducial observers is then demonstrated in Figure 6.3. A larger mass value lowers the y-intercept of this curve. If the mass is sufficient for this factor to change sign, a trapping region within which outgoing photons must propagate towards decreasing radial parameter r will be present. This occurs because outgoing photons follow null trajectories $\frac{dr_\gamma}{dct} = 1 - \frac{\sqrt{R_M(r)}}{r}$. If the mass m is below the threshold value where the sign change occurs, there will be no trapped region. For the self-generating mass, this threshold value is given by $m \simeq 0.63 M_P$. Masses smaller than this will not generate a black hole.

A conformal density plot for this energy distribution is presented in Figure 6.4. The center $r = 0$ is the left boundary of the diagram and is everywhere a time-like curve. In contrast to the Minkowski space-time diagram in Figure 2.11, the center

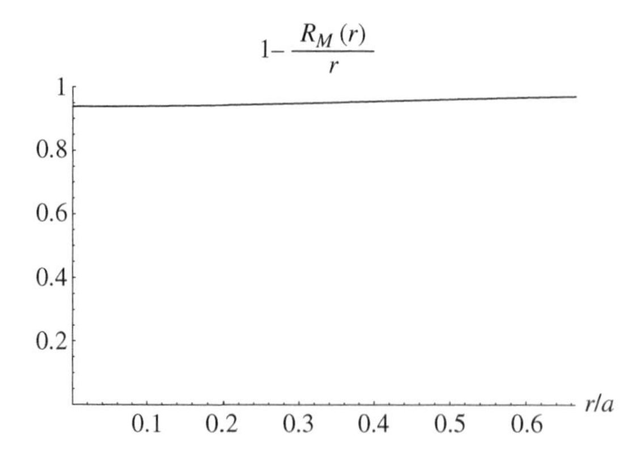

Figure 6.3 Metric factor $-g_{00}$ and g_{rr}^{-1} .

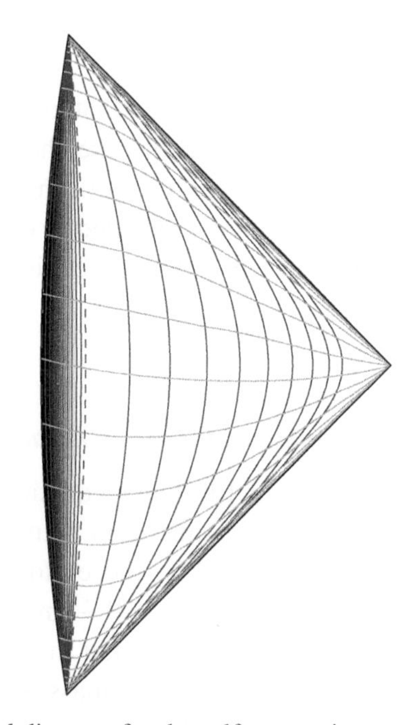

Figure 6.4 Conformal diagram for the self-generating mass distribution. Vertical curves represent surfaces of fixed radial coordinate, and horizontal curves represent surfaces of fixed temporal coordinate for geometrically stationary observers. The dashed curve represents the outermost surface of non-vanishing energy density.

is not a vertical line due to the influence of the energy density. Fixed-time surfaces are graded in units of the surface scale $R_{surface}$ and are everywhere space-like. Fixed radial coordinate curves are originally graded in tenths, then in units of the surface scale.

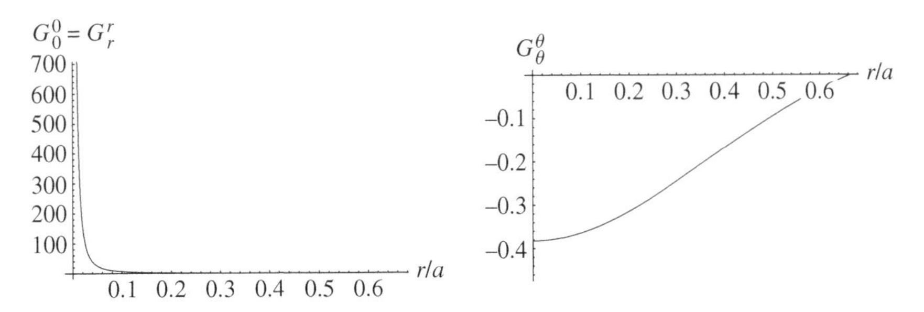

Figure 6.5 Einstein tensor components for self-generating mass.

In the region of non-vanishing energy density, the non-vanishing components of the Einstein tensor $G^{ct}_{ct} = G^r_r$ and $G^\theta_\theta = G^\phi_\phi$ are demonstrated in Figure 6.5. The Einstein tensor vanishes in the vacuum region $r > R_{surface}$, $\frac{r}{a} > 0.64$.

Energy conditions

Numerical values testing the energy conditions can be obtained for the self-gravitating densities. The null and weak energy conditions are satisfied everywhere for these solutions. Likewise, the dominant energy condition is also satisfied everywhere for observers with arbitrary radial motions. However, the dominant energy condition for rapid azimuthal motions of an observer is violated in the region just inside of the surface $r \lesssim R_{surface}$, due to coherence and gravitational binding from the interior mass distribution. Rapid azimuthal motions are those exceeding the condition:

$$r\, U_\theta > \frac{2R'_M(r)}{\sqrt{(r\, R''_M(r))^2 - (2R'_M(r))^2}}, \tag{6.46}$$

when defined. Exterior to the region of coherence, as well as proximal to the center, all energy conditions are satisfied.

6.3.4 Co-gravitating masses

Macroscopic bodies constructed from large numbers of constituents maintain mean gravitational fields while those constituents undergo varying states of coherence. Therefore, such bodies should be described using density matrices, allowing the majority of the constituents to be disentangled. However, because of the non-linear nature of Eq. 2.87, one cannot simply add solutions of self-gravitating systems to obtain a composite system. One can gain insight by numerically examining an example system of co-gravitating masses.

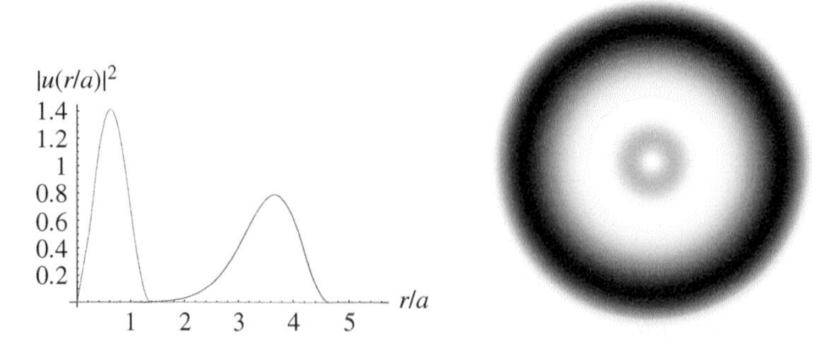

Figure 6.6 Probability densities (*left*) and mass densities (*right*) of an $\ell = 0$ and seven $\ell = 3$ co-gravitating masses of mass m.

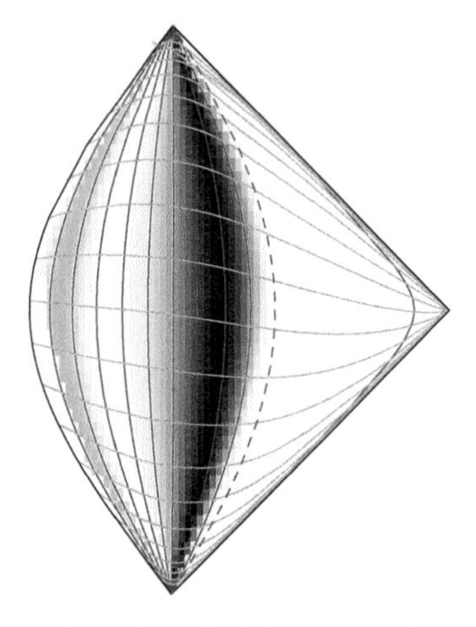

Figure 6.7 Conformal density plot of eight co-gravitating masses. The dashed curve represents the radial scale of the outermost surface of non-vanishing energy density.

General ℓ states

A general eigenstate of angular momentum with $\ell > 0$ has angular dependence and is thus not spherically symmetric. However, if one sums the modulus squared of the spherical harmonics in an expression like 6.40 over the quantum number ℓ_z, the resultant form has no angular dependence.

Co-gravitating quanta with a single mass m with $\ell = 0$, and $2\ell + 1 = 7$ masses m with $\ell = 3$ will be examined, as was explored in Figure 2.9. However, in this case, the central mass will be chosen to be of the type that has vanishing probability density at the center and is thus suitable to satisfy Fermi–Dirac statistics. A solution to 2.87 is plotted in Figure 6.6. The solution represents an incoherent sum of the coherent mass states, each of which gravitates in the collective mean field consistent with the principle of equivalence.

A conformal density plot of the global causal structure of the geometry of co-gravitating masses is presented in Figure 6.7. For this example, the metric approaches the Minkowski form both asymptotically as well as near $r \rightarrow 0$. Also, all components of the Einstein tensor are everywhere finite. The diagonal components of the Einstein tensor are equal, but do not vanish at the center of the distribution.

7

The physics of horizons and trapping regions

One of the principles of modern physics that is most adhered to is the expectation that models constructed to describe the phenomena of the physical universe should not depend upon any absolute frame of reference. The discovery of the CMB radiation perhaps demonstrates a counter example to this supposition, due to the preferred frame at rest relative to the energy content of the universe during its initial phase of expansion, as will be discussed in the next chapter. However, for most phenomena, the co-variance of the laws modeling those phenomena is consistent with the expectation of independence of the fundamental physics from the particular frame of reference utilized by the observer. This principle is embodied in the concept of *complementarity* [23] in the description of black holes. In its most direct expression, complementarity simply states that no observer should ever witness a violation of a law of nature. In particular, one expects that for a freely falling observer, there should be no local effects of gravitation as espoused by the principles of equivalence and relativity.

In this treatment, a *horizon* will always be a light-like surface that globally separates causally disconnected regions of space-time. Since light itself is characterized by both classical and quantum properties, geometries with horizons offer insights into the subtle relationships between general relativity and quantum physics. Generally, horizons can be only *globally* (not locally) defined, which means that local experiments performed by freely falling, inertial observers cannot detect the presence of such horizons. Yet, because of their light-like nature, no finite mass system can possibly follow the same trajectories. On a conformal diagram, a horizon is *always* represented by an ingoing or outgoing line with slope ± 1. On the other hand, the geometry of a *trapping surface* is defined by the *local* behaviors of light-like particles at a given time. Such surfaces need not be light-like, yet likewise, no finite mass system can follow such a trajectory. A horizon is a light-like trapping surface.

A *black hole* in particular represents an attractive geometry with an outgoing light-like horizon, from within which no communication can reach an exterior observer. As previously mentioned, a static spherically symmetric geometry with its mass within the Schwarzschild radius $2G_N M/c^2$ is an example of black hole geometry, with the radial scale of the horizon given by the Schwarzschild radius. One should note that that there can be no observational verification of a black hole short of measuring the properties of its horizon. Otherwise, the object might just be some manifestation of gravitationally condensed matter whose properties are yet to be explored in present accelerators using present astrophysical bounds or might be a horizonless transient black object.

It is well-known that coherent particles have enigmatic characteristics in terms of the predictions of classical general relativity (just as the coherent physics of stationary atoms violate many of the predictions of classical electromagnetism). For instance, the horizon of a black hole with mass M, spin $J \equiv R_J \, M \, c$, and charge $Q^2 \equiv \frac{R_Q^2 c^4}{G_N}$ can be shown to have radial coordinate $R_H = R_S \pm \sqrt{R_S^2 - (R_J^2 + R_Q^2)}$ on the equator, where $R_S \equiv \frac{2G_N M}{c^2}$. The term under the radical is unphysical for an electron, $R_S^2 - (R_J^2 + R_Q^2) = 4(\frac{m_e}{M_P})^2 L_P^2 - ((\frac{\hbar}{2m_e c})^2 + \alpha L_P^2) \cong -(\frac{\hbar}{2m_e c})^2$, being dominated by the quantized angular momentum. Therefore (as was also shown in Section 5.4), if an electron satisfies the equations of classical general relativity as a "point" particle, it exhibits a *naked singularity*, which refers to the singularity at $r = 0$ that is exposed to the exterior in the absence of a horizon. One thus expects that quantum behaviors likely substantially modify classical general relativity in the regions of scale comparable to that of the horizon.

Supplement: Nuclear density black hole estimation

Strong gravitational fields will couple to all microscopic interactions. Nuclear forces saturate at close distances, resisting densities that would compress constituent nucleons to have radii less than about $r_o \approx 1$ fermi $= 10^{-15}$ m. This results in close packing of nucleons resulting in a nuclear radius of $R = A^{1/3} r_o$, where A is the mass number of the nucleus. One can thus estimate what size nucleus would match its Schwarzschild radius $R_S = \frac{2G_N M}{c^2} = 2\frac{M}{M_P} L_P$. The two scales match at the critical mass number satisfying $A_{crit}^{1/3} r_o = 2A_{crit} \frac{m_n}{M_P} L_P$, or:

$$A_{crit}^{2/3} \simeq \frac{1}{2} \frac{r_o}{L_P} \frac{M_P}{m_n},$$

where $m_n \approx m_p$ is the mass of a nucleon. Since a solar mass consists of about $1.2 \times 10^{57} m_n$, this sets the critical mass number at about $A_{crit} \approx 10 \times 10^{57}$, or 8–9 solar masses. For nuclear matter more massive than this, gravitational physics requires the use of quark dynamics and Higgs-scale microscopic physics.

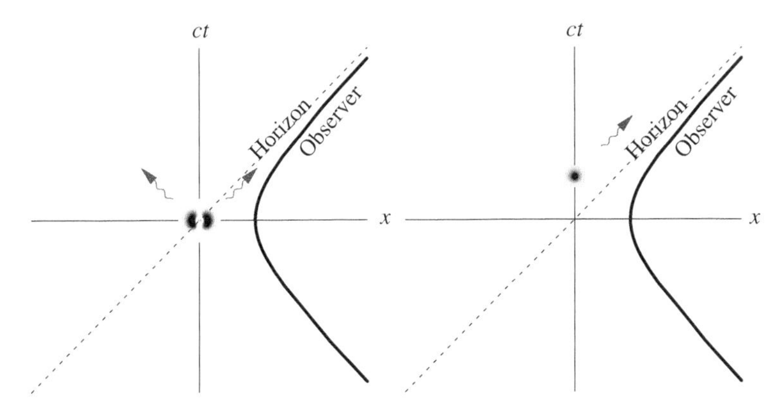

Figure 7.1 Quantum information from an inertial system viewed by an accelerating observer. The hyperbolic trajectory represents the motion of the uniformly accelerating observer. The coherent pair emit photons at $t = 0$ to form a new state, whose properties remain hidden from the observer.

As was demonstrated in Section 2.4, the inclusion of quantum mechanics in a special relativistic system with a horizon also introduces thermodynamic effects. Since a quantum system will maintain its coherence across the horizon of a uniformly accelerating observer, there will be an information loss associated with the coordinate description developed by that accelerating observer. An example is given by the coherent particle pair illustrated in Figure 7.1. The effective temperature associated with the horizon was developed in Section 2.4. Clearly, there are no special physical properties associated with the horizon in the Minkowski space-time background of the accelerating system. To an inertial observer, the horizon of this accelerating observer is just another light-like surface. Certainly, this horizon in flat space-time should have negligible thermal effects upon an inertial observer as that horizon traverses the inertial observer. However, for the accelerating observer, this light-like surface represents a boundary beyond which information concerning subtle quantum correlations becomes lost forever (if the acceleration is perpetual). This is an expression of complementarity in flat space-time. The only modifications to the laws of physics should be those appropriate to the accelerations of one's frame of reference. The relationship of these flat-space expectations to curvilinear expectations will be developed in the next section.

7.1 Static horizons in Rindler, Schwarzschild, and radially stationary geometries

The Schwarzschild geometry defined by the metric in Eq. 5.64 undergoes peculiar behavior when the quantity in the brackets vanishes: $[1 - \frac{R_S}{r}] \rightarrow 0$. The Schwarzschild radius $R_S \equiv 2MG_N/c^2$ represents the radial coordinate closest to the center from which a light beam could escape to infinity, and a region where the coordinates become anomalous. This radial coordinate is often called the *event horizon* of a Schwarzschild *black hole*, since any light emitted from a point closer to the center cannot be seen outside.

Although R_S represents a breakdown in the coordinates of a distant Schwarzschild observer, there are *no* local physical distinctions at this coordinate. The curvature of the space-time is finite, and usual laws of physics locally apply at $r = R_S$. However, the center $r = 0$ is both a coordinate *and* a physical singularity for the geometry, exerting unbounded forces in its vicinity.

Supplement: Schwarzschild radii of astrophysical objects of interest

Using Earth parameters $M_{Earth} \simeq 6 \times 10^{24}$ kg, $G_N \simeq 6.7 \times 10^{-11}$ N (m/kg)^2, and $c \simeq 3 \times 10^8$ m/s, one calculates a Schwarzchild radius for the Earth of about 9×10^{-3} m $= 9$ mm. Notice that this is not a proper distance, just a radial coordinate. Even if the Earth were a black hole, objects at a radial coordinate equal to the radius of the Earth would experience normal gravitational effects with accelerations $g \simeq 10\,\mathrm{m/sec}^2$ due to Birkhoff's theorem, or Newton's assertion that the gravity on the Earth's surface is as if all mass were concentrated at the center.

The Schwarzschild radii of other astrophysical bodies can be calculated as a factor of that of the Earth. For instance, the Schwarzschild radius of the Sun is about 3 km. There is also a gravitationally condensed massive body at the center of the galaxy, referred to as Sagittarius A, that is a strong candidate as an astrophysical black hole. The Schwarzschild radius of Sagittarius A is about 10 million km.

It is of interest to estimate the forces on a body in the vicinity of the Schwarzschild radius. The tidal forces in a geometry are proportional to the local Riemann curvature components associated with the specific orientation of the object. Those components have dimensions of inverse length squared and are factors of the form $R^{\alpha}_{\beta\mu\nu} \sim \frac{R_S}{r^3}$. Tidal accelerations of objects of scale δL are therefore of the order $a_S \sim \frac{R_s}{r^3} c^2\, \delta L$, proportional to the Schwarzschild radius. For a solar-sized black hole, the human body could withstand pressures and tidal effects up to about $100\, R_{S\odot}$, nowhere near the solar horizon itself. However, a human could traverse the horizon of Sagittarius A without hardly noticing any effects foretelling of impending doom.

Another interesting characteristic of the Schwarzschild geometry is that light doesn't traverse finite coordinate displacements in r and t at the "speed" of light in a flat space-time c. Upon inspecting null radial geodesics ($ds^2 = 0$), outgoing/ingoing light-like trajectories satisfy:

$$\frac{dr}{dt_S} = \pm\left(1 - \frac{R_S}{r}\right) \times c. \tag{7.1}$$

The closer the beam is emitted from the Schwartzchild radius, the more pronounced the difference. Yet, since the laws of physics are still valid in this geometry, locally inertial (freely falling) observers *will* measure the speed of light to be c. This measurement, $\frac{ds_r}{d\tau_r} = c$, is also true for local fiducial observers using rulers of fixed proper-length and the proper times of fiducial clocks (which *do not* display the Schwarzschild time, t_S).

One can calculate the amount of Schwarzschild time t that it would take for a mass to fall from rest at radial coordinate $r = r_o$ to the edge of the event horizon $r = R_S$. In terms of the Schwarzschild time coordinate t_S, the mass would take infinitely long to reach this radius, while the local proper time τ measured by a clock falling with the particle will be finite (e.g., see Reference [23]):

$$c(\Delta\tau_{horizon}) = \frac{r_o}{2}\sqrt{\frac{r_o}{R_S}}\left[\cos^{-1}\left(\frac{2R_S}{r_o} - 1\right) + 2\sqrt{\frac{R_S}{r_o}\left(1 - \frac{R_S}{r_o}\right)}\right]. \tag{7.2}$$

The freely falling observer's clock continues to tick unaltered as it traverses the Schwarzschild horizon. The observer will reach the singularity at $r = 0$ in a finite proper time $c\tau = \frac{\pi}{2} r_o (\frac{r_o}{R_S})^{1/2}$.

Fiducial observers measure proper time intervals $d\tau_r$ given by fixing the coordinate (r, ϑ, φ) in the metric: $(d\tau_r)^2 = [1 - \frac{R_S}{r}](dt_S)^2$. In this expression, dt_S represents an infinitesimal time interval measured by a distant Schwarzschild observer far from the gravitating source mass. Generally, the time intervals and tick frequencies of stationary clocks for different fiducial observers in a gravitational field satisfy the ratio:

$$\frac{d\tau_2}{d\tau_1} = \frac{\nu_1}{\nu_2} = \sqrt{\frac{g_{00}(\vec{r}_2)}{g_{00}(\vec{r}_1)}} \Rightarrow \sqrt{\frac{1 - \frac{R_S}{r_2}}{1 - \frac{R_S}{r_1}}}. \tag{7.3}$$

A clock very near the horizon ticks very slowly relative to one further away, and it must experience tremendous forces in order to keep it from falling through the horizon.

7.1.1 Stationary quantum black hole

Before exploring near horizon phenomena, several characteristics of trapping surfaces that are independent of any singular behavior of the geometry can be explored.

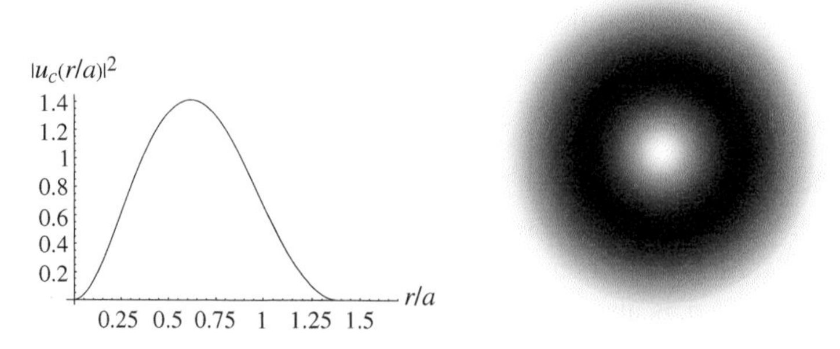

Figure 7.2 Self-generating, non-singular, Planck mass black hole.

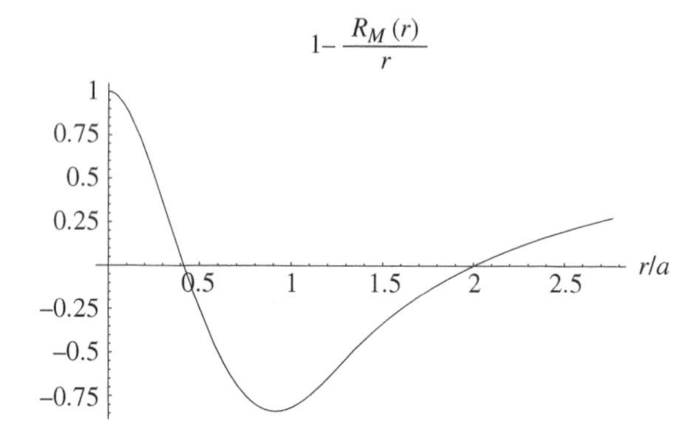

Figure 7.3 Plot of $-g_{00}$ for non-singular black hole geometry of stationary Planck mass.

As an example of a non-singular quantum black hole, consider the self-generating s-wave solution of Eq. 6.44 corresponding to a Planck mass, $m = M_P$. The mass distribution of the Planck mass is represented in Figure 7.2. All components of the curvature tensor are non-singular everywhere for this distribution.

The plot of g_{00} in Figure 7.3 clearly demonstrates the existence of a trapping region in this geometry. The trapping surfaces bound the region of space-time for which $R_M(r) > r$. The behavior of light-like trajectories in the geometry displayed in Figure 7.4 clearly demonstrate this trapping behavior. The figures chart out the trajectories of outgoing and ingoing photons that are evenly spaced at $t = 0$. All outgoing null trajectories in the diagram on the left, within the trapping region, have decreasing radial coordinates, while those interior and exterior to this region have increasing radial coordinates. The two vertical trajectories represent the inner and outer trapping surfaces of the geometry. Within the trapping region, all fixed radial coordinate surfaces are space-like, forbidding the existence of fiducial observers.

The conformal diagram for this geometry is demonstrated in Figure 7.5. The diagram is bounded on the left by the center $r = 0$. It should be noted that the center is

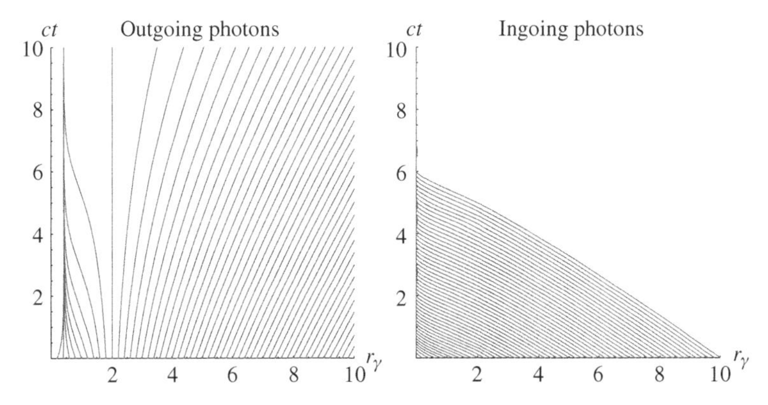

Figure 7.4 Standard space-time plots of outgoing (*left*) and ingoing (*right*) light-like trajectories in stationary black hole geometry.

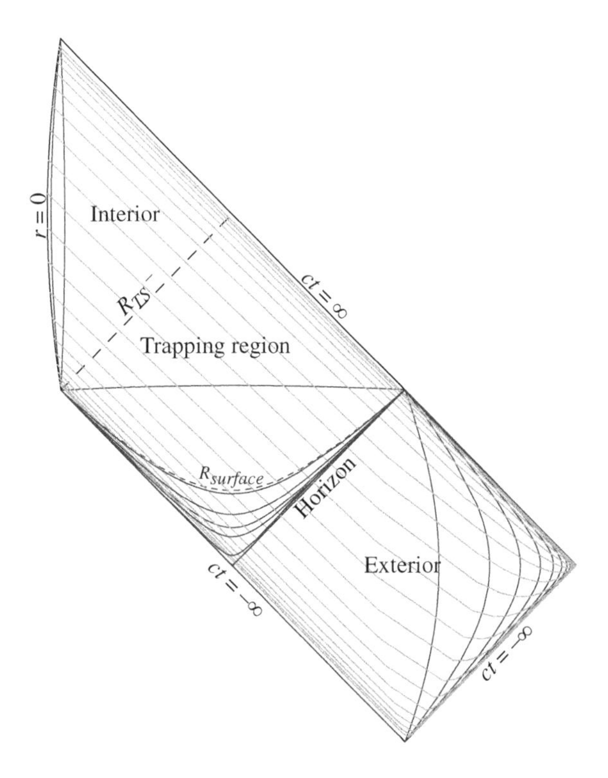

Figure 7.5 Conformal diagram of Planck mass, non-singular black hole.

time-like, allowing fiducial objects in this non-singular interior (one typically does not expect *observers* to be Planck scale). However, no process in the interior can send an outgoing communication to the exterior, despite receiving all ingoing communications from the exterior. All fixed temporal coordinate surfaces are space-like everywhere. The trapping region is bounded by two outgoing light-like surfaces, the inner trapping surface $r = R_{TS}^-$ and the horizon. Distant fiducial observers in the exterior will maintain minimal awareness of the black hole. One should note that the surface of the mass distribution $r = R_{surface}$, which is represented by the dashed curve, lies within the trapping region and is space-like, in contrast to the time-like surface of an exterior fixed radial coordinate. This should not be too surprising, since any potential evaporation must involve space-like transfers of energy densities to the exterior.

7.1.2 Schwarzschild black holes

If the Schwarzschild geometry defined by the metric in Eq. 5.64 holds throughout space-time, this defines a Schwarzschild black hole. This section will characterize such geometries.

Near-horizon coordinates

Very near the horizon of a Schwarzschild black hole, the radial coordinate can be treated to be essentially R_S to zeroth order. An alternative radial coordinate ρ_r characterizing the proper radial distance from the horizon will prove useful. This proper distance satisfies:

$$d\rho_r = \frac{dr}{\sqrt{1 - \frac{R_S}{r}}} \cong \sqrt{R_S}\frac{dr}{\sqrt{r - R_S}} \Rightarrow \rho_r \cong 2\sqrt{R_S\,(r - R_S)}. \tag{7.4}$$

Therefore, near the horizon the Schwarzschild metric can be approximated using:

$$ds^2 \cong -\rho_r^2 \frac{(dct_S)^2}{4R_S^2} + d\rho_r^2 + r^2(\rho_r)(d\vartheta^2 + \sin^2\vartheta\, d\varphi^2). \tag{7.5}$$

It is perhaps not too surprising that the near-horizon Schwarzschild coordinates are related to the Rindler space-time coordinates in the metric form of Eq. 1.41 developed for a uniformly accelerating observer in Minkowski space-time. Very near the horizon, the tangential coordinates appear essentially flat, and the radial scale is close to R_S. The acceleration $a_S = \frac{c^2}{2R_S}$ that would be associated with this Rindler space-time is directly related to the temperature of the horizon. However, note that a_S is not associated with the proper acceleration of a fiducial observer in the Rindler space-time.

Thus, the coordinate singularity associated with the Schwarzschild radius is conveniently parameterized using the following identifications:

$$\begin{aligned}
\omega_t &\equiv \frac{ct_S}{2R_S} = \frac{c^3}{4G_N M}\,t_S && \text{dimensionless Rindler time,} \\
a_{proper} &= \frac{c^2}{\rho_r} && \text{proper acceleration,} \\
ds^2 &= -\rho_r^2\, d\omega_t^2 + d\rho_r^2 + |\mathbf{dx}_\perp|^2 && \text{Rindler metric.}
\end{aligned} \tag{7.6}$$

The coordinate singularity $\rho_r = 0$ representing the horizon is seen to be the hyperbolic analog of the polar coordinate singularity representing angular ambiguity in polar coordinates at $r = 0$. In the transformation from Rindler time, the Schwarzschild time t_S rescales as if there were an asymptotic acceleration given by:

$$\frac{a_\infty}{c^2}\rho_r\, dct_S = \rho_r\, d\omega_t = \frac{c^2}{4G_N M}\rho_r\, dct_S \Rightarrow a_\infty = \frac{c^4}{4G_N M}. \tag{7.7}$$

Since there is actually no asymptotic proper acceleration, this temporal rescaling results in an asymptotic temperature that the distant Schwarzschild observer associates with the horizon of the black hole.

7.1.3 Temperature, entropy, and geometry

This section will examine the subtleties of space-times with finite-scaled horizons. In particular, the existence of causally disjoint regions in a space-time within which quantum coherent processes are occurring has implications with regards to information loss, as previously illustrated in Figure 7.1. Since any description of a coherent process using information exterior to the horizon must sum over consistent interior states, the existence of a static horizon motivates its statistical/thermal description in terms of exterior microphysics. For the uniformly accelerating observer in Minkowski space-time, differences in the representation of the particle vacuum states of an accelerating observer versus those an inertial observer have been used to explain the emission of radiations associated with the horizon. The temperature assigned to the horizon will be demonstrated using various arguments. The form of the entropy will then be inferred by presuming the validity of the first law of thermodynamics.

A fundamental characteristic of a horizon is that it delineates causally disjoint regions of the geometry, generating an information deficit about the region causally excluded by the horizon. Several quantum states beyond the horizon might correspond to a given state in the causally accessible region. Such states must be handled statistically with regards to the physics describing the accessible region. Statistical physics assigns an entropy associated with the number of microscopic configurations that can correspond to a given "coarse" (or macroscopic) measurement. This entropy parameterizes the disordered internal energy describing the heat in the first law of thermodynamics, and the non-decreasing degree of randomness in the second law of thermodynamics.

In Chapter 2, a relationship was found between the acceleration in a (flat) Rindler space-time and the temperature of a thermal bath, expressed in Eq. 2.99. The effective acceleration of the Schwarzschild observer relative to the horizon in

Eq. 7.7 already results in an apparent temperature of:

$$k_B T = \frac{\hbar a_\infty}{2\pi c} = \frac{\hbar c}{4\pi R_S}, \tag{7.8}$$

associated with the horizon. The development of such a temperature for the geometry portends the existence of thermodynamic parameters describing average values of macroscopic state variables. Thermodynamic parameters are generally obtained by statistically averaging over the possible non-coherent configurations associated with a given coarse measurement. As discussed in Section 2.2.3, the thermodynamic (average) energy, free energy, temperature, and entropy in the canonical ensemble are related by $\langle \hat{H} \rangle = F - T\frac{\partial F}{\partial T} = F + TS$. The partition function Z is the statistical factor that normalizes the probability distributions, $Z = e^{-F/k_B T} = Tr\, e^{-\hat{H}/k_B T}$. Using energy eigenstates to evaluate the trace, the partition function is generally given by:

$$Z = \int dE\, \eta(E)\, e^{-E/k_B T}, \tag{7.9}$$

where the *density of states* factor $\eta(E)$ counts the number of microstates associated with a given energy configuration in converting the sum over energy eigenstates into an integral. The thermal average of a quantum operator involves the trace over arbitrary basis states of that operator with the thermal density operator $\hat{\rho} = \frac{e^{-\hat{H}/k_B T}}{Z} = e^{(F-\hat{H})/k_B T}$, where $Tr\hat{\rho} = 1$. Alternative derivations of Eq. 7.8 will be demonstrated using the statistical physics of the horizon.

Periodicity and temperature

For quantum state vectors, the time translation operator for Heisenberg representation states is given by $e^{-\frac{i}{\hbar}t\hat{H}}$. As an operation, this bears a direct resemblance to the thermal weight factor from Boltzmann statistics $e^{-\frac{1}{k_B T}\hat{H}} \equiv e^{-\theta_T \hat{H}}$ if one makes an analytic continuation into imaginary time. The space-time metric likewise transitions from a Minkowski to a Euclidean form under the transformation $\frac{i}{\hbar}t \rightarrow \theta_T \equiv \frac{\omega_E}{\hbar c}$, when the time-like coordinate t becomes a space-like coordinate ω_E:

$$ds^2 = -(dct)^2 + dx^2 + dy^2 + dz^2 \Rightarrow ds^2_{Euclidean} = (d\omega_E)^2 + dx^2 + dy^2 + dz^2. \tag{7.10}$$

An arbitrary thermal representation operator $\hat{Q}(\theta) \equiv e^{\theta\hat{H}}\hat{Q}e^{-\theta\hat{H}}$ is the analog of a time-translated Heisenberg operator. The thermal average of a thermal representation operator is given by $\langle \hat{Q}(\theta) \rangle_{thermal} \equiv Tr\hat{\rho}\,\hat{Q}(\theta)$. Because of the cyclicity of the trace operation $Tr(\mathbf{ABC}) = Tr(\mathbf{CAB})$, thermal averages of thermal representation operators obey a periodicity condition:

$$\langle \hat{Q}(\theta) \rangle_{thermal} = Tr \hat{\rho}\, \hat{Q}(\theta) = \langle \hat{Q}\,(\theta + 1/k_B T) \rangle_{thermal} = \langle \hat{Q}\,(\theta + \theta_T) \rangle_{thermal}. \tag{7.11}$$

Therefore, thermal representation operators are periodic in the thermal parameter conjugate to the Hamiltonian on the interval $\theta : 0 \to \frac{1}{k_B T}$.

This periodicity in $\theta_{Rindler}$ relates directly to the calculation of the temperature of the horizon. Just transitioning the horizon, the Rindler metric takes a Euclidean form with $\omega_t \to i\theta_R$, and:

$$ds^2_{Euclidean} = \rho_r^2\, d\theta_R^2 + d\rho_r^2 + |\mathbf{dx}_\perp|^2. \tag{7.12}$$

The *conical angle* θ_R is periodic in this metric, varying from 0 to 2π. This inverse Rindler temperature is the rotational angle around an axis whose proper distance measure is ρ_r. The relationship of the Rindler time to the Schwarzschild time using Eq. 7.6 likewise relates the angle θ_R conjugate to the Rindler time with the angle θ_S conjugate to the Schwarzschild time. Thus the periodicity of the two measures must be related:

$$\theta_R = \frac{\hbar c}{2R_S}\theta_S \quad \Rightarrow \quad 2\pi = \frac{\hbar c}{2R_S}\frac{1}{k_B T_H}. \tag{7.13}$$

The last expression directly calculates the *Hawking temperature* [84] T_H that a distant Schwarzschild observer associates with the black hole horizon:

$$k_B T_H = \frac{\hbar c}{4\pi R_S} = \frac{\hbar c^3}{8\pi G_N M}. \tag{7.14}$$

Once a temperature is associated with the black hole, the conjugate entropy is determined assuming the validity of the first law of thermodynamics:

$$dE = dMc^2 = T_H\, dS. \tag{7.15}$$

It should be noted that any volume energy associated with the work of expansion against the otherwise empty space is neglected in this expression. A direct substitution using Eq. 7.14 gives the entropy of the black hole:

$$S = \frac{4\pi G_N M^2}{\hbar c} k_B + \text{constant} \Rightarrow S = \left(\frac{k_B c^3}{\hbar}\right)\frac{4\pi R_S^2}{4G_N} = \left(\frac{k_B c^3}{\hbar}\right)\frac{Area}{4G_N}. \tag{7.16}$$

Several features of this answer are of interest:

- Entropy is proportional to the *area* of the horizon in Planck units, whereas typically entropy is an extensive variable proportional to the *volume*. More precisely, $S = k_B \frac{Area}{4L_P^2}$. This motivates many of the concepts in holographic analyses in cosmology, since information is an extensive variable on the surface of the horizon, rather than within a bulk volume.

- Entropy cannot be calculated perturbatively in Planck's constant $\hbar \to 0$. The entropy of a black hole is inherently a quantum phenomenon.
- A finite temperature requires that the system has thermal radiations (*Hawking radiation*), which means that it cannot be static, as was assumed in the derivations. At best, the system can be stationary, if infalling radiations can be arranged to precisely match any outgoing thermal radiations.
- A freely falling observer will not detect a horizon or temperature without violating the principle of equivalence. Complementarity requires that no observer witness any violation of the laws of physics.
- The entropy has no obvious interpretation as to the form of any micro-states that are counted in order to generate this entropy. In particular, the entropy is proportional to $\frac{1}{\hbar}$, so that any quantum contributions must somehow generate this inverse relationship.

One might be concerned with the puzzle of who detects the thermal radiations in asymptotic Schwarzschild space-time. A fiducial asymptotic observer is expected to detect the radiations, while a freely falling observer should not. Some consider the distinction to be ambiguous in Schwarzschild space-time, while as shown in the supplement on page 240, it is not ambiguous using radially stationary (non-orthogonal) coordinates. Thus, one concludes that the radiations are detected by any observer that maintains fiducial coordinates against the gravitational flows of the geometry.

Thermal translations and temperature

The quantum states of the region interior to the horizon consistent with a given exterior measurement must be statistically summed over. Therefore, expectation values of any quantum operator in the exterior region can be expressed using traces of the microcanonical density matrix:

$$\rho_{exterior}(b, b') = \sum_{a_{in}} \psi^*(a_{in}, b)\psi(a_{in}, b'), \tag{7.17}$$

where a_{in} represents the interior quantum states consistent with the exterior measurement. This sum implies an information deficit due to the coarse graining of the local measurement.

The form of the density matrix in Eq. 7.17 can be evaluated using the near-horizon Rindler coordinates. Generic exterior fields χ_{OUT} and interior fields χ_{IN} will commute across space-like regions, due to microscopic causality. However, interior and exterior states can be related through thermal translation (analogous to

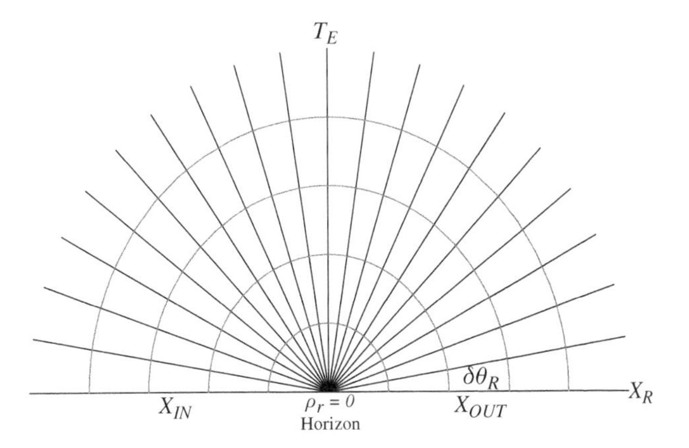

Figure 7.6 Rindler geometry for Euclidean form of metric.

relating in/out scattering states through temporal translations):

$$|\chi(\theta)\rangle = e^{-\theta \hat{H}} |\chi\rangle. \tag{7.18}$$

In particular, the Euclidean form of the Rindler metric 7.12 has the geometry represented in Figure 7.6. In this figure, T_E is the Euclidean Rindler component $T_E = \rho_r \sin\theta_R$, while $x_R = \rho_r \cos\theta_R$ is the original spatial direction of the acceleration. χ_{OUT} fields in the exterior on the fixed T_E surface are to the right of the horizon $\rho_r = 0$, while χ_{IN} fields in the interior are to the left of the horizon. The Euclidean angular translation $\delta\theta_R$ is directly related to the Rindler temporal coordinate ω_t.

The overlap amplitude in Eq. 7.17 of the interior field to the exterior field is then given by:

$$\psi(a_{in}, b) = \frac{1}{\sqrt{Z}} \langle \chi_{a_{in}} | \chi_b(\theta_R = \pi) \rangle = \frac{1}{\sqrt{Z}} \langle \chi_{a_{in}} | e^{-\pi \hat{H}_R} | \chi_b \rangle, \tag{7.19}$$

where Z is the normalization factor for the amplitude. Substitution into Eq. 7.17 gives an expression for the Rindler space density matrix:

$$\rho_{exterior}(b', b) = \frac{1}{Z} \sum_{a_{in}} \langle \chi_{b'} | e^{-\pi \hat{H}_R} | \chi_{a_{in}} \rangle \langle \chi_{a_{in}} | e^{-\pi \hat{H}_R} | \chi_b \rangle = \langle \chi_{b'} | \frac{e^{-2\pi \hat{H}_R}}{Z} | \chi_b \rangle. \tag{7.20}$$

This gives a unique form for the Rindler temperature $T_R = \frac{1}{2\pi}$, which is dimensionless (as is the Rindler time ω_t).

The temperature associated with the density operator in Schwarzschild space-time is obtained by relating the Rindler Hamiltonian to the Schwarzschild Hamiltonian. This is done in terms of the conjugate temporal parameters $\omega_t = \frac{ct_S}{2R_S}$.

This relates the Hamiltonians according to $\hat{H}_R = \frac{2R_S}{c} \hat{H}_S$. Therefore, the density operator in Schwarzschild space-time is given by:

$$\hat{\rho}_{exterior} = \frac{e^{-2\pi \hat{H}_R}}{Z} = \frac{e^{-2\pi \left(\frac{4G_N M}{\hbar c^3}\right)\hat{H}_S}}{Z} = \frac{e^{-\hat{H}_S/k_B T_H}}{Z}. \tag{7.21}$$

This again verifies the previous forms of the temperature of the horizon.

The prior calculations all obtain the temperature of the horizon and then calculate other thermal parameters from the temperature. A direct calculation of the entropy of a Schwarzschild horizon is given in Appendix F.1.

Entropy and holography

The entropy of a system is a direct measure of the number of microscopic configurations, or number of states, that would yield the same macroscopic state, which is related to the information that can be known about the system as discussed in Section 2.2.4. Systems with the maximum entropy provide one with minimum information, and conversely, a system with maximum information is in a known state and thus has vanishing entropy. Typically, the number of states of a system, and thus its entropy, is additive within a *volume*. However, for a black hole, the entropy is proportional to its *area* in Planck units:

$$S_{\text{Black Hole}} = k_B \, \frac{1}{4} \frac{Area}{L_P^2}. \tag{7.22}$$

This considerably decreases the number of degrees of freedom of a black hole, when compared to typical physical systems.

The *holographic principle* results if one asserts that a black hole represents a physical system about which the minimum information can be known, and thus such a system has the maximum entropy that a physical system can attain. To understand the implications of this assertion, consider a spherical region that contains an energy distribution within a given volume $\mathcal{V}$, bounded by a spherical area $A = \partial \mathcal{V}$, which minimizes the surface to volume ratio. The holographic principle then implies that the total entropy contained in that region cannot exceed that of a black hole with area A. This is true because one can arrange a collapsing spherical shell of light-like energy with just the right amount of additional energy to create a black hole once it crosses the bounding area, thus forming a black hole with that area. By the second law of thermodynamics, this process cannot have decreased the overall entropy, and thus the previous entropy *must* have been less than that of the final black hole. Thus the holographic principle establishes a limit upon the entropy of a physical system in terms of its (space-like) area:

$$S_{\text{arbitrary system}} \leq k_B \, \frac{1}{4} \frac{Area}{L_P^2}. \tag{7.23}$$

All known systems, including the observable universe as a whole, fall within these bounds.

The holographic principle then allows one to use light-like projections in appropriate space-times to establish entropy bounds within a given closed region. This can be done because light bends towards regions of attractive gravity, bending more in larger gravitational fields. For instance, consider a gravitationally lensed image projected upon a screen. The rays from a bundle of light that passes near a gravitational source become more spread out as they pass by the source, increasing their cross-sectional area. Suppose a surface image of a gravitational source is projected onto a screen away from that source. Then, if the trajectories of the rays are reversed at the screen, those rays will focus back onto the object by reciprocity or time-reversal invariance. This means that the actual surface area of emission from the source is less than its projected area, $A_{source} \leq A_{image}$.

If light is emitted away from the horizon of a black hole, because of the extreme gravitational bending, even rays originating from the rear area of that black hole can be imaged on a screen. However, the image area cannot be smaller than the area of the horizon, due to the anti-focusing of the rays as they move away from the black hole. This then puts a holographic bound on the entropy per unit area σ projected on the screen:

$$\sigma \equiv \frac{S_{source}}{A_{image}} \leq \frac{1}{4}\frac{A_{source}}{A_{image}}\frac{k_B}{L_P^2} \leq \frac{1}{4}\frac{k_B}{L_P^2}. \tag{7.24}$$

There is a caveat. The discussion assumed that the background cosmology itself does not focus light beams independent of their nearness to attractive sources. Within such cosmologies, the previous arguments concerning light-like projections are not valid.

7.2 Dynamic spherically symmetric black holes

The thermal radiations of a static black hole horizon have demonstated that the geometry cannot really be static, at best it can only be stationary. One further suspects that some subtle aspects of gravitating quantum systems might be difficult to ascertain using static backgrounds, since quantum mechanics incorporates dynamic measurability constraints (such as the energy-time uncertainty principle, etc.). This motivates the reader to examine spatially coherent, dynamic geometries as backgrounds for quantum systems, which will be the primary topic of the next sections. The geometric dynamics will be chosen so that quantum behaviors can be described in a convenient manner. The Penrose diagrams of spherically symmetric evaporating and accreting black holes will be developed as a straightforward dynamic background upon which quantum behaviors can be studied. These diagrams will explicitly and unambiguously demonstrate the horizons and coordinate anomalies associated with the geometries, allowing clear physical interpretation of the large-scale causal structures.

7.2.1 Spatially coherent excreting black holes

Radially stationary coordinates provide a temporal coordinate that is everywhere time-like, (i.e., fixed coordinate time surfaces are everywhere space-like). A spatially coherent geometry is one whose radial mass scale everywhere depends only upon this time coordinate shared by geometrically stationary observers. The metric will be described using [85]:

$$ds^2 = -\left(1 - \frac{R_M(ct)}{r}\right)(dct)^2 + 2\sqrt{\frac{R_M(ct)}{r}}dct\, dr + dr^2 + r^2\left(d\theta^2 + sin^2\theta\, d\phi^2\right). \tag{7.25}$$

There are no curvature singularities anywhere in the geometry away from that at $r = 0$. The location of the horizon is not apparent from the metric, however, one can immediately examine the (null) trajectories of outgoing light-like particles. These trajectories satisfy $\frac{dr_\gamma}{dct} = -\sqrt{\frac{R_M(ct)}{r}} + 1$. Therefore outgoing photons are momentarily stationary at the location $R_{TS} = R_M(ct)$, defining the trapping surface within which all future seeking trajectories must have decreasing radial coordinate.

The mixed components of the Einstein tensor can be directly calculated:

$$((G^\mu{}_\nu)) = \begin{pmatrix} 0 & 0 & 0 & 0 \\ -\frac{\dot{R}_M(ct)}{r^2} & -\frac{\dot{R}_M(ct)}{r^2\sqrt{\zeta}} & 0 & 0 \\ 0 & 0 & -\frac{\dot{R}_M(ct)}{4r^2\sqrt{\zeta}} & 0 \\ 0 & 0 & 0 & -\frac{\dot{R}_M(ct)}{4r^2\sqrt{\zeta}} \end{pmatrix}, \tag{7.26}$$

where the dimensionless parameter $\zeta \equiv \frac{R_M(ct)}{r}$. This tensor clearly vanishes in the static limit $\dot{R}_M \to 0$. The dynamic black hole generated will serve as a space-time background that will have a dynamic horizon, as well as other features of interest, which provide surfaces near which quantum phenomena cannot be neglected. The quantum behaviors of gravitating systems in this macroscopically generated space-time can be examined in a straightforward manner.

7.2.2 Form of conformal coordinates

For a black hole satisfying Eq. 7.25 with a constant rate of mass accretion/excretion $\ddot{R}_M = 0$, the conformal coordinates can be determined directly. The geometry will

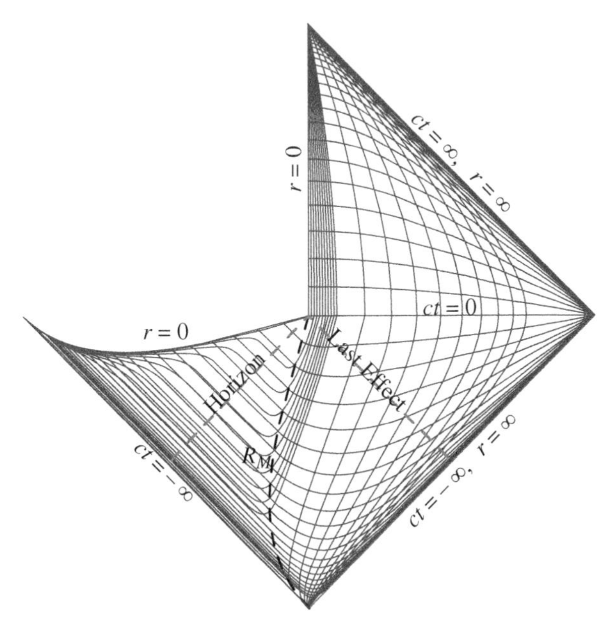

Figure 7.7 Conformal diagram for a spatially coherent black hole that evaporates steadily to zero mass at $ct = 0$. Curves of constant r are displayed in tenths of units, units of length, and decades of units, and are time-like exterior to R_M, and space-like interior to this dynamic scale. Curves of constant ct are successive in units of length, and are everywhere space-like. The horizontal solid line on the right represents the end of excretion $ct = 0$. The dashed curve labeled R_M represents the radial mass scale ($\zeta_M = 1$).

be assumed to excrete steadily from the distant past until $t = 0$. The conformal coordinates are given by:

$$ct_* = \frac{r}{2}\left(exp\left[\int^{\frac{R_M(ct)}{r}} \frac{(1+\sqrt{\zeta'})\,d\zeta'}{\{\zeta'(1+\sqrt{\zeta'})+\dot{R}_M\}}\right] - exp\left[\int^{\frac{R_M(ct)}{r}} \frac{(1-\sqrt{\zeta'})\,d\zeta'}{\{\zeta'(1-\sqrt{\zeta'})-\dot{R}_M\}}\right]\right)$$

$$r_* = \frac{r}{2}\left(exp\left[\int^{\frac{R_M(ct)}{r}} \frac{(1+\sqrt{\zeta'})\,d\zeta'}{\{\zeta'(1+\sqrt{\zeta'})+\dot{R}_M\}}\right] + exp\left[\int^{\frac{R_M(ct)}{r}} \frac{(1-\sqrt{\zeta'})\,d\zeta'}{\{\zeta'(1-\sqrt{\zeta'})-\dot{R}_M\}}\right]\right). \tag{7.27}$$

A non-integral form solution for these coordinates can be found, if desired [86]. These coordinates allow the conformal diagram to be directly constructed in Figure 7.7. The singularity of the black hole, represented by the space-like curve $r = 0$ bounding the left-hand region of the diagram from above, vanishes at $t = 0$. No communication to the left of the light-like horizon can escape hitting the singularity. Likewise, no communication to the right of the light-like incoming surface of last effect can communicate with the singularity. The radial mass scale $R_M(ct) = 2G_N M(ct)/c^2$ traces a time-like surface where radially outgoing light-like trajectories remain temporarily stationary in coordinate r as they are traversed by this scale.

One should note that the integrals in Eq. 7.27 are singular if $\dot{R}_M \rightarrow 0$ and the region $r \rightarrow \infty$ is in the domain of integration. This non-analytic behavior in the static limit demonstrates that the conformal coordinates smoothly corresponding with the asymptotic Minkowski forms are qualitatively different from what might be obtained using one's intuitions about Schwarzschild geometry. Fixed time surfaces are seen in Figure 7.7 to remain space-like all the way to the singularity, unlike those in Schwarzschild space-time.

7.2.3 Geodesic motion

The behaviors of freely falling systems on a given space-time background can be directly determined only if that motion has negligible effect upon the background. For classical particles, the description of the particle trajectory is completely defined in terms of its motion on any particular boundary surface.

Four-velocities

The radial component of (dimensionless) four-velocities on the metric 7.25 is given by:

$$U^r = -\zeta^{1/2}\, U^{ct} \pm \sqrt{(U^{ct})^2 - \Theta_m}, \tag{7.28}$$

where the mass factor $\Theta_m \equiv \begin{cases} 0 & m = 0 \\ 1 & m \neq 0. \end{cases}$ In particular, geometrically stationary massive particles satisfy $U^{ct} = 1$, $U^r = -\zeta^{1/2}$. The geodesic equations for radial motions in this geometry are given by:

$$\frac{dU^{ct}}{d\lambda} = -\frac{\zeta^{1/2}}{2r}(\zeta^{1/2} U^{ct} + U^r)^2, \tag{7.29}$$

and:

$$\frac{dU^r}{d\lambda} = -\frac{\dot{R}_M}{2r\zeta^{1/2}}(U^{ct})^2 - \Theta_m \frac{\zeta}{2r}. \tag{7.30}$$

The horizon $\zeta = \zeta_H$ and the surface of last effect $\zeta = \zeta_I$ define the geodesics of particular outgoing and ingoing massless particles that pass through the point $(ct, r) = (0, 0)$.

The four-velocities of freely gravitating particles can be reparameterized in terms of the single dimensionless variable ζ in the black hole geometry. This reparameterization involves the transformation:

$$\frac{d\zeta}{d\lambda} = \frac{\dot{R}_M U^{ct} - \zeta U^r}{r}. \tag{7.31}$$

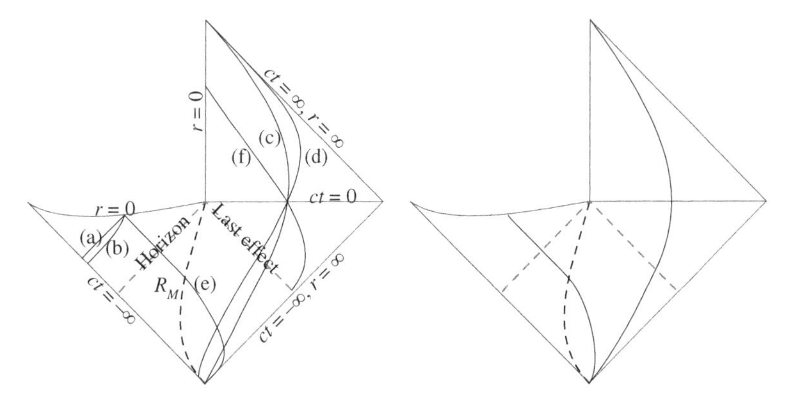

Figure 7.8 Classical trajectories for various massive gravitating particles. The outgoing trajectories in the diagram on the left terminate on the singularity at ($ct = -r_o, r = 0$) [(a) and (b)], and pass through ($ct = 0, r = +r_o$) [(c) and (d)] at various speeds. The ingoing trajectories likewise terminate at ($ct = -r_o, r = 0$) [(e)] and pass through ($ct = 0, r = +r_o$) [(f)]. The diagram on the right represents the trajectories of the geometrically stationary observers passing through those points.

Since the affine derivatives can be expressed in terms of the substantive derivative $\frac{d}{d\lambda} = U^\beta \partial_\beta$, Eqs. 7.29 and 7.30 allow the geodesic four-velocities to be expressed as functions of the single parameter ζ, i.e., $U^\beta = U^\beta(\zeta)$, eliminating the separate dependence on the radial coordinate r.

The four-velocities of outgoing particles have singular behavior at the horizon $r = R_H(ct)$, while the ingoing particles have singular behavior at the surface of last effect $r = R_I(ct)$. The radial components of the four-velocity of outgoing particles change sign at the radial mass scale $\zeta_M = 1$.

Classical geodesic trajectories

The classical trajectories of freely falling bodies can be directly represented on the conformal diagram. Massive particle trajectories are demonstrated in Figure 7.8. In the diagram, outgoing trajectories (a) and (b) are faster/slower observers that hit the singularity at the same time, $ct = -r_o$. Likewise, outgoing trajectories (c) and (d) reach an exterior point $r = r_o$ at $t = 0$. Trajectory (e) is an ingoing mass that hits the singularity at the same time as trajectories (a) and (b), while trajectory (f) is an ingoing mass that passes through the point ($ct = 0, r_o$) at the same time as trajectories (c) and (d). Notice that an ingoing trajectory that does not hit the singularity must originate exterior to the surface of last effect. The diagram on the right plots the trajectories of geometrically stationary masses that pass through these same points. The motions of all of these observers is, everywhere, time-like.

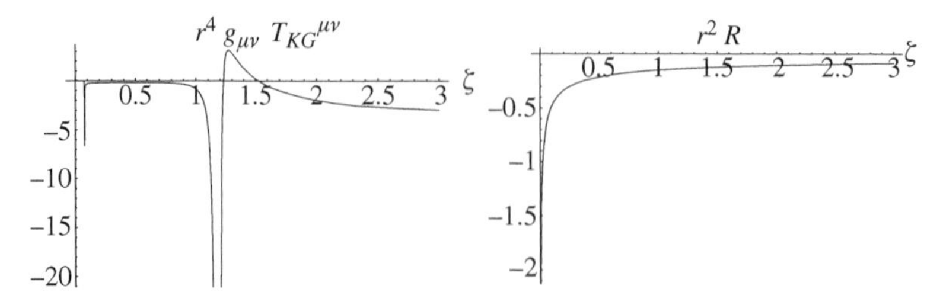

Figure 7.9 Plots of invariant scalar densities associated with gravitating quanta on the spatially coherent black hole background. The left diagram plots r^4 times the trace of the energy-momentum tensor of the Klein–Gordon field, while the diagram on the right plots r^2 times the Ricci scalar of the dynamic black hole. The horizontal axis for both diagrams is the dimensionless parameter $\zeta = R_M(ct)/r$.

7.2.4 Klein–Gordon field

The behaviors of a Klein–Gordon scalar field on the spatially coherent black hole will be explored. Since a massless field introduces no additional scale, the massless form of this equation will be solved:

$$-\frac{\partial^2\psi_\ell}{(\partial ct)^2} + \frac{\partial}{\partial ct}\left[\sqrt{\frac{R_M}{r}}\left(\frac{\partial\psi_\ell}{\partial r}\right)\right] + \frac{\partial}{\partial r}\left[\sqrt{\frac{R_M}{r}}\left(\frac{\partial\psi_\ell}{\partial ct}\right)\right]$$
$$+\frac{\partial}{\partial r}\left[\left(1-\frac{R_M}{r}\right)\left(\frac{\partial\psi_\ell}{\partial r}\right)\right] + -\left[\frac{\ell(\ell+1)+\frac{R_M}{r}}{r^2} + \frac{1}{r}\frac{\partial}{\partial ct}\left(\sqrt{\frac{R_M}{r}}\right)\right]\psi_\ell = 0. \tag{7.32}$$

For a system that uniformly excretes ($\dot{R}_M$ = constant), it is very convenient that the dynamics of the fields in Eq. 7.32 can be expressed in terms of the single dimensionless parameter ζ:

$$(-\zeta + \zeta^{3/2} + \dot{R}_M)(\zeta + \zeta^{3/2} + \dot{R}_M)\frac{d^2\psi_\ell}{d\zeta^2} + 2\zeta^2\left(\zeta(-2+3\zeta) + 3\sqrt{\zeta}\,\dot{R}_M\right)\frac{d\psi_\ell}{d\zeta}$$
$$+\left(2\ell(\ell+1)\zeta^2 + 2\zeta^3 + \zeta^{3/2}\dot{R}_M\right)\psi_\ell = 0. \tag{7.33}$$

Given appropriate boundary conditions, this second-order differential equation can be solved in a straightforward manner.

S-wave ($\ell = 0$) solutions of Eq. 7.32 will be demonstrated. Boundary conditions at $ct = 0$ (or $\zeta = 0$) require that the field ψ_0 and its derivatives must appropriately match those of a free-space Minkowski Klein–Gordon field $\ddot{\psi}_M - \psi''_M = 0$ near that region. The matching Minkowski space solution for the field away from $r = ct$

is given by:

$$\psi_M(ct, r) = \left[A + B\, log \left(\frac{r + ct}{r - ct} \right) \right] \Theta(r - ct). \tag{7.34}$$

Plots of the scalar density of s-wave solutions to Eq. 7.33 on the black hole background are compared to the Ricci scalar of the background in Figure 7.9. The physical scalar density invariant is given by $g_{\mu\nu} T^{\mu\nu}$ (i.e., the trace of the scalar field's energy-momentum tensor). If the Klein–Gordon field self-gravitates, this would be directly related to the Ricci scalar $\mathcal{R}$ of the space-time. In the diagram, $r = 0$ corresponds to $\zeta = \infty$, and $\zeta = 0$ corresponds to $r = \infty$ or $t = 0$. The Klein–Gordon density invariant is singular at the horizon and surface of last effect, while the Ricci scalar is not. The non-linear form of the Klein–Gordon energy-momentum tensor in derivatives makes superpositions of co-gravitating Klein–Gordon fields clumsy, and makes the dependency of the density invariant upon the radial coordinate r irreparably different from that of the Ricci scalar. Derivative powers can only be matched using length scales within the Lagrangian, combined with the Planck length in terms involving gravity.

7.3 Macroscopic co-gravitation of quanta

The foundation for developing co-gravitating quanta consistent with a given space-time will next be established. Co-gravitation requires that the mutually gravitating constituents can be cluster disentangled. In order to avoid the extreme non-linear manifestation of the dependence of geometric curvature upon the sources, the dynamics of the mutually gravitating quantum fields can be expressed in terms of the affine parameters (like proper times) that describe local motions of a given quantum type on the space-time. Algebraic equations that relate the energy-momentum densities of the quantum fields to Einstein's tensor can then be developed. The relevant quantum dynamics is likewise conveniently expressed in terms of the affine parameter associated with the gravitating particles/fields. The affine or proper-time derivative of local physical parameters can be expressed in terms of the substantive derivative $\frac{d}{d\lambda} = U^\beta \frac{\partial}{\partial x^\beta} \equiv U^\beta \partial_\beta$, where the affine parameter λ for a massive particle corresponds to the proper time $c\tau$. This then makes the incorporation of the principles of equivalence/relativity straightforward, resulting in substantive quantum flows on the geometry. Much of the collective non-linearity associated with co-gravitation gets incorporated in the affine flow parameters U^β, essentially defining the analogous Lorentz frame for any off-shell quantum dynamics.

Insights into the geometry of quantum flows can be gained by examining suitably simple, well-known quantum systems. As mentioned in Chapter 2, macroscopic quantum fluids maintain persistent quantum flows that are directly coupled to the local geometrodynamics. Fluid continuity directly follows from the equations of motion resulting from Lagrangians constructed using substantive derivatives. Superfluid flows then result from the local gauge invariance of those Lagrangian forms. Such substantive flows can directly be generalized for linear spinor fields using Eq. 6.18.

Lagrangians that utilize substantive derivatives most directly incorporate the spirit of the principle of equivalence. Therefore, the chosen form for the Lagrangian of a gravitating non-interacting cluster a in the geometry will be taken to depend on the derivative of the field through the form $\psi_{a,c\tau} \equiv U^\beta \partial_\beta \psi_a$. Writing the fields as complex parameters $\psi_a = e^{i\xi_a} |\psi_a|$, the Euler–Lagrange equations 6.14, 6.15, and 6.16 for a Lagrangian holonomic in the phase then ensure probability conservation for variations in the phases ξ_a, and phase coherence for variations in the moduli $|\psi_a|$. Therefore, this general class of systems satisfies the dynamics expected of inertial quantum fields.

The form of the contribution of the gravitating quanta to the energy-momentum tensor in Eq. 3.51 can be calculated using the dynamics of a multi-component field χ (separating moduli and phases, or real and imaginary parts). The energy-momentum contribution of quantum a to the overall system then depends linearly upon derivatives of the phase, directly linking the local phase of the field ψ_a to its contribution to the local energy-momentum of the system, consistent with experiment. Any collective field that vanishes in the absence of gravitation should also contribute to the local energy densities. The superposable forms can be exploited to develop macroscopic co-gravitating structures built of non-interacting components.

7.3.1 Co-gravitating quanta

Co-gravitating quanta should mutually contribute to a particular space-time configuration as follows:

- Each component should contribute to the overall gravitation independent of the coherence state of other disentangled components;
- The energy densities of the components should additively contribute to the overall energy density used to drive the Einstein equation for the gravitating environment in the $G \to 0$ limit;
- Some components should be able to maintain quantum coherence and satisfy probability flux conservation and expected energy-momentum/time-space phase relationships;
- Gravitating quantum systems should be able to maintain observed linearity as previously disentangled components interact, forming changing configurations of those systems;

- The collective gravitational field should incorporate the principle of equivalence with regards to the gravitation of each component.

This makes a density matrix approach ideal for describing a co-gravitating system. Using such an approach, the contribution of each component (cluster) of the incoherent sum is independent of the coherence state of the other components.

For the spatially coherent black hole geometry, one can establish useful derivatives of the geodesic four-velocities using Eqs. 7.29, 7.30 and 7.31 given by:

$$\partial_\beta U^\beta = \frac{\Theta_m}{r} \frac{\zeta^{1/2}}{2} \frac{\dot{R}_M + \zeta^{3/2}}{\dot{R}_M u^{ct} - \zeta u^r}, \tag{7.35}$$

and:

$$\partial_{ct} U_r - \partial_r U_{ct} = 0. \tag{7.36}$$

Thus, for massless geodesics, the standard form of both the divergence and the curl of the geodesic four-velocities vanish.

Cluster disentangled quanta

The Lagrangian of a gravitating non-interacting cluster a in the geometry will be taken to be:

$$\mathcal{L}_a \equiv \sqrt{-g}\ \mathrm{L}_a = -\sqrt{-g}\left[\frac{i\hbar c}{2} U_a^\beta (\psi_a^* \partial_\beta \psi_a - (\partial_\beta \psi_a^*)\psi_a) - m_a c^2 \psi_a^* \psi_a\right]. \tag{7.37}$$

Expressing the fields as complex parameters $\psi_a \equiv |\psi_a| e^{i\xi_a}$, the Euler–Lagrange equations ensure probability conservation:

$$\frac{1}{\sqrt{-g}} \partial_\beta (\sqrt{-g}\, |\psi_a|^2 U_a^\beta) = 0, \tag{7.38}$$

and phase coherence:

$$U_a^\beta \partial_\beta \xi_a = -\frac{m_a c}{\hbar}. \tag{7.39}$$

The extremum Lagrangian vanishes $\mathcal{L}_a[\psi_{a,extremal}, \partial_\mu \psi_{a,extremal}] = 0$. This means that these quanta cannot contribute to a fluid pressure. These quanta can describe non-interacting radiating particles, since the conservation principles satisfied by such quanta remain valid asymptotically, independent of the geometry.

An energy parameter E_m can be defined for any constituent as the standard form invariant of the little group of transformations from Chapter 4 (i.e., mc^2 for massive particles, E_0 for massless particles). In terms of this invariant, a general solution to

Eq. 7.39 must satisfy:

$$\partial_{ct}\xi = \frac{E_m}{\hbar c}(U_{ct} + Q_\xi\, U^r) = \frac{E_m}{\hbar c}\left[-(1-\zeta)U^{ct} + (\sqrt{\zeta} + Q_\xi)U^r\right], \tag{7.40}$$

and

$$\partial_r\xi = \frac{E_m}{\hbar c}(U_r - Q_\xi\, U^{ct}) = \frac{E_m}{\hbar c}\left[(\sqrt{\zeta} - Q_\xi)U^{ct} + U^r\right], \tag{7.41}$$

where the covariant forms of the four-velocities are given by lowering the contravariant index:

$$\begin{aligned} U_{ct} &= (\zeta - 1)U^{ct} + \zeta^{1/2}U^r, \\ U_r &= \zeta^{1/2}\, U^{ct} + U^r. \end{aligned} \tag{7.42}$$

Since the four-velocities depend only on the the dimensionless parameter ζ, $U^\beta = U^\beta(\frac{R_M(ct)}{r})$, the dimensionless functions $Q_\xi = Q_\xi(\zeta)$ must likewise depend only upon the dimensionless parameter ζ.

An equation for the additional phase factor Q_ξ can be found by examining the analytic properties of the phase. The integrability of the phase $[\partial_r, \partial_{ct}]\,\xi(ct, r) = 0$ implies that the function Q_ξ must satisfy:

$$U^\beta \partial_\beta Q_\xi = \partial_{ct}U_r - \partial_r U_{ct} - (\partial_\beta U^\beta)\, Q_\xi. \tag{7.43}$$

This equation will appropriately simplify using the relations 7.35 and 7.36. In particular, Q_ξ is constant along the trajectory of a massless particle.

Dimensional analysis aids considerably in establishing solutions for the previous differential equations. The solutions must generally have dimensions that can be reduced in terms of the parameter $\lambda_{E_m} \equiv \frac{\hbar c}{E_m}$. The quanta must generate probability densities in terms of $|\psi(ct, r)|^2 \equiv \frac{|\tilde{\psi}(\zeta)|^2}{\lambda_{E_m} r^2}$, where $\tilde{\psi}(\zeta)$ is dimensionless. The phases must satisfy $\xi(ct, r) \equiv r\tilde{\xi}(\zeta)$, where the reduced phase $\tilde{\xi}(\zeta)$ carries the dimension of inverse Compton wavelength λ_{E_m}. For massless particles, one can immediately determine using Eqs. 7.38 and 7.43 that the parameters $|\tilde{\psi}(\zeta)|^2$ and $Q(\zeta)$ are constants. Furthermore, using Eq. 3.51, the energy-momentum tensor of quantum a satisfies:

$$\mathcal{T}_{a}{}^{\beta}{}_{\mu} = \sqrt{-g}\; T_{a}{}^{\beta}{}_{\mu} = \sqrt{-g}\,(\hbar c)\, U_a^\beta\, (\partial_\mu \xi_a)\, |\psi_a|^2. \tag{7.44}$$

This energy-momentum density already has the correct functional dependency in the variables (ct, r) needed to directly contribute to the Einstein tensor. In addition, there is a linear connection of the phase dynamics to the energy-momentum tensor. This means that the microscopic energy-momentum associated with the phase of such a coherent field can algebraically contribute to the overall local energy-momentum tensor that is connected to the geometry through Einstein's equation.

The energy-momentum tensor 7.44 need not be conserved, due to the background gravitation. The covariant divergence of the energy-momentum tensor of a field ψ_a is given by:

$$T_{a\ \mu;\beta}^{\beta} = -\Gamma^{\lambda}_{\mu\beta} \frac{\partial \mathrm{L}_a}{\partial(\partial_\beta \psi_a)} \partial_\lambda \psi_a + \text{complex conjugate}. \tag{7.45}$$

In addition, the form of this tensor is not necessarily symmetric (except for massless quanta), satisfying:

$$T_a^{ct\,r} - T_a^{r\,ct} = \Theta_{m_a} E_{m_a} \left|\psi_a\right|^2 Q_{\xi_a}. \tag{7.46}$$

Since the Einstein tensor is both geometrically conserved and symmetric, the composite energy-momentum tensor driving the Einstein equation must likewise be geometrically conserved and symmetric, even if individual constituent tensors do not share these properties.

7.3.2 The core gravitating field

The cluster decomposable co-gravitating fields will superpose as contributors to the given background space-time metric. The remaining contribution to the energy-momentum of the geometry should be due to a real, core gravitating field directly associated with the collective system of co-gravitating quanta. This should be analogous to an electric field being generated by a collection of charges. This *core gravitating field* should incorporate the overall symmetries of the system, rather than those of the individual constituents. Once a functional form for the core gravitating field has been established, the form of the constituent contributions from the collection of gravitating scalars to the overall energy-momentum tensor in Einstein's equation can be self-consistently solved algebraically using:

$$G_\mu^\beta = -8\pi \frac{L_P^2}{\hbar c} \left(T_{core\,\mu}^{\ \ \beta} + \sum_a T_{a\ \mu}^{\beta} \right). \tag{7.47}$$

Since the core gravitating field is generated by the gravitating collective, this field should vanish as $G_N \to 0,\ L_P \to 0$.

The space-time for the spatially coherent black hole is spherically symmetric, so that all source velocities should have vanishing angular components $u^\theta = 0 = u^\phi$. For such sources, since the Lagrangian densities L_a in Eq. 7.37 vanish for the extremal fields that satisfy the equations of motion, their energy-momentum tensors cannot contribute to the non-vanishing components of the Einstein tensor G_θ^θ and G_ϕ^ϕ. Thus, the core gravitating field must incorporate those symmetries transverse to the four-velocity contributions of those quanta.

To provide maximal flexibility in establishing the core gravitating field, the Euler–Lagrange equation for this field will be chosen to be identically satisfied by the core quantum type. Such a choice gives complete flexibility towards satisfying the angular components of the Einstein tensor. A general form for the Lagrangian of a real core field that locally gravitates with geodesic four-velocity components U_c^β is given by:

$$\mathcal{L}_{General\,Core} = \mp\sqrt{-g}\,\hbar c\;\psi_c\left[U_c^\beta\,\partial_\beta\psi_c + \frac{1}{2}\left(U_c^\beta\,\partial_\beta(\log\sqrt{-g}) + \partial_\beta U_c^\beta\right)\psi_c\right], \tag{7.48}$$

where the sign determines the signature of the energy density of the core field in the causal patch. One expects a negative energy density for a core field associated with gravitational binding, as was the case with $\rho_w(r)$ in Figure 5.5. A field with negative energy density cannot create free asymptotic quanta. Its quanta can be detected only through exchanges in the near zone.

The invariant general core Lagrangian 7.48 is proportional to the four-divergence of the probability current density of the core field $\mathcal{L}_{General\,Core} = \mp\sqrt{-g}\,(\frac{\hbar c}{2})(\psi_c^2 U_c^\beta)_{;\beta}$, where the factors in the second term on the right-hand side combine to give the covariant four-divergence of the four-velocity. The Euler–Lagrange equations are identically satisfied for this Lagrangian due to the geodesic equations of the four-velocity $\vec{U}_c$. For the spatially coherent black hole, one can calculate an effective interaction of the form $U_c^\beta\,\partial_\beta\log(\sqrt{-g}) = 2\frac{U_c^r}{r}$, where the components U_c^β satisfy the appropriate geodesic equations 7.29 and 7.30 for the core field quantum type. The form of the core field ψ_c is therefore completely determined by Einstein's equations.

For the spatially coherent black hole, the core gravitating field will be constructed from massless quanta. Using Eq. 7.48, the Lagrangian form:

$$\begin{aligned}\mathcal{L}_{core} = -\sqrt{-g}\,\hbar c\Big[&\psi_{c+}(U_{c+}^\beta\partial_\beta\psi_{c+} + \tfrac{1}{2}U_{c+}^\beta\partial_\beta(\log\sqrt{-g})\psi_{c+}) + \\ &-\psi_{c-}(U_{c-}^\beta\partial_\beta\psi_{c-} + \tfrac{1}{2}U_{c-}^\beta\partial_\beta(\log\sqrt{-g})\psi_{c-})\Big].\end{aligned} \tag{7.49}$$

has Euler–Lagrange equations for real fields:

$$\sqrt{-g}\,\psi_{c\pm}(\partial_\beta U_{c\pm}^\beta) = 0 \tag{7.50}$$

that from 7.35 are indeed trivially satisfied for massless core quanta, with arbitrary functional forms $\psi_{c\pm}$. The core field energy-momentum tensor for this Lagrangian is given by:

$$T_{core}{}_\mu^\beta = -\hbar c\left[\psi_{c+}U_{c+}^\beta\partial_\mu\psi_{c+} - \psi_{c-}U_{c-}^\beta\partial_\mu\psi_{c-}\right] - \delta_\mu^\beta\mathcal{L}_{core}/\sqrt{-g}. \tag{7.51}$$

Figure 7.10 Conformal density plots of the dimensionless densities $|\tilde{\psi}|^2$ for the massless Klein–Gordon field (*left*), a geometrically stationary scalar field (*middle*), and the core gravitating field (*right*), on the spatially coherent evaporating black hole geometry.

Einstein's equation then constrains the form of the core gravitating field to satisfy:

$$-\frac{\dot{R}_M}{4r^2\sqrt{\zeta}} = G^{\theta}{}_{\theta} = -8\pi\frac{L_P^2}{\hbar c}T_{core}{}^{\theta}{}_{\theta} = 8\pi\frac{L_P^2}{\hbar c}\mathcal{L}_{core}/\sqrt{-g}, \tag{7.52}$$

which yields the equation satisfied by the core gravitating fields:

$$\frac{\dot{R}_M}{16\pi L_P^2 r^2\sqrt{\zeta}} = \left(U_{c+}^{\beta}\partial_{\beta}\psi_{c+}^2 + \frac{2U^r}{r}\psi_{c+}^2\right) - \left(U_{c-}^{\beta}\partial_{\beta}\psi_{c-}^2 + \frac{2U^r}{r}\psi_{c-}^2\right). \tag{7.53}$$

This represents a first-order differential equation in the core fields.

Using dimensional analysis, the core gravitating field must be of the form $\psi_c^2(ct,r) \equiv \frac{\tilde{Y}_c^2(\zeta)}{(L_P^2 r)}$, which can be compared to the previously mentioned form for the radiations $|\psi_a(ct,r)|^2 \equiv |\tilde{\psi}_a(\zeta)|^2/(\lambda_{E_a} r^2)$. One can see that the core gravitating field is of a qualitatively different type from the co-gravitating quanta, directly relating the core field to the collective rather than individual component behaviors. Furthermore, the energy-momentum tensor can be factored into a dimensionless form satisfying $\tilde{T}_{core}^{\mu\nu} \equiv \frac{L_P^2 r^2}{\hbar c}T_{core}^{\mu\nu}$. It is instructive to note that the core field contains an extra dimensional factor of the Planck length L_P when expressed in terms of the dimensionless reduced function, as compared with the co-gravitating fields. This extra factor is a manifestation of the gravitational nature of the core field, ensuring that it vanishes in the absence of gravity.

Given a solution for the core gravitating field, the co-gravitating scalar fields then algebraically satisfy Einstein's equation. Since the overall energy-momentum tensor consists of a sum of otherwise linearly independent contributors, such solutions are always possible. Macroscopic solutions for a given set physical boundary conditions can thus be constructed [87]. The diagrams in Figure 7.10 demonstrate the square of reduced dimensionless fields $|\tilde{\psi}|^2$ on the geometry. The massless Klein–Gordon field has solutions that have dominant density near the horizon, while the geometrically stationary scalar is much more spread out in the exterior and Minkowski regions of the geometry. The core gravitating field is singular at the center (since the geometry itself has this singularity), but away from that singular behavior has dominant density near the radial mass scale. The middle diagram represents the analogous quantum stationary distribution compared to the classical trajectories depicted on the right in Figure 7.8.

7.4 Temporally transient black objects

A Schwarzschild black hole has a space-like center $r = 0$ that implies the existence of a horizon. The horizon delineates the outermost boundary of a region of space-time within which all future-trending trajectories must ultimately hit that singular center. However, it is possible to construct a geometry that manifests as a transient black hole in the exterior but has an innermost boundary to the region of the space-time for which outgoing trajectories have decreasing radial coordinate. The innermost region then serves as a "depository" that temporarily stores any information that falls through the trapping region. The examination of a singularity-free and spherically symmetric transient black object whose center remains always time-like will be the subject of this section.

The exterior geometry of a transient black object is similar to that of a long-lived transient black hole, with a few subtle differences. For a temporally transient black hole, the existence of a horizon means that all in-falling objects are not seen to vanish by an external observer until the horizon itself vanishes. For a transient black object, a sturdy enough in-falling particle might in principle appear severely redshifted temporarily, then emerge near the center after the object has evaporated.

The information dynamics of strong field gravitation is also a topic of some interest. The strong gravitational field at the core of a gravitationally condensed object presumably couples *all* gravitating particles. That coupling must be associated with the production of radiations for black objects. In the exterior, the co-gravitating radiations are expected to be mostly massless quanta (whether gravitational or electro-weak) that collectively carry away gravitational curvature. Thus, the model will presume that strong gravity will allow baryon-number non-conservation, since no known massless quanta carry a baryon number. This means that much of the information that falls into a black object must transmute in form into that carried by the massless radiations. For a non-singular transient black object, the dynamics of both generic "standard" communications as well as entangled communications can be directly studied.

Form of the metric

The dynamic black object will be modeled using a time-dependent metric of the form:

$$ds^2 = -\left(1 - \frac{R_M(ct,r)}{r}\right)(dct)^2 + 2\sqrt{\frac{R_M(ct,r)}{r}}\, dct\, dr$$
$$+ dr^2 + r^2 d\vartheta^2 + r^2 sin^2\vartheta\, d\varphi^2, \tag{7.54}$$

where a non-vanishing radial mass scale $R_M(ct,r) \equiv 2G_N M(ct,r)/c^2$ is the length scale associated with the mass-energy content of the black object. Substitution of this form into Einstein's equations relates this mass to an energy density through $G^0{}_0 = \frac{1}{r^2}\frac{\partial}{\partial r} R_M(ct,r) = -\frac{8\pi G_N}{c^4} T^0{}_0$. The metric takes the form of Minkowski space-time both asymptotically ($r \gg R_M$) as well as when the radial mass scale vanishes.

As before, radial trajectories for test particles of mass m have four-velocity components that satisfy:

$$U^r = -\sqrt{\frac{R_M}{r}} U^{ct} \pm \sqrt{(U^{ct})^2 - \Theta_m}, \quad \Theta_m \equiv \begin{cases} 1 & m \neq 0 \\ 0 & m = 0 \end{cases}, \tag{7.55}$$

where the $\pm$ defines the trajectories of "outgoing"/"ingoing" test particles on the geometry. Geometrically stationary trajectories satisfy $U^{ct}_{(GS)} = 1$, $U^r_{(GS)} = -\sqrt{\frac{R_M}{r_{GS}}}$. The radial coordinate r can be interpreted as the proper distance between a geometrically stationary observer with coordinates (ct, r) and the center $r = 0$ at the same value of t.

Trapping surfaces

As can be seen from the form of the metric Eq. 7.54 and the four-velocity Eq. 7.55, outgoing massless particles at the surfaces R_{TS} instantaneously given by solutions to:

$$1 = \sqrt{\frac{R_M(ct, R_{TS})}{R_{TS}}}, \tag{7.56}$$

are momentarily stationary in the radial coordinate. Fixed radial coordinate curves labeled by r are space-like between these two surfaces and time-like exterior to these surfaces. If solutions to 7.56 exist, the surfaces bound the trapping regions of the space-time. There can be no observers with stationary or increasing radial coordinate within a trapping region. If transient trapping regions exist in the absence of a horizon, the geometry contains a *black object*, whereas if there is a horizon, the geometry contains a *black hole*.

7.4.1 Modeling a temporally transient black object

A transient black object will form from the collapse of material and evolve through the transmutation of that material into radiations that eventually eliminate any trapping region [88]. Exterior to the trapped matter and radiations, the geometry is that of a radially stationary geometry. Solutions for light-like trajectories on a stationary geometry with a background radial mass scale given by R_S^{bg} can be found in closed forms, given by:

$$ct - ct_o = r_\gamma(ct) - r_o + 2\sqrt{R_S^{bg}}\left(\sqrt{r_\gamma(ct)} - \sqrt{r_o}\right) + 2R_S^{bg}\, log\left(\frac{\sqrt{r_\gamma(ct)} - \sqrt{R_S^{bg}}}{\sqrt{r_o} - \sqrt{R_S^{bg}}}\right), \tag{7.57}$$

for outgoing photons, and:

$$ct_o - ct = r_\gamma(ct) - r_o - 2\sqrt{R_{So}}\left(\sqrt{r_\gamma(ct)} - \sqrt{r_o}\right) + 2R_{So}\, log\left(\frac{\sqrt{r_\gamma(ct)} + \sqrt{R_{So}}}{\sqrt{r_o} + \sqrt{R_{So}}}\right), \tag{7.58}$$

for ingoing photons.

The functional form of the radial mass scale $R_M(ct, r)$ will be chosen to preclude any singular behavior at $r = 0$ itself. If such a non-singular geometry forms a horizon, energy densities near the space-like center must necessarily be exotic, since their constituents cannot be time-like. The exotic nature of the densities is quantified using the energy conditions. One is motivated to fulfill the energy conditions over the broadest possible region of space-time. The dominant energy condition for this geometry is satisfied as long as:

$$0 \leq \frac{\partial}{\partial ct} R_M(ct, r) \leq 2\sqrt{\frac{R_M(ct, r)}{r}} \frac{\partial}{\partial r} R_M(ct, r). \tag{7.59}$$

The transient black object developed in this section will remain time-like everywhere, eliminating complications of exotic physics near the center.

Pressureless collapse

A solution to the non-linear relation 7.59 for dynamic radial mass scales satisfying the energy conditions is given by:

$$R_M^{EC}(ct, r) = \frac{4}{9} \frac{r^3}{(ct_B - ct)^2}, \tag{7.60}$$

where ct_B is an arbitrary constant. Substitution of this form into Einstein's equation describes a pressure-less collapse of matter whose edge will be defined as

$r_{collapse}(ct)$. During collapse, the geometry exterior to this edge has vanishing T_ν^μ. The exterior surface is described by the equation:

$$R_M^{EC}(ct, r_{collapse}(ct)) = R_{So} \equiv \frac{2G_N M_o}{c^2}, \tag{7.61}$$

where M_o denotes the total mass in the cosmology.

Therefore, during the collapse the radially mass scale will be given by:

$$R_M^{collapse}(ct, r) = \begin{cases} R_M^{EC}(ct, r) & \text{interior, } R_M^{EC}(ct, r) < R_{So} \\ R_{So} & \text{exterior, static vacuum.} \end{cases} \tag{7.62}$$

The exterior surface $r_{collapse}(ct)$ is geometrically stationary. The collapse will continue until the material reaches a critical density, for which quantum non-locality effects are presumed to dominate the dynamics.

Evaporation

Excretion or evaporation involves a local decrease in the radial mass scale of the geometry. Regions with $\frac{\partial}{\partial ct} R_M(ct, r) < 0$ will obviously violate the energy condition 7.59, so the goal is to minimize the region within which the non-local effects needed for evaporation are significant.

Once a sufficient mass has collapsed within its Schwarzschild radius, a trapping region defining the black object will develop. The time of the onset of a trapping region will be denoted t_{Dark}. Evaporation due to micro-gravitational dynamics is presumed to begin at this time. However, the collapse continues until it reaches the critical density. The center $r = 0$ remains non-singular and time-like.

After evaporation begins $t > t_{Dark}$, the exterior region is no longer delineated by the collapse surface $r_{collapse}(ct)$; rather, it is delineated by the outermost trapped surface $R_S(ct) \equiv R_{TS}^+$, which is defined in terms of the energy scale of the black object that has yet to evaporate. The model describing the evaporation of this *radial surface scale* $R_S(ct)$ will be motivated using the thermal evaporation rates expected from the quasi-static geometry. Generally, the rate of interior mass change is expected to be of the form:

$$\dot{M}c^2 = \frac{\text{energy}}{\text{emitted quantum}} \times \frac{\text{number of quanta emitted}}{\text{unit time}}, \tag{7.63}$$

where the dot refers to derivatives with respect to ct. The energy of the emitted quantum is expected to be defined by the coherence length scale of the geometry, i.e., the spatial extent of the radial surface scale $R_S(ct)$. This scale likewise defines the inverse temperature of the quasi-static geometry $R_S = \frac{\hbar c}{4\pi k_B T_H}$. The characteristic energy of the emitted quanta is thus of the order $\frac{\hbar c}{R_S(ct)}$. The rate of emission of the quanta for fiducial observers in the static geometry is expected to be of the order of one quantum per unit Rindler time. This rate, $\frac{\kappa}{R_S(ct)}$, is likewise

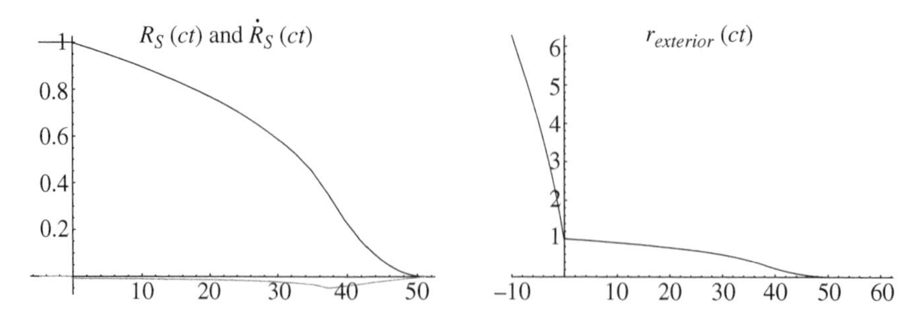

Figure 7.11 Plots of $R_S(ct)$ and $\dot{R}_S(ct)$ (*left*), and $r_{exterior}(ct)$ (*right*) in units of R_{So}. The black curve on the left plot represents the radial surface scale, while the negative gray curve represents its derivative.

inversely proportional to the radial surface scale. Thus, multiplying 7.63 by $\frac{2G_N}{c^2}$, the dynamics of the surface scale will be taken to satisfy:

$$\dot{R}_S(ct) = -\frac{2L_P^2}{R_S(ct)}\frac{\kappa}{R_S(ct)}, \tag{7.64}$$

where κ is a dimensionless number of order one. This form of evaporation will continue until quantum measurement and/or geometric consistency constraints are no longer met.

Gravitational evaporation ceases once there is no longer a trapping region. This occurs when the radial surface scale reaches the surface of the core of critical density $R_S(ct_{remnant}) = r_{core}(ct_{remnant})$. Once there is no longer any trapping region, the geometry ceases to contain a black object. For this example, however, the remnant of gravitational evaporation will be assumed to continue to undergo microscopic decay consistent with geometric and quantum measurement constraints. The radial surface scale and the exterior surface scale are demonstrated in Figure 7.11. In the exterior region, changes in the interior energy density will be communicated via collections of energy-carrying and geometry-changing massless quanta. The quanta collectively carry sufficient energy to modify the interior radial surface scale by $\delta R_S < 0$, themselves propagating through a stationary affine space with lesser interior radial surface scale $R_S - |\delta R_S|$. This guarantees that all energy conditions will be satisfied in the exterior during evaporation.

A massless geometry-changing quantum itself propagates through a stationary affine space, which allows a functional form to be developed for its trajectory. An exterior outgoing quantum emitted at (ct_{emit}, r_{emit}) that is located at $r_\gamma(ct)$ propagates through a background geometry satisfying Eq. 7.57, with $R_S^{bg}(ct_{emit}) = R_S(ct_{emit}) - |\delta R_S(ct_{emit})|$. The parameter δR_S represents the radial scale of the energy carried by that quantum. The radial mass scale at an arbitrary exterior point (ct, r) can be calculated from the retarded time of emission $(ct_{ret}, r_{exterior}(ct_{ret})) = (ct_{emit}, r_{emit})$ corresponding to the source of the geometry-changing quanta traversing that exterior point. During evaporation, the exterior surface of emission will be parameterized in terms of the outer trapping surface using $r_{exterior}(ct_{ret}) = R_S(ct_{ret}) + \delta_{stretch}$, where $\delta_{stretch}$ can be chosen to be arbitrarily small, since the quanta collectively carry geometry-changing energies consistent with the particular stage of evaporation. The radial mass scale at a general exterior point is therefore given in terms of this retarded time by:

$$R_M(ct, r) \equiv R_S(ct_{ret}(ct, r)), \quad \text{exterior region.} \tag{7.65}$$

This form then ensures a causal propagation of the evaporation of energy from the interior.

Therefore, the local radial mass scale $R_M(ct, r)$ will incorporate the following characteristics:

- The geometry-changing quanta emit from the radial surface scale R_S, but propagate through a geometry that is slightly flatter by δR_S precisely due to the energy transported by the quanta;
- Since the geometry-changing massless quanta communicate curvature changes, all energy conditions are satisfied in the exterior;
- The quantum degenerate core region is chosen to maintain its critical density during evaporation, which satisfies all energy conditions;
- The region between the exterior and the degenerate core will be assumed to maintain the spatially coherent form $R_S(ct)$. The forms smoothly match the behaviors during the transition from collapse to evaporation and define the edge of the core region $r_{core}(ct)$.

This form for the radial mass scale has been chosen so as to satisfy energy conditions in the broadest region of the space-time. Standard space-time plots of the density $T^0{}_0$ for this geometry are displayed in Figure 7.12. The in-falling pressureless matter reaches critical density, then slowly evaporates/decays away. The diagram on the right enhances the evaporation, whose density is much less than the critical density. This diagram begins as evaporation commences. Decay ends as the exterior surface vanishes.

Quantum measurement constraints and geometric consistency

A dynamic geometry should satisfy quantum measurements constraints with regards to dynamic surfaces. For this dynamic black object, energy changes are related to the energy carried per emission, given by $\delta R_S = \dot{R}_S \, \delta ct \Rightarrow \delta E = -\frac{c^2}{2G_N}\delta R_S \simeq \frac{\hbar c}{R_S(ct)}$. The uncertainty principle $\delta E \;\; \delta t \geq \frac{\hbar}{2}$ then implies that a decrease in the radial surface scale has a lower limit given by $|\delta R_S| \, \delta ct_{emissions} \gtrsim L_P^2$.

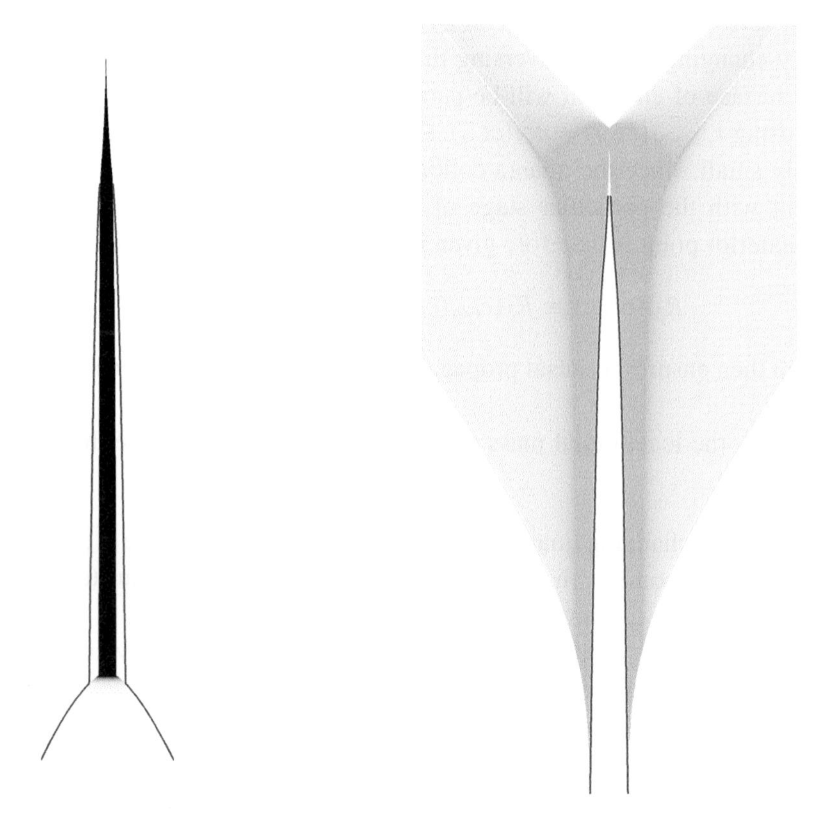

Figure 7.12 Energy density plots for a transient black object, with time ct progressing upwards, and a spatial coordinate x horizontal. The diagram on the left represents the overall energy density generating the geometry, while that on the right screens out the density of the collapsing/collapsed matter and enhances that of evaporating quanta. The (darker gray) surface represented by the merging curves delineates the exterior from the interior.

There is also an upper limit upon the rate of evaporation/decay, due to *geometric consistency*. The rate of evaporation must be dynamically consistent in a manner that guarantees that the energy-carrying quanta do not leave space-time so flat that subsequent quanta eventually catch up. Outgoing geometry changing massless quanta propagate through a background geometry satisfying Eq. 7.57, where the background geometry of successive quanta are modified by the previous quanta. The geometric consistency limit is reached if the trajectories of two successive radial emissions coincide as $t \to \infty$.

7.4.2 Geometric characterizations of the geometry

A plot of a snapshot of the radial mass scale at a fixed time during evaporation is demonstrated in Figure 7.13. The diagram depicts the dependence of the radial

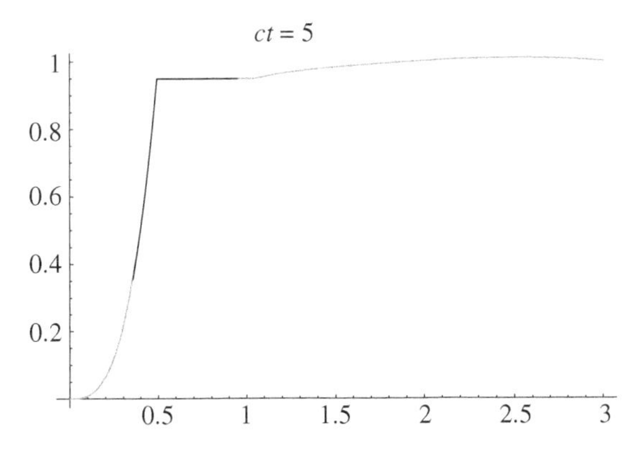

Figure 7.13 Snapshot of radial mass scale of evaporating black object $R_M(5, r)$ in units of R_{So}.

mass scale on the radial coordinate at time $ct = 5R_{So}$. The gray curve represents the spatial dependence of the radial mass scale at the given time. The trapping region is represented by a black region superposed upon the radial mass scale curve. The flat region of the curve beyond $r \gtrsim 2R_{So}$ is in the stationary exterior yet to be affected by evaporation. The portion of the curve that is flat in the interior represents a spatially coherent region with vanishing ${T^0}_0$. Interior to this region lies a core of degenerate critical density.

Light-like curves on the dynamic geometry

Light-like surfaces for the given metric 7.54 define conformal coordinates, which are labeled (v, u), based upon their correspondence on reference hypersurfaces. Here, $v = ct_* + r_*$ and $u = ct_* - r_*$, where (ct_*, r_*) are the conformal time-space coordinates, and:

$$\dot{r}_v = -1 - \sqrt{\frac{R_M}{r_v}}, \quad \dot{r}_u = 1 - \sqrt{\frac{R_M}{r_u}}. \tag{7.66}$$

The light-like trajectories are depicted on standard space-time (ct, r) diagrams in Figure 7.14. All trajectories are sourced from or terminate on the center $r = 0$, spaced in units of R_{So}, the Schwarzschild radius of the geometry. One should note that photons emitted just prior to the formation of the black object initially propagate with increasing radial coordinate and then have decreasing radial coordinate as the trapped region forms. Once the trapping region forms, none of the interior outgoing photons can traverse the interior trapping surface R_{TS}^- while that trapping region is present. However, for this transient black object, *all* outgoing photons will eventually reach light-like future infinity. The particular outgoing light-like surface communicating the beginning of evaporation joined with the collapsed surface

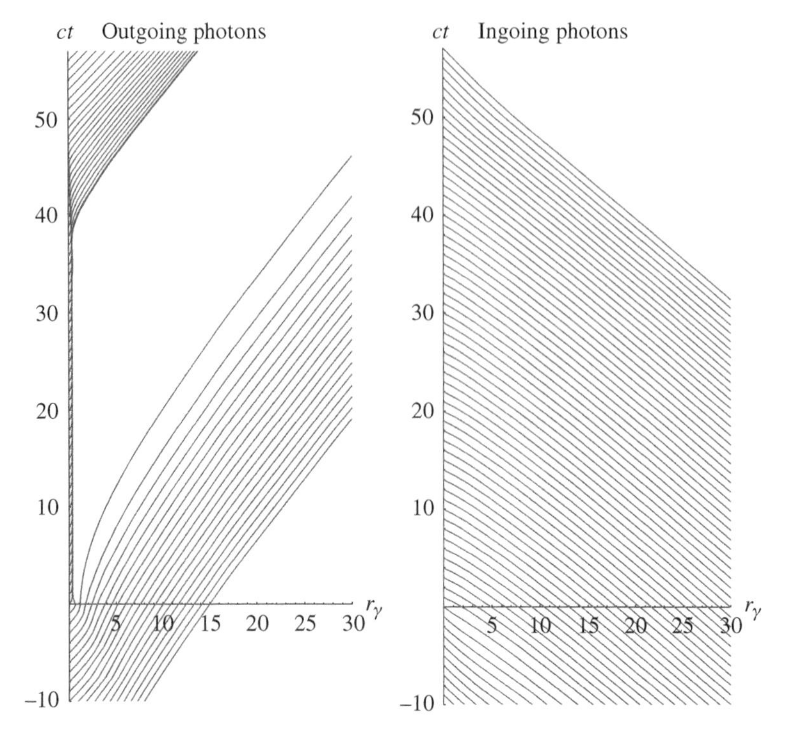

Figure 7.14 Standard space-time diagrams of light-like trajectories in units of R_{So}. Outgoing trajectories are depicted on the left, ingoing trajectories on the right.

prior to evaporation serve as the exterior surface of correspondence for assigning the parameter v on conformal diagrams.

7.4.3 Conformal diagram of the transient black object

The significant features of the geometry are demonstrated on the conformal diagrams in Figure 7.15. The diagrams are bounded from the left by the center $r = 0$, which as always time-like, from the lower right by past light-like infinity $skri^-$, and from the upper right by future light-like infinity $skri^+$. The past and future light-like infinities are consistent with those of Minkowski space-time. The static radial mass scale (Schwarzschild radius) of the geometry, which is depicted by the (gray) curve labeled R_{So}, scales the diagram. The time-like dashed curve labeled by R_X represents the exterior surface of the collapse prior to its crossing R_{So}, and the exterior boundary of the trapping region R_S afterwards. There are no horizons on the diagram. The center of the conformal coordinates $(Y_\rightarrow = 0, Y_\uparrow = 0)$ is chosen to correspond with the coordinate $(ct = 0, r = R_{So})$. Evaporation begins on the space-like surface $ct = ct_{Dark}$, when the collapsing surface crosses its Schwarzschild radius. Critical density is reached at the time

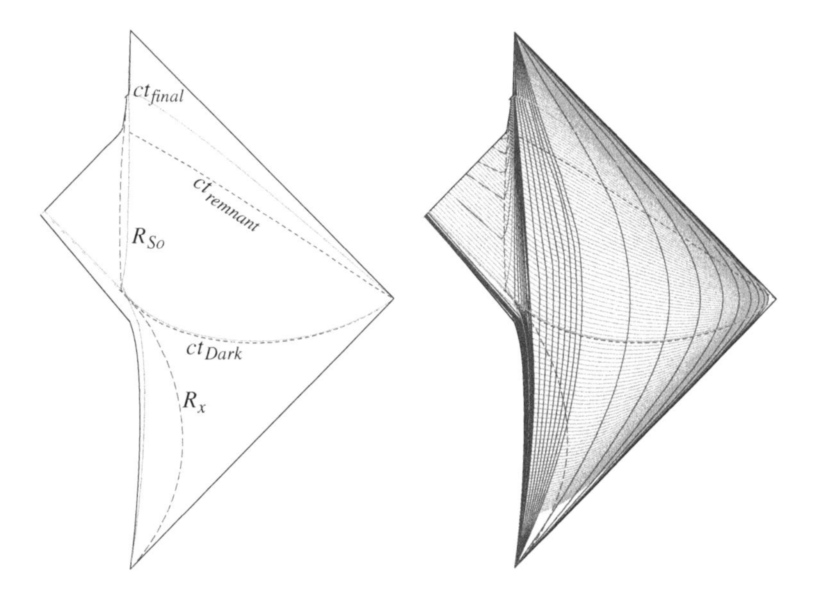

Figure 7.15 Conformal diagrams of the transient black object. The diagram on the left demonstrates dynamic surfaces of interest, while that on the right displays surfaces of fixed (ct, r) coordinates.

represented by the solid space-like surface just barely distinguishable after ct_{Dark} as the collapse terminates. The dashed space-like surface labeled $ct_{remnant}$ depicts the end of the black object and thermal evaporation, and the beginning of decay of the thermal remnant. The solid space-like curve labeled ct_{final} depicts the end of decay of the thermal remnant. After the communication of the end of this decay, the space-time is flat everywhere.

Fixed time coordinate curves in the right diagram of Figure 7.15, which are everywhere space-like, are graded in units of R_{So}. Curves of fixed radial coordinate r are graded in units of $0.1R_{So}$ from the center out to $2R_{So}$, then integral values of this unit, then decades of this unit. Fixed radial coordinate surfaces are seen to be space-like within the trapping region. Those fixed radial coordinate surfaces $r < R_{So}$ that are crossed by the trapping surface have a slope of $+1$ on the diagram at that instant, as they transition from time-like to space-like surfaces.

7.4.4 Appearance of trapping surface traversing observers

Only that information carried by a system entering a black object which can be communicated to an exterior observer is relevant to that observer's description of the system. Since the geometry has no horizon, there are subtle differences in the observable dynamics of falling through the trapping region of a black object versus a black hole. A horizon is a light-like surface, thus no system can be seen to vanish through a temporally transient horizon until the horizon itself vanishes. It is of interest to examine these dynamics for a temporally transient black object.

An exterior geometrically stationary observer shares proper time τ with that of the asymptotic observer t, and thus can serve as a convenient platform for observing in-falling systems. For illustrative purposes, a geometrically stationary observer that remains $r = R_{So}$ after the remnant has completed its decay will be chosen as such an observational platform. This inertial observer will remain near enough to the black object that observations will be timely, while escaping any direct experience of strong gravitational effects.

Since the center remains always time-like, and matter within the central core satisfies all energy conditions, in-falling energies can, in principle, remain intact throughout the lifetime of the black object. Communications from such an in-falling system that remains intact will be examined. A sturdy, freely falling, geometrically stationary probe that reaches $r = R_{So}$ at time $ct = 10\, R_{So}$ will be observed as it traverses the interior of the black object. The probe is presumed to be robust enough to withstand the evaporation mechanism within the life cycle of the black object. It has proper specifications that emit standard frequency outgoing photons at a rate of c per R_{So} directly outwards towards the geometrically stationary observer. The photon trajectories are displayed in Figure 7.16. The in-falling probe is depicted by the innermost bold trajectories in the diagrams, and the observational platform by the outermost bold trajectories.

The diagrams demonstrate that there is an extended period of time, between the communication of the probe crossing into the trapped region and the communication of the end of the black object, during which any photons observed from the emitter are extremely redshifted. All outgoing communications from the probe are temporarily trapped during this period, as they collect within the inner surface of the trapping region. This collection of trapped communications is rapidly released after the trapping region vanishes. Eventually, all emitted communications will be received by the observer. Once the space-time becomes flat, the rate of observation of signals and frequency of the observed emissions will become identical to the proper specifications of the probe.

Frequency shifts of quanta emitted by in-falling emitters

The redshift of the emitted quanta can be calculated from the geodesic equation:

$$\frac{dU^\beta}{d\lambda} + \Gamma^\beta_{\mu\nu} U^\mu U^\nu = 0, \tag{7.67}$$

where λ is the affine parameter characterizing the trajectory. Once four-velocities for outgoing massless quanta are calculated, the temporal component of the

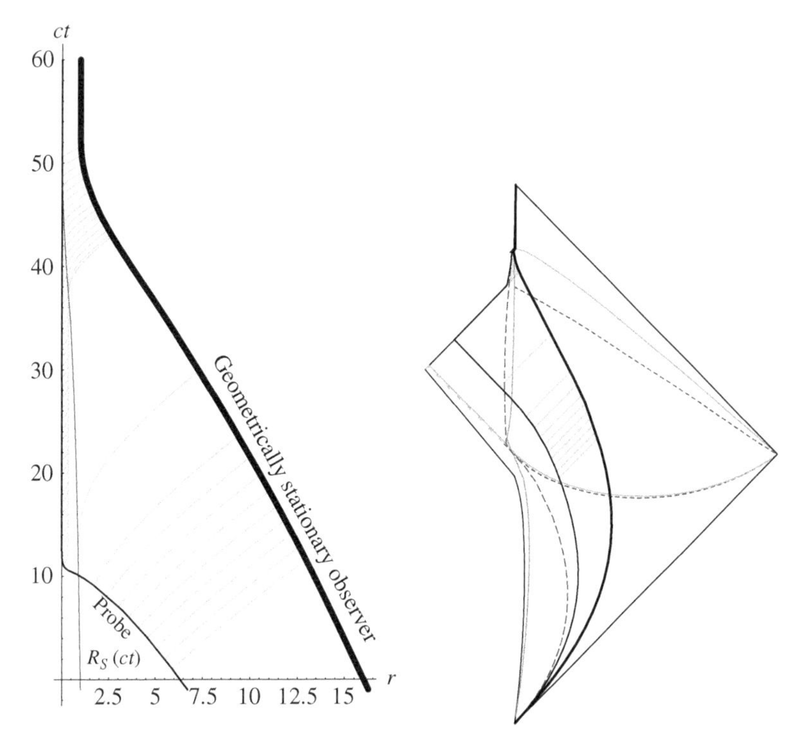

Figure 7.16 Standard photons are emitted by an in-falling geometrically stationary probe outwards to be detected by a nearby exterior observer. A standard space-time diagram is represented on the left, and the conformal diagram is represented on the right.

four-velocity of a gravitating photon $\vec{U}_\gamma$ observed by an observer with four-velocity $\vec{U}_o$, given by $-\vec{U}_o \cdot \vec{U}_\gamma$, is expected to satisfy:

$$-\vec{U}_o \cdot \vec{U}_\gamma = \left(U_o^{ct} \pm \sqrt{(U_o^{ct})^2 - 1} \right) U_\gamma^{ct}, \tag{7.68}$$

where U_o^{ct} is the temporal component of an ingoing/outgoing observer's four-velocity. This quantity is directly proportional to the observed energy of the photon. For a geometrically stationary observer with $U_o^{ct} = 1$, the observed photon temporal component is seen to be simply the component U_γ^{ct} calculated from the geodesic equation.

The relationship 7.68 can likewise be used to express the temporal component of the four-velocity of the photon in terms of its emitted form in the proper frame * of the probe:

$$U_\gamma^{ct} = \frac{(U_\gamma^{ct})_{emitted}}{U_*^{ct} \pm \sqrt{(U_*^{ct})^2 - 1}}, \tag{7.69}$$

where the $\pm$ sign refers to outgoing/ingoing probes, and U_*^{ct} is the temporal component of the (dimensionless) four-velocity of the probe. Since both the observer and the probe are geometrically stationary, then $U_o^{ct} = 1 = U_*^{ct}$. Thus, the geodesic equation directly calculates the observed energy $E_\gamma \propto U_\gamma^{ct}$ of the gravitating photons. The associated frequency is directly related to the rate of observation in Figure 7.16, prior to the termination of the trapping region. The rate of observation of photons becomes severely redshifted as the probe traverses the interior of the black object. However, once the black object evaporates away, communications from the probe re-emerge. This is in complete contrast to observations from a temporally transient black hole, since no normal energies can remain untransmuted upon encountering the space-like center of an analogous transient black hole. As the remnant decays, the rate of reception is more rapid than the rate of emission, but each emission continues to be individually redshifted, until the remnant completely decays away.

7.4.5 Entanglement of massless quanta

The absence of a space-like center for the transient black object also allows a direct exploration of the trajectories of entangled photons through the black object. Consider a massive, unstable particle, initially coincident with the previous in-falling geometrically stationary probe, that decays into an entangled pair of photons just as the particle encounters the outer trapping surface $R_S(ct_{decay})$. One of the photons is emitted radially outward, while the other photon is emitted radially inward to conserve microscopic momentum. One observer (Alice) is a geometrically stationary observer with final location $x = R_{So}$, while the other observer (Bob) is a geometrically stationary observer located diametrically opposite the former observer with final location $x = -R_{So}$. This arrangement is demonstrated in Figure 7.17. The in-falling unstable particle is represented by the dashed curve, while the entangled photons which are the products of the decay are represented as right-moving (R) and left-moving (L) light-like trajectories, ultimately detected by Alice and Bob.

In the space-time diagram on the left, the interior mass scale $R_S(ct)$ is also demonstrated as a solid (gray) time-like curve slowly approaching zero. The decay occurs just as the unstable particle encounters the trapping surface, so that the right-moving photon trajectory that ultimately reaches Alice initially has a vertical slope, moving outwards as the surface of the trapping region shrinks away. The left-moving photon crosses the center but remains trapped as it becomes an outgoing photon, until the trapping region vanishes due to the evaporation of the black object. After evaporation, this photon is ultimately detected by Bob. In the conformal diagram on the right of Figure 7.17, Alice has polar location $\varphi = 0$, while Bob has polar location $\varphi = \pi$. These observers are actually space-like separated, which

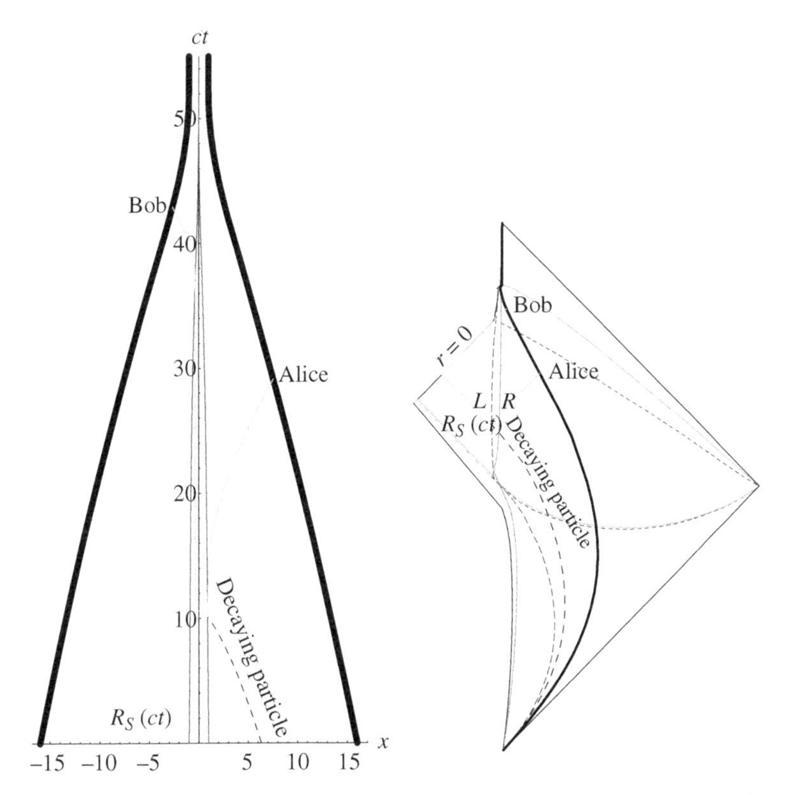

Figure 7.17 Trajectories of an entangled photon pair emitted by an unstable particle just as it crosses into the trapped region. The left diagram is a standard space-time diagram using (x, ct), while the right diagram utilizes radial conformal coordinates $(Y_{\rightarrow}, Y_{\uparrow})$.

is not depicted on the radially symmetric conformal plot which overlays their trajectories. In contrast to the situation in flat space-time, Alice and Bob likely detect the photons during differing epochs, and with differing energies. Thus, the space-time and energy-momentum entanglement information measured by the different observers is both temporally shifted and redshifted in energies.

This example directly demonstrates that the loss of entanglement information is only temporary for this geometry and that standard laws of physics demonstrate no obvious instances of violation, consistent with complementarity. One can generally assert that any information that falls through a general black object is *not* lost to the universe as a whole. It might, however, get mixed up and globalized through subtle quantum correlations of the radiations that result from the energy transmutations during black object evaporation.

8

Cosmology

8.1 A synopsis of Big Bang cosmology

One of the fundamental principles guiding the expected behaviors of the cosmology is a generalization of ideas of Copernicus, known as the *cosmological principle*. This principle presumes that no non-rotating observer at rest to the CMB radiation is more special than any other. Since the observed universe has large-scale uniformity, the cosmological principle imposes an overall homogeneity and isotropy to the universe. In addition, the dynamics of most of the aggregate features in the universe can be described assuming that the energy-momentum content of the cosmology is consistent with being an ideal fluid.

The Friedmann–Lemaitre equations 8.6, which describe the dynamics of an ideal fluid cosmology, are spatially scale invariant (if the cosmological constant is negligible), but not temporally scale invariant. The form of those equations that govern a spatially flat ($\kappa = 0$) expansion satisfies spatial scale invariance (at least to a very good approximation), due to the fact that the energy densities that drive the dynamics are intensive thermodynamic variables. However, there *is* apparently a beginning time $t_o \approx 13.7(\pm 0.2)$ billion years ago, which represents the earliest backwards-looking extrapolation of the standard model expansion called the *Big Bang*. The physics during these earliest moments is an active field of research. Thus, the cosmological principle does not refer to the temporal evolution of the universe.

There is relatively recent evidence for the existence of *dark energy*, driving an acceleration of the rate at which galactic clusters separate. Such geometrodynamic behaviors are contrary to the usual attractive nature of gravitation. Persistent dark energy implies that far off regions of the universe will remain forever beyond observation, defining a horizon *within* which these regions will have no causal influence. One should note that the existence of any horizon due to remnant dark energy *does* introduce a persistent spatial scale to the macroscopic cosmological dynamics that becomes substantial during the later stage of cosmological evolution.

Evidence for the existence of persistent dark energy comes from several independent observations. The luminosities of type Ia supernovae show that the rate of expansion of the universe was decelerating in the distant past (consistent with the gravitational attraction of the dispersing energy density) but has been accelerating for the past 6×10^9 years [89]. These supernovae explode when the mass from a binary companion of a white dwarf star accretes onto the white dwarf until it reaches the *Chandrasekhar limit*. The Chandrasekhar limit is associated with gravitational saturation of the Fermi degeneracy pressure of the electrons in the white dwarf that keep the star from collapsing (via the Pauli exclusion principle). The supernovae explode with a well-known luminosity profile, which allows them to serve as standard candles for measuring distance. Distance measurements are crucial in describing cosmological evolution. More recent supernovae appear more faint than they should (~25%), yet this faintness cannot be due to intervening dust, which would change the color profile, as sunset does to the Sun.

This acceleration is independently supported by analysis of CMB radiation [90,91]. The CMB radiation, which is extremely isotropic, images the final snapshot of the hot dense plasma of the Big Bang as hydrogen atoms are formed, and the universe becomes transparent. This snapshot images the universe when it was about 300 000 years old, at a black body temperature of about 0.3 eV (~3000 K). At that time, cosmological scales were 1100 times smaller than today. The background has cooled to the 2.74 K black body spectrum observed today, some of which was commonly seen as a portion of the thermal "snow" on analog broadcast television. However, CMB radiation is not completely uniform, exhibiting very slight, yet regular, temperature fluctuations. The fluctuations in CMB radiation are indicative of acoustic waves that propagated on the primordial plasma prior to the decoupling of the photons that make up this radiation. The observed locations of the acoustic peaks on these waves are consistent with a spatially flat background. If there were spatial curvature, these peaks would appear to have propagated a different distance since their formation than that observed. Also, the height of the acoustic waves is sensitive to the normal (gravitationally attractive) energy density of the universe. Careful measurements of these waves are consistent with normal and dark matter density making up only about 27% of the present total energy density of the universe. *Dark matter* is a type of matter that does not (strongly) couple to photons (which couple charges) and thus cannot be seen, but contributes to gravitational sourcing in the same manner as *baryonic* matter, the material of stars, planets, and humans. The presence of dark matter is independently measured by examining the orbits of stars within galaxies.

The luminosity of standard candles (like the type Ia supernovae), the fine structure of the CMB radiation, and other independent measures (like the deuterium/hydrogen ratio), are all in quantitative agreement to a cosmology with a non-vanishing (positive) cosmological constant fit to the data. The energy density resulting from this fit defines a *dark energy* responsible for the acceleration of distant galactic clusters during late cosmological times. This repulsive characteristic was Einstein's original motivation for including a cosmological constant in the equations of a steady-state universe. The existence of a persistent dark energy density that might be described in terms of a cosmological constant defines a length scale for the global structure of the cosmology.

Whole-sky, deep field observations yield additional interesting properties of the universe. The (luminal) *horizon problem* expresses the paradox of observed large-scale homogeneity and isotropy of macro-physical properties of the universe beyond regions of causal influence. For instance, during the time that photons in CMB radiation have traveled since decoupling from the primordial plasma, they have traversed 100 times the distance that they could have traveled prior to that decoupling. The decoupling occurred when the universe was over a thousand times smaller than it is today. This means that the CMB radiation image should contain $\sim 10^6$ luminally disconnected regions. However, the temperature of the CMB radiation is quite uniform across the whole sky, despite the expectation that opposing directions have yet to be in causal contact, since those photons are just reaching the halfway point at Earth. In addition, angular correlations of the fluctuations in CMB radiation have been accurately measured by several experiments [90]. These correlations provide evidence for space-like coherent phase associations among the cosmological fluctuations reflected in the CMB anisotropies. It is expected that during the earliest of epochs, the quantum coherence of gravitating subsystems qualitatively altered the dynamics of the cosmology from what is presently observed. The entangled nature of co-gravitating quantum states with space-like separations should manifest as some form of spatial coherence in the large-scale structure of the geometrodynamics.

Supplement: Cosmological features and parameters

The observation of standard candles, galactic rotation curves, features of the CMB, and other phenomena, imply a standard cosmological model that evolves into the relative densities for photons, baryons, matter = baryons + dark matter, and dark energy observed today given by $\Omega_\gamma \sim 4.6 \times 10^{-5}$, $\Omega_b \sim 0.04$, $\Omega_m \sim 0.267$, and $\Omega_\Lambda \sim 0.73$. The present-day energy densities are usually expressed as ratios $\Omega_X \equiv \frac{\rho_X}{\rho_{crit}}$ relative to the maximum energy density that would *not* close the observed universe, in terms of the present Hubble expansion rate: $\rho_{crit} \approx 0.5 \times 10^{-5}$ GeV/cm^3. A universe without dark energy with this critical density would manifest an indefinite, slowing expansion that just barely never collapses. The ratios Ω_X all must add to unity for a spatially flat cosmology.

The measured values are consistent with a cosmology consisting of a thermal fluid with remnant pressureless matter (both baryons and dark matter), leftover CMB radiation, and dark energy, which manifests the evolution of a hot Big Bang that initiated about 13.7 billion years ago. The present prevalence of Big Bang photons Ω_γ yields a ubiquitous number density of these primordial photons of about $n_\gamma \approx 413/\text{cm}^3$, resulting in an entropy density of about $s_\gamma \approx 2905\, k_B/\text{cm}^3$. The night sky is dark only because the CMB radiation has cooled to 2.74 K due to the expansion of the universe, well below visible frequencies, thus resolving Olbers' paradox.

However, the universe is quite hot with regards to the remnant baryons. Quantitatively, the universe overall has $\frac{1}{k_B}\frac{entropy}{baryon} \sim 10^{10}$, as compared to a value of $\sim 10^{-2}$ for the Sun, and ~ 1 for a neutron star. This implies that the baryons are a small remnant of a large number of annihilations during a very hot primordial period.

The observed large-scaled structure of the universe grew from the fluctuations from uniformity in the CMB at last scattering (decoupling) of the order $\frac{\delta\rho_{CMB}}{\rho_{LS}} \sim 10^{-5}$. This value compares to the corresponding present day values for superclusters $\frac{\delta\rho_{superclusters}}{\rho_o} \sim 1$, galactic clusters $\frac{\delta\rho_{clusters}}{\rho_o} \sim 10^2$, galaxies $\frac{\delta\rho_{galaxies}}{\rho_o} \sim 10^5$, and stars $\frac{\delta\rho_{stars}}{\rho_o} \sim 10^{30}$.

8.1.1 The equations of standard cosmology

The equations of general relativity applied to the universe as a whole define a cosmology. If one assumes that the large-scale distribution of the matter and space-time of the universe is essentially homogeneous (the same independent of location) and isotropic (the same independent of the orientation of perspective), then the large-scale metric behavior of the universe should not depend upon angular orientation. The metric can be temporally dependent, but no center can be (globally) special. The space-time invariant interval must, therefore, be described by the Robertson–Walker metric:

$$ds^2 = -(dct)^2 + a^2(ct)\left(\frac{dr_{RW}^2}{1-\kappa\, r_{RW}^2} + r_{RW}^2\left(d\vartheta^2 + \sin^2\vartheta\, d\varphi^2\right)\right). \tag{8.1}$$

The parameter $a(ct)$ is a spatial distance *scale factor* for cosmological versus microscopic proper spatial scales, taking units of length. The Robertson–Walker radial coordinate r_{RW} is typically dimensionless. The sign and value of the factor κ defines the *spatial curvature* of the cosmology, and a *spatially flat* cosmology is described by $\kappa = 0$. Observers with fixed coordinate r_{RW} can be shown to be *co-moving* inertial observers sharing the same time, $\frac{dt}{d\tau} = 1$. These observers are geometrically stationary in this cosmology. Thus, the coordinate time t *is* a proper time for these observers. Since the universe is presently expanding, some define the time $t = 0$ as corresponding to the vanishing of the scale factor $a(ct = 0) = 0$. This perspective presumes the dominance of a classical description of the temporal progression of the dynamics all the way back to this time. Such a perspective introduces a metric singularity at $t = 0$, which will not be the perspective taken in the descriptions presented later.

The rate of scale expansion is directly related to the observed Hubble rate of the expansion H,

$$H = \frac{\frac{da(ct)}{dt}}{a(ct)}, \tag{8.2}$$

which measures the recession of distant galactic superclusters away from the Earth. This then matches up the parameter t in the universal metric with the time of a standard clock at the center of the galactic cluster orbited by the Milky Way galaxy, which is quite close to the time of Earth-based clocks. Indeed, the temporal parameter in the cosmological metric requires minimal reinterpretation, in contrast to Schwarzschild time, which is the proper time only for a distant observer.

As previously discussed, the inclusion of a cosmological constant into Einstein's equation:

$$\mathcal{G}_{\mu\nu} \equiv \mathcal{R}_{\mu\nu} - \frac{1}{2} g_{\mu\nu} \mathcal{R} = -\left(\frac{8\pi G_N}{c^4} \mathbf{T}_{\mu\nu} + \Lambda \, g_{\mu\nu} \right) \tag{8.3}$$

provides a convenient parameterization for the phenomena described using static dark energy as a substantial constituent of the energy content of the universe [92]. If the geometric and dynamic conservation principles are strictly valid, the constant Λ cannot have evolved from a primordial parameter, or be evolving towards a remnant value.

An ideal fluid (one with negligible viscosity) has the energy-momentum tensor reproduced from Eq. 5.72:

$$\mathbf{T}_{\mu\nu} = P \, g_{\mu\nu} + (\rho + P) U_\mu U_\nu. \tag{8.4}$$

Substitution into Einstein's field equations result in the Friedmann–Lemaitre equations for cosmology given by:

$$\begin{aligned} \left(\frac{\frac{da}{dct}}{a} \right)^2 &= \frac{8\pi G_N}{3c^4} \rho + \left(\frac{1}{R_\Lambda} \right)^2 - \frac{\kappa}{a^2} = \frac{8\pi G_N}{3c^4} (\rho + \rho_\Lambda) - \frac{\kappa}{a^2}, \\ \frac{\frac{d^2 a}{dct^2}}{a} &= -\frac{4\pi G_N}{3c^4} (\rho + 3\,P) + \left(\frac{1}{R_\Lambda} \right)^2 = -\frac{4\pi G_N}{3c^4} (\rho + 3\,P - 2\,\rho_\Lambda), \end{aligned} \tag{8.5}$$

where $\Lambda \equiv \frac{3}{R_\Lambda^2} = \frac{8\pi G_N}{c^4}\rho_\Lambda$. These equations exhibit spatial scale invariance in the scale factor a if the term with the spatial curvature κ is absent. Spatial scale invariance means that the equations are equally valid for scale a as for scale $\lambda \times a$, where λ is any real number. The term involving R_Λ introduces relative cosmological scale without breaking the overall scaling behavior in a, the term P is the cosmological pressure of the ideal fluid cosmology, ρ_Λ is the *dark energy* density associated with scale R_Λ, and G_N is Newton's gravitational constant from classical physics.

During the early thermal expansion, the rate of expansion on the left-hand side of Eq. 8.6 neglects the relatively small but constant density on the right-hand side associated with the cosmological constant ρ_Λ. As the density of the energy content of the universe becomes more dilute (as ρ decreases), the expansion rate of the universe $\left(\frac{da}{dct}\right)$ slows down. Radiation (thermal energy of a type for which any mass scale can be ignored, like photons, neutrinos, and very hot matter) has a form so that its associated wavelength expands as the universe expands. This means that it dilutes in a manner proportional to the inverse quartic power of the scale factor $\rho_{radiation} \propto a^{-4}$, since energy and each of the three components of momentum decrease. Pressureless matter dilutes only because of the spatial increase in the volume scale of the universe $\rho_{matter} \propto a^{-3}$, since most of the energy of each particle is contained in its fixed mass, which does not redshift. This means that during the thermal evolution of the universe, initially hot radiation during the *radiation dominated* epoch (which dilutes *most* rapidly) eventually becomes more dilute than the remnant pressureless matter during the *matter dominated* epoch. During either of these epochs, the expansion rate decreases due to the decreasing energy density content ρ. In late times like the present, both radiation and matter have become so dilute that the small but constant *dark energy* term ρ_Λ dominates, causing an increasingly rapid expansion of the universe. The late time behavior of the universe will be examined in Section 8.2.

8.1.2 de Sitter geometry

A universe defined by a positive cosmological constant is called a *de Sitter* space-time. This static space-time is specified by a constant radial length scale R_{dS} in the metric given by:

$$ds^2 = -\left(1 - \frac{r_{dS}^2}{R_{dS}^2}\right)(dct_{dS})^2 + \frac{(dr_{dS})^2}{1 - \frac{r_{dS}^2}{R_{dS}^2}} + r_{dS}^2 d\vartheta^2 + r_{dS}^2 \sin^2\vartheta\, d\varphi^2. \tag{8.6}$$

The coordinates $(ct_{dS}, r_{dS}, \vartheta, \varphi)$ represent proper-time, radial, and angular coordinates for an inertial observer at $r_{dS} = 0$. The mixed Einstein tensor generated

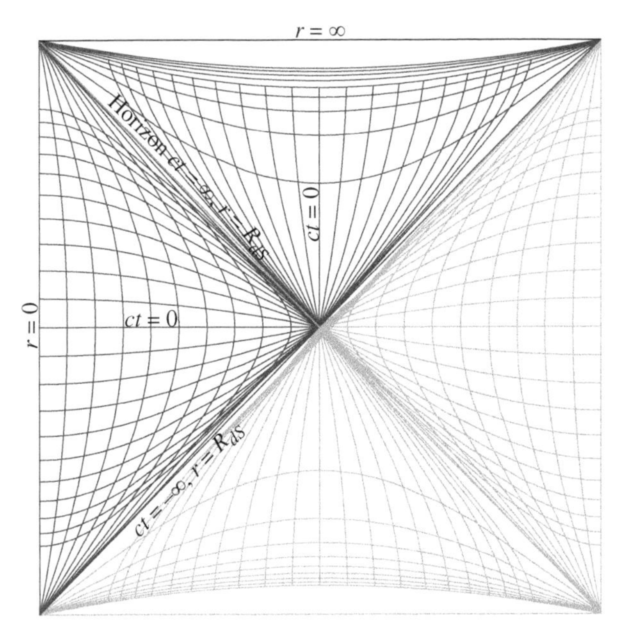

Figure 8.1 Conformal diagram for a de Sitter geometry.

by this metric is of the form $\mathcal{G}^{\mu}{}_{\nu} = \frac{3}{R_{dS}} \delta^{\mu}_{\nu}$, resulting in a cosmology with energy density ρ_{Λ} and pressure $P = -\rho_{\Lambda}$, where the cosmological constant is given by $\Lambda = \frac{3}{R_{dS}^2}$.

For conformal temporal parameter $t^* = t_{dS}$, a conformal form for the radial coordinate satisfies:

$$r^* = \begin{cases} R_{dS} \tan^{-1}(r_{dS}/R_{dS}), & r_{dS} < R_{dS}, \\ -\frac{R_{dS}}{2} \log\left(\frac{r_{dS}-R_{dS}}{r_{dS}+R_{dS}}\right), & r_{dS} > R_{dS}. \end{cases} \tag{8.7}$$

This transformation allows a direct construction of the conformal diagram for this geometry, demonstrated in Figure 8.1. The diagram is bounded on the left and right by time-like surfaces $r_{dS} = 0$ and on the bottom and top by space-like surfaces $r_{dS} = \infty$. Fixed radial coordinates are initially graded in tenths (left and right regions), then units (top and bottom regions) of the de Sitter scale R_{dS}. Fixed radial coordinate curves go from being time-like for $r_{dS} < R_{dS}$ to being space-like for $r_{dS} > R_{dS}$ at the de Sitter radial scale R_{dS}, which is an *ingoing* horizon for an observer at $r_{dS} = 0$. Fixed temporal coordinate curves (which are graded in tenths of the de Sitter scale R_{dS}/c) go from being space-like for $r_{dS} < R_{dS}$ to being time-like for $r_{dS} > R_{dS}$ at the scale R_{dS}. Time progresses from bottom to top in the left-most region of the diagram, and from right to left in the top-most region. The de Sitter horizon $r_{dS} = R_{dS}$ originates at the bottom-right corner of the diagram, and terminates at the upper-left corner. As was the case with Schwarzschild space-time, the horizon has temporal coordinate $t_{dS} = \infty$. However, unlike Schwarzschild space-time, de Sitter space has no physical singularity.

There is a similar outgoing light-like surface delineating regions which can be impacted upon by a system on the left surface $r_{dS} = 0$, its *particle out horizon*. This surface is a $t_{dS} = -\infty$ surface also with $r_{dS} = R_{dS}$ originating at the bottom-left corner of the diagram terminating at the upper-right corner. The gray region of the diagram to the right of this horizon cannot be causally affected by a system at $r_{dS} = 0$. The diagram is temporally symmetric about the point of interception of the de Sitter horizon with the particle out horizon. The triangular region on the far-left is completely causally disjoint from the triangular region on the far-right.

8.2 Dynamic de Sitter cosmology

The observed cosmology is certainly not static, yet it seems to be evolving towards (and perhaps from) de Sitter-like conditions. The classical form of a Roberson–Walker cosmology breaks down at an earliest scale for which $a \to 0$, at which time, the densities described in Eq. 8.6 become singular. Singularities are usually inconvenient mathematically and unrealizable physically. However, to construct a non-singular cosmology, there needs to be some initial degenerate state of finite density. It will be assumed that such a state will be defined by (or define) the micro-physical parameters of the cosmology. The dynamics of that cosmology will be developed in this section.

8.2.1 Fluid cosmology

General fluid cosmologies are consistent with accepted observational evidence during intermediate and late times, and they need not have singular behavior in primordial times. If the primordial behavior is non-singular, the primordial fluid itself provides a non-vanishing geometric scale for the cosmology. This primordial fluid begins the evolution of the cosmology through its dissolution into the other components of the cosmology. Thus, a description of cosmological dynamics in terms of its fluid content is both general and convenient.

Consistent with observation, the large scale cosmological description should assume isotropic flows $U_\vartheta = 0 = U_\varphi$, and global homogeneity. One can eliminate the geometric factors in terms of fluid variables using Einstein's equations for an ideal fluid to describe a fluid consistent geometry:

$$\begin{aligned} P &= \mathbf{T}^r_r = \mathbf{T}^\theta_\theta = \mathbf{T}^\phi_\phi = -\tfrac{c^4}{8\pi G_N}\mathcal{G}^\theta_\theta, \\ \rho &= 3P + \tfrac{c^4}{8\pi G_N}\mathcal{G}^\mu_\mu. \end{aligned} \tag{8.8}$$

The fluid flow velocities can likewise be determined to satisfy:

$$(U_0)^2 = (T_{00} - g_{00}P)/(\rho + P) = -\left(\frac{c^4}{8\pi G_N}G_{00} + g_{00}P\right)/(\rho + P),$$
$$(U_r)^2 = (T_{rr} - g_{rr}P)/(\rho + P) = -\left(\frac{c^4}{8\pi G_N}G_{rr} + g_{rr}P\right)/(\rho + P). \tag{8.9}$$

Thus, given any metric form for a fluid consistent cosmology, the fluid parameters are defined by Eq. 8.8.

As has been demonstrated for black holes and black objects, dynamic geometries manifest qualitatively different behaviors from their corresponding static analog. Since the standard description of the cosmology utilizes a Robertson–Walker geometry that evolves towards a final state de Sitter form, it is convenient to develop a dynamic metric that incorporates the properties of these two geometries. A metric with a dynamic scale that diagonalizes towards a Robertson–Walker form when maintaining the temporal coordinate, and towards a de Sitter-like form when maintaining the radial coordinate, is given by [93]:

$$ds^2 = -\left(1 - \frac{r^2}{R_\rho^2(ct)}\right)(dct)^2 - 2\frac{r}{R_\rho(ct)}dct\,dr + dr^2 + r^2 d\vartheta^2 + r^2\sin^2\vartheta\, d\varphi^2. \tag{8.10}$$

The fluid scale $R_\rho(ct)$ represents a type of dynamic de Sitter scale, defining a *dynamic de Sitter* cosmology. As is common for radially dynamic cosmologies described in this manner, any fixed temporal surface represents a space-like volume *everywhere* in the geometry, even across any horizons.

The coordinates $(ct, r, \vartheta, \varphi)$ represent proper-temporal, radial, and angular coordinates for an observer at $r = 0$. Proper-temporal intervals for this geometrically stationary observer are given by dt, while the proper-length intervals of measuring rods are given by dr. Thus, these coordinates are consistent with those associated with describing the micro-physical dynamics of an observer at $r = 0$. However, utilizing the cosmological principle, any given center $r = 0$ is no more special than any other. There are forms of the spatial coordinates that can be freely translated and rotated.

The metric form Eq. 8.10 introduces a microscopic fluid scale R_ρ that can be directly related to the usual Robertson–Walker scale. The coordinate transformation that diagonalizes this metric into a Robertson–Walker form while maintaining the same temporal coordinate $t_{RW} = t$ must give angular isotropy in either metric:

$$r = a(ct)\, r_{RW}, \tag{8.11}$$

and also satisfy the differential form $dr = a(ct)[\frac{r_{RW}}{R_\rho(ct)}dct + dr_{RW}]$ obtained through algebraic diagonalization. This results in relationships between the geometric Robertson–Walker scale $a(ct)$ and the fluid scale:

$$\frac{1}{R_\rho^2} \equiv \frac{8\pi G_N}{3c^4}\rho, \quad \frac{\dot{a}}{a} = \frac{1}{R_\rho} \quad \Rightarrow \quad a(ct) = R_\rho(0) \exp\left(\int_0^{ct} \frac{dct'}{R_\rho(ct')} \right). \quad (8.12)$$

Since the metric is diagonal in terms of the Robertson–Walker coordinates with $g_{00} = -1$, all observers with fixed spatial coordinate r_{RW} (absent angular motions) have the same proper time intervals dt. These geometrically stationary observers are co-moving systems with fixed Robertson–Walker radial coordinate $r_{RW} = constant$. However, the proper radial distance between these co-moving observers is not fixed, because of the factor $a(ct)$ in Eq. 8.11.

The metric 8.10 has several features of interest. It incorporates a fluid scale that can evolve away from/towards a cosmological pseudo-constant in Einstein's equation. This means that it is straightforward to describe an early inflationary period that evolves into a final dark energy–dominated epoch through an intermediate Big Bang cosmology. The metric allows the dynamic evolution of dark energy and quantum evolution of the cosmology without a need to introduce any true cosmological *constants*. The cosmology will thus be assumed to evolve in the absence of any true cosmological constant $\Lambda_{true} = 0$. Rather, there will be a dynamic parameter $R_\rho(ct)$ that becomes observationally consistent in intermediate and late times with a cosmological constant.

Einstein's equations for 8.10 allow the geometrodynamics to be expressed solely in terms of the energy content:

$$\frac{d\rho}{dct} = -\sqrt{\frac{24\pi G_N \rho}{c^4}}\,(P + \rho). \quad (8.13)$$

This equation is equivalent to the Friedmann–Lemaitre equations 8.6 of standard cosmology, written completely only in terms of the physical densities ρ and P.

8.2.2 Multi-fluid cosmology

The standard cosmology described at the beginning of this section models various epochs during which the dynamics are dominated by fluids with differing equations of state. The primordial (earliest) state gives rise to a hot, dense ultra-relativistic plasma (through so-called *reheating*), which behaves as a *radiation-dominated* cosmology. During the radiation epoch, the temperature is so extreme that the masses of the proliferous particles and antiparticles is negligible, so all particles exhibit ultra-relativistic motions. As the radiation cools, the quarks condense into baryons, and baryons condense into light nuclei within the first few minutes. The radiation eventually cools to a point dominated by leftover (pressureless) dust consisting of baryonic and dark matter, which was produced in minuscule quantities during the earlier stages, through processes which are yet to be understood. During matter domination, the plasma of electrons and nuclei condense into atoms, making the universe essentially transparent to the thermal photons that had been in equilibrium with the plasma. The present CMB radiation is the snapshot of this decoupling. For the last 6 billion years, the cosmology has been settling into a dark energy–dominated equation of state. This overall evolution can be developed using the coupled-state dynamics of a multi-fluid cosmology.

As a specific example, the cosmology will be modeled using physical densities that obey the equations of state for dark energy, radiation, and dust:

$$\rho = \rho_{DE} + \rho_{rad} + \rho_{dust}, \tag{8.14}$$

where primordial and remnant dark energies will be included in the term $\rho_{DE} = \rho_{primordial} + \rho_{remnant}$. Equation 8.13 can then be decomposed into coupled-rate equations.

A general form for the component rate equation for the dissolution of the primordial energy density is given by:

$$\frac{d\rho_{DE}}{dct} = -\Gamma_{DE\rightarrow rad}(ct) - \frac{3}{R_\rho}(P_{DE} + \rho_{DE}), \tag{8.15}$$

where radiation in some sense "precipitates" from the primordial energy component in early times. The term represented by $-\Gamma_{DE\rightarrow rad}$ is a generic rate of the dissolution of the primordial energy into radiation and remnant dark energy, whose detailed form is determined by micro-physical processes. The second term is a general form that incorporates the equation of state of the primordial energy. As previously mentioned, this period of conversion from significant quantities of primordial energy density into other components is commonly referred to as reheating.

During the Big Bang, radiation has its energy and momentum components redshifted during an expansion. The radiation density scales with the inverse fourth power of the Robertson–Walker scale $a(ct)$, since each component of the energy-momentum has frequency-wavelength freely expanding with the cosmology. Because the inverse fluid scale $1/R_\rho$ is the logarithmic derivative of the Robertson–Walker metric scale a from Eq. 8.12, the rate equation describing the radiation can be expected to take the form:

$$\frac{d\rho_{rad}}{dct} = \Gamma_{DE\rightarrow rad}(ct) - \frac{4}{R_\rho}\rho_{rad} - \Theta(\rho_{rad} - \rho_{threshold})\frac{\rho_{rad}}{c\tau_{r\rightarrow d}}. \tag{8.16}$$

The first term on the right of Eq. 8.16 incorporates the dissolution of primordial dark energy into radiation, and the second incorporates the appropriate redshift, and equation of state $P_{rad} = \frac{1}{3}\rho_{rad}$, for radiation. This results in a factor

$-\frac{3}{R_\rho}(P_{rad}+\rho_{rad})=-\frac{4}{R_\rho}\rho_{rad}$. The third term incorporates the generation of remnant dust with a rate constant $1/\tau_{r\to d}$. It assumes that above a threshold density, microscopic asymmetries during the non-equilibrium expansion (or some other processes) generate this remnant dust directly from the radiation.

The energy components of the constituents of the pressureless dust are not expected to redshift during the expansion. Therefore, the dust density scales with the inverse third power of a, yielding a rate equation of the form:

$$\frac{d\rho_{dust}}{dct}=\Theta(\rho_{rad}-\rho_{threshold})\frac{\rho_{rad}}{c\tau_{r\to d}}-\frac{3}{R_\rho}\rho_{dust}. \tag{8.17}$$

The first term on the right of Eq. 8.17 generates the remnant dust in early times directly from the radiation consistent with 8.16, while the second term ensures proper scaling of this density during expansion due to its equation of state $P_{dust}=0$.

The forms of Eqs. 8.15–8.17 ensure that the evolution of the total density given by Eq. 8.13 is consistent with the summed density components, as long as the pressure content is appropriate. For a thermal black body, the radiation component has an equation of state $P_{rad}=\frac{1}{3}\rho_{rad}$, and the dust has been assumed not to contribute to the pressure $P_{dust}=0$. The primordial form of the dark pressure can be generally taken to be that of a macroscopically coherent system undergoing the phase transition to a hot dense plasma. The equation of state for the remnant dark energy is experimentally consistent with $P_{DE}=-\rho_{DE}$. For present purposes, this equation of state will be assumed to be maintained for the primordial fluid. Thus, the form of the overall pressure can be taken as:

$$P=P_{dust}+P_{rad}+P_{DE}=\frac{1}{3}\rho_{rad}-\rho_{DE}. \tag{8.18}$$

This defines all of the parameters needed to describe the evolution of the cosmology.

Adiabatic expansion

Equation 8.13, which describes the dynamics of the expansion of the energy content of the cosmology, can be re-expressed in terms of the Robertson–Walker scale factor in the following way:

$$\frac{d}{dct}(\rho\, a^3)=-P\,\frac{d}{dct}a^3. \tag{8.19}$$

This has been interpreted to describe an adiabatic expansion of the energy contained in the volume enclosed by a co-moving spherical surface with a fixed Robertson–Walker coordinate $r_{RW}=1$, using the first law of thermodynamics [94]. The density ρ in Eq. 8.19 represents the sum of all component energy densities in the cosmology and expends no heat in expanding the co-moving volume.

Moving all terms involving the primordial fraction of energy to the right-hand side of the equation and substituting Eq. 8.15, the expression 8.19 becomes:

$$\frac{d}{dct}[(\rho_{rad} + \rho_{dust})a^3] = \Gamma_{DE \to rad}(ct)\, a^3 - (P_{rad} + P_{dust})\, \frac{d}{dct} a^3. \tag{8.20}$$

This equation takes the form of the first law of thermodynamics for the comological fraction consisting of radiation and dust. Therefore, the rate of reheating in the co-expanding volume is directly described by the micro-physical details of dissolution:

$$T \frac{dS_{reheat}}{dt} = \Gamma_{DE \to rad}\, V. \tag{8.21}$$

Given a micro-physically derived rate of the dissolution of primordial material into radiation, the dynamics of reheating can be immediately calculated using this equation.

8.2.3 Temporal evolution of scales

The Robertson–Walker scale is given by an integral over the fluid scale of the form:

$$\log\left[\frac{a(ct)}{a(ct_o)}\right] = \int_{ct_o}^{ct} \frac{dct'}{R_\rho(ct')}. \tag{8.22}$$

This means that over any interval for which the fluid scale remains essentially unchanged, the evolution is *inflationary*:

$$a(ct) = a(ct_o)\, e^{H_o (t - t_o)} \quad \text{during inflationary epochs.} \tag{8.23}$$

This exponential evolution describes the universe dominated by any remnant dark energy, as well as during "flat" expansion of any primordial inflation.

The energy density of radiation ρ_γ varies as $\frac{1}{a^4}$. Substitution of this behavior into the rate equation $H^2 = c^2 \left(\frac{\dot{a}}{a}\right)^2 = \frac{8\pi G_N}{c^2}\rho$ during radiation domination results in the relations:

$$a\, \dot{a} = \lambda_r,\; R_\rho(ct) = \frac{a^2(ct)}{\lambda_r}, \tag{8.24}$$

where λ_r is a constant with dimensions of length. Thus,

$$a^2(ct) = a^2(ct_o) + 2\,\lambda_r\,(ct - ct_o) \quad \text{during the radiation epoch.} \tag{8.25}$$

One might note that during radiation domination, Eq. 8.24 implies that $\dot{R}_\rho = 2$, i.e., the fluid scale uniformly expands with time as $R_\rho(ct) = R_\rho(ct_o) + 2\,(ct - ct_o)$.

Near radiation-matter equality, the scale satisfies the hybrid form:

$$\frac{t}{t_{eq}} = \frac{2 + \left(\frac{a}{a_{eq}} - 2\right)\left(\frac{a}{a_{eq}} + 1\right)^{1/2}}{2 - \sqrt{2}}. \tag{8.26}$$

During matter domination, the energy density varies inversely with the spatial volume $\rho_m \sim \frac{1}{a^3}$. This results in scale evolution satisfying:

$$[a(ct)]^{3/2} = a_m^{3/2} + \frac{3}{2} a_m^{1/2} \dot{a}_m [ct - ct_m] \quad \text{during the matter epoch.} \tag{8.27}$$

These four evolutionary descriptions are consistent with present observational evidence.

The *particle horizon* is the incoming light-like surface that delineates the furthest region of the universe that can have communicated since the beginning from those regions that have yet to be seen. The proper distance to the particle horizon d_H is most directly obtained from the diagonal Robertson–Walker form of the metric $d_H = a(ct) \times r_H$, where r_H was the radial coordinate of a photon that was emitted at $t = 0$ and is just arriving at $r = 0 = r_{RW}$. Null motions satisfy $dr_{RW} = -\frac{dct}{a(ct)}$, so that the proper distance to the particle horizon is given by:

$$d_H = a(ct) \int_0^{ct} \frac{dct'}{a(ct')}. \tag{8.28}$$

Substitution of the temporal dependence of the scale a during the various epochs demonstrates how the particle horizon varies with time:

$$d_H \simeq \begin{cases} \frac{c}{H_I}\left(e^{H_I(t-t_I)} - 1\right) + d_I & \text{during inflationary epochs} \\ 2\,c(t - t_r) + d_r & \text{during radiation epoch} \\ 3\,c(t - t_m) + d_m & \text{during matter epoch} \end{cases} \tag{8.29}$$

It should be noted that one cannot simply multiply the age of the universe by the speed of light to determine the size of the observable universe.

8.2.4 Temperature in a dynamic cosmology

The evolution of the temperature in a quasi-equilibrium thermal universe depends upon the dominant equation of state during the various epochs of evolution. The primordial state prior to thermalization during "reheating" is expected to have long-range coherence and a high degree of order. Temperature evolution during this transition critically depends upon the micro-physics of the dissolution and will not be examined here.

After reheating, the early universe is in a state of very hot radiation, where, for any known particles, $k_B T \gg m_a c^2$. The equation of state for black body photons in thermal equilibrium is well-known:

$$\rho_\gamma = g_\gamma \frac{\pi^2}{30} \frac{(k_B T)^4}{(\hbar c)^3}, \quad P_\gamma = \frac{1}{3}\rho_\gamma, \quad s_\gamma = \frac{4}{3}\frac{\rho_\gamma}{T}, \quad n_\gamma = g_\gamma \frac{\zeta(3)}{\pi^2}\left(\frac{k_B T}{\hbar c}\right)^3, \tag{8.30}$$

where the degeneracy factor for photons spin states is given by $g_\gamma = 2$, ρ_γ is the energy density, P_γ the pressure, s_γ the entropy density, and n_γ the number density. Micro-physical couplings will generate a proliferation of ultra-relativistic particle–anti-particle pairs, giving an overall fluid density for the universe during radiation domination of:

$$\rho = \left(\sum_{bosons\ B} g_B + \frac{7}{8} \sum_{fermions\ F} g_F \right) \frac{\pi^2}{30} \frac{(k_B T)^4}{(\hbar c)^3} \equiv g_\rho(T) \frac{\pi^2}{30} \frac{(k_B T)^4}{(\hbar c)^3}, \tag{8.31}$$

where the sums are over the dimensionless degeneracies of all ultra-relativistic particle types at the given temperature, and the difference in the factors is due to the differing quantum statistics.

After decoupling (last scattering) as the plasma becomes atomic, the dust becomes transparent to photons, and thus these components fall out of equilibrium. Prior to that time, the neutrinos, which interact very weakly, had already fallen out of equilibrium with the plasma. The temperature of the CMB radiation continued to redshift with a cosmological scale factor dominated by the pressureless matter, whose energy component does not redshift. In terms of the temperature of the CMB radiation at some time during the matter domination epoch T_{MD}, and the density of pressureless dust at that time ρ_{MD}, the matter-dominated energy density of the universe satisfies:

$$\rho_m \equiv \rho_b + \rho_{dm} = \rho_{MD} \left(\frac{T}{T_{MD}} \right)^3. \tag{8.32}$$

During the present late stage, the expansion is dominated by dark energy, whose nature is a problem of current interest.

Cosmological redshift

A cosmological redshift of freely propagating photons is a direct consequence of geodesic motion in an expanding universe. The *redshift parameter* z is defined in terms of the frequency of a photon measured by an observer compared to that when it was emitted:

$$z \equiv \frac{\nu_{emit} - \nu_{observed}}{\nu_{observed}}. \tag{8.33}$$

Heuristically, one expects the wavelength of massless quanta to expand with the scale of the cosmology, giving $\frac{\lambda_1}{\lambda 2} = \frac{a(ct_1)}{a(ct_2)} = \frac{\nu_2}{\nu_1}$. This can be rigorously demonstrated using the geodesic equation for radially propagating massless quanta:

$$\frac{dU^{ct}}{d\lambda} = \frac{1}{R_\rho}\left(\frac{r}{R_\rho}U^{ct} - U^r\right)^2, \tag{8.34}$$

and substituting the requirement that null radial geodesics satisfy $U^r = (\frac{r}{R_\rho} \pm 1)\,U^{ct}$. Thus, using 8.12:

$$\frac{d}{dct}\log U^{ct} = -\frac{1}{R_\rho} = -\frac{d}{dct}\log a \quad \Rightarrow \quad \frac{U^{ct}(ct_2)}{U^{ct}(ct_1)} = \frac{\nu_2}{\nu_1} = \frac{a(ct_1)}{a(ct_2)}. \tag{8.35}$$

This means that during any observational epoch,

$$z + 1 = \frac{a(ct_{observed})}{a(ct_{emit})}. \tag{8.36}$$

This redshift parameter serves as a convenient measure of the cosmological time at which any observed radiation was emitted. It is particularly useful when examining the known spectral characteristics of that radiation as they appear today.

Geodesic motion of massive particles

For completeness, the equations for the geodesic motions of massive particles will also be demonstrated. It is of particular interest to examine motions relative to the geometrically stationary, co-moving observers. In terms of the Robertson–Walker scale factor, the four velocity of a massive particle must satisfy:

$$\frac{dct}{dc\tau} = \sqrt{1 + a^2\left|\frac{d\mathbf{r}_{RW}}{dc\tau}\right|^2}. \tag{8.37}$$

The spatial coordinates of the Robertson–Walker geometry are dimensionless. Proper distance measurements in the metric requires multiplication by the dimensioned scale factor, yielding $d\mathbf{L} = a(ct)\,d\mathbf{r}_{RW}$. Thus, the proper spatial component of the four-velocity calculated by a co-moving observer includes this scale factor, with its units of length:

$$\mathbf{U}_* = a(ct)\frac{d\mathbf{r}_{RW}}{dc\tau}, \tag{8.38}$$

where from Eq. 8.37, $U^{ct} = \sqrt{1 + |\mathbf{U}_*|^2}$. One should note that these are not the spatial components of the coordinate four-velocity $u^\beta = \frac{dx^\beta}{d\tau}$, whose dimensions follow those of the coordinates.

The temporal component of the four-velocity satisfies:

$$\frac{dU^{ct}}{dc\tau} + a(ct)\dot{a}(ct)\left|\frac{d\mathbf{r}_{RW}}{dc\tau}\right|^2 = \frac{dU^{ct}}{dc\tau} + \frac{\dot{a}(ct)}{a(ct)}|\mathbf{U}_*|^2 = 0. \tag{8.39}$$

This equation can be directly solved for the magnitude of the spatial components of the *peculiar velocity* $\mathbf{U}_*$ that parameterizes proper motions relative to the co-moving system:

$$|\mathbf{U}_*(ct)| = |\mathbf{U}_*(ct_o)| \, \frac{a(ct_o)}{a(ct)}. \tag{8.40}$$

The velocity $|\mathbf{U}_*(ct)|$ is the magnitude of the spatial component of the four-velocity determined by a local co-moving observer at the time t. One should note that for a cosmology that expands indefinitely, all inertial motions eventually come to rest with respect to some co-moving location.

This form for the inertial motions has physical significance. For instance, consider a random walk of a co-moving mass m at time t_o caused by elastic scattering with a thermal photon of energy $E_{\gamma o} = \kappa \frac{\hbar c}{a(ct_o)}$. The three-momentum $m|\mathbf{U}_*(ct_o)|c$ imparted upon the mass $2\frac{E_{\gamma o}}{c}$ will result in motion of the mass relative to the co-movers given in terms of the proper displacement velocity $|\mathbf{U}_*(ct)|c = 2\frac{\kappa}{m}\frac{\hbar}{a(ct)}$. A thermal photon elastically striking at a later time from the opposite direction has precisely the momentum needed to bring this mass back to rest at a different co-moving coordinate. Thus, on average the net effects of random walks do not contribute energy to the co-moving systems. The walks do, however, disperse non-uniform concentrations towards uniformity, as well as generate fluctuations, in a manner analogous to classical random walks.

Supplement: Classical random walk

An unbiased random walk involves a system undergoing random steps from a given position, $x_{s+1} = x_s + \Pi_s \, \delta L$, where $\Pi_s = \pm 1$ is random. The motions can be thought of as being caused by random collisions with unseen particles from either direction with equal likelihood. The average position of an ensemble of random walkers vanishes, but the average distance traveled does not: $\langle x_N^2 \rangle = N\,(\delta L)^2$. If the interval between the steps is constant (at least on the average), then the total number of steps satisfies $N = \frac{t}{\delta t}$. Thus, after a time interval t, an ensemble of random walkers originally at $x = 0$ will have diffused into a region with average squared distance:

$$\langle x^2(t) \rangle = \frac{(\delta L)^2}{\delta t} t \equiv 2\,D\,t \overset{3-dim}{\Rightarrow} \langle \mathbf{r}^2(t) \rangle = 6\,D\,t,$$

where the constant $D = \frac{(\delta L)^2}{2\,\delta t}$ is defined as the *diffusion constant*. This demonstrates how the fluctuations causing random walks result in a dispersion of the walkers.

The effect of the mechanism causing the random walks also results in dissipations and viscosity. Suppose that the random walks are caused by collisions with unseen particles of mass m and average speed v. A walker moving with a drift velocity v_d in such an environment will experience larger momentum transfers from collisions in the forward direction than those from the receding direction. This difference in the momentum transfers results in a viscous force proportional to the drift velocity:

$$\left\langle \frac{\delta p}{\delta t} \right\rangle = \langle F \rangle = -\frac{2m}{\delta t}\, v_d \equiv -\zeta\, v_d,$$

for non-relativistic motions, where the *viscous coefficient* is given by $\zeta \equiv \frac{2m}{\delta t}$. The unseen particles travel the distance $L = v_x \delta t$ between the collisions, which means that the diffusion constant can be written as $D = \frac{v_x^2 \delta t}{2}$. This means that the product of the diffusion constant with the viscous coefficient depends only upon the average energy of the unseen particles:

$$D\zeta = \frac{1}{3} m v^2 \propto k_B T.$$

This well-known expression, known as *Einstein's relation*, states that the fluctuation measured by D times the dissipation measured by ζ should have a simple temperature dependence.

An illustrative cosmology

To gain further insight into cosmological geometrodynamics, a model will be constructed to demonstrate the features of a dynamic de Sitter cosmology. For illustrative purposes, the parameters will be chosen for convenience and clarity in the diagrammatic representation of various transitional epochs in the cosmological evolution in Figure 8.2. The actual cosmological transition scales would be indistinguishable on a diagram with known phenomenological parameters. The primordial energy density was chosen to roll away from an initially stationary state, generating hot radiation. Once this radiation reaches a threshold density, some of its energy density gets transmuted into pressureless dust. The dust (matter) eventually dominates the cosmology until the remnant dark energy density $\rho_\Lambda(ct \to \infty)$ dominates during late times.

For the illustrated cosmological parameters, the redshift parameter $z_\Lambda + 1$ for observations at the time of dark energy dominance, and the expansion rate $\dot{a}$, are displayed in Figure 8.3. The expansion rate for $ct \gg ct_\Lambda$ becomes exponential, consistent with remnant dark energy.

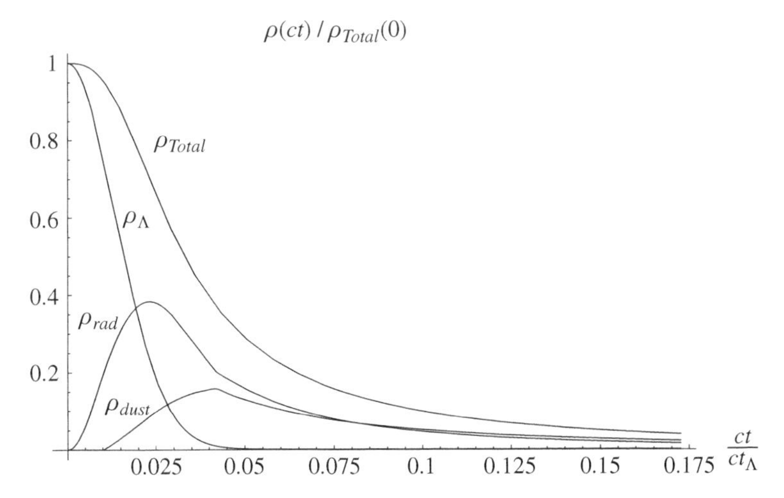

Figure 8.2 Primordial, radiation, and dust energy densities during early modeled transition to thermal cosmology. The temporal scale has been chosen to be relative to the time of remnant dark energy dominance.

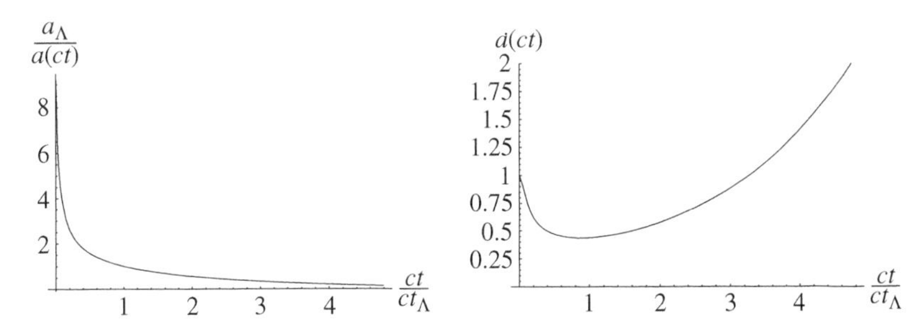

Figure 8.3 Redshift parameter measured during transition to dark energy domination $\frac{a(ct_\Lambda)}{a(ct)} \equiv z_\Lambda + 1$ and Robertson–Walker scale rate $\dot{a}$ for the illustrated dynamic de Sitter cosmology.

8.2.5 Conformal diagram of dynamic de Sitter cosmology

The conformal diagram for this geometry is illustrated in Figure 8.4. The diagram has several features of interest. Foremost is that it has only three boundaries, an initial space-like volume $t = 0$, a final space-like future infinity, and a time-like center $r = 0$. Some of these characteristics are also present in the static de Sitter diagram, Figure 8.1, although the initial surface on that diagram lies in the infinite past. The conformal coordinates of the diagram have been chosen to vanish at the point of intersection of the late-stage de Sitter-like horizon and the particle out-horizon. These horizons delineate the left and right regions in the cosmology that cannot have *any* causal influence upon each other. The dynamic fluid scale R_ρ coincides

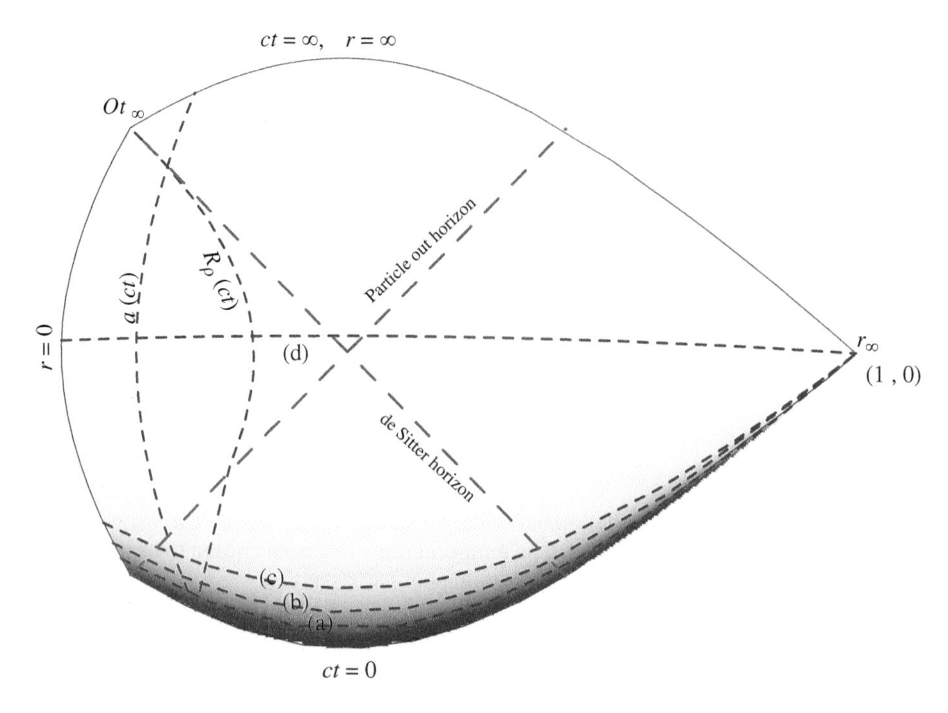

Figure 8.4 Global causal structure of a dynamic de Sitter cosmology. The diagram displays features such as the horizons, the Robertson–Walker scale $a(ct)$, and the fluid scale R_ρ, as overlays upon the total energy density $\rho_{Total}(ct)$.

with the horizon at a unique future infinity point on the diagram, emulating a de Sitter-like future. The Robertson–Walker scale $a(0)$ has been chosen to initially coincide with the primordial fluid scale $R_\rho(0)$, making the geometry non-singular (although this need not be the case). The cosmology begins in a state with primordial fluid density $\rho_\Lambda(ct = 0)$. Radiation first dominates the dynamics of the cosmology on the space-like volume at time labeled (a). The pressureless dust (matter) reaches a maximum density at time (b) and eventually dominates the cosmology after the time labeled (c). Finally, the remnant dark energy density $\rho_\Lambda(ct \to \infty)$ dominates the dynamics of the cosmology after the time labeled (d). As was the case for a de Sitter space-time, there are regions in this cosmology that are completely causally disjoint. One should note that the dynamic de Sitter horizon has a finite radial coordinate at $t = 0$ on the diagram.

The conformal diagrams in Figure 8.5 exhibit fixed-coordinate surfaces in the dynamic geometry. Curves of fixed coordinates (ct, r) for the dynamic de Sitter metric 8.10 are displayed in the diagram on the left. Curves of fixed r (transverse area) are all time-like to the left of the fluid scale R_ρ, which represents a dynamic surface where *ingoing* light-like trajectories are momentarily stationary in the

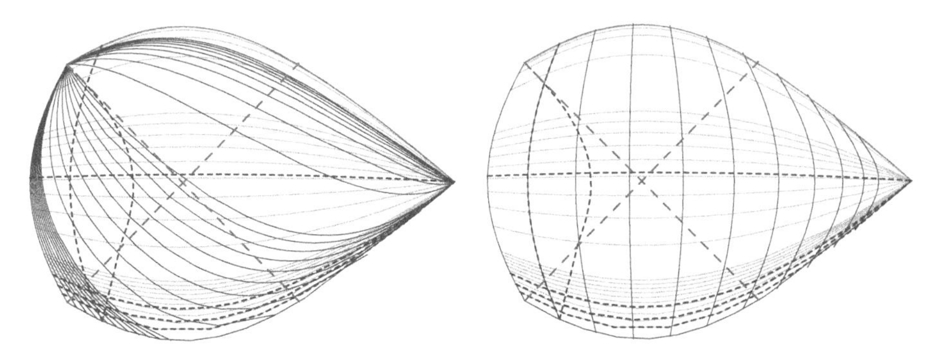

Figure 8.5 Conformal diagrams displaying fixed dynamic de Sitter coordinates (*left*) and fixed Robertson–Walker coordinates (*right*).

radial coordinate (a trapping surface expelling ingoing radiations). The fixed-radial coordinate curves, which everywhere represent surfaces of fixed transverse area, are initially graded in tenths, then units, and decades of the given scale. Fixed area curves all initiate at various points on the volume $t = 0$ and terminate at the future infinity point shared by the fluid scale and the horizon. The upper boundary of the diagram is a fixed infinite area surface $r \to \infty$. Curves of fixed temporal coordinates, initially graded in tenths, units, decades, then centuries of the given scale, all initiate on the time-like surface $r = 0$ and terminate at the extremal point (1,0), which is the *only* extremal conformal coordinate on the diagram. These fixed-coordinate time curves remain space-like volumes throughout the cosmology, parameterizing the proper times of co-moving (geometrically stationary) observers.

The corresponding Robertson–Walker coordinates $(ct, r_{RW} = \frac{r}{a(ct)})$ are displayed in the right diagram in Figure 8.5. Curves of the fixed Robertson–Walker radial coordinate r_{RW} remain time-like surfaces throughout the space-time. Each of these surfaces, graded in units of the given scale, initiate on the $t = 0$ surface and terminate on the static infinite area surface $r = \infty$. Observers with fixed Robertson–Walker radial coordinate represent inertial co-moving centers in this cosmology. One should note that there are no surfaces on the geometry connecting two extremal conformal coordinates $Y_{\rightarrow} = \pm 1$ or $Y_{\uparrow} = \pm 1$ on the diagram, in contrast to previous conformal plots demonstrated.

Transient black hole on a dynamic de Sitter cosmology

For completeness, a local geometry will be embedded within the dynamic cosmology to examine the overall modifications induced. A transient, spatially coherent black hole can be chosen as a local system with strong gravitational fields [95]. The conformal diagram of the combined cosmology is demonstrated in Figure 8.6. In the figure, the black hole singularity is denoted by the space-like bold curve

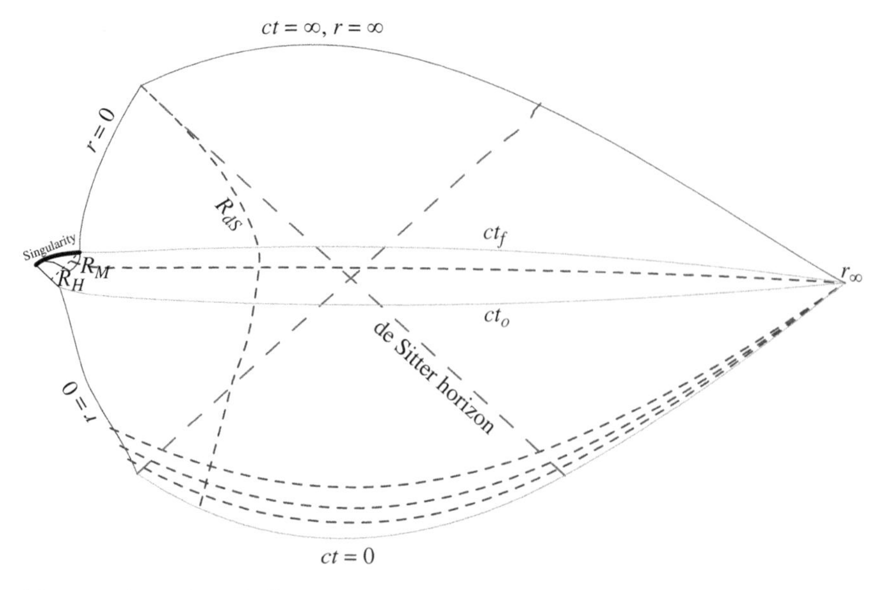

Figure 8.6 Conformal diagram of a temporally transient black hole in a dynamic de Sitter cosmology. The black hole at $r = 0$ persists for times satifying $ct_o \leq ct \leq ct_f$.

at $r = 0$, the radial mass scale of the black hole is labeled R_M, and the black hole horizon is labeled R_H. The black hole has been chosen to persist during the transition of the cosmology from matter to dark energy domination.

The radial coordinates of the horizons of both the dynamic de Sitter cosmology and the transient black hole are very slightly modified in the combined cosmology. The black hole horizon is very slightly smaller than it would be in a Minkowski background, since escape from the vicinity of the black hole horizon is slightly less difficult due to the expanding background. Likewise, distant incoming photons hit the center $r = 0$ sooner than they would had the black hole never been present, *very* slightly expanding the de Sitter horizon. The temporary existence of the black hole *does* make the center $r = 0$ unique in this case. One should note that, since general relativity is a *local* theory, the black hole indeed dominates the geometry in its vicinity during its persistence, regardless of the background cosmology. The cosmology does not shred the black hole asunder, regardless of the period of its persistence.

8.3 Co-gravitating quanta on dynamic cosmology

It is instructive to examine the behavior of a massive scalar field on the geometry. The solution to the geodesic equation 8.39 for the Robertson–Walker metric can be expressed:

$$U^r(ct) = U^r(ct_o)\left(\frac{a(ct_o)}{a(ct)}\right)^2, \qquad U^{ct} = \sqrt{1 + a^2(ct)(U^r(ct))^2}. \tag{8.41}$$

These components can be used to develop the Lagrangian for a substantive scalar field. Expressing the field in the Euler–Lagrange equations 6.14 and 6.15 in terms of $\chi = \frac{\psi(ct)}{r_{RW}}$, the magnitude of this field is of the form:

$$\begin{aligned} |\psi(ct)| &= \frac{a(ct_o)((U^r(ct_o))^2 a^4(ct_o) + a^2(ct_o))^{1/4}}{a(ct)((U^r(ct_o))^2 a^4(ct_o) + a^2(ct))^{1/4}} |\psi(ct_o)| \\ &\overset{U^r \to 0}{\Longrightarrow} \left(\frac{a(ct_o)}{a(ct)}\right)^{3/2} |\psi(ct_o)|, \end{aligned} \tag{8.42}$$

while its phase satisfies:

$$\xi(ct) = \xi(ct_o) - \frac{mc}{\hbar}\int_{ct_o}^{ct} \frac{dct'}{U^{ct}(ct')} \overset{U^r \to 0}{\Longrightarrow} \xi(ct_o) - \frac{mc^2(t - t_o)}{\hbar}. \tag{8.43}$$

In these expressions, the right-most terms describe co-moving fields, $U^r = 0$. The phase is precisely what is expected for an inertial massive quantum. The quantum generates an energy density given by:

$$-T^{ct}{}_{ct} = \rho_m(ct, r_{RW}) = mc^2 \, |\chi(ct, r_{RW})|^2 \overset{\text{co-moving}}{\Longrightarrow} mc^2 \, |\chi(ct_o, r_{RW})|^2 \left(\frac{a(ct_o)}{a(ct)}\right)^3, \tag{8.44}$$

which gives the expected redshift of a massive system in the cosmology, whose energy density falls off with the cube of the scale factor.

To complete this section, the Lagrangian for a core gravitating field will be displayed. A core gravitating field is a collective, real field constructed to vanish in the $G_N \to 0$ limit that provides maximal flexibility in contributing to Einstein's equations, since the Euler–Lagrange equations are automatically satisfied on the geometry. For the Robertson–Walker form of the metric,

$$\mathcal{L}_c = -\sqrt{-g}\,\hbar c\,\psi_c \left[U_c^\beta \partial_\beta \psi_c + \left(\frac{1}{2}\partial_\beta U_c^\beta + \frac{3}{2}\frac{\dot{a}}{a} U_c^{ct} + \frac{U_c^{r_{RW}}}{r_{RW}}\right)\psi_c \right], \tag{8.45}$$

while for the dynamic de Sitter form,

$$\mathcal{L}_c = -\sqrt{-g}\,\hbar c\,\psi_c \left[U_c^\beta \partial_\beta \psi_c + \left(\frac{1}{2}\partial_\beta U_c^\beta + \frac{U_c^r}{r}\right)\psi_c \right]. \tag{8.46}$$

This field can produce a non-linear, collective energy-momentum term generated by the co-gravitating, disentangled contributors to the geometry that is consistent with Einstein's equation, as was done in Section 7.3.

8.4 Cosmological fluctuations

Density fluctuations in the cosmology introduce small local curvatures upon the background that create self-interacting clusters. The clusters are then expected to co-gravitate as thermal enclaves on the large-scale cosmological background. These clusters are presumed to evolve into the observed structure of the universe. The basics of the evolution of fluctuations will be examined in this section.

8.4.1 Acoustic waves

If the universe were exactly uniform, no structure (galaxies, stars, planets, etc.) would have formed, since these structures imply density inhomogeneities. A close examination of the almost ideal CMB radiation reveals temperature fluctuations on the order of $\frac{\delta\rho_{fluc}}{\rho} \approx 5 \times 10^{-5}$ at the time of photon decoupling [96]. Small primordial fluctuations are believed to have propagated as acoustic waves on the hot plasma during radiation domination, which are seen in the snapshot of last scattering as the CMB radiation sourced during early matter domination, as well as the observed distribution of galactic clusters. During radiation domination, the charges in the plasma reached statistical quasi-equilibrium with the photons through proliferous Compton scatterings, essentially behaving as particles with non-vanishing chemical potentials, since Compton scattering (as well as ultra-relativistic pair creations and annihilations) conserve particle number.

Because photon number was essentially conserved, the usual form of the continuity equation describing photon number density expressed in terms of the Robertson–Walker scale is given by:

$$\frac{\partial n_\gamma}{\partial ct} + \sum_j \frac{1}{a(ct)} \frac{\partial}{\partial x^j_{RW}} (n_\gamma \, U_j) = 0. \tag{8.47}$$

Since the Robertson–Walker spatial coordinates are dimensionless, it is quite convenient to rewrite this equation in terms of the dimensionless *conformal time* η, defined by:

$$d\eta \equiv \frac{dct}{a(ct)}. \tag{8.48}$$

The photon continuity equation then becomes:

$$\frac{\partial n_\gamma}{\partial \eta} + \sum_j \frac{\partial}{\partial x^j_{RW}} (n_\gamma \, U^j_\gamma) = 0. \tag{8.49}$$

The Euler equation for fluid dynamics is likewise of the form:

$$(\rho_\gamma + P_\gamma) \frac{\partial U^j_\gamma}{\partial \eta} = -\frac{\partial P_\gamma}{\partial x^j_{RW}}. \tag{8.50}$$

These equations will provide the dynamics of temperature fluctuations on the primordial plasma.

Examining the thermodynamic parameters for a photon gas from Eq. 8.30, the number density is proportional to the cube of the temperature, the energy density is proportional to the fourth power of the temperature, and the pressure satisfies $P_\gamma = \frac{1}{3}\rho_\gamma$. The temperature fluctuations can be expressed in terms of its Fourier components:

$$k_B T(\eta, \mathbf{r}_{RW}) = k_B T(\eta) \left(1 + \int dk \, \Theta_k(\eta) \, e^{i \, k_j \, x^j_{RW}} \right), \tag{8.51}$$

where the dimensionless temperature perturbations $\Theta_k \ll 1$ are assumed to be small. Substitution of this form into Eqs. 8.49 and 8.50, and assuming that the dominant component of the observed wave velocity will be in the direction of the propagation mode vector k, will result in a wave equation for the temperature perturbations:

$$3\Theta''_k(\eta) + k^2 \, \Theta_k(\eta) = 0. \tag{8.52}$$

This indicates that the conformal speed of sound of the acoustic wave is given by $1/\sqrt{3}$. Thus the perturbations have a proper speed of sound given by $c/\sqrt{3}$, which is relativistic.

The observed location of the first acoustic peak in the CMB radiation is quite near that expected due to its speed of propagation. If the overall cosmology had spatial curvature that would result in an open universe, the features would appear to be closer than they should be, and conversely, for a closed universe, the features would appear to be further than they should. This means that to a great deal of accuracy, the observed cosmology is spatially flat.

Growth of primordial density perturbations

Perturbations on the smooth cosmological background grow with time. From the equation for cosmological fluid dynamics in 8.13 and the definition of the fluid scale R_ρ in Eq. 8.12, the dynamics expressed in terms of the geometric scales satisfies:

$$\dot{\rho} = -3 \, \frac{\rho + P}{R_\rho} = -3 \, (\rho + P) \frac{\dot{a}}{a}. \tag{8.53}$$

Small spatial perturbations in the density will locally modify the scale factor. A small density perturbation over a (time-like) co-moving trajectory δct will satisfy $\delta\rho(ct, \mathbf{r}) = \rho(ct, \mathbf{r}) \, \delta ct$. Using 8.53, this results in scale variations satisfying:

$$\frac{\delta a}{a} = \frac{\dot{a} \, \delta ct}{a} = \frac{\dot{a}}{a} \frac{\delta\rho}{\dot{\rho}} = -\frac{1}{3} \frac{\delta\rho}{\rho + P}. \tag{8.54}$$

This useful expression directly relates geometric perturbations to energy density perturbations. A more detailed examination of density perturbations can be found in the literature [96].

For a spatially flat cosmology, this growth can be estimated in a straightforward manner. During the early universe, the Hubble expansion rate from 8.6 for a spatially flat universe satisfies $H^2 = \frac{8\pi G_N}{c^2}\rho$. A small positive fluctuation in energy density will locally close the universe, with a positive curvature κ that must be consistent with the expansion rate: $H^2 = \frac{8\pi G_N}{c^2}\tilde{\rho} - \frac{\kappa c^2}{a^2}$. Likewise, a small negative fluctuation in energy density will give negative local curvature. If these two equations are subtracted, the density perturbations are estimated to grow according to:

$$\frac{\delta\rho}{\rho} = \frac{3c^4}{8\pi G_N}\left(\frac{\kappa}{\rho\, a^2}\right) = \begin{cases} \delta_r \left(\frac{a(ct)}{a(ct_r)}\right)^2 & \text{radiation domination} \\ \delta_m \left(\frac{a(ct)}{a(ct_m)}\right) & \text{matter domination} \end{cases} \tag{8.55}$$

where $\delta_r \equiv \frac{\delta\rho(ct_r)}{\rho(ct_r)}$ is a constant describing the density fluctuations at a time of typical radiation domination, while $\delta_m \equiv \frac{\delta\rho(ct_m)}{\rho(ct)}$ is a constant describing the density fluctuations at a time of typical matter domination. This result is consistent with Eq. 8.54.

8.4.2 Quantum measurement and geometry

One can next examine the constraints that quantum measurement and geometric consistency place upon primordial perturbations. The fluid scale $R_\rho(ct)$ defines the inner boundary of a trapping surface. This can be seen by examining null geodesics generated by the metric 8.10, which satisfy:

$$\dot{r}_\gamma^\pm = \frac{r_\gamma^\pm}{R_\rho} \pm 1. \tag{8.56}$$

This makes the fluid scale the maximum proper (co-moving) length that a measuring rod held rigid by microscopic forces can attain, since even ingoing photons beyond this scale $r_\gamma^- > R_\rho$ *must* have increasing radial scale $\dot{r}_\gamma^- > 0$ for a fixed time in Eq. 8.56. However, as was likewise the case with radially stationary black holes, the surface scale R_ρ is not a horizon, since it is not a light-like surface, only a surface where light is temporarily stationary in the radial coordinate r. It does represent a maximum limit of the size scale of coordinated extended co-moving systems. In this sense, R_ρ represents the IR (infrared) long-distance limit for exchanges of thermal energies. Likewise, it represents the UV (ultraviolet) short-distance limit for coherent gravitational dynamics during thermalization. Quantum phenomena, such as zero-point motions, are typically coherent over space-like regions, so the scales $R_\rho(0) = a(0)$ in early times parameterizes the limits on regions of thermal and coherent correlations.

Other relationships are satisfied by the cosmological scale parameters. Since the fluid scale initially coincides with the Robertson–Walker scale, Eq. 8.12 guarantees that $\dot{a}(ct = 0) = 1$. Also, one might note that Eq. 8.12 directly connects the fluid scale to the energy density $\frac{1}{R_\rho^2} \sim \rho$. This implies that:

$$\frac{\dot{R}_\rho}{R_\rho} = -\frac{1}{2}\frac{\dot{\rho}}{\rho}. \tag{8.57}$$

This relationship is true throughout any cosmological epoch or transition period. The dynamic de Sitter fluid scale is directly related to the energy density sourcing the geometry.

The constraints of quantum measurement and geometric consistency provide meaningful bounds upon the physics of the primordial perturbations. In particular, the zero-point energies of a quantum system of cosmological scale should manifest as non-trivial contributors to subsequent dynamics, as the thermal densities redshift. Some intriguing results from considerations of dark energy as a manifestation of zero-point quantum behaviors will be briefly explored in what follows.

8.4.3 Dark energy partitions

One of the most profound outstanding problems in cosmology today is that of the fundamental nature of dark energy. Since the dark energy seems to be associated (at least approximately) with a cosmological constant, its scale has been somehow frozen in, unlike the redshifting energy scales of the other constituents of the cosmos. The dark energy observed today should be modeled to be frozen in as a result of a phase transition involving the source of that energy. Postulating that the thermal de-coherence of aggregate cosmological energies from the zero-point energies during this phase transition drives statistical variations in the energy density specify a class of cosmological models for which the CMB fluctuation amplitude at last scattering is approximately the observed value of 10^{-5} [97]. Gravitational de-coherence at the UV scale defined by this dark energy generates a fixed microscopic scale of cosmological significance, which explains the observed space-like/supraluminal correlations of large-scale cosmological structures. Loosely speaking, one could conjecture that the zero-point kinetic motions during decoherence should drive the density fluctuations, while the zero-point potential energies remain at the original, frozen-in microscopic scale. Since these energies are inherently a quantum effect, one expects the fluctuations to exhibit the space-like correlations consistent with a quantum phenomenon.

A weakly interacting sea of the quantum fluctuations due to zero-point motions is expected to exhibit local statistical variations in the energy. These variations can be quantified using simple counting arguments for a sufficiently well-defined state. A sea of fluctuations can be locally partitioned. Defining the statistical weight of the zero-point motions of the sources associated with a partition A having energy E_A as $\Omega(E_A)$ allows the probability of such a partitioning to be described by $P(E_A) = \frac{\Omega(E_A)}{\Omega_{tot}}$. If one requires that the most likely configuration of energy partitions results when (the logarithm of) this probability is maximized with respect to the energy partition E_A, the distribution has uniform zero-point energy. This result is analogous to the zeroth law of thermodynamics, which requires uniformity of temperature for systems in thermodynamic equilibrium. Subsequently using arguments analogous to those that establish the second law of thermodynamics, $\log \Omega(E_\Lambda)$ is likewise expected to be a non-decreasing function if previously isolated systems are placed in mutual contact.

The canonical ensemble of thermodynamics results from examining small partitions A in a thermal bath of fluctuations. Examination of the logarithm of the lowest-order fluctuations of the reservoir from uniformity due to the sea of zero-point fluctuations results in the relation:

$$\left\langle (\delta E)^2 \right\rangle = E_\Lambda^2 \frac{d}{dE_\Lambda} < E >, \tag{8.58}$$

for a bath of fluctuations characterized by ("temperature") E_Λ. Typically, an equation of state will connect the extensive variable $< E >$ to the extensive variable $\mathcal{N}$ that counts the number of available degrees of freedom of the system in the form $< E > = \mathcal{N} \epsilon \, \zeta(\frac{E_\Lambda}{\epsilon})$, for some micro-physical energy ϵ and dimensionless function ζ. Substitution of such an equation of state into Eq. 8.58 demonstrates that $\frac{\langle (\delta E)^2 \rangle}{<E>^2} = \frac{E_\Lambda}{<E>} \times (z \frac{d}{dz} \log \zeta(z))|_{z=\frac{E_\Lambda}{\epsilon}} \sim \frac{1}{\mathcal{N}}$. The scaling of these fluctuations with the inverse square root of the number of degrees of freedom is typical of systems in statistical physics. Expressed in terms of the densities, the fluctuations satisfy:

$$\frac{\left\langle (\delta E)^2 \right\rangle}{\langle E \rangle^2} = \frac{\left\langle (\delta \rho)^2 \right\rangle}{\rho^2} \sim \frac{\rho_\Lambda}{\rho}. \tag{8.59}$$

This result is analogous to the manner that background thermal energy $k_B T$ drives the fluctuations of thermal systems. The amplitude of relative fluctuations is expected to be of the order:

$$\Delta_{PT} \equiv \left(\frac{\rho_\Lambda}{\rho} \right)^{1/2}, \tag{8.60}$$

where ρ_{PT} is the cosmological energy density at the time of any phase transition that decouples from the dark energy.

Consider next a phase transition into the radiation epoch. The densities at radiation-matter equality $\rho_M(z_{eq}) = \rho_{rad}(z_{eq})$ can be used to extrapolate back to the phase transition period to determine the redshift parameter z at the time of the transition. The (non-relativistic) baryon-electron plasma (dust) is expected to scale using $\rho_m(z) = \rho_{m_o}(1+z)^3$ in terms of the presently observed dust density ρ_{m_o}, whereas the radiation scales during the early expansion as $\rho_{rad}(z) = \rho_{PT}(\frac{1+z}{1+z_{PT}})^4$. Dividing these equations at radiation-matter equality results in the relation:

$$1 + z_{PT} \simeq \left[\frac{\rho_{PT}}{\rho_{m_o}} (1 + z_{eq}) \right]^{\frac{1}{4}}. \tag{8.61}$$

This expresses the redshift at the time of the phase transition z_{PT} in terms of the energy density at that time ρ_{PT}.

Equation 8.55 can next be used to evolve the perturbations after they have formed. The scale of fluctuations at last scattering should be related to fluctuations during the phase transition by:

$$\Delta_{LS} = \left(\frac{a_{LS}}{a_{eq}} \right) \left(\frac{a_{eq}}{a_{PT}} \right)^2 \Delta_{PT} = \frac{(1 + z_{PT})^2}{(1 + z_{eq})(1 + z_{LS})} \Delta_{PT}. \tag{8.62}$$

Combining Eqs. 8.61, 8.62, and 8.60, the amplitude at last scattering is given by:

$$\Delta_{LS} = \frac{(1 + z_{PT})^2}{(1 + z_{eq})(1 + z_{LS})} \left(\frac{\rho_\Lambda}{\rho_{PT}} \right)^{1/2} \simeq \frac{1}{1 + z_{LS}} \sqrt{\frac{\Omega_{\Lambda_o}}{(1 - \Omega_{\Lambda_o})(1 + z_{eq})}}$$

$$\cong 2.5 \times 10^{-5}, \tag{8.63}$$

where a spatially flat cosmology and a transition during, or into, radiation domination has been assumed. The values taken for the phenomenological parameters are given by $\Omega_{\Lambda o} \cong 0.73$, $z_{eq} \cong 3500$, and $z_{LS} \cong 1100$. Surprisingly, this estimate is independent of the density during the phase transition ρ_{PT} and is of the order observed for the fluctuations in the CMB (e.g., see Reference [91], section 23.2 page 221). Calculations for a phase transition that occurs during matter domination prior to last scattering yield a similar result of $\Delta_{LS} \approx 2 - 4 \times 10^{-5}$. Thus, this approach is particularly robust in explaining the observed scale of the cosmological perturbations.

Zero-point energies and thermalization

The zero-point energies describing the uncertainties of a pseudo-stable equilibrium state can typically be expressed in terms of massless modes using $E_0 = \sum_{modes} \frac{1}{2}\hbar\omega_k$. For sufficiently dense states, this sum can be very accurately expressed in terms of an integral using the density of states, which can be shown to be independent of the shape of the boundary region [98] and is proportional to its volume. Thus, if one defines the fraction $f_\Lambda : 0 \rightarrow 1$ to be that proportion of the zero-point energy that remains after driving the cosmological anisotropies to establish remnant dark energy, the dark energy density should satisfy:

$$\rho_\Lambda = f_\Lambda \frac{g_{zp}}{4\pi^2} \int_0^{k_{UV}} (\hbar k v_p) k^2 dk = f_\Lambda \frac{g_{zp}}{16\pi^2} \hbar v_p k_{UV}^4, \tag{8.64}$$

where v_p is the phase velocity of the modes, and g_{zp} is the degeneracy of the quanta. In this equation, ρ_Λ should essentially be the observed remnant dark energy density.

As previously mentioned, the fluid scale R_ρ (which parameterizes the trapping surface) defines the short-distance gravitational coherence limit $k_{UV} = \frac{2\pi}{2R_\rho}$ (or perhaps $k_{UV} = \frac{\pi}{2R_\rho}$, depending upon periodicity arguments). This means that during the thermal phase transition, the energy density is expected to be of the scale:

$$\rho_{PT} = \frac{3c^4}{8\pi G_N} \left(\frac{1}{R_\rho^{PT}} \right)^2 \sim \frac{3M_P^2 c^4}{2\pi^2 \hbar c} \sqrt{\frac{\rho_\Lambda}{f_\Lambda g_{zp} \hbar v_p}} \ll \rho_{Planck} \equiv \frac{(M_P c^2)^4}{(\hbar c)^3}. \tag{8.65}$$

Such a phase transition occurs at a scale far below the Planck density, at thermal energies of a thousand or so GeVs, and a redshift $z_{PT} \sim 10^{16}$. Thus, the use of R_ρ as the UV length scale for zero-point perturbations from uniformity gives indications for physics of significant cosmological importance at TeV energy scales. It is intriguing that such arguments favor spatial flatness, as well as hint at manifestations of Higgs-scale micro-physics in cosmological macro-physics [99].

8.5 Time in cosmology

As has been discussed in Part I, in classical relativity t is a geometric parameter labeling well-localized surfaces, while quantum systems with well-defined energies undergo a type of temporal oscillation. Experiments using cold atoms have measured temporal oscillations consistent with the maintenance of quantum coherence in a gravitational field. However, it is of interest to consider temporal parameterizations of the cosmological initial or final state. In this section, a few notions will be submitted for the reader's perusal.

8.5.1 Beginning of operational time

A universe in a true quantum ground state would not be temporally localized (i.e., it would be in an "eternal" state of vanishingly small entropy and fixed energy). However, in quantum mechanics, if there is *any* possibility of a transition that conserves all appropriate parameters, a quasi-stationary system can eventually be measured to be in that altered state. If the altered state involves many such measurements and local interrelations, spatial and temporal relationships can then be established. During this "initial" (phase) transition, a macroscopic quantum system has both significant quantum *and* thermal fractions.

Thus, the temporal behavior of a universe in a macroscopic quantum coherent state poses a fundamental question. In a coherent state, the temporal behavior of the state vector that is descriptive of the likelihood of the measurement of specific temporal relationships is oscillatory and non-local. However, the state vector represents only what predictive information an observer *can* deduce, not a defined outcome from a measurement. A quantum system either *will* or *will not* congeal into a particular measured state. A single system does not partially congeal into the multiple states described by the state vector in the appropriate proportions, as would be the case for a density matrix describing an evolving incoherent sum of disentangled coherent systems. That a system is not measured by a single act to have a value given by prediction is not unusual in statistical sciences, when probabilities or continuous distributions describe discrete occurrences. What is different about a coherent quantum state is that there can be interference relationships unattainable in the classical situation (for which, although the state has yet to be measured, its state is just hidden, not undefined). This is true whether a spatial or temporal relationship is being examined. Until an interaction or interrelation has occurred, the temporal relationship of a coherent aspect is not only unknown, but unknowable. Einstein's equations presume classical behavior of the universe as a whole, and classical descriptions typically arise from expectation (averaged) values over quantum ensembles. As a transitioning quantum system becomes thermal, density matrix descriptions *do* become appropriate, as has been utilized in describing the Big Bang.

A primordial quantum state should exhibit quasi-stationary, non-localized behavior, absent classical notions of temporal locality or time flow, as would a universe consisting of a single atom in its ground state. More accurately, a universe in a quasi-stationary state (like a radioactive nucleus) can have considerable temporal extension (non-locality) relative to the natural time scales associated with the inherent interactions in the decayed state. Nucleons shuffle among the many micro-configurations consistent with the macroscopic nucleons. The fewer the possible configurations, the fewer the shufflings prior to decay. If there are few configurations, few temporal relationships can be established. In this manner, the $t = 0$ moment becomes expanded (or perhaps more appropriately, *contracted* as a measure of fewer interrelations). Until time can be measured, it is problematic to ascertain its meaning.

As such, it can be argued that there is no meaningful parameterization of time before it can be operationally described. The geometric time t parameterized in cosmological equations is expected to be a classical correspondence limit of the *operational* time t_{op} described by ticks and rhythms of clocks on a grid. In terms of the operational time, Einstein's equations are expected to be valid when the energy densities are statistically averaged $\mathcal{G}_{\mu\nu}^{op} = -\frac{8\pi G_N}{c^4}\langle \hat{\mathbf{T}}_{\mu\nu}\rangle$. Local coherence is (at least partially) broken whenever there is a "tick" (usually involving at least two distant interrelations) of a quantum clock, which is clearly a *discontinuous* procedure. An operational time parameter can be defined in terms of the large-scale averaging of interrelational operations. In late times, the energy content of local observers is dominated by the local energies. However, in very early times, macroscopic coherence could modify the expected temporal dynamics. In the absence of incoherent processes, temporal progression of the operational time as described is absent.

8.5.2 End of cosmological time

The existence of a cosmological horizon presents particularly intriguing possibilities. Because of the finite entropy associated with the horizon, the universe could undergo fluctuations that would significantly alter established configurations. Such extremely rare occurrences will be described in this section.

Cosmological entropy

A universe with remnant dark energy eventually expands so rapidly that the global geometry manifests a horizon through which any object that crosses can never return. If this truly describes the universe, the Earth will eventually be effectively alone in communications with the other constituents of the local gravitationally bound galactic super-cluster that the Milky Way presently orbits. The existence of a finite-sized horizon surrounding an observer defines a finite entropy. This can be understood by recognizing that a horizon "hides" a region of the universe, requiring the quantum-correlated information contained in the unseen region to be handled in a statistical way. The entropy of a system is a measure of the number of ways that hidden parameters (or parameters that are too "fine" to be discerned by the "coarse graining" of the observation) can result in the observed state. Quite generally, the entropy one associates with a horizon can be shown to be proportional to its area:

$$S = \frac{k_B c^3}{\hbar}\frac{A}{4G_N}. \tag{8.66}$$

This is so because the subtle (quantum) correlations that connect the observable to the hidden regimes must match behavior on the boundary (the horizon), and only on the boundary. Therefore a finite-horizon area implies a finite entropy. The horizon itself is a light-like surface with regards to a particular traversing motion. For a universe with remnant dark energy, incoming light is stationary at the horizon.

In thermodynamics, a conjugate temperature is associated with entropy. This temperature can be calculated from the energy using the first law of thermodynamics. Writing the energy within the horizon in terms of the integral over the space-like volume defined by a fixed co-moving time, $E = \int_0^{R_\Lambda} \rho_\Lambda \, 4\pi r^2 dr = \frac{4}{3}\pi R_\Lambda^3 \rho_\Lambda$. Expressing the dark fluid density in terms of the dark fluid scale using $\rho_\Lambda = \frac{3c^4}{8\pi G_N} \frac{1}{R_\Lambda^2}$ allows the energy to be written as $E = (\frac{\hbar c}{2\pi R_\Lambda}) \times S$. This results in a temperature that an observer at the center associates with the horizon given by $k_B T_\Lambda = \frac{\hbar c}{2\pi R_\Lambda}$, which is consistent with the result obtained by directly calculating the temperature from the de Sitter metric 8.6. To get an idea of the horizon temperature, a value for its radial coordinate can be estimated from the observed dark energy density to be about $R_d \approx 1.5 \times 10^{28}$ cm $\approx$ 16 billion light years. The expected temperature of such a horizon is approximately $T_\Lambda \approx 2 \times 10^{-30}$ K above absolute zero. However, the entropy of the horizon $S_\Lambda = k_B \frac{\pi R_\Lambda^2}{L_P^2} \approx 3 \times 10^{122} \, k_B$ is quite large, easily dominating any contribution from other energy densities in the cosmology [23].

Poincaré recurrences

Although the entropy of the horizon is quite large, it is none the less finite. Since the system has finite entropy, there are a fixed number of states that the fluctuating system shuffles through. A finite entropy furthermore requires that the system must have a discrete spectrum of states that can be counted. The entropy in some sense counts the available configurations ($S = k_B \log \Omega \Rightarrow \Omega = e^{S/k_B}$ in the microcanonical ensemble). If a typical micro-reshuffle occurs on temporal scales of the order $\Delta t_{resuffle}$, then one can estimate that within a time of the order $\Delta t_{recurrence} \approx \Delta t_{resuffle} (e^{S/k_B})^p$ the system will likely undergo a (Poincaré) recurrence into a previous microstate, where p is some unitless power associated with the dimension of the system. Since the entropy of the horizon is quite large, the number of configurations associated with this observable universe as a whole is huge, since this large entropy is then exponentiated to calculate this number. Therefore, if the universe has a remnant dark energy density with the presently observed value, a recurrence is not likely anytime soon. However, such a recurrence would drastically modify any *local* information content established in such a system.

There are further implications with regards to being able to establish well-defined boundary states for quantum systems, as well as the expectation of having a system manifest decreasing correlation with its previous state as it progressively evolves in a thermal environment. As a concrete example, consider a closed system of finite entropy described by a thermal density matrix and a thermal correlator of an observable $\hat{A}(t)$ in the form [25]:

$$F(t) \equiv \langle \hat{A}(0) \, \hat{A}(t) \rangle_{thermal} = \frac{1}{Z} Tr \, e^{-\beta \hat{H}} \, \hat{A}(0) \, e^{it\hat{H}/\hbar} \hat{A}(0) \, e^{-it\hat{H}/\hbar}. \tag{8.67}$$

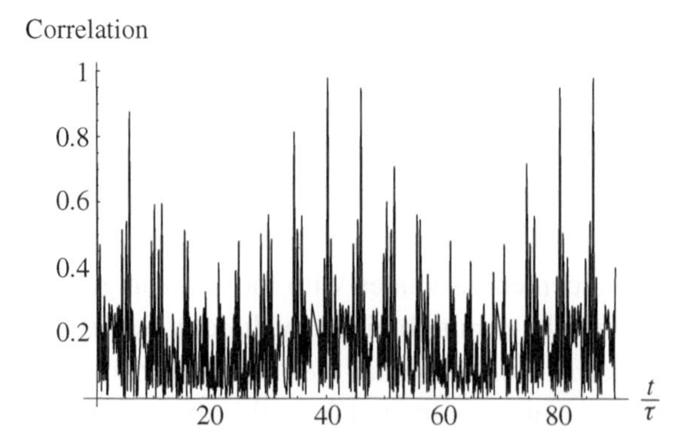

Figure 8.7 Absolute value of a thermal correlator F(t).

Expansion in terms of a complete set of (discrete) energy eigenstates gives:

$$F(t) = \frac{1}{Z} \sum_{jk} e^{-\beta E_j} e^{-it(E_k - E_j)/\hbar} |A_{jk}|^2. \tag{8.68}$$

A plot of the evolution of the absolute value $|F(t)|$ for an example system is shown in Figure 8.7. In the figure, the time is normalized relative to $\tau \equiv \Delta t_{micro}\, e^{S/k_B}$. The operator $\hat{A}$ has been chosen to have no matrix elements connecting states of equal energy. This then implies that the time average of the thermal correlator $F(t)$ vanishes.

Next, consider the long time average of the product $F(t)F^*(t)$, defined by:

$$L \equiv \lim_{T \to \infty} \frac{1}{2T} \int_{-T}^{+T} dt\, F(t)\, F^*(t). \tag{8.69}$$

Using the expression for $F(t)$ in terms of matrix elements, the long-time average can be rewritten as:

$$L = \frac{1}{Z^2} \sum_{ijkm} e^{-\beta(E_i + E_k)} |A_{ij}|^2 |A_{km}|^2\, \delta_{E_j - E_m + E_k - E_i}, \tag{8.70}$$

where the delta parameter is defined to vanish if the argument $(E_j - E_m + E_k - E_i)$ is non-zero and take unit value if the argument is zero. From Eq. 8.70, the long-time average L is obviously non-zero and positive. Thus it is not possible for the correlator $F(t)$ to tend to zero as the time tends to infinity. This is certainly contrary to the usual notions that correlations in systems with random influences should eventually fall to zero over long enough times. Were the entropy infinite, the long-time average *would* vanish.

The value of the long-time average for such finite systems can be estimated, and it is typically of the order of some power of e^{-S/k_B} (for large entropies). As

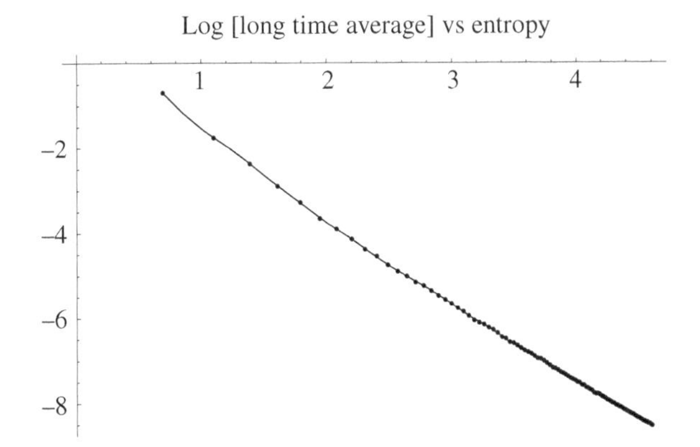

Figure 8.8 Relation between long time average, L, and entropy, S.

a concrete example, the long-time average has been calculated for a numerically tractable system generated by a random matrix Hamiltonian in Figure 8.8. In the figure, the horizontal axis represents the entropy, while the vertical axis represents the logarithm of the long-time average $\log L$. For "large" entropies, the graph displays a fixed slope $\tilde{p}$, defining the power in the long time average $L \sim (e^{-S/k_B})^{\tilde{p}}$. For a finite-entropy system, it is likely that the correlations fluctuate stochastically in the long time limit. From Figure 8.8, the large time behavior of the correlator is random "noise", with a long-time average given by $L \sim e^{-\tilde{p}\,S/k_B}$.

9

Gravitation of interacting systems

Although the composite system of internally interacting constituents might undergo geodesic motion, the individual interacting components themselves generally do not. This final chapter will briefly examine common methods of introducing microscopic interactions into gravitational physics.

9.1 A charged geometry

Electromagnetism is considered by many to be the most well understood of fundamental interactions. The theory manifests local gauge invariance, which is reflected by phase symmetry in its quantum version. Most micro-physical models of more complicated interactions build upon gauge symmetries as fundamental characteristics of viable theories. Therefore, as a means of gaining insights into the interplay of gravitational and microscopic interactions, the simplest of electrostatic charged systems will be examined in this section.

The form of the energy-momentum tensor of an electromagnetic field is well understood from Maxwell's equations. Using the gravitational generalization of Eq. 1.31,

$$T_{\mu\nu}^{EM} = \frac{1}{4\pi} \left(F^{\alpha}{}_{\mu} F^{\beta}{}_{\nu} \, g_{\alpha\beta} - \frac{1}{4} \, F_{\alpha\beta} F^{\alpha\beta} \, g_{\mu\nu} \right), \tag{9.1}$$

where the space-time indices on the field strengths are raised and lowered using the metric $g_{\mu\nu}$. Recall that the field strengths satisfy $F_{\mu\nu} = \partial_{\mu} A_{\nu} - \partial_{\nu} A_{\mu} = A_{\nu\,;\mu} - A_{\mu\,;\nu}$, where the second equality is true because of the anti-symmetry properties of the definition, and the symmetry properties of the affine connections, as long as the potential A_{μ} transforms as a vector, and its covariant derivative can be defined. The electromagnetic energy-momentum tensor 9.1 will later be used in Einstein's equations to develop a charged geometry.

9.1.1 Electromagnetism on curved space-times

The covariant field components for the electric field $\mathcal{E}_k$ and the magnetic field $\mathcal{B}_k$ will be defined by:

$$((F_{\mu\nu})) = ((\partial_\mu A_\nu - \partial_\nu A_\mu)) = \begin{pmatrix} 0 & -\mathcal{E}_1 & -\mathcal{E}_2 & -\mathcal{E}_3 \\ \mathcal{E}_1 & 0 & \mathcal{B}_3 & -\mathcal{B}_2 \\ \mathcal{E}_2 & -\mathcal{B}_3 & 0 & \mathcal{B}_1 \\ \mathcal{E}_3 & \mathcal{B}_2 & -\mathcal{B}_1 & 0 \end{pmatrix}. \tag{9.2}$$

These fields satisfy the usual homogeneous gauge field equations:

$$\sum_{km} \epsilon_{jkm}\, \partial_k\, \mathcal{E}_m + \partial_0\, \mathcal{B}_j = 0, \quad \sum_j \partial_j\, \mathcal{B}_j = 0, \tag{9.3}$$

resulting in Faraday's law of induction (and the Lorentz force equation), and excluding magnetic monopoles, for any integrable field.

The gauge field Lagrangian for the vector potential is of the general form

$$\mathcal{L}_A = -\sqrt{-g}\left(\frac{1}{16\pi} F^{\mu\nu} F_{\mu\nu} - \frac{1}{c} J^\beta A_\beta\right). \tag{9.4}$$

Stationarity of the action under variations of the functional form of the vector potential results in the inhomogeneous gauge field equations,

$$F^{\mu\beta}_{\ ;\beta} = \frac{1}{\sqrt{-g}}\partial_\beta\left(\sqrt{-g}F^{\mu\beta}\right) = \frac{4\pi}{c}J^\mu \quad \Rightarrow \quad \partial_\mu\left(\sqrt{-g}J^\mu\right) = 0, \tag{9.5}$$

which incorporate Gauss'/Coulomb's Law sourcing electric fields and the Ampere/Maxwell relation for magnetic fields. The relations in 9.5 are true due to the anti-symmetry and integrability of the field strengths, requiring $\Gamma^\beta_{\mu\alpha} F^{\mu\alpha} = 0$ and $\partial_\mu\, \partial_\beta(\sqrt{-g}\, F^{\mu\beta}) = 0$. The vector J^β does not guarantee that for any fixed, closed space-like two-surface, any change in the interior charge is due solely to the flux of charge current through that surface, since $J^\beta_{;\beta} = \partial_\beta J^\beta + \Gamma^\beta_{\mu\beta} J^\mu = 0$. The second term in this expression need not vanish. However, conservation of charge is guaranteed when generated by a current $\mathcal{J}^\beta \equiv \sqrt{-g}\, J^\beta$, even with non-vanishing space-time curvatures. For instance, a point charge is represented by a charge density that must satisfy:

$$Q = \int d^4x' \sqrt{-g(\vec{x}')}\, Q\, \frac{\delta^4(\vec{x} - \vec{x}')}{\sqrt{-g(\vec{x}')}} = \int d^3x'\, Q\, \delta^3(\mathbf{x} - \mathbf{x}'). \tag{9.6}$$

Since the current density is expressed in terms of the standard velocity of the charge $\mathbf{v} = \mathbf{u}/u^0$, the current density of such a distribution is given by:

$$J^\beta(\vec{x}) = \int dc\tau\, Q\, u^\beta_Q(c\tau) \frac{\delta^4(\vec{x} - \vec{x}_Q(c\tau))}{\sqrt{-g(\vec{x})}}. \tag{9.7}$$

Thus the current $\mathcal{J}^{\beta} = \sqrt{-g}\, J^{\beta}$ represents the usual electromagnetic current density [100].

If one prefers to work with vector potentials in a generalized Lorentz gauge $A^{\beta}_{;\beta} = 0$, then a direct substitution of Eq. 9.2 into Eq. 9.5 yields a form of the inhomogeneous Maxwell equations in terms of the gauge fields given by:

$$A^{\mu;\beta}_{;\beta} + R^{\mu}{}_{\beta} A^{\beta} = -\frac{4\pi}{c} J^{\mu}. \tag{9.8}$$

The extra curvature terms result from the non-vanishing of $g^{\mu\nu}(A^{\beta}_{;\beta;\nu} - A^{\beta}_{;\nu;\beta})$, since covariant derivatives do not commute.

The deviation of the motion of a mass m and charge q from geodesic motion due to an electromagnetic field is given by:

$$m\left(\frac{du^{\mu}}{d\tau} + \Gamma^{\mu}_{\alpha\beta} u^{\alpha} u^{\beta}\right) = \frac{q}{c} F^{\mu\nu} u_{\nu}, \tag{9.9}$$

where u^{β} are components of the tangent four-velocity of the particle trajectory. This reproduces the usual Lorentz force equation, with the inclusion of gravitational influences on the system, consistent with the principle of equivalence. More generally, the four-divergence of the electromagnetic energy density:

$$(T^{\mu\nu}_{EM})_{;\nu} = -\frac{1}{c} F^{\mu}{}_{\nu} J^{\nu}, \tag{9.10}$$

describes the interplay of electromagnetism with other forces.

For the convenience of the reader, the form of Maxwell's equations expressed in a geometry with a metric whose coordinates are orthogonal will be displayed. A general diagonal metric can be expressed in the form:

$$ds^2 = -(h_0)^2(dx^0)^2 + (h_1)^2(dx^1)^2 + (h_2)^2(dx^2)^2 + (h_3)^2(dx^3)^2. \tag{9.11}$$

For this metric, the electric components of the contravariant field strengths are given by $F^{0j} = \frac{\mathcal{E}_j}{h_0^2 h_j^2}$, while the magnetic components are given by $F^{jk} = \epsilon_{jkm} \frac{\mathcal{B}_m}{h_j^2 h_k^2}$. The inhomogeneous Maxwell equations then take the form:

$$\frac{1}{h_0 h_1 h_2 h_3} \sum_j \frac{\partial}{\partial x^j}\left(\frac{h_{j-} h_{j+}}{h_0 h_j} \mathcal{E}_j\right) = \frac{4\pi}{c} J^0, \tag{9.12}$$

for the contravariant electric field strength components, and:

$$\frac{1}{h_0 h_1 h_2 h_3}\left(\sum_{km} \epsilon_{jkm} \frac{\partial}{\partial x^k}\left(\frac{h_0 h_m}{h_j h_k} \mathcal{B}_m\right) - \frac{\partial}{\partial x^0}\left(\frac{h_{j-} h_{j+}}{h_0 h_j} \mathcal{E}_j\right)\right) = \frac{4\pi}{c} J^j, \tag{9.13}$$

for the contravariant magnetic field strength components. The three spatial scale factors h_k, h_{k-}, h_{k+} can be used to relate the derivatives in these equations to the standard spatial divergence and curl operations if desired [101]. The inhomogeneous Maxwell equations of course remain unchanged, since they define the covariant components of the field strengths in terms of the electric and magnetic fields. For general non-orthogonal coordinates, it is usually less complicated to work directly with the field strengths $F_{\mu\nu}$ rather than the electric and magnetic fields.

9.1.2 Stationary charged geometry

Next, the geometry of a radially stationary charge will be developed. The spherically symmetric, radially stationary geometry will be described by the metric:

$$ds^2 = -(1-\beta^2)(dct)^2 - 2\beta\, dct\, dr + dr^2 + r^2(d\vartheta^2 + \sin^2\vartheta\, d\varphi^2). \tag{9.14}$$

A charge Q will generate spherically symmetric fields with components $A_r = 0 = A_\vartheta = A_\varphi$, and $A_0 = A_0(r)$. The field strengths defined by these fields satisfy $F_{0r} = -\mathcal{E}_r = -\partial_r A_0$ and $F^{0r} = \mathcal{E}_r$. Gauss' law can then be solved for the form of the electric field:

$$F^{0\beta}_{;\beta} = 4\pi\rho_Q \Rightarrow \mathcal{E}_r = \frac{Q}{r^2},\ A_0 = -\frac{Q}{r}, \tag{9.15}$$

where the charge Q has units of esu. This solution is reminiscent of Coulomb's law in flat space-time.

The energy-momentum tensor developed from this solution will have components:

$$T_{00}^{EM} = -\frac{1}{8\pi}(1-\beta^2)\left(\frac{Q}{r^2}\right)^2,\ T_{0r}^{EM} = -\frac{1}{8\pi}\beta\left(\frac{Q}{r^2}\right)^2,\ T_{rr}^{EM} = \frac{1}{8\pi}\left(\frac{Q}{r^2}\right)^2. \tag{9.16}$$

Using this energy-momentum tensor in Einstein's equations for the metric 9.14 results in the condition:

$$\frac{d}{dr}\left[r\,\beta^2(r)\right] = \frac{G_N}{c^4}\frac{Q^2}{r}. \tag{9.17}$$

Far from the charge, the geometry is expected to have the radially stationary form of Schwarzschild geometry, i.e., $r\,\beta^2(r) \overset{r\to\infty}{\Rightarrow} \frac{2G_N M}{c^2}$. Therefore, the function $\beta(r)$ in the metric 9.14 is expected to satisfy:

$$\beta(r) = -\sqrt{\frac{2G_N M}{c^2 r} - \frac{G_N Q^2}{c^4 r^2}}. \tag{9.18}$$

This radially stationary solution becomes problematic if it remains unaltered in a region for which the argument of the square root is negative. One can use related solutions with spherically symmetric charge distributions as long as the term denoting the charge distribution never becomes large enough to make the argument negative.

The static, diagonal form of this metric alternatively gives the Reissner–Nordstrom metric:

$$ds_{RN}^2 = -\left(1 - \frac{2G_N M}{c^2 r} + \frac{G_N Q^2}{c^4 r^2}\right)(dct_{RN})^2 + \frac{dr^2}{1 - \frac{2G_N M}{c^2 r} + \frac{G_N Q^2}{c^4 r^2}} + r^2(d\vartheta^2 + \sin^2\vartheta\, d\varphi^2). \tag{9.19}$$

This is the geometry of a stationary "bare" mass M and charge Q located at its center $r = 0$. It is of interest to examine the analytic and coordinate singularities of the geometry. The Ricci scalar for the geometry is given by $\mathcal{R} = -\frac{1}{r^2}\frac{d}{dr}[2r\,\beta^2 + r^2\,\frac{d}{dr}\beta^2]$. This means that if the vacuum solution holds everywhere, there is a time-like physical singularity at $r = 0$. This is in contrast to the space-like singularity at $r = 0$ in Schwarzschild geometry. It is somewhat perplexing that the addition of a single charge, no matter how small, to a Schwarzschild geometry would so drastically alter the global causal structure of geometry. There is no smooth transitioning of the global causal structures as $Q \to 0$. Of course, one must take care when using the intuitions one gains studying static systems to dynamic circumstances, as has been previously demonstrated.

There are also coordinate singularities in 9.14 and 9.19 when $(1 - \frac{2G_N M}{c^2 R_{(\pm)}} + \frac{G_N Q^2}{c^4 R_{(\pm)}^2}) = 0$. The surfaces $r = R_{(\pm)}$ that satisfy this condition are given by:

$$R_{(\pm)} = \frac{G_N M}{c^2}\left[1 \pm \sqrt{1 - \frac{Q^2}{G_N M^2}}\right]. \tag{9.20}$$

Fixed-radial parameter surfaces change from being time-like in regions for which $|\beta(r)| < 1$ to being space-like in regions for which $|\beta(r)| > 1$. Thus, for this stationary geometry, the surfaces corresponding to $\beta^2 = 1$ should be light-like surfaces. Light-like trajectories on the geometry satisfy:

$$\frac{dr_\gamma^\pm}{dct} = \beta(r_\gamma^\pm) \pm 1, \tag{9.21}$$

where for this geometry, $\beta \leq 0$ everywhere. Thus, for outgoing light-like trajectories $\beta(R_\pm) = -1$, which implies that $\frac{dr_\gamma^+}{dct}\Big|_{r_\gamma^+ = R_{(\pm)}} = 0$. Since the surfaces $R_{(\pm)}$ persist indefinitely, they represent horizons on the charged geometry, defining the system as a black hole.

If $Q^2 > G_N M^2$, the metric has a time-like singularity with no horizon, a so-called *naked singularity*, since there is no real solution to Eq. 9.20. Elementary particles like an electron satisfy this property, but one expects quantum dynamics to modify the geometry for such particles. For an *extremal* black hole defined by $Q^2 = G_N M^2$, the two horizons are coincident at $R_{(\pm)} = \frac{G_N M}{c^2}$. For an extremal geometry, $|\beta(r)| \leq 1$ everywhere, so for all regions in the space-time outgoing photons have non-decreasing radial coordinates, $\frac{dr_\gamma^+}{dct} \geq 0$. Therefore, fiducial observers can, in principle, maintain a fixed radial coordinate away from the horizon, even in the interior.

9.2 Self-gravitating charged canonical proper-time systems

The complexities of self-gravitating interacting systems can be explored using the canonical proper-time formulation with $Q \neq 0$. To begin, consider a quantum system with mass m and charge q propagating in a radially stationary electric field. The equations of motion for a momentarily stationary charge q from Eq. 9.9 is given by:

$$m \left(\frac{du^j}{d\tau} + \Gamma^j_{00}(u^0)^2 \right) = -\frac{q}{c} u_\nu F^{\nu j}, \tag{9.22}$$

where the radial component of the four-velocity vanishes $u^r = 0$, and its covariant components are given by $u_0 = g_{00}\, u^0$ and $u_r = g_{0r}\, u^0$. For the metric form 9.14, the needed affine connection can be calculated using $\Gamma^r_{00} = -\frac{1}{2} g^{r\mu}\, \partial_\mu\, g_{00}$, where $g_{00} = -(1-\beta^2(r)) = -g^{rr}$, $g_{0r} = -\beta(r)$, and $u^0 = \frac{c}{\sqrt{1-\beta^2}}$. The electrostatic force term also requires the factor $u_\nu F^{\nu r} = u_0 \mathcal{E}_r$. Substitution of these functional forms into 9.22 gives the equation of motion for the charge that must be reproduced by the canonical proper energy operator $\hat{K}$:

$$\frac{du^r}{d\tau} = \frac{1}{2} c^2 \partial_r \beta^2 + \frac{q}{m} \sqrt{1-\beta^2}\, \mathcal{E}_r = -\partial_r \Phi(r). \tag{9.23}$$

The proper potential Φ can be obtained from this equation, given the metric function $\beta(r)$.

One can utilize a metric form analogous to $\beta(r)$ from 9.18 for a geometry with a charge distribution (or even more generally, use a metric of the form displayed in Eq. 5.71), producing an electric field $\mathcal{E}_r = \frac{Q(r)}{r^2}$, where $Q(r)$ represents the total charge contained within a spherical region parameterized by radial coordinate r. Defining a normalized distribution $\mathcal{P}(r) \equiv \int_0^r \psi^2(\zeta)\, d\zeta$, with $\mathcal{P}(\infty) = 1$, the mass and charge distributions are expected to satisfy:

$$M(r) = m\, \mathcal{P}(r),\; Q(r) = q\, \mathcal{P}(r). \tag{9.24}$$

From Einstein's equation, $T_0^0 = \frac{c^4}{8\pi G_N r^2} \frac{d}{dr}[r\,\beta^2(r)]$, the gravitational source mass lying within a spherical region of radius r is given by:

$$M_T(r)\,c^2 \equiv -\int_0^r T^0{}_0\, 4\pi \tilde{r}^2 d\tilde{r} = \frac{c^4}{2G_N}\, r\,\beta^2(r). \tag{9.25}$$

The electromagnetic contribution to this mass can be calculated from its energy-momentum tensor:

$$(T_{EM})_0^0 = g^{0\nu}(T_{EM})_{\nu 0} = \frac{1}{8\pi}\left[\frac{Q(r)}{r^2}\right]^2, \tag{9.26}$$

which yields a metric with a β of the form:

$$\beta^2(r) = \frac{2G_N}{c^2 r} M_T(r) = \frac{2G_N}{c^2 r}\left(M(r) + \frac{1}{2c^2}\int_0^r d\tilde{r}\, \frac{Q^2(\tilde{r})}{\tilde{r}^2}\right). \tag{9.27}$$

This gives the previous form 9.18 in the exterior of a regularized charge distribution.

Substitution of the metric factor 9.27 into Eq. 9.23:

$$\Phi(r) = -\frac{1}{2}\beta^2(r)c^2 + \frac{q}{m}\int_0^r d\tilde{r}\, \frac{Q(\tilde{r})}{\tilde{r}^2}\sqrt{1-\beta^2(\tilde{r})} \tag{9.28}$$

yields a canonical proper potential form that is quite complicated, since the factors $Q(r)$ and $\beta(r)$ contain integrals over the distribution $\mathcal{P}(r)$ whose dynamics are determined by the canonical proper equation of motion. However, certain characteristic features of this equation are evident. The factor β^2 is first-order in the Newtonian coupling G_N, while the second term in 9.28 has a contribution that is zeroth-order in this coupling. Thus, the potential is repulsive unless the charge is quite weak $\frac{q^2}{\hbar c} \lesssim \left(\frac{m}{M_P}\right)^2$. This condition is not met if one substitutes the mass of any known charged elementary particle as the parameter m. One should note that the bare mass term M in 9.27 gets modified by the electromagnetic self-energy, as should be expected if mass is ultimately to be self-consistently generated by its interactions.

9.3 Gravitation of interacting spinors

To complete the chapter, a form of the Lagrangian for gravitating linear spinor fields with local gauge symmetries will be displayed. As previously shown in Eq. 6.18, one can define spinor-valued four-velocities $\mathbf{U}^\beta \equiv \mathbf{\Gamma}^{\hat{\mu}} \mathcal{V}^\beta_{\hat{\mu}}$. The gauge covariant form of Eq. 6.19 is then given by:

$$L_m = \frac{1}{2\Gamma}\frac{\hbar c}{i}\left[\bar{\mathbf{\Psi}}^{(\Gamma)}_{(\gamma)} \mathbf{U}^\beta \left(\partial_\beta - \frac{q}{\hbar c} A^r_\beta\, i\mathbf{G}_r\right)\mathbf{\Psi}^{(\Gamma)}_{(\gamma)} - (\text{c. c.})\right] + \frac{(\gamma)}{\Gamma} mc^2\, \bar{\mathbf{\Psi}}^{(\Gamma)}_{(\gamma)}\mathbf{\Psi}^{(\Gamma)}_{(\gamma)}, \tag{9.29}$$

where (c. c.) specifies the complex conjugate of the previous expression, and the matrix $\mathbf{G}_r$ is a hermitian generator of the local gauge group of symmetries for the linear spinor field $\mathbf{\Psi}^{(\Gamma)}_{(\gamma)}$. This form is quite similar to the general form for substantive quantum flows and will have the same cluster decomposition properties.

9.3.1 Lie transformation algebra for spinor fields

Linear spinor fields are useful for describing the micro-physics of gravitating systems for two reasons. The first has been their cluster decomposition properties that allow straightforward combinations of systems to have arbitrary degrees of quantum entanglements at varying times. The second property is that the *group* metric generated by the operators $\hat{\Gamma}^{\tilde{\mu}}$ is the Minkowski metric. Because operators like $\hat{\Gamma}^{\tilde{\mu}}\hat{P}_{\tilde{\mu}}$ are invariant under the Lorentz subgroup of transformations, the components of the momenta likewise transform in a manner consistent with this metric defining *group* invariants. There is no analogous group metric for the Poincaré group, since there are no non-abelian operators in that group to generate this metric. Since the four-momentum operators $\hat{P}_{\tilde{\mu}}$ transform like basis vectors, the invariance of $\hat{P}_{\tilde{\mu}}\,\eta^{\tilde{\mu}\tilde{\nu}}\,\hat{P}_{\tilde{\nu}}$ will have significance defining the *space-time* metric. The relevant group properties of the extended Poincaré group will be explored further in this section.

From Chapter 4, the general extended Poincaré group parameters are parameterized by $\mathcal{P}_X \equiv \{\vec{\omega}, \mathbf{U}, \mathbf{\Theta}, \vec{a}, \alpha\}$, with generators $\{\hat{\Gamma}^{\mu}, \hat{K}_j, \hat{J}_k, \hat{P}_\nu, \hat{\mathcal{M}}_T\}$ representing Dirac boosts, Lorentz boosts, rotations, space-time translations, and affine translations. For brevity, the mutually commuting extended translations will be labeled by a barred parameter $\bar{a} \equiv \{\alpha, \vec{a}\}$, while the extended Lorentz group parameters will be underlined $\underline{a} = \{\mathbf{\Theta}, \mathbf{U}, \vec{\omega}\}$. The overall group structure will be examined for the product transformation of pure translations with pure-extended Lorentz group transformation defined by the convention $\hat{U}(\mathcal{P}_X) \equiv \hat{X}(\bar{a})\hat{W}(\underline{a})$. A reversal of this order results in a representation that is a similarity transformation on some of the elements.

The individual subgroups will have defined group operations within each subgroup, while the overall group will have a group operation based upon the convention:

$$
\begin{aligned}
\hat{X}(\bar{b})\,\hat{X}(\bar{a}) &\equiv \hat{X}(\bar{\phi}_x(\bar{b};\bar{a})),\\
\hat{W}(\underline{b})\,\hat{W}(\underline{a}) &\equiv \hat{W}(\bar{\phi}_w(\underline{b};\underline{a})),\\
\hat{U}(\mathcal{P}_X{}')\,\hat{U}(\mathcal{P}_X) &\equiv \hat{U}(\Phi(\mathcal{P}_X{}';\mathcal{P}_X)).
\end{aligned}
\tag{9.30}
$$

Using group properties, one can generally show that the translational group operation is independent of the initial extended Lorentz group parameter $\underline{a}$,

$\bar{\Phi}_x(\bar{b}, \underline{b}; \bar{a}) = \bar{\phi}_x(\bar{b}; \bar{\Phi}_x(\bar{I}, \underline{b}; \bar{a}))$, where I is the identity element. Likewise, the extended Lorentz group operation is independent of the final translation $\bar{b}$ using this convention, $\underline{\Phi}_w(\underline{b}; \bar{a}, \underline{a}) = \underline{\phi}_w(\underline{\Phi}_w(\underline{b}; \bar{a}, \underline{b}^{-1}); \underline{\phi}_w(\underline{b}; \underline{a}))$. The inverse group element to $\mathcal{P}_X = \{\underline{a}, \bar{a}\}$, which can be calculated using $\hat{W}^{-1}\hat{X}^{-1} = \hat{W}^{-1}\hat{X}^{-1}\hat{W}\hat{W}^{-1}$, is given by $\mathcal{P}_X^{-1} = \{\underline{\Phi}_w(\underline{a}^{-1}; \bar{a}^{-1}, \underline{I}), \bar{\Phi}_x(\bar{I}, \underline{a}^{-1}; \bar{a}^{-1})\}$. Group associativity is expressed in the relationships:

$$\bar{\Phi}_x(\bar{c}, \underline{c}; \bar{\Phi}_x(\bar{b}, \underline{b}; \bar{a})) = \bar{\Phi}_x(\bar{\Phi}_x(\bar{c}, \underline{c}; \bar{b}), \underline{\Phi}_w(\underline{c}; \bar{b}, \underline{b}); \bar{a}), \tag{9.31}$$

$$\underline{\Phi}_w(\underline{c}; \bar{\Phi}_x(\bar{b}, \underline{b}; \bar{a}), \underline{\Phi}_w(\underline{b}; \bar{a}, \underline{a})) = \underline{\Phi}_w(\underline{\Phi}_w(\underline{c}; \bar{b}, \underline{b}); \bar{a}, \underline{a}). \tag{9.32}$$

In particular, the translationally independent group of transformations:

$$\bar{\Phi}_x(\bar{I}, \underline{c}; \bar{\Phi}_x(\bar{I}, \underline{b}; \bar{x})) = \bar{\Phi}_x(\bar{I}, \underline{\Phi}_w(\underline{c}; \bar{I}, \underline{b}); \bar{x}),$$

forms a Lie transformation group satisfying Eq. A.5, with $\bar{x}' \equiv \bar{f}(\bar{x}; \underline{b}) = \bar{\Phi}_x(\bar{I}, \underline{b}; \bar{x})$. One should note that the full group of transformations is generally beyond that of a traditional Lie transformation group.

Given the group operation, the complete set of group parameters (like structure constants, Lie structure matrices, transformation matrices, etc.) can be constructed. A particular set of matrices will be useful for present calculations. The matrices that define how the group transformations mix the generators of the group are calculated through Eq. A.26 to be of the form:

$$\begin{aligned}
\oplus_r^{\;s}(\mathcal{P}) &= \frac{\partial}{\partial \mathcal{P}^{r\prime}} \Phi^s(\mathcal{P}^{-1}; \Phi(\mathcal{P}'; \mathcal{P}))\bigg|_{\mathcal{P}' \to I} \\
&= \frac{\partial \Phi^m(\mathcal{P}'; \mathcal{P}))}{\partial \mathcal{P}^{r\prime}}\bigg|_{\mathcal{P}'=I} \frac{\partial \Phi^s(\mathcal{P}^{-1}; \mathcal{P}'))}{\partial \mathcal{P}^{m\prime}}\bigg|_{\mathcal{P}'=\mathcal{P}},
\end{aligned} \tag{9.33}$$

where it is convenient to define $\mathbf{O}_m{}^s(\mathcal{P}) \equiv \frac{\partial \Phi^s(\mathcal{P}^{-1}; \mathcal{P}'))}{\partial \mathcal{P}^{m\prime}}\Big|_{\mathcal{P}'=\mathcal{P}}$, and the parameter $\Theta_r{}^m(\mathcal{P}) \equiv \frac{\partial \Phi^m(\mathcal{P}'; \mathcal{P}))}{\partial \mathcal{P}^{r\prime}}\Big|_{\mathcal{P}'=I}$ as defined in Appendix A.1.1. With these definitions, the group transformation matrices satisfy $\oplus_r{}^s(\mathcal{P}) = \Theta_r{}^m(\mathcal{P}) \mathbf{O}_m{}^s(\mathcal{P})$.

The transformations $\bar{\phi}_x(\bar{b}; \bar{a}) = \bar{\Phi}_x(\bar{b}, \underline{I}; \bar{a})$ form an abelian subgroup of translations. Since the extended translations all commute with each other and are mixed among each other by the extended Lorentz transformations, a general operation of the form $F(\xi^\Lambda \hat{P}_\Lambda)$ will transform under extended Lorentz transformations according to:

$$\hat{U}(\{\underline{a}, \bar{I}\})\, F(\xi^\Lambda \hat{P}_\Lambda)\, \hat{U}^{-1}(\{\underline{a}, \bar{I}\}) = F(\xi^\Lambda \oplus_\Lambda{}^\Gamma(\underline{a}^{-1}) \hat{P}_\Gamma), \tag{9.34}$$

where the capital Greek indices sum over the five parameters including the space-time coordinates and the affine coordinate conjugate to the transverse mass. The fact that the translations are abelian allows a very useful transformation of the translation group parameters, which of course are associated with space-time coordinates. If one utilizes the factor $\mathbb{O}_\Gamma{}^\Lambda(\bar{x}) \equiv \frac{\partial}{\partial x^\Gamma}\phi_x^\Lambda(\bar{x}^{-1};\bar{x})$, one can note that the associativity condition 9.31 implies that:

$$\mathbb{O}_\Gamma^\Lambda(\bar{x}) \oplus_\Lambda^\Upsilon(\bar{a}) = \frac{\partial\phi_x^\Lambda(\bar{x};\bar{a})}{\partial x^\Gamma}\mathbb{O}_\Lambda^\Upsilon(\bar{\phi}_x(\bar{x};\bar{a}))$$
$$\Rightarrow dx^\Gamma\mathbb{O}_\Gamma^\Lambda(\bar{x}) \oplus_\Lambda^\Upsilon(\bar{a}) = d\phi_x^\Lambda(\bar{x};\bar{a})\mathbb{O}_\Lambda^\Upsilon(\bar{\phi}_x(\bar{x};\bar{a})). \tag{9.35}$$

Therefore, if one defines the special set of coordinates that are defined by:

$$\frac{\partial\xi^{\tilde{\Upsilon}}}{\partial x^\Gamma} \equiv \mathbb{O}_\Gamma^{\tilde{\Upsilon}}(\bar{x}), \tag{9.36}$$

then these coordinates have the property that:

$$\xi^{\tilde{\Upsilon}}(\bar{a}) = \int_{\bar{I}}^{\bar{a}} dx^\Gamma\mathbb{O}_\Gamma^\Upsilon(\bar{x}), \tag{9.37}$$

$$\xi^{\tilde{\Gamma}}(\bar{b}) \oplus_{\tilde{\Gamma}}^{\tilde{\Upsilon}}(\bar{a}) = \int_{\bar{a}}^{\bar{\phi}_x(\bar{b};\bar{a})} d\phi_x^\Gamma\,\mathbb{O}_\Gamma^{\tilde{\Upsilon}}(\bar{\phi}_x). \tag{9.38}$$

This means that these coordinates satisfy $\xi^{\tilde{\Upsilon}}(\bar{\phi}_x(\bar{b};\bar{a})) = \xi^{\tilde{\Gamma}}(\bar{b}) \oplus_{\tilde{\Gamma}}^{\tilde{\Upsilon}}(\bar{a}) + \xi^{\tilde{\Upsilon}}(\bar{a})$. Furthermore, one can demonstrate that for the abelian subgroup, $\oplus_\Lambda^\Upsilon(\bar{a}) = \delta_\Lambda^\Upsilon$. Thus, the special coordinates translate according to:

$$\xi^{\tilde{\Upsilon}}(\bar{\phi}_x(\bar{b};\bar{a})) = \xi^{\tilde{\Upsilon}}(\bar{b}) + \xi^{\tilde{\Upsilon}}(\bar{a}), \tag{9.39}$$

where the coordinate transformation is related to the group operation via:

$$\frac{\partial\xi^{\tilde{\Upsilon}}(\bar{x})}{\partial x^\Gamma} = \mathcal{V}_\Gamma^{\tilde{\Upsilon}}(\bar{x}) = \left.\frac{\partial\Phi_x^{\tilde{\Upsilon}}(\bar{x}^{-1},\underline{I};\bar{x}')}{\partial x'^\Gamma}\right|_{\bar{x}'=\bar{x}}. \tag{9.40}$$

This equation directly relates the tetrads $\mathcal{V}_\beta^{\tilde{\mu}}$ (connecting the curvilinear metric and the Minkowski metric from Eq. 5.58) to the extended Poincaré group operation.

More generally, the special coordinates satisfy:

$$\xi^{\tilde{\Upsilon}}(\bar{\phi}_x(\bar{x}_2,\underline{a}_2;\bar{x}_1)) = \xi^{\tilde{\Upsilon}}(\bar{x}_2) + \xi^{\tilde{\Lambda}}(\bar{x}_1) \oplus_{\tilde{\Lambda}}{}^{\tilde{\Upsilon}}(\underline{a}_2^{-1}), \tag{9.41}$$

or perhaps more suggestive:

$$\bar{\phi}_x(\bar{x}_2,\underline{a}_2;\bar{x}_1) = \bar{x}_3(\overline{\xi(\bar{x}_2)} + \overline{\xi(\bar{x}_1) \oplus (\underline{a}_2^{-1})}). \tag{9.42}$$

These relationships follow from expressions like $\oplus_\Gamma{}^r(\bar{a},\underline{I}) = 0 = \oplus_s{}^\Lambda(\bar{I},\underline{a})$ and $\oplus_s^\Lambda(\bar{a},\underline{a}) = -\xi^{\tilde{\Gamma}}(\bar{a})\,(g_s)_{\tilde{\Gamma}}^\Upsilon \oplus_\Upsilon{}^\Lambda(\underline{a})$ that result directly from the properties of the abelian subgroup. The expression 9.42 demonstrates a direct mapping of tangent space coordinates that satisfy the usual Minkowski space-time characteristics into curvilinear coordinates, consistent with the principle of equivalence. The forms of many of the group structure elements for the extended Poincaré algebra can be found in references 70,102.

Appendix A

Addendum for Chapter 1

A.1 Groups and special relativity

A.1.1 Fundamentals of group theory

Properties of groups

A *group* is a set of elements that have the following properties:

- Contains the identity transformation $\mathbf{1}$;
- If E and E' are elements in the group, then there exists a group operation (generically called "multiplication") that always produces an element of the group, $E'' \equiv E' \cdot E$ (closure);
- For every transformation element E, there exists in the group an inverse element E^{-1}, where $E^{-1} \cdot E = \mathbf{1}$;
- The group operation is associative, $E'' \cdot (E' \cdot E) = (E'' \cdot E') \cdot E$.

A particular type of group satisfies an additional property. For an *abelian* group, the group operation yields the same answer regardless of the order, $E' \cdot E = E \cdot E'$. Often, a subset of the elements within a group satisfies all four group properties. This subset is referred to as a *subgroup*.

Generally, two different groups are *isomorphic* if there is a one-to-one relationships between the elements of the groups with regards to group operations. More generally, if a set of elements in one group are in direct relationship with one element in another, the groups are *homomorphic*.

A particular class of groups is quite useful for describing transformations in quantum physics. Invertible $N \times N$ matrices (i.e., matrices with non-vanishing determinants) form the set of *linear groups*. For a general group $\mathcal{G}$, a *representation* of that group is a linear group $\mathbf{\Gamma}$ for which $\mathcal{G}$ has a homomorphic mapping with $\mathbf{\Gamma}$. If there is no similarity transformation that can bring a representation into block diagonal form, that representation is *irreducible*. When all elements of the representation are simply connected to the identity transformation, that group is referred to as the *covering group* of $\mathcal{G}$.

If a representation is irreducible, there must be some (off-diagonal) mixing between some of the matrices in the representation. *Shur's lemma* formally states this by asserting that any matrix that commutes with *all* of the matrices of an irreducible representation is a multiple of the unit matrix. In general, given an irreducible representation of $\mathcal{G}$ denoted by $\mathbf{D}(\mathcal{G})$,

$$\text{if for all elements } E,\ \mathbf{D}^{-1}(E)\,\mathbf{\Lambda}\,\mathbf{D}(E) = \mathbf{\Lambda} \quad \Rightarrow \quad \mathbf{\Lambda} = \lambda \mathbf{1}. \tag{A.1}$$

This property is useful for developing orthogonality conditions for finite groups. For instance, consider the matrix:

$$\mathbf{\Lambda} \equiv \sum_{E} \mathbf{D}^{-1}(E)\,\mathbf{X}\,\mathbf{D}(E), \tag{A.2}$$

where the sum is over all elements of the group, and $\mathbf{X}$ is an arbitrary matrix with the same dimensions as the representation. Perform a similarity transformation on this form using element $\mathbf{D}(E')$. Since the sum is over all elements of the group, it is likewise over all transformed elements of the group. Therefore, by Shur's lemma:

$$\sum_{E} \mathbf{D}^{-1}(E)\,\mathbf{X}\,\mathbf{D}(E) = \lambda \mathbf{1}, \tag{A.3}$$

and the value of λ depends only on the choice of $\mathbf{X}$. This relationship can be used to establish orthogonality conditions between matrix elements of the representation.

Lie groups

A continuous set of R parameters $\underline{a} \equiv \{a^1, a^2, \ldots, a^r\}$ forms an R-dimensional parameter space. If this space forms a manifold with group properties, it is called a *Lie group*. The parameter group properties satisfy:

- Identity element I;
- Group operation $\underline{c} = \underline{b} \cdot \underline{a}$, or $c^r = \phi^r(\underline{b};\underline{a})$ is a differentiable function of all b^s and a^s;
- $\phi(\underline{a}^{-1};\underline{a}) = \underline{I} = \phi(\underline{a};\underline{a}^{-1})$;
- Associativity, $\phi(\underline{c};\phi(\underline{b};\underline{a})) = \phi(\phi(\underline{c};\underline{b});\underline{a})$.

Lie transformation groups

An R-parameter Lie group of transformations is a group on an N-dimensional space given by:

$$x'^j = f^j(x^1, \ldots, x^N : a^1, \ldots, a^R), \tag{A.4}$$

or symbolically, $\vec{x}' = \vec{f}(\vec{x} : \underline{a})$. The functional form satisfies the following group properties:

- $\vec{f}(\vec{x} : \underline{I}) = \vec{x}$;
- If $\vec{x}' = \vec{f}(\vec{x} : \underline{a})$ and $\vec{x}'' = \vec{f}(\vec{x}' : \underline{b})$, then

$$\vec{x}'' = \vec{f}(\vec{x} : \phi(\underline{b};\underline{a})) = \vec{f}(\vec{f}(\vec{x} : \underline{a}) : \underline{b}); \tag{A.5}$$

- Invertibility implies that the Jacobian is non-vanishing $\frac{\partial(f^1,\ldots,f^N)}{\partial(x^1,\ldots,x^N)} \neq 0$.

Generators of infinitesimal transformations

Consider a variation of coordinates $\vec{x}$ under an infinitesimal group transformation parameterized by $\delta\underline{a}$. Assuming that repeated indices are summed,

$$\begin{aligned} f^j(\vec{x} : \delta\underline{a}) &\cong x^j + \delta a^r \left.\frac{\partial f^j(\vec{x}:\underline{a})}{\partial a^r}\right|_{\underline{a}=\underline{I}}, \text{ or} \\ \delta x^j &= \delta a^r\, u_r^j(\vec{x}), \quad \text{where} \quad u_r^j(\vec{x}) \equiv \left.\frac{\partial f^j(\vec{x}:\underline{a})}{\partial a^r}\right|_{\underline{a}=\underline{I}}. \end{aligned} \tag{A.6}$$

A general scalar function $F(\vec{x})$ then varies under a group transformation as follows:

$$\delta F(\vec{x}) = \delta a^r \, u_r^j(\vec{x}) \frac{\partial}{\partial x^j} F(\vec{x}). \tag{A.7}$$

The generator for infinitesimal group transformations $\mathbf{G}_r(\vec{x})$ (which is the same as the infinitesimal generators for quantum field transformations up to factors of i and $\hbar$) is defined by:

$$\mathbf{G}_r(\vec{x}) \equiv u_r^j(\vec{x}) \frac{\partial}{\partial x^j}. \tag{A.8}$$

The transformed group parameter associated with an infinitesimal transformation satisfies:

$$\phi^r(\underline{\delta a}; \underline{a}) \cong a^r + \delta a^s \left. \frac{\partial \phi^r(\underline{b}; \underline{a})}{\partial b^s} \right|_{\underline{b}=\underline{I}} \equiv a^r + \delta a^s \, \Theta_s^r(\underline{a}). \tag{A.9}$$

The last relationship gives the differential change in the group parameter resulting from an infinitesimal group transformation $da^r = \delta a^s \, \Theta_s^r(\underline{a})$. Next, consider the group property Eq. A.5. If the parameter $\underline{b}$ is infinitesimal, one obtains:

$$u_s^j(\vec{f}(\vec{x} : \underline{a})) = \Theta_s^n(\underline{a}) \frac{\partial}{\partial a^n} f^j(\vec{x} : \underline{a}). \tag{A.10}$$

If one next takes the derivative of this relationship with respect to a^r, and then sets $\underline{a}$ to the identity element, one obtains the *group algebra*:

$$[\mathbf{G}_r(\vec{x}), \mathbf{G}_s(\vec{x})] = c^n_{r\,s} \mathbf{G}_n(\vec{x}), \tag{A.11}$$

where:

$$c^n_{r\,s} \equiv - \left[\frac{\partial}{\partial b^s}, \frac{\partial}{\partial a^r} \right] \phi^n(\underline{b}; \underline{a})|_{\underline{b}=\underline{I}=\underline{a}}, \tag{A.12}$$

are defined to be the *structure constants* of the Lie group.

The *Jacobi identity* is an algebraic identity for any arbitrary operators $\mathcal{O}_s$. For the group generators, the Jacobi identity constrains the structure constants:

$$\begin{gathered} [\mathbf{G}_r, [\mathbf{G}_s, \mathbf{G}_n]] + [\mathbf{G}_n, [\mathbf{G}_r, \mathbf{G}_s]] + [\mathbf{G}_s, [\mathbf{G}_n, \mathbf{G}_r]] = 0, \\ c_{s\,n}{}^m c^k_{r\,m} + c_{r\,s}{}^m c^k_{n\,m} + c_{n\,r}{}^m c^k_{s\,m} = 0. \end{gathered} \tag{A.13}$$

If one defines the matrices $(\mathbf{A}_n)_s^m \equiv c_{s\,n}{}^m$, then:

$$[\mathbf{A}_s, \mathbf{A}_n] = c_{s\,n}{}^m \mathbf{A}_m, \tag{A.14}$$

i.e., the group constants themselves satisfy the group algebra. This is known as the *adjoint representation* of the Lie ring. By defining the *Lie ring metric*:

$$g_{s\,n} \equiv c_{s\,r}{}^m c_{n\,m}{}^r, \tag{A.15}$$

the upper index of the structure constants can be lowered, giving $c_{s\,r\,n} \equiv c_{s\,r}{}^m g_{m\,n}$. This lowered form of the structure constants is antisymmetric in all indices.

A *semi-simple* Lie group has no abelian subgroups, which implies that the matrix of Lie ring metric elements has a non-vanishing determinant and is therefore invertible. For a semi-simple group, an operator that commutes with all generators can be defined by:

$$\mathbf{C} \equiv g^{r\,s} \mathbf{G}_r \mathbf{G}_s . \tag{A.16}$$

That this operator, known as the *Casimir operator* commutes with the generators is due only to the symmetry of the metric inverse under exchange of indices. Therefore, by Shur's lemma, for irreducible representations the Casimir operator must be a multiple of the identity operator in that subspace, and that multiple can be used to label the representation.

Invariant integration

For continuous groups, a left invariant measure $d\tau_A$ is one that satisfies:

$$\int_{\mathcal{M}} d\tau_A F(A) = \int_{B\mathcal{M}} d\tau_{BA} F(B^{-1}A), \tag{A.17}$$

where $\mathcal{M}$ is the whole group manifold. What is needed is $d\tau_A = d\tau_{BA}$, or:

$$d\phi^1(\underline{b};\underline{a}) \wedge \ldots \wedge d\phi^R(\underline{b};\underline{a}) = \left| \frac{\partial(\phi^1, \ldots, \phi^R)}{\partial(a^1, \ldots, a^R)} \right| da^1 \wedge \ldots \wedge da^R . \tag{A.18}$$

Using the associativity condition, one can show that $\frac{\partial \phi^r(\underline{b};\underline{a})}{\partial a^m} \Theta^m_s(\underline{a}) = \Theta^r_s(\underline{\phi}(\underline{b};\underline{a}))$, which allows the definition of an invariant measure:

$$d\tau_{\underline{a}} = \frac{da^1 \wedge \ldots \wedge da^R}{det(\Theta(\underline{a}))} = \frac{d\phi^1 \wedge \ldots \wedge d\phi^R}{det(\Theta(\underline{\phi}))} . \tag{A.19}$$

A.1.2 Group theoretic conventions

To establish the conventions used in this manuscript, a few formal aspects of group theory consistent with those in quantum mechanics will be briefly developed. In terms of the group elements $\mathcal{M}$, a vector representation of that group S satisfies:

$$S(\mathcal{M}_2)\, S(\mathcal{M}_1) = S(\Phi(\mathcal{M}_2; \mathcal{M}_1)) \tag{A.20}$$

where $\Phi(\mathcal{M}_2; \mathcal{M}_1)$ is the group composition rule. The generators of infinitesimal transformations are given by:

$$i\,\mathbf{X}_r \equiv \left. \frac{\partial}{\partial \mathcal{M}^r} S(\mathcal{M}) \right|_{\mathcal{M}=\mathcal{I}} . \tag{A.21}$$

Lie structure matrices can be defined by:

$$\Theta^s_r(\mathcal{M}) \equiv \left. \frac{\partial \Phi^s(\mathcal{M}'; \mathcal{M})}{\partial \mathcal{M}'^r} \right|_{\mathcal{M}'=\mathcal{I}} , \tag{A.22}$$

from which one can obtain the group structure constants:

$$c^m_{sn} \equiv \left. \frac{\partial}{\partial \mathcal{M}^n} \Theta^m_s(\mathcal{M}) \right|_{\mathcal{M}=\mathcal{I}} - \left. \frac{\partial}{\partial \mathcal{M}^s} \Theta^m_n(\mathcal{M}) \right|_{\mathcal{M}=\mathcal{I}} \tag{A.23}$$

for the Lie algebra:

$$[\mathbf{X}_r, \mathbf{X}_s] = -i c^m_{rs} \mathbf{X}_m . \tag{A.24}$$

The generators transform under the representations of the group as given by the relation:

$$S(\mathcal{M}^{-1})\mathbf{X}_r\, S(\mathcal{M}) = \boldsymbol{\Theta}_r^s(\mathcal{M})\,\mathbf{X}_s, \tag{A.25}$$

where the matrices $\oplus$ given by:

$$\boldsymbol{\Theta}_r^s(\mathcal{M}) \equiv \left.\frac{\partial}{\partial \mathcal{M}'^r}\Phi^s(\mathcal{M}^{-1};\, \Phi(\mathcal{M}';\, \mathcal{M}))\right|_{\mathcal{M}'=\mathcal{I}}, \tag{A.26}$$

form a fundamental representation of the group.

A.1.3 Group structure of transformations in special relativity

The element composition rule and group structure (structure constants, generators for infinitesimal transformations, covering group) of the Lorentz group for arbitrary boosts and rotations will be demonstrated. The Lorentz group is the subgroup of the Poincaré group (which includes space-time translations) of rotations $\boldsymbol{\theta}$ and velocity boosts $\mathbf{u}$. First, the elements of the rotation subgroup of the Lorentz group will be examined. For the Lorentz subgroup, the generators $X = \{\mathbf{K}, \mathbf{J}\}$ can be shown using Eq. A.25 to transform according to its *fundamental representation* as follows:

$$\mathbf{S}(\mathbf{u}, \boldsymbol{\theta})\mathbf{X}_s\mathbf{S}^{-1}(\mathbf{u}, \boldsymbol{\theta}) = \boldsymbol{\Theta}_s^r(\mathbf{u}, \boldsymbol{\theta})\mathbf{X}_r. \tag{A.27}$$

The transformation properties of the generators of the Lorentz group will next be developed.

The rotation group

The rotation group, often referred to as the group of orthogonal transformations in three-dimensions O[3] transforms vectors $\vec{v}$ according to $v_j = R_{jk}v_k$, where $\mathbf{R}^T\mathbf{R} = \mathbf{1}$. Pure rotations are a subgroup of the Lorentz group. This group is closely related to the group of transformations of complex 2×2 matrices with unit determinant SU[2] by defining the matrix $\mathbf{v} \equiv v_j\boldsymbol{\sigma}_j$, in terms of the Pauli spin matrices $\boldsymbol{\sigma}_j$:

$$\boldsymbol{\sigma}_x = \begin{pmatrix} 0 & 1 \\ 1 & 0 \end{pmatrix}, \quad \boldsymbol{\sigma}_y = \begin{pmatrix} 0 & -i \\ i & 0 \end{pmatrix}, \quad \boldsymbol{\sigma}_z = \begin{pmatrix} 1 & 0 \\ 0 & -1 \end{pmatrix}. \tag{A.28}$$

The determinant of this matrix is invariant under similarity transformations. Note, $det(\mathbf{v}) = -\underline{v} \cdot \underline{v}$, so that invariance of the SU[2] matrices under similarity transformations corresponds to invariance of vector length under rotations. Each similarity transformation corresponds to a rotation, meaning that SU[2] is homomorphic to O[3]. From the Pauli spin matrix relation $Tr(\boldsymbol{\sigma}_j\boldsymbol{\sigma}_k) = 2\delta_{jk}$, one can extract the component of the vector from its corresponding matrix using $v_j = \frac{1}{2}Tr(\boldsymbol{\sigma}_j\mathbf{v})$. Therefore, the homomorphic mapping must satisfy:

$$R_{jk} = \frac{1}{2}Tr(\boldsymbol{\sigma}_j\mathbf{S}^{-1}\boldsymbol{\sigma}_k\mathbf{S}). \tag{A.29}$$

The homomorphism satisfies the following important points:

- There are two matrices **S** and -**S** in SU[2] that correspond to a given single matrix in O[3].
- Both groups are *compact*, i.e., the volume of the group space is finite. This volume corresponds to the integral over the angles that parameterize the groups.
- The manifold of rotations can be represented by an angle θ and an axis of rotation $\hat{n}$. This manifold can be visualized as the set of all points inside a sphere of radius

π. Since a point on the surface of the sphere corresponds to the same rotation as that diametrically opposite, these points have to be identified together. Therefore, the group manifold of O[3] is not singly connected, since not every closed curve on the manifold can be topologically shrunk to a point.

- There is no identification of distant transformations of the linear group SU[2], which makes this group simply connected.

The simply connected group that is homomorphic to a multiply connected group is known as its *covering group*. The local properties (like group algebra) of the groups are identical, and the groups only differ in global properties (like degree of connectivity). Thus, SU[2] is the covering group of O[3]. Quantized systems undergoing transformations under the group SU[2] also admit states with half-integral spins. Therefore, in this treatment, the rotation group will be parameterized using SU[2].

The rotation subgroup representation is given by:

$$\mathbf{R}(\boldsymbol{\theta}_2)\mathbf{R}(\boldsymbol{\theta}_1) = \mathbf{R}(\boldsymbol{\theta}_{(R)}(\boldsymbol{\theta}_2;\boldsymbol{\theta}_1)) \tag{A.30}$$

in terms of the group composition element $\boldsymbol{\theta}_{(R)}(\boldsymbol{\theta}_2;\boldsymbol{\theta}_1)$. The inverse element is given by $\{\boldsymbol{\theta}\}^{-1} = \{-\boldsymbol{\theta}\}$.

The group structure of the fundamental representation for SU[2] can be determined using the rule:

$$\mathbf{R}(\boldsymbol{\theta}) = e^{i\boldsymbol{\theta}\cdot\boldsymbol{\sigma}/2} = \mathbf{1}\cos\left(\frac{\theta}{2}\right) + i\hat{\theta}\cdot\boldsymbol{\sigma}\sin\left(\frac{\theta}{2}\right). \tag{A.31}$$

The composition rule then satisfies:

$$\begin{aligned}
\cos\left(\frac{\theta_{(R)}(\boldsymbol{\theta}_2;\boldsymbol{\theta}_1)}{2}\right) &= \cos\left(\frac{\theta_2}{2}\right)\cos\left(\frac{\theta_1}{2}\right) - \hat{\theta}_2\cdot\hat{\theta}_1\sin\left(\frac{\theta_2}{2}\right)\sin\left(\frac{\theta_1}{2}\right)\\
\hat{\theta}_{(R)}\sin\left(\frac{\theta_{(R)}(\boldsymbol{\theta}_2;\boldsymbol{\theta}_1)}{2}\right) &= \hat{\theta}_2\sin\left(\frac{\theta_2}{2}\right)\cos\left(\frac{\theta_1}{2}\right) + \hat{\theta}_1\cos\left(\frac{\theta_2}{2}\right)\sin\left(\frac{\theta_1}{2}\right)\\
&\quad + \hat{\theta}_2\times\hat{\theta}_1\sin\left(\frac{\theta_2}{2}\right)\sin\left(\frac{\theta_1}{2}\right).
\end{aligned} \tag{A.32}$$

By direct substitution, Eq. A.31 gives the fundamental representation:

$$\begin{aligned}
\boldsymbol{\Theta}_{J_k}^{J_m}(\mathbf{0},\boldsymbol{\theta}) &= \cos(\theta)\delta_{k,m} + (1-\cos(\theta))\hat{\theta}_k\hat{\theta}_m + \sin(\theta)\hat{\theta}_j\epsilon_{jkm}\\
\boldsymbol{\Theta}_{K_k}^{K_m}(\mathbf{0},\boldsymbol{\theta}) &= \cos(\theta)\delta_{k,m} + (1-\cos(\theta))\hat{\theta}_k\hat{\theta}_m + \sin(\theta)\hat{\theta}_j\epsilon_{jkm}.
\end{aligned} \tag{A.33}$$

The Lie structure matrices from Eq. A.22 can also be calculated using Eq. A.32 by examining infinitesimal $\boldsymbol{\theta}_2$:

$$\Theta_r^{(R)s}(\boldsymbol{\theta}) = \delta_{r,s} + \frac{\theta_k}{2}\epsilon_{krs} + \left(\frac{\theta}{2}\cot\left(\frac{\theta}{2}\right) - 1\right)(\delta_{r,s} - \hat{\theta}_r\hat{\theta}_s). \tag{A.34}$$

Lorentz velocity boosts

General pure Lorentz velocity boosts do not form a subgroup of the Lorentz group, since sequential velocity boosts will generally give a resultant system that is both boosted and rotated relative to the initial frame of reference. General sequential pure Lorentz boosts can be written in terms of a single pure Lorentz boost and a pure rotation:

$$\mathbf{L}(\mathbf{u}_2)\,\mathbf{L}(\mathbf{u}_1) \equiv \mathbf{L}(\mathbf{u}_{(L)}(\mathbf{u}_2\,;\,\mathbf{u}_1))\,\mathbf{R}(\boldsymbol{\theta}_{(L)}(\mathbf{u}_2\,;\,\mathbf{u}_1)). \tag{A.35}$$

The covering group of the whole Lorentz group whose parameter space is simply connected is the linear group of complex 2×2 matrices with unit determinant, SL[2,C]. The composition rule for SL[2,C] can be expressed:

$$\mathbf{L}_\beta = e^{\beta\cdot\sigma/2} = \mathbf{1}\,\cosh\left(\frac{\beta}{2}\right) + \hat{\beta}\cdot\boldsymbol{\sigma}\,\sinh\left(\frac{\beta}{2}\right). \tag{A.36}$$

Defining the four-velocity $\vec{u}$ using $u^0 = \sqrt{1+|\mathbf{u}|^2} \equiv \cosh(\zeta)$ and $\mathbf{u} \equiv \hat{u}\sinh\zeta$, the Lorentz boost takes the form:

$$\mathbf{L}(\mathbf{u}) = \sqrt{\frac{u^0+1}{2}}\,\mathbf{1} + \sqrt{\frac{u^0-1}{2}}\,\hat{u}\cdot\boldsymbol{\sigma}. \tag{A.37}$$

The composition rule for pure boosts can be obtained by direct multiplication of the matrix representation:

$$\begin{aligned}
\tan\left(\frac{\theta_{(L)}(\mathbf{u}_2;\mathbf{u}_1)}{2}\right) &= \frac{|\mathbf{u}_2\times\mathbf{u}_1|}{(u_2^0+1)(u_1^0+1)+\mathbf{u}_2\cdot\mathbf{u}_1}\\
\hat{\theta}_{(L)}(\mathbf{u}_2;\mathbf{u}_1) &= \frac{\mathbf{u}_2\times\mathbf{u}_1}{|\mathbf{u}_2\times\mathbf{u}_1|}\\
u^0_{(L)}(\mathbf{u}_2;\mathbf{u}_1) &= u_2^0 u_1^0 + \mathbf{u}_2\cdot\mathbf{u}_1\\
\left(\hat{u}_{(L)}(\mathbf{u}_2;\mathbf{u}_1)\cos\left(\frac{\theta_{(L)}}{2}\right) - \hat{u}_{(L)}(\mathbf{u}_2;\mathbf{u}_1)\times\hat{\theta}_{(L)}\sin\left(\frac{\theta_{(L)}}{2}\right)\right)&\sqrt{\frac{u^0_{(L)}-1}{2}}\\
&= \hat{u}_2\sqrt{\frac{u_2^0-1}{2}}\sqrt{\frac{u_1^0+1}{2}} + \hat{u}_1\sqrt{\frac{u_2^0+1}{2}}\sqrt{\frac{u_1^0-1}{2}}.
\end{aligned} \tag{A.38}$$

General Lorentz transformations

General Lorentz boosts and rotations form a group of transformations whose representation will be given by the convention $\mathcal{L}(\mathbf{u},\boldsymbol{\theta}) \equiv \mathbf{L}(\mathbf{u})\mathbf{R}(\boldsymbol{\theta})$. The group composition behavior will be represented:

$$\mathcal{L}(\mathbf{u}_2,\boldsymbol{\theta}_2)\,\mathcal{L}(\mathbf{u}_1,\boldsymbol{\theta}_1) = \mathbf{L}\left(\mathbf{u}_{(L)}(\mathbf{u}_2;R(\boldsymbol{\theta}_2)\mathbf{u}_1)\right)\,\mathbf{R}\left(\boldsymbol{\theta}_{(R)}(\boldsymbol{\theta}_{(L)}(\mathbf{u}_2;R(\boldsymbol{\theta}_2)\mathbf{u}_1);\boldsymbol{\theta}_{(R)}(\boldsymbol{\theta}_2;\boldsymbol{\theta}_1))\right). \tag{A.39}$$

The inverse element of this representation of the Lorentz group satisfies:

$$\{\mathbf{u},\boldsymbol{\theta}\}^{-1} = \{-R(-\boldsymbol{\theta})\mathbf{u},\,-\boldsymbol{\theta}\}. \tag{A.40}$$

The fundamental representation matrix elements can be constructed using Eq. A.25 to obtain:

$$\begin{aligned}
\boldsymbol{\Theta}_{J_k}^{J_m}(\mathbf{u},\mathbf{0}) &= u^0\delta_{k,m} + (1-u^0)\hat{u}_k\hat{u}_m \\
\boldsymbol{\Theta}_{J_k}^{K_m}(\mathbf{u},\mathbf{0}) &= u_j\epsilon_{jkm} \\
\boldsymbol{\Theta}_{K_k}^{J_m}(\mathbf{u},\mathbf{0}) &= -u_j\epsilon_{jkm} \\
\boldsymbol{\Theta}_{K_k}^{K_m}(\mathbf{u},\mathbf{0}) &= u^0\delta_{k,m} + (1-u^0)\hat{u}_k\hat{u}_m.
\end{aligned} \tag{A.41}$$

The general matrix is then constructed from a rotation and sequential boost using:

$$\boldsymbol{\Theta}_r^s(\mathbf{u},\boldsymbol{\theta}) = \boldsymbol{\Theta}_r^m(\mathbf{u},\mathbf{0})\,\boldsymbol{\Theta}_m^s(\mathbf{0},\boldsymbol{\theta}) = \boldsymbol{\Theta}_r^{(L)m}(\mathbf{u})\,\boldsymbol{\Theta}_m^{(R)s}(\boldsymbol{\theta}). \tag{A.42}$$

In addition, the Lie structure matrices from Eq. A.22 can be shown to be given by:

$$\begin{aligned}
\Theta_{\theta_k}^{(\mathcal{L})\theta_j}(\mathbf{u},\boldsymbol{\theta}) &= \Theta_k^{(R)j}(\boldsymbol{\theta}) \\
\Theta_{u_k}^{(\mathcal{L})\theta_j}(\mathbf{u},\boldsymbol{\theta}) &= \Theta_{u_k}^{(L)\theta_m}(\mathbf{u})\Theta_m^{(R)j}(\boldsymbol{\theta}) \\
\Theta_{u_k}^{(L)\theta_j}(\mathbf{u}) &= -\tfrac{u_m}{u^0+1}\,\epsilon_{mkj} \\
\Theta_{\theta_k}^{(\mathcal{L})u_j}(\mathbf{u},\boldsymbol{\theta}) &= \Theta_{\theta_k}^{(L)u_j}(\mathbf{u}) = u_m\,\epsilon_{mkj} \\
\Theta_{u_k}^{(\mathcal{L})u_j}(\mathbf{u},\boldsymbol{\theta}) &= \Theta_{u_k}^{(L)u_j}(\mathbf{u}) = u^0\,\delta_{k,j}.
\end{aligned} \tag{A.43}$$

To complete this section, a convention for the Lorentz transformation matrices on four-vectors will be established. Define $\mathcal{R}_\mu^\nu$ and $\mathcal{L}_\mu^{\ \nu}$, which act on (covariant) four-vectors according to $\Lambda_\mu^{\ \nu}\omega_\nu = \omega'_\mu$, where $\mathcal{R}$ will rotate four-vectors, and $\mathcal{L}$ will velocity boost those four-vectors. The form of those transformation matrices will be taken as:

$$\begin{aligned}
\mathcal{R}_k^m(\underline{\theta}) &= \cos(\theta)\delta_{k,m} + (1-\cos(\theta))\hat{\theta}_k\hat{\theta}_m + \sin(\theta)\hat{\theta}_j\epsilon_{jkm} \\
\mathcal{R}_0^0(\underline{\theta}) &= 1 \\
\mathcal{L}_k^m(\underline{u}) &= \delta_{k,m} - (1-u^0)\hat{u}_k\hat{u}_m \\
\mathcal{L}_0^m(\underline{u}) &= -u_m = \mathcal{L}_m^0(\underline{u}) \\
\mathcal{L}_0^0(\underline{u}) &= u^0.
\end{aligned} \tag{A.44}$$

The form of the infinitesimal four-Lorentz generators:

$$\begin{aligned}
i\,(\mathcal{J}_m)_\mu^\nu &\equiv \frac{\partial}{\partial\theta_m}\mathcal{R}_\mu^\nu(\boldsymbol{\theta})\Big|_{\boldsymbol{\theta}=\mathbf{0}} \\
i\,(\mathcal{K}_m)_\mu^\nu &\equiv \frac{\partial}{\partial u_m}\mathcal{L}_\mu^\nu(\mathbf{u})\Big|_{\mathbf{u}=\mathbf{0}},
\end{aligned} \tag{A.45}$$

has non-vanishing elements given by:

$$\begin{aligned}
i\,(\mathcal{J}_m)_j^k &= \epsilon_{mjk} \\
i\,(\mathcal{K}_m)_0^k &= -\delta_{m,k} = i\,(\mathcal{K}_m)_k^0.
\end{aligned} \tag{A.46}$$

Appendix B

Addendum for Chapter 2

B.1 Interactions and non-inertial quantum effects

B.1.1 Gravitating charged mass

The probability distribution discussed in Eq. 2.89 involves gravitating mass only. If the mass has commensurate charge q, the interaction form has an additional term:

$$U_q = q\frac{Q(r)}{r}. \tag{B.1}$$

This results in a modification of the second term on the left-hand side of Eq. 2.89 by simple multiplication by a factor $(1 - \frac{q^2}{G_N m^2}) \equiv (1 - \alpha\frac{M_P^2}{m^2})$. Solutions to this equation can be explored in various ways. One examination involves leaving the mass scale $a = \frac{\hbar^2}{G_N m^3}$ invariant, which likewise fixes the distribution form in $\zeta = r/a$. Modifications in the binding energy of an originally self-generating mass due to charge are parameterized in Figure B.1. In the figure, the mass and charge distribution remain as indicated in Figure 2.6. The change in the binding energy as a function of the dimensionless charge parameter $\frac{q^2}{\hbar c}\frac{M_P^2}{m^2}$. The self-repulsion of the charge generates negative binding, as indicated in the figure.

B.1.2 Quantum accelerating oscillator

The accelerating proper time oscillator of Section 1.3.2 can be examined for quantized oscillations. For a sufficiently stiff oscillator $\omega^2 > \left(\frac{a}{c}\right)^2$, the quantum form of Eq. 1.72 generates an annihilation operator $\hat{B}_\Omega$ for an oscillator with modes of frequency $\Omega = \sqrt{\omega^2 - (a/c)^2}$ that is modified from that of the oscillator in an inertial frame $\hat{b}_\omega$ from Eq. 2.31 with modes of frequency ω:

$$\hat{B}_\Omega \equiv \hat{b}_\omega - \sqrt{\frac{2m\Omega}{\hbar}}\frac{a}{2\Omega^2}\hat{1} \equiv \hat{b}_\omega - \alpha_\Omega\hat{1}. \tag{B.2}$$

One can directly examine relationships between the eigenstates of the "inertial" oscillator $\hat{b}_\omega^\dagger\hat{b}_\omega|n_\omega\rangle = n_\omega|n_\omega\rangle$ and the accelerating form $\hat{B}_\Omega^\dagger\hat{B}_\Omega|N_\Omega >= N_\Omega|N_\Omega >$. In particular, the ground state $|n_\omega = 0\rangle$ is represented by a Poisson distribution of excitations of the form with the additional acceleration:

$$|\langle N_\Omega|n_\omega = 0\rangle|^2 = \frac{\alpha_\Omega^{2N_\Omega}}{N_\Omega!}e^{-\alpha_\Omega^2}. \tag{B.3}$$

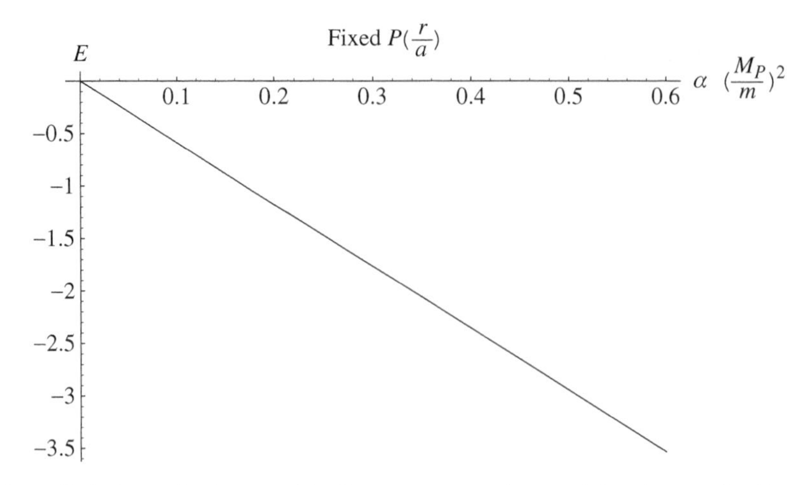

Figure B.1 Canonical proper binding energy vs. charge of fixed mass distribution $a = constant$.

As expected, the ground state of one system is composed of a superposition of excitations of the other.

The canonical proper energy takes the form:

$$\hat{K} = \left(\hat{B}_\Omega^\dagger \hat{B}_\Omega + \frac{1}{2}\right)\hbar\Omega + m\left(c^2 - \frac{a}{2\Omega^2}\right)\hat{1}. \tag{B.4}$$

Since the average number in a Poisson distribution satisfies $\bar{N}_\Omega = \alpha_\Omega^2$, the average canonical proper energy of the state $n_\omega = 0$ still satisfies $\bar{K} = mc^2 + \frac{1}{2}\hbar\Omega$, despite the acceleration.

B.1.3 Thermal properties of accelerating coordinates

An alternative derivation of the temperature associated with accelerations makes use of the normalization of outgoing massless radiations [103]. Consider the metric form from the accelerating coordinate system described by Eq. 1.37, $ds^2 = -(dc\tau_a)^2 + 2\,\frac{c\tau_a}{x_a}\,dc\tau_a\,dx_a + (1 - (\frac{c\tau_a}{x_a})^2)\,(dx_a)^2$. A massless scalar field using these coordinates satisfies the Klein–Gordon equation given by $\partial_\mu\psi\, g^{\mu\nu}\,\partial_\nu\psi = 0$. One can construct outgoing/ingoing solutions at fixed coordinates x_a of the form:

$$\psi^\pm(x_a, \tau_a) = \mathcal{N} exp\left[i\left(k_x x_a \mp \int^{\tau_a} \omega_\pm(\tau')\,d\tau'\right)\right]. \tag{B.5}$$

Substituting this form into the Klein–Gordon equation yields frequencies that must satisfy:

$$\omega_\pm(\tau_a) = \frac{k_x c}{\frac{c\tau_a}{x_a} \pm 1 - i0^+} = \frac{\omega_\infty}{\frac{c\tau_a}{x_a} \pm 1 - i0^+}. \tag{B.6}$$

The causal factor $i0^+$ has been included so that integrals over any singular values are properly retarded. Notice that the wave number takes asymptotic value of the frequency $\omega_\infty \equiv k_x c$.

Consider now the phase of an outgoing wave. Radiations sourced from times satisfying $c\tau_o < -x_a$ will require the causal factor in the denominator of Eq. B.6, giving:

$$\int_{\tau_o}^{\tau_a} \omega_+(\tau)d\tau = \mathcal{P}\int_{\tau_o}^{\tau_a} \omega_+(\tau)d\tau - i\pi\frac{x_a}{c}\omega_\infty = \mathcal{P}\int_{\tau_o}^{\tau_a} \omega_+(\tau)d\tau - i\pi\frac{c}{a}\omega_\infty, \tag{B.7}$$

where the proper acceleration satisfies $a = c^2/x_a$. Performing the integral using the form of the frequency in Eq. B.6, exterior radiations are presumed to take the form:

$$\psi^+_{exterior}(x_a, \tau_a) = \mathcal{N}_{source} exp\left\{i\frac{\omega_\infty}{c}x_a\right\}\left(\frac{\frac{c\tau_a}{x_a}+1}{\frac{c\tau_o}{x_a}+1}\right)^{-i\frac{\omega_\infty}{c}x_a}, \tag{B.8}$$

which is normalized over radiations sourced beyond the horizon. However, this form excludes any possible contribution from non-propagating modes in the interior. General radiations will take the form:

$$\psi^+_{norm}(x_a, \tau_a) = exp\left\{i\frac{\omega_\infty}{c}x_a\right\}\left(\frac{\frac{c\tau_a}{x_a}+1}{\left|\frac{c\tau_o}{x_a}+1\right|}\right)^{-i\frac{\omega_\infty}{c}x_a}$$

$$\left[\mathcal{N}_{source}e^{-\pi\frac{c}{a}\omega_\infty}\,\Theta(-c\tau_o - x_a) + \mathcal{N}_{exterior}\,\Theta(c\tau_o + x_a)\right] \tag{B.9}$$

$$\equiv \psi^+_{int}(x_a, \tau_a)\,\Theta(-c\tau_o - x_a) + \psi^+_{ext}(x_a, \tau_a)\,\Theta(c\tau_o + x_a). \tag{B.10}$$

Matching the normalizations of the radiations $|\psi_{exterior}|^2 = |\psi_{norm}|^2$ gives the relations between the constants $|\mathcal{N}_{source}e^{-\pi\frac{c}{a}\omega_\infty}|^2 + |\mathcal{N}_{exterior}|^2 = |\mathcal{N}_{source}|^2$ Therefore, using Eq. B.10, the ratio of the interior normalization to the exterior propagating normalization for general outgoing radiations at the exterior position $(c\tau_a, x_a)$ is given by:

$$\frac{|\psi^+_{int}(x_a, \tau_a)|^2}{|\psi^+_{ext}(x_a, \tau_a)|^2} = \frac{|\mathcal{N}_{source}|^2 e^{-2\pi\frac{c}{a}\omega_\infty}}{|\mathcal{N}_{source}|^2\left(1 - e^{-2\pi\frac{c}{a}\omega_\infty}\right)} = \frac{1}{e^{\beta\epsilon_\infty} - 1}, \tag{B.11}$$

where the inverse temperature satisfies $\beta \equiv \frac{2\pi c}{\hbar a}$. This form matches that of Section 2.4.

Appendix C

Addendum for Chapter 3

C.1 The counting of bound states

This appendix demonstrates techniques in counting discrete eigenvalues in homogeneous integral equations. Conditions will be established that determine when bound states exist, and when the number of such states proliferates.

A general homogeneous integral equation with real, discrete eigenvalues λ_r can be expressed:

$$\lambda_r \mathcal{T}(M;\lambda_r) = \int dM' \mathcal{K}(M|M')\mathcal{T}(M';\lambda_r), \tag{C.1}$$

where it is assumed that the kernel can be made symmetric $\mathcal{K}(M|M') = \mathcal{K}(M'|M)$ by suitable redefinition of the eigenstate $\mathcal{T}(M;\lambda_r)$. Because of the symmetry of the kernel, the eigenstates are orthogonal. This is demonstrated by multiplying (C.1) by $\mathcal{T}(M;\lambda_s)$ and integrating, giving:

$$(\lambda_r - \lambda_s)\int dM\mathcal{T}(M;\lambda_s)\mathcal{T}(M;\lambda_r) = 0 \overset{\text{normalize}}{\Longrightarrow} \int dM\mathcal{T}(M;\lambda_s)\mathcal{T}(M;\lambda_r) = \delta_{sr}. \tag{C.2}$$

The normalized eigenstates allow the construction of functions whose traces count the states. The functions defined by:

$$\begin{aligned}\mathcal{N}(M'|M) &\equiv \sum_{r=1}^{N} \mathcal{T}(M';\lambda_r)\mathcal{T}(M;\lambda_r),\\ \Lambda(M'|M) &\equiv \sum_{r=1}^{N} \lambda_r\, \mathcal{T}(M';\lambda_r)\mathcal{T}(M;\lambda_r),\end{aligned} \tag{C.3}$$

satisfy the following integral equation:

$$\Lambda(M'|M) = \int dM''\, \mathcal{K}(M'|M'')\mathcal{N}(M''|M). \tag{C.4}$$

The trace of the function $\mathcal{N}(M'|M)$ counts the number of eigenstates, and the Schwartz inequality from (C.4) results in a condition on the eigenvalues:

$$N = \int dM\, \mathcal{N}(M|M), \quad \sum_{r=1}^{N} \lambda_r^2 \leq \int dM' \mathcal{K}^2(M|M'). \tag{C.5}$$

This approach can be utilized to examine the bound states generated by interacting particles.

The scattering equations for discrete bound state energies below the scattering threshold $E_r \equiv (M_{threshold} - M_r)c^2 > 0$ in Eq. 3.7 have the form:

$$\mathcal{T}(M, \alpha; E_r) = \int dM' \frac{k(M|M', \alpha)}{\beta(M', \alpha) + E_r} \mathcal{T}(M', \alpha; E_r), \tag{C.6}$$

where again the kernel can always be made symmetric $k(M|M', \alpha) = k(M'|M, \alpha)$, $\{E_r\}$ are the binding energies, and α measures the strength of the interaction. This equation can be transformed into a convenient form by defining the function $\mathcal{V}(M, \alpha; E_r) \equiv \frac{\mathcal{T}(M,\alpha;E_r)}{\beta(M',\alpha)+E_r}$, giving the equation:

$$E_r \mathcal{V}(M, \alpha; E_r) = \int dM' k(M|M', \alpha) \mathcal{V}(M', \alpha; E_r) - \beta(M, \alpha) \mathcal{V}(M, \alpha; E_r). \tag{C.7}$$

Again, this equation implies that the eigenstates are orthogonal and can be normalized according to the condition $\int dM \, \mathcal{V}(M, \alpha; E_s) \mathcal{V}(M, \alpha; E_r) = \delta_{sr}$.

Motivated by (C.3), a general moment of order s will be defined by:

$$\Lambda_\alpha^{(s)}(M'|M) \equiv \sum_{r=1}^{N_\alpha} (E_r)^s \, \mathcal{V}(M', \alpha; E_r) \mathcal{V}(M, \alpha; E_r), \tag{C.8}$$

where N_α is the number of eigenstates associated with an interaction of strength α. The trace of the moments sum powers of the eigenvalues:

$$\int dM \Lambda_\alpha^{(s)}(M|M) = \sum_{r=1}^{N_\alpha} (E_r)^s, \; N_\alpha = \int dM \Lambda_\alpha^{(0)}(M|M). \tag{C.9}$$

The moments satisfy the following integral equations:

$$\Lambda_\alpha^{(s+1)}(M'|M) = \int dM'' k(M|M'', \alpha) \Lambda_\alpha^{(s)}(M''|M) - \beta(M, \alpha) \Lambda_\alpha^{(s)}(M'|M),$$
$$\Lambda_\alpha^{(n+s)}(M'|M) = \int dM'' \Lambda_\alpha^{(n)}(M'|M'') \Lambda_\alpha^{(s)}(M''|M). \tag{C.10}$$

By completing the square on the form of the integral equation for $\Lambda_\alpha^{(2)}(M'|M)$, the following upper limits for bound state energies can be derived:

$$\sum_{r=1}^{N_\alpha} (E_r) \leq \frac{1}{2} \left(\int dM' \int dM \left[k(M|M', \alpha) \right]^2 + 1 \right),$$
$$\sum_{r=1}^{N_\alpha} (E_r)^2 \leq \int dM' \int dM \left[k(M|M', \alpha) \right]^2. \tag{C.11}$$

These expression limit the range of bound state energies.

Similarly, by completing the square on the form of the integral equation for $\Lambda_{\alpha}^{(1)}(M'|M)$, the following upper limits on the number of bound states can be derived:

$$N_{\alpha} \leq \frac{1}{2}\left(\int dM' \int dM \left[k(M|M',\alpha)\right]^2 + \sum_{r=1}^{N_{\alpha}} \frac{1}{E_r^2}\right),$$

$$N_{\alpha} \leq \int dM' \int dM \frac{\left[k(M|M',\alpha)\right]^2}{\beta(M',\alpha)} - 2\sum_{r=1}^{N_{\alpha}} E_r. \tag{C.12}$$

These conditions place the following constraints upon the number and energies of the states:

- There will be no bound states if the right-hand sides of (C.12) imply that $N_{\alpha} < 1$;
- There are analytic limits on the energies of a finite number of bound states;
- The number of bound states can become unbounded if any of the bound state energies become vanishingly small;
- The large N_{α} behavior is due to the small $\beta(M,\alpha)$ behavior in the weak binding limit within the domain of integration, where α is a measure of the strength of the interaction.

The specific analytic forms of $\beta(M,\alpha)$ and $k(M|M',\alpha)$ can likewise imply *lower* limits upon the number of bound states [46]. A proliferation of the number of three-particle bound states for weak two-particle binding is a prediction of the Efimov effect, discussed in Section 3.2.2.

C.2 Three-particle relativistic kinematics

The specific form of the connected three-particle scattering amplitudes will be developed in this appendix. The off-shell invariant energy will be specified by $\zeta_{(a)} \equiv \frac{Z-e_a}{-\vec{u}\cdot\vec{u}_{(a)}}$, and the off-diagonal invariant energies will be defined by $\omega_{(a)} \equiv \frac{M-e_a}{-\vec{u}\cdot\vec{u}_{(a)}}$. The connected amplitude W_{ab} from Eq. 3.19 satisfies the integral equation:

$$\begin{aligned}
&W_{ab}(M, \mathbf{u}_{(a)}, \hat{\mathbf{q}}_{(a)}|M_o, \mathbf{u}_{(b)o}, \hat{\mathbf{q}}_{(b)o}; Z) \\
&= W_{ab}^D(M, \mathbf{u}_{(a)}, \hat{\mathbf{q}}_{(a)}|M_o, \mathbf{u}_{(b)o}, \hat{\mathbf{q}}_{(b)o}; Z) + \\
&- \sum_d \bar{\delta}_{ad} \int \frac{d^3u'_{(d)}}{u^{0'}_{(d)}} \tau_{(a)}(\omega_{(a)}, \hat{\mathbf{q}}_{(a)}|\omega'_{(a)}, \hat{\mathbf{q}}'_{(a)}; \zeta_{(a)}) F_{ad}(\mathbf{u}_{(a)}, \mathbf{u}'_{(d)}, \mathbf{u}_{(a)} \cdot \mathbf{u}'_{(d)}) \\
&\times \frac{1}{M'-Z} W_{db}(M', \mathbf{u}'_{(d)}, \hat{\mathbf{q}}'_{(d)}|M_o, \mathbf{u}_{(b)o}, \hat{\mathbf{q}}_{(b)o}; Z),
\end{aligned} \tag{C.13}$$

where the function F_{ad} contains the Jacobian factors resulting from changing from invariant integration over individual momenta to appropriate system-wide kinematic parameters. This function can be determined by evaluating the intermediate integration over angles and invariant energies, given by:

$$F_{ad}(\mathbf{u}_{(a)}, \mathbf{u}'_{(d)}, \mathbf{u}_{(a)} \cdot \mathbf{u}'_{(d)}) \equiv \int dM' \, d^2\hat{\mathbf{q}}'_{(d)} \frac{u^0_{(a)} \delta^3(\mathbf{u}_{(a)} - \mathbf{u}'_{(a)}) \rho^{(3)}_{(d)}(M', u^{0'}_{(d)})}{\sqrt{\rho^{(3)}_{(a)}(M', u^{0'}_{(a)}) \rho^{(3)}_{(d)}(M', u^{0'}_{(d)})}}. \tag{C.14}$$

The explicit form of this integration is derived in Reference [48] and is expressed in terms of the function B_{ad} defined by:

$$B_{ad}(\mathbf{u}_{(a)}, \mathbf{u}'_{(d)}, \mathbf{u}_{(a)} \cdot \mathbf{u}'_{(d)}) \equiv 1 + \mathbf{a}_{(ad)} \cdot \mathbf{u}_{(a)} + \mathbf{b}_{(ad)} \cdot \mathbf{u}'_{(d)} + \left(\mathbf{a}_{(ad)} \cdot \mathbf{b}_{(ad)}\right)\left(\mathbf{u}_{(a)} \cdot \mathbf{u}'_{(d)}\right) - \left(\mathbf{a}_{(ad)} \cdot \mathbf{u}'_{(d)}\right)\left(\mathbf{b}_{(ad)} \cdot \mathbf{u}_{(a)}\right), \tag{C.15}$$

where the parameters $\mathbf{a}_{(ad)}$ and $\mathbf{b}_{(ad)}$ are defined in terms of momentum gradients $(\boldsymbol{\nabla}_{(d)})_j \equiv \frac{\partial}{\partial q'_{(d)j}}$ and the energy of spectator a in the (d) pair center of momentum system $\epsilon_a^{(d)}$:

$$\mathbf{a}_{(ad)}(\mathbf{u}_{(a)}, \mathbf{u}'_{(d)}) \equiv -\boldsymbol{\nabla}_{(d)} M'_{(a)}, \quad \mathbf{b}_{(ad)}(\mathbf{u}_{(a)}, \mathbf{u}'_{(d)}) \equiv \frac{\mathbf{u}'_{(d)}}{u^{0'}_{(d)} + 1} - \boldsymbol{\nabla}_{(d)} \epsilon_a^{(d)}. \tag{C.16}$$

The Jacobian factor F_{ad} can then be expressed:

$$F_{ad}(\mathbf{u}_{(a)}, \mathbf{u}'_{(d)}, \mathbf{u}_{(a)} \cdot \mathbf{u}'_{(d)}) = \left[\frac{\epsilon_a^{(a)} \epsilon_d^{(d)}}{|\mathbf{q}'_{(a)}||\mathbf{q}'_{(d)}|}\right]^{\frac{1}{2}} \frac{M'^2_{(a)} M'^2_{(d)}}{\epsilon'_d \epsilon_a^{(d)} \epsilon_{d-}^{(d)} u^0_{(a)}} \frac{1}{B_{ad}(\mathbf{u}_{(a)}, \mathbf{u}'_{(d)}, \mathbf{u}_{(a)} \cdot \mathbf{u}'_{(d)})}, \tag{C.17}$$

where $\epsilon_a^{(a)} \equiv M' u^0_{(a)} - M'_{(a)}$. Clearly, the relativistic kinematics needed for a unitary, cluster-decomposable description is somewhat complicated, as expected.

The driving term of the fully connected amplitude is given by:

$$W^D_{ab}(M, \mathbf{u}_{(a)}, \hat{\mathbf{q}}_{(a)} | M_o, \mathbf{u}_{(b)o}, \hat{\mathbf{q}}_{(b)o}; Z) = \tau_{(a)}(\omega_{(a)}, \hat{\mathbf{q}}_{(a)} | \omega'_{(a)}, \hat{\mathbf{q}}'_{(a)}; \zeta_{(a)}) \frac{F_{ab}(\mathbf{u}_{(a)}, \mathbf{u}_{(b)o}, \mathbf{u}_{(a)} \cdot \mathbf{u}_{(b)o})}{M - Z} \tau_{(b)}(\omega'_{(b)}, \hat{\mathbf{q}}'_{(b)} | \omega_{(b)o}, \hat{\mathbf{q}}_{(b)o}; \zeta_{(b)o}) \tag{C.18}$$

where the primed variables are defined by the kinematic constraints.

Appendix D

Addendum for Chapter 4

D.1 Group theoretic considerations for quantum representations

As was seen for the Galilean group of transformations, the phase invariance of quantum probabilities allow that quantum states transform under unitary ray representations in general. A few of the relevant properties of such representations will be discussed in this appendix.

D.1.1 Unitary ray representations

States of quantum systems $|\psi\rangle$ transform in a unitary manner $|\psi'\rangle = U(G)|\psi\rangle$ such that probability densities are invariant $\langle\phi'|\psi'\rangle = \langle\phi|\psi\rangle$. This requires the unitary operation to satisfy $U^\dagger(G)\,U(G) = \mathbf{1}$. Therefore, under subsequent transformations, unitary transformations generally have an additional phase:

$$U(G')\,U(G) = e^{i\,\xi(G';G)}U(G'\cdot G). \tag{D.1}$$

The factor ξ is a real phase called the *local exponent*. A general quantum mechanical representation is a representation up to this phase, called a *ray representation*. If $\xi(G';G) = 0$ for all transformations, then the representation is called a *vector representation*.

Properties of ray representations

The local exponent of a ray representation has the following properties:

- Since $U(I) = \mathbf{1}$, then $\xi(I;G) = 0 = \xi(G;I)$;
- $U(G^{-1})U(G) = e^{i\,\xi(G^{-1};G)}\mathbf{1}$, which implies that $U(G^{-1}) = e^{i\,\xi(G^{-1};G)}U^\dagger(G)$, with $\xi(G^{-1};G) = \xi(G;G^{-1})$;
- Group associativity $G_3\cdot(G_2\cdot G_1) = (G_3\cdot G_2)\cdot G_1$ requires that:

$$\xi(G_3;G_2\cdot G_1) + \xi(G_2;G_1) = \xi(G_3;G_2) + \xi(G_3\cdot G_2;G_1);$$

- If each operator $U(G)$ is multiplied by an arbitrary phase, one obtains a new representation $U'(G) \equiv e^{i\,\zeta(G)}U(G)$. The local exponents of this new representation satisfy:

$$\xi'(G_2;G_1) = \xi(G_2;G_1) + \zeta(G_2) + \zeta(G_1) - \zeta(G_2\cdot G_1).$$

  Therefore, if phases $\zeta(G)$ can be found such that $\xi'(G_2;G_1) = 0$, then the representation $U(G)$ is said to be *equivalent* to a vector representation.

Consider a finite N-dimensional representation of $U(G)$ with matrix elements $S_{mn}(G) \equiv \langle m|U(G)|n\rangle$. Such representations can be shown to be always equivalent to a vector representation. The $N \times N$ matrices $\mathbf{S}$ satisfy $\mathbf{S}(G_2)\mathbf{S}(G_1) = e^{i\,\xi(G_2;G_1)}\mathbf{S}(G_2 \cdot G_1)$. The determinant of this expression satisfies $\det \mathbf{S}(G_2) \det \mathbf{S}(G_1) = e^{i\,N\xi(G_2;G_1)} \det \mathbf{S}(G_2 \cdot G_1)$, which allows an equivalence transformation of the form $\mathbf{S}'(G) \equiv \frac{\mathbf{S}(G)}{[\det \mathbf{S}(G)]^{1/N}}$. This new representation forms a vector representation of the group $\mathbf{S}'$ that satisfies $\mathbf{S}'(G_2)\mathbf{S}'(G_1) = \mathbf{S}'(G_2 \cdot G_1)$. Therefore, finite-dimensional ray representations are always equivalent to vector representations. One can similarly show that ray representations of one-parameter groups and abelian groups are always equivalent to vector representations. However, as shown in Section 4.1, one cannot find a representation for the inhomogeneous Galilean group that is equivalent to a vector representation.

A group is referred to as being *compact* if the volume of the measure over the whole group manifold is finite. For compact groups, the left- and right-invariant measures are identical. One can show that for compact groups:

- Every representation is equivalent to a unitary representation;
- The representations obey orthogonality conditions;
- Every representation is fully reducible to a sum of irreducible representations, all of finite dimensions.

D.2 Group properties of extended Lorentz group

For convenience, define ladder operators $\Delta_k^{(\pm)}$ as follows:

$$\Delta_k^{(\pm)} \equiv \Gamma^k \,(\pm)\, i\, K_k. \tag{D.2}$$

Then the complete algebra can be expressed in terms of raising and lowering operators:

$$\left[\Gamma^0, J_k\right] = 0 \tag{D.3}$$

$$\left[\Gamma^0, \Delta_k^{(\pm)}\right] = (\pm)\,\Delta_k^{(\pm)} \tag{D.4}$$

$$[J_z, J_\pm] = \pm J_\pm \tag{D.5}$$

$$[J_+, J_-] = 2J_z \tag{D.6}$$

$$\left[J_z, \Delta_z^{(\pm)}\right] = 0 \tag{D.7}$$

$$\left[J_z, \Delta_\pm^{(\pm)}\right] = \pm\Delta_\pm^{(\pm)} \tag{D.8}$$

$$\left[J_\pm, \Delta_\pm^{(\pm)}\right] = 0 \tag{D.9}$$

$$\left[J_\pm, \Delta_\mp^{(\pm)}\right] = \pm 2\Delta_z^{(\pm)} \tag{D.10}$$

$$\left[J_\pm, \Delta_z^{(\pm)}\right] = \mp\Delta_\pm^{(\pm)} \tag{D.11}$$

$$\left[\Delta_z^{(+)}, \Delta_z^{(-)}\right] = -2\Gamma^0 \tag{D.12}$$

$$\left[\Delta_z^{(\pm)}, \Delta_\pm^{(\pm)}\right] = 0 \tag{D.13}$$

$$\left[\Delta_z^{(+)}, \Delta_\pm^{(-)}\right] = \mp 2J_\pm \tag{D.14}$$

$$\left[\Delta_z^{(-)}, \Delta_\pm^{(+)}\right] = \mp 2 J_\pm \tag{D.15}$$

$$\left[\Delta_+^{(\pm)}, \Delta_-^{(\pm)}\right] = 0 \tag{D.16}$$

$$\left[\Delta_+^{(+)}, \Delta_+^{(-)}\right] = 0 \tag{D.17}$$

$$\left[\Delta_+^{(\pm)}, \Delta_-^{(\mp)}\right] = -4\left[J_z(\pm)\Gamma^0\right], \tag{D.18}$$

which will be convenient forms from which to construct the spinor representation.

Spinor Equations

Eigenstates of the Casimir operator Eq. 4.48 along with the commuting operators J_z and Γ^0 will next be constructed. To develop a basis of states, it is convenient to construct an operator that raises and lowers eigenvalues of the operator Γ^0 analogous to the angular momentum raising and lowering operators. This operator is given by:

$$\Delta_J^{(\pm)} \equiv \underline{J} \cdot \underline{\Delta}^{(\pm)}. \tag{D.19}$$

The relevant commutation relations needed to construct the basis are given by:

$$[J_z, J_\pm] = \pm J_\pm \tag{D.20}$$

$$\left[\Gamma^0, \Delta_J^{(\pm)}\right] = (\pm)\Delta_J^{(\pm)} \tag{D.21}$$

$$\left[J_z, \Delta_J^{(\pm)}\right] = 0 \tag{D.22}$$

$$\left[\Gamma^0, J_\pm\right] = 0 \tag{D.23}$$

$$\left[J_z, \Gamma^0\right] = 0 \tag{D.24}$$

$$\left[J^2, \underline{J}\right] = 0 \tag{D.25}$$

$$\left[J^2, \Gamma^0\right] = 0 \tag{D.26}$$

$$\left[J^2, \Delta_J^{(\pm)}\right] = 0. \tag{D.27}$$

D.2.1 Construction of $\Gamma = 1$ spinor states

The forms of the matrices corresponding to $\Gamma = 1$ are expected to have dimensionality $N_1 = 10$. The normalization of the spinors will be chosen to satisfy:

$$\overline{\left(\chi_a^{(A)}\chi_b^{(B)}\right)}\left(\chi_{a'}^{(A')}\chi_{b'}^{(B')}\right) = \frac{1}{2}\left(\delta^{AA'}\delta_{aa'}\delta^{BB'}\delta_{bb'} + \delta^{AB'}\delta_{ab'}\delta^{BA'}\delta_{ba'}\right) \tag{D.28}$$

which results in a state normalization of the form:

$$\overline{\psi_{\gamma,M}^{\Gamma,J}}\,\psi_{\gamma',M'}^{\Gamma,J'} = (-)^{1-\gamma}\,\delta^{JJ'}\delta_{\gamma\gamma'}\delta_{MM'}. \tag{D.29}$$

The spinor states satisfying this normalization are given by:

$$\begin{aligned}
\psi_{0,0}^{1,0} &= \chi_+^{(+)}\chi_-^{(-)} - \chi_-^{(+)}\chi_+^{(-)} \\
\psi_{1,1}^{1,1} &= \chi_+^{(+)2} \\
\psi_{1,0}^{1,1} &= \sqrt{2}\,\chi_+^{(+)}\chi_-^{(+)} \\
\psi_{1,-1}^{1,1} &= \chi_-^{(+)2} \\
\psi_{0,1}^{1,1} &= -\sqrt{2}\,\chi_+^{(+)}\chi_+^{(-)} \\
\psi_{0,0}^{1,1} &= -\chi_+^{(+)}\chi_-^{(-)} - \chi_-^{(+)}\chi_+^{(-)} \\
\psi_{0,-1}^{1,1} &= -\sqrt{2}\,\chi_-^{(+)}\chi_-^{(-)} \\
\psi_{-1,1}^{1,1} &= \chi_+^{(-)2} \\
\psi_{-1,0}^{1,1} &= \sqrt{2}\,\chi_+^{(-)}\chi_-^{(-)} \\
\psi_{-1,-1}^{1,1} &= \chi_-^{(-)2}.
\end{aligned} \tag{D.30}$$

Matrix representation of $\Gamma = 1$ systems

Matrix elements of the operators given by Eqs. 4.56–4.60 can be constructed using the states given in Eq. D.30 to obtain the matrix representation for $\Gamma = 1$ systems:

$$\begin{aligned}
&\mathbf{\Gamma^0} = \begin{pmatrix} 0 & \underline{0} & \underline{0} & \underline{0} \\ \underline{0} & \mathbf{1} & \mathbf{0} & \mathbf{0} \\ \underline{0} & \mathbf{0} & \mathbf{0} & \mathbf{0} \\ \underline{0} & \mathbf{0} & \mathbf{0} & -\mathbf{1} \end{pmatrix} \qquad
\mathbf{\Gamma}^k = \frac{\sqrt{2}}{2}\begin{pmatrix} 0 & \underline{w}_k^T & \underline{0} & \underline{w}_k^T \\ -\underline{w}_k & \mathbf{0} & \mathcal{J}_k & \mathbf{0} \\ \underline{0} & -\mathcal{J}_k & \mathbf{0} & \mathcal{J}_k \\ -\underline{w}_k & \mathbf{0} & -\mathcal{J}_k & \mathbf{0} \end{pmatrix} \\
&\mathbf{J_k} = \begin{pmatrix} 0 & \underline{0} & \underline{0} & \underline{0} \\ \underline{0} & \mathcal{J}_k & \mathbf{0} & \mathbf{0} \\ \underline{0} & \mathbf{0} & \mathcal{J}_k & \mathbf{0} \\ \underline{0} & \mathbf{0} & \mathbf{0} & \mathcal{J}_k \end{pmatrix} \qquad
\mathbf{K_k} = -\frac{\sqrt{2}}{2}i\begin{pmatrix} 0 & -\underline{w}_k^T & \underline{0} & \underline{w}_k^T \\ -\underline{w}_k & \mathbf{0} & \mathcal{J}_k & \mathbf{0} \\ \underline{0} & \mathcal{J}_k & \mathbf{0} & \mathcal{J}_k \\ \underline{w}_k & \mathbf{0} & \mathcal{J}_k & \mathbf{0} \end{pmatrix} \\
&\mathbf{g} = \begin{pmatrix} -1 & \underline{0} & \underline{0} & \underline{0} \\ \underline{0} & \mathbf{1} & \mathbf{0} & \mathbf{0} \\ \underline{0} & \mathbf{0} & -\mathbf{1} & \mathbf{0} \\ \underline{0} & \mathbf{0} & \mathbf{0} & \mathbf{1} \end{pmatrix}
\end{aligned} \tag{D.31}$$

where:

$$\begin{aligned}
&\underline{w}_x \equiv \frac{\sqrt{2}}{2}\begin{pmatrix} -1 \\ 0 \\ 1 \end{pmatrix} \qquad \underline{w}_y \equiv -i\frac{\sqrt{2}}{2}\begin{pmatrix} 1 \\ 0 \\ 1 \end{pmatrix} \qquad \underline{w}_z \equiv \begin{pmatrix} 0 \\ 1 \\ 0 \end{pmatrix} \\
&\mathcal{J}_x \equiv \frac{\sqrt{2}}{2}\begin{pmatrix} 0 & 1 & 0 \\ 1 & 0 & 1 \\ 0 & 1 & 0 \end{pmatrix} \quad \mathcal{J}_y \equiv \frac{\sqrt{2}}{2i}\begin{pmatrix} 0 & 1 & 0 \\ -1 & 0 & 1 \\ 0 & -1 & 0 \end{pmatrix} \quad \mathcal{J}_z \equiv \begin{pmatrix} 1 & 0 & 0 \\ 0 & 0 & 0 \\ 0 & 0 & -1 \end{pmatrix}
\end{aligned} \tag{D.32}$$

Since not all of the generators are hermitian in this representation, the representation is seen to be finite-dimensional, but not unitary (the same as for the $\Gamma = \frac{1}{2}$ system). It should be noted that in contrast to the $\Gamma = \frac{1}{2}$ system, for the $\Gamma = 1$ system the matrix forms for the spinor metric and $\mathbf{\Gamma^0}$ are *not* proportional.

The representation transforms a scalar, a vector particle ($J = 0$) and its adjoint, and a self-adjoint vector particle. The 10-spinors are normalized such that for standard states labeled by γ:

$$\bar{\mathbf{\Psi}}^{(1)}_{(\gamma')}(\vec{p}, J', s'_z)\, \mathbf{\Psi}^{(1)}_{(\gamma)}(\vec{p}, J, s_z) = (-)^{1-\gamma}\, \delta_{\gamma'\gamma}\, \delta_{J'J}\, \delta_{s'_z s_z}, \tag{D.33}$$

$$\mathbf{\Psi}^{(1)}_{(\gamma)}(\vec{p}, J, s_z) \times \bar{\mathbf{\Psi}}^{(1)}_{(\gamma)}(\vec{p}, J, s_z) = (-)^{1-\gamma}\, \mathbf{\Pi}^{(1)}_{(\gamma)}(\vec{p}, J, s_z), \tag{D.34}$$

where $\bar{\mathbf{\Psi}} \equiv \mathbf{\Psi}^\dagger \mathbf{g}$, and the $\mathbf{\Pi}^{(1)}_{(\gamma)}(\vec{p}, J, s_z)$ are 10×10 projection matrices $\mathbf{\Pi}^2 = \mathbf{\Pi}$. The spinors satisfy:

$$\mathbf{\Gamma}^\mu \hat{P}_\mu \mathbf{\Psi}^{(1)}_{(\gamma)}(\vec{p}, J, s_z) = -\gamma mc\, \mathbf{\Psi}^{(1)}_{(\gamma)}(\vec{p}, J, s_z). \tag{D.35}$$

The normalizations follow those of the Dirac spinors corresponding to $\Gamma = \frac{1}{2}$.

D.3 Group properties of the extended Poincaré group

The local exponents of the extended Poincaré algebra will be developed in this appendix. Unitary ray representations of the group will be shown always to be equivalent to vector representations.

D.3.1 Local factors

As previously demonstrated, for a general ray representation of a quantum mechanical system group transformations on a quantum state vector can introduce additional phase factors;

$$U(\underline{b})\, U(\underline{a}) = e^{i\zeta(\underline{b};\underline{a})} U(\underline{\phi}(\underline{b};\underline{a})), \tag{D.36}$$

where ζ is the local exponent and $\underline{\phi}(\underline{b};\underline{a}) = \underline{b} \cdot \underline{a}$ represents the general group multiplication element. The behavior of the local exponent under group transformations can in general introduce local factors that are c-numbers into the algebra of the generators. These local factors can have physical significance, as was the case of the Galilean group for non-relativistic transformations [60]. Any local factors for this extended Poincaré group will be examined next.

If non-vanishing commutation relations such as $[\hat{A}, \hat{B}] = i(\hat{C} + \xi_{A,B}\hat{1})$ exist between operators, one can generally eliminate the additional c-number introduced by the ray representation by simply redefining the operator $\hat{C}' \equiv \hat{C} + \xi_{A,B}\hat{1}$, which will not effect any commutation relations involving $\hat{C}$ or $\hat{C}'$. Quite often, such redefinitions are sufficient to eliminate all of the local factors in the commutation relations. All local factors for this group can be shown to vanish using this reasoning except those in the following list of commutators:

- Eqs. 4.40 and 4.42, where the terms on the right-hand sides are shifted by $K_m - \xi_{\Gamma^m,\Gamma^0}$;
- Eqs. 4.44 and 4.46, where the terms on the right-hand sides are shifted by $\Gamma^m + \xi_{K_m,\Gamma^0}$;
- Eq. 4.47, where the term on the right-hand side is shifted by $\Gamma^0 + \xi_{K_z,\Gamma^z}$;
- Eqs. 4.90, 4.91, and 4.94, where the terms on the right-hand sides are shifted by $P_m + \xi_{\Gamma_m,\mathcal{M}_T}$;

- Eqs. 4.92 and 4.94, where the terms on the right-hand sides are shifted by $-P_0 + \xi_{\Gamma^0, \mathcal{M}_T}$;
- Eq. 4.93, where the term on the right-hand side is shifted by $\mathcal{M}_T + \xi_{\Gamma^0, P_0}$.

The local factors ξ appear in precisely the appropriate manner such that they can be absorbed into re-definitions of K_m, Γ^μ, P_μ, and $\mathcal{M}_T$, eliminating their appearance in the commutation relations. Henceforth, it will be assumed that the representation is one in which all local factors have been eliminated.

D.4 Hilbert space for massless particles

The Hilbert space of the generators of Lorentz transformations that leave the standard state vector for massless particles invariant is constructed by examining the commutation relations between the generators:

$$\mathbf{k}_x \equiv \begin{pmatrix} 0 & 0 & i & 0 \\ 0 & 0 & 0 & 0 \\ i & 0 & 0 & -i \\ 0 & 0 & i & 0 \end{pmatrix}, \quad \mathbf{k}_y \equiv \begin{pmatrix} 0 & i & 0 & 0 \\ i & 0 & 0 & -i \\ 0 & 0 & 0 & 0 \\ 0 & i & 0 & 0 \end{pmatrix}, \quad \mathbf{j}_z \equiv \begin{pmatrix} 0 & 0 & 0 & 0 \\ 0 & 0 & i & 0 \\ 0 & -i & 0 & 0 \\ 0 & 0 & 0 & 0 \end{pmatrix}. \tag{D.37}$$

The generators satisfy the algebra:

$$[\mathbf{k}_x, \mathbf{k}_y] = 0, \ [\mathbf{j}_z, \mathbf{k}_x] = i\mathbf{k}_y, \ [\mathbf{j}_z, \mathbf{k}_y] = -i\mathbf{k}_x. \tag{D.38}$$

Raising/lowering matrices can be defined by $\mathbf{k}_\pm \equiv \mathbf{k}_x \pm i\mathbf{k}_y$, which have commutation relations with the helicity generator $[\mathbf{j}_z, \mathbf{k}_\pm] = \pm\mathbf{k}_\pm$.

For eigenstates of the operator $\hat{j}_z$ satisfying $\hat{j}_z|\lambda\rangle = \lambda|\lambda\rangle$, the raising/lowering operators must satisfy $\hat{k}_\pm|\lambda\rangle = \beta_\pm(\lambda)|\lambda \pm 1\rangle$. Since $\hat{k}_+\hat{k}_-|\lambda\rangle = \hat{k}_-\hat{k}_+|\lambda\rangle$, the factors $\beta_\pm$ must satisfy $\beta_+(\lambda - 1)\,\beta_-(\lambda) = \beta_-(\lambda + 1)\,\beta_+(\lambda)$. Action on the minimum helicity state by $\hat{k}_-$, as well as action on the maximum helicity state by $\hat{k}_+$ imply that $\beta_\pm(\lambda) = 0$ for all helicities. Thus there is no action in the Hilbert space of massless particles by ladder operators that can change the value of the helicity by integral units. For little group elements, only the rotation about the standard direction has non-vanishing action on the state vector. This also implies that a particle that transforms under a massless representation of this algebra can have only one of the two extremal helicity values associated with its spin.

D.5 Extended Poincaré Lorentz invariants

For completeness, the non-vanishing commutation relations involving generators of the extended Poincaré group and Lorentz invariant bilinear products are listed below:

$$[\hat{\Gamma}^k, \eta_{\alpha\beta}\hat{\Gamma}^\alpha\hat{\Gamma}^\beta] = i\sum_{mn} \epsilon_{kmn}(\hat{J}_m\hat{\Gamma}^n + \hat{\Gamma}^n\hat{J}_m) + i(\hat{K}_k\hat{\Gamma}^0 + \hat{\Gamma}^0\hat{K}_k) \tag{D.39}$$

$$[\hat{\Gamma}^0, \eta_{\alpha\beta}\hat{\Gamma}^\alpha\hat{\Gamma}^\beta] = i\sum_j(\hat{K}_j\hat{\Gamma}^j + \hat{\Gamma}^j\hat{K}_j) \tag{D.40}$$

$$[\hat{\mathcal{M}}_T, \eta_{\alpha\beta}\hat{\Gamma}^\alpha\hat{\Gamma}^\beta] = -i\left(\hat{P}_\beta\hat{\Gamma}^\beta + \hat{\Gamma}^\beta\hat{P}_\beta\right) \tag{D.41}$$

$$[\hat{P}_\mu, \eta_{\alpha\beta}\hat{\Gamma}^\alpha\hat{\Gamma}^\beta] = i\eta_{\mu\beta}\left(\hat{\mathcal{M}}_T\hat{\Gamma}^\beta + \hat{\Gamma}^\beta\hat{\mathcal{M}}_T\right) \tag{D.42}$$

$$[\hat{\Gamma}^\mu, \eta^{\alpha\beta}\hat{P}_\alpha\hat{P}_\beta] = -2i\eta^{\mu\beta}\hat{\mathcal{M}}_T\hat{P}_\beta. \tag{D.43}$$

D.6 Mixing of massless eigenstates

Suppose that massless particles of finite-transverse or conformal mass mix to form flavor eigenstates in a manner consistent with the single helicity states giving V-A couplings in weak interactions, and analogous to the quark mixing that suppresses neutral, strangeness-changing currents. The transverse mass eigenstates will be labeled by $|m_j\rangle$, while the flavor eigenstate-defining generation a will be labeled $|f_a\rangle$. The mixing, due to the dynamics, is expected to unitarily relate the states:

$$|f_a\rangle = \sum_j |m_j\rangle\, U_{j\,a}. \tag{D.44}$$

Since finite affine parameter transformations for massless particles takes the form $e^{-\frac{i}{\hbar}\lambda \hat{\mathcal{M}}_T}$, the transition amplitude for propagating massless flavor eigenstates $f_a \rightarrow f_b$ is given by:

$$A(f_b \leftarrow f_a) = \sum_j U^*_{j\,b}\, e^{-\frac{i}{\hbar} m_j c\, \Delta\lambda}\, U_{j\,a}. \tag{D.45}$$

The affine parameter is given by the spatial/temporal distance of the null particle trajectory $\Delta\lambda = L = cT$.

The transition probability is given by $\mathcal{P}(f_b \leftarrow f_a) = |A(f_b \leftarrow f_a)|^2$, which results in mixing satisfying:

$$\mathcal{P}(f_b \leftarrow f_a) = \delta_{b\,a} + -2\sum_{j<k}\left\{2Re[\Upsilon_{j\,k}(b,a)]sin^2\left(\frac{\delta m_{jk}\, c\, L}{2\hbar}\right)\right.$$
$$\left. + Im[\Upsilon_{j\,k}(b,a)]sin\left(\frac{\delta m_{jk}\, c\, L}{\hbar}\right)\right\}, \tag{D.46}$$

where $\Upsilon_{j\,k}(b,a) \equiv U_{j\,b}\, U^*_{j\,a}\, U^*_{k\,b}\, U_{k\,a}$. Thus, massless particles with differing transverse mass eigenvalues $\delta m_{jk} = m_j - m_k$ can indeed allow dynamical mixing of flavor eigenstates.

D.7 Finite extended parameter transforms

Finite transformations in the extended group parameters will be examined in this appendix. To examine transformations on the generators that define the minimal Poincaré extension of the extended Lorentz group, we will express the group parameter conjugate to the operators $\hat{\Gamma}^\mu$ (*Gamma boosts*) in terms of a magnitude and direction, $\omega_\mu \equiv \omega\, n_\mu$. One can directly demonstrate that the commutation relations Eqs. 4.93 and 4.94 imply that for time-like $(\vec{n}\cdot\vec{n} = -1)$ or space-like $(\vec{n}\cdot\vec{n} = +1)$ transformations:

$$e^{i\omega_\mu \hat{\Gamma}^\mu}\, \hat{\mathcal{M}}_T e^{-i\omega_\nu \hat{\Gamma}^\nu} = \cos(\sqrt{-\vec{\omega}\cdot\vec{\omega}})\, \hat{\mathcal{M}}_T + \frac{\sin(\sqrt{-\vec{\omega}\cdot\vec{\omega}})}{\sqrt{-\vec{\omega}\cdot\vec{\omega}}}\omega_\mu \eta^{\mu\nu}\, \hat{P}_\nu \tag{D.47}$$

$$e^{i\omega_\mu \hat{\Gamma}^\mu}\, \hat{P}_\beta e^{-i\omega_\nu \hat{\Gamma}^\nu} = \omega_\beta \frac{\sin(\sqrt{-\vec{\omega}\cdot\vec{\omega}})}{(\sqrt{-\vec{\omega}\cdot\vec{\omega}})}\hat{\mathcal{M}}_T + \left[\delta^\nu_\beta + \frac{\omega_\beta \omega^\nu}{\vec{\omega}\cdot\vec{\omega}}(\cos(\sqrt{-\vec{\omega}\cdot\vec{\omega}}) - 1)\right]\hat{P}_\nu. \tag{D.48}$$

One can see that for eigenstates of $\hat{\Gamma}^0$ such as the finite-dimensional representations discussed previously, the time-like transformations mix eigenvalues of $\hat{P}_0$ (masses) with eigenvalues of $\hat{\mathcal{M}}_T$. The mass values oscillate under variations in the parameter ω. If the group elements ω are discrete, then the discrete set of masses will mix into each other and eigenvalues of $\mathcal{M}_T$ under operations using finite transformations in group elements conjugate to Γ^0. There will be a discrete set of transformations that will mix a massive state into an allowable massless state with null four-momentum. In this sense, the operator $\mathcal{M}_T$ which is necessary for group closure can be interpreted as a transverse mass operator.

If the Gamma boosts are light-like ($\vec{n}\cdot\vec{n}=0$) , the transformations are given by:

$$e^{i\omega_\mu \hat{\Gamma}^\mu}\hat{\mathcal{M}}_T e^{-i\omega_\nu \hat{\Gamma}^\nu} = \hat{\mathcal{M}}_T - \omega n_\mu \,\eta^{\mu\nu}\,\hat{P}_\nu \tag{D.49}$$

$$e^{i\omega_\mu \hat{\Gamma}^\mu}\hat{P}_\beta e^{-i\omega_\nu \hat{\Gamma}^\nu} = -\omega n_\beta\,\hat{\mathcal{M}}_T + \left(\delta_\beta^\nu + \frac{\omega^2}{2} n_\beta\, n^\nu\right)\hat{P}_\nu . \tag{D.50}$$

The transverse mass of massless eigenstates remains invariant under these transformations.

D.8 General chiral projections

The matrix $\mathbf{\Gamma}_E$ from Section 4.4.6 can be combined with the spinor metric to form a convenient projector. By examining the commutation relations given in Eq. 4.80, 4.84, 4.284, and 4.285, the matrix $\mathbf{g}^{(\Gamma)}\,\mathbf{\Gamma}_E$ commutes with the generators for Lorentz transformations:

$$\mathbf{g}^{(\Gamma)}\mathbf{\Gamma}_E\,\mathbf{J}_k^{(\Gamma)} = \mathbf{J}_k^{(\Gamma)}\,\mathbf{g}^{(\Gamma)}\mathbf{\Gamma}_E, \quad \mathbf{g}^{(\Gamma)}\mathbf{\Gamma}_E\,\mathbf{K}_k^{(\Gamma)} = \mathbf{K}_k^{(\Gamma)}\,\mathbf{g}^{(\Gamma)}\mathbf{\Gamma}_E, \tag{D.51}$$

while it anti-commutes with the gamma matrices:

$$\mathbf{g}^{(\Gamma)}\mathbf{\Gamma}_E\,\mathbf{\Gamma}^\beta = -\mathbf{\Gamma}^\beta\,\mathbf{g}^{(\Gamma)}\mathbf{\Gamma}_E. \tag{D.52}$$

For the Dirac representation, $\mathbf{g}^{(\frac{1}{2})}\,\mathbf{\Gamma}_E = \gamma_5$, while for the 10-dimensional $\Gamma=1$ representation, this matrix takes the form:

$$\mathbf{g}^{(1)}\mathbf{\Gamma}_E^{(1)} = \begin{pmatrix} -1 & \mathbf{0} & \mathbf{0} & \mathbf{0} \\ 0 & \mathbf{0} & \mathbf{0} & \mathbf{1} \\ 0 & \mathbf{0} & \mathbf{1} & \mathbf{0} \\ 0 & \mathbf{1} & \mathbf{0} & \mathbf{0} \end{pmatrix}, \tag{D.53}$$

where the bold-faced matrices $\mathbf{1}$ represent 3×3 identity matrices. This means that the generalized chirality projection $\Pi_\pm^{(\Gamma)}$ defined by:

$$\Pi_\pm^{(\Gamma)} \equiv \frac{\mathbf{1}\pm\mathbf{g}^{(\Gamma)}\,\mathbf{\Gamma}_E}{2} \tag{D.54}$$

is a Lorentz invariant. For the Dirac representation, this operator defines the standard chirality operation that gives the helicity of the standard state. These matrices indeed form mutually orthogonal projections:

$$\Pi_\pm^{(\Gamma)}\,\Pi_\pm^{(\Gamma)} = \Pi_\pm^{(\Gamma)}, \quad \Pi_\pm^{(\Gamma)}\,\Pi_\mp^{(\Gamma)} = 0.$$

The operation leaves the helicity operator invariant:

$$\Pi_\pm^{(\Gamma)}\,\mathbf{W}_\alpha(\vec{p}) = \mathbf{W}_\alpha(\vec{p})\,\Pi_\pm^{(\Gamma)}, \tag{D.55}$$

where $\mathbf{W}_\alpha(\vec{p}) = i\epsilon_{\alpha\beta\mu\nu}\mathbf{\Gamma}^\beta\mathbf{\Gamma}^\mu\eta^{\nu\lambda}p_\lambda$. By direct substitution, the projected spinors can be seen to satisfy:

$$\mathbf{\Gamma}^\beta p_\beta\left[\Pi_\pm^{(\Gamma)}\mathbf{u}_{(\gamma)}^{(\Gamma)}(\vec{p}, m, J, s_z)\right] = -(\gamma)mc\left[\Pi_\mp^{(\Gamma)}\mathbf{u}_{(\gamma)}^{(\Gamma)}(\vec{p}, m, J, s_z)\right]. \tag{D.56}$$

This demonstrates that chiral projected states can only satisfy the linear spinor equations for massless systems. The two projected states reflect the two possible helicity states of these systems.

Appendix E

Addendum for Chapter 5

E.1 Structure equations for general relativity

It is often most expedient to simply calculate the non-vanishing connections directly from the metric and substitute into Eq. 5.33 to calculate the curvature tensor components. An alternative method, which will be presented here, is also sometimes convenient to perform these calculations.

E.1.1 Exterior derivatives

A bit of formal structure will be developed for calculating curvatures. First will be to define an *exterior derivative*. The exterior derivative is an anti-symmetric operation useful for examining integrals over oriented surfaces. The exterior derivative of a scalar function (or 0-form) gives a *1-form*:

$$\mathbf{d}f \equiv \partial_\beta f \, \mathbf{d}x^\beta. \tag{E.1}$$

The exterior derivative of a general p-form $\mathbf{F}$ defines an *exact* (p+1)-form $\mathbf{dF}$. If the exterior derivative of a form vanishes, it is referred to as a *closed form*. The exterior derivative of a scalar multiple of an exact form satisfies:

$$\mathbf{d}(f\,\mathbf{dF}) \equiv \mathbf{d}f \wedge \mathbf{dF} \quad \text{where} \quad \mathbf{d}x^\alpha \wedge \mathbf{d}x^\beta = -\mathbf{d}x^\beta \wedge \mathbf{d}x^\beta. \tag{E.2}$$

This establishes a convention for placing the additional one-form $\mathbf{d}f$ in constructing the resultant (p + 2)-form. One can then imply that $\mathbf{dd}(f\,\mathbf{dF}) = \frac{1}{2}([\partial_\alpha, \partial_\beta] f)\mathbf{d}x^\alpha \wedge \mathbf{d}x^\beta \wedge \mathbf{dF} = 0$, i.e., $\mathbf{ddF} = 0$ when acting on arbitrary p-forms. This means that $\mathbf{dF}$ is always a closed form.

Quite often it is convenient to use a non-coordinate basis in order to calculate curvatures. For a non-coordinate basis, the basis forms $\mathbf{e}^\beta$ are not generally exterior differentials of a coordinate. Since the inner product between basis forms and basis tangent vectors should be invariant, for the bilinear exterior derivative operation $\langle \mathbf{e}^\beta, \vec{e}_\mu \rangle = \delta^\beta_\mu$. This then implies that $\langle \mathbf{de}^\beta, \vec{e}_\mu \rangle = -\langle \mathbf{e}^\beta, \mathbf{d}\vec{e}_\mu \rangle$. Eq. E.1 is just a re-expression of the chain rule $\mathbf{d} = \mathbf{e}^\beta \times \vec{e}_\beta$.

E.1.2 Rotation 1-forms

The basis tangent vectors are 0-forms. This means that an exterior differential operation on a basis tangent vector must result in a linear superposition of basis vectors and basis 1-forms. The coefficients of this superposition are the affine connections:

$$\mathbf{d}\vec{e}_\alpha = \vec{e}_\lambda \otimes \Gamma^\lambda_{\alpha\beta} \, \mathbf{d}x^\beta. \tag{E.3}$$

The *rotation 1-forms* are defined as $\mathbf{\Omega}^\lambda{}_\alpha \equiv \Gamma^\lambda_{\alpha\beta}\,\mathbf{d}x^\beta$. The derivative of the metric can be expressed in terms of the rotation 1-forms:

$$\mathbf{d}\left(\vec{e}_\alpha \cdot \vec{e}_\beta\right) = (\mathbf{d}\vec{e}_\alpha)\cdot \vec{e}_\beta + \vec{e}_\alpha \cdot \left(\mathbf{d}\vec{e}_\beta\right) \;\Rightarrow\; \mathbf{d}g_{\alpha\beta} = \mathbf{\Omega}^\lambda{}_\alpha\, g_{\lambda\beta} + g_{\alpha\lambda}\,\mathbf{\Omega}^\lambda{}_\beta. \tag{E.4}$$

Also, exterior derivatives of the rotation 1-forms satisfy:

$$\mathbf{d\Omega}^\lambda{}_\alpha = (\partial_\mu \Gamma^\lambda_{\alpha\beta})\,\mathbf{d}x^\mu \wedge \mathbf{d}x^\beta. \tag{E.5}$$

These derivatives will contribute to the calculation of the curvature.

E.1.3 Curvature 2-forms

By their nature, tangent vectors are not objects that lie *on* the manifold. The integrability condition does not require that two exterior derivatives acting on a tangent vector need vanish. However, by the definition of exterior derivatives, two such derivatives will be antisymmetric.

Therefore, an exterior derivative of (Eq. E.3) gives an anti-symmetric differential of the basis vector proportional to the Riemann curvature tensor:

$$\mathbf{d}(\mathbf{d}\vec{e}_\alpha) = \mathbf{d}\vec{e}_\lambda \wedge \mathbf{\Omega}^\lambda_\alpha + \vec{e}_\lambda \otimes \mathbf{d\Omega}^\lambda_\alpha = \vec{e}_\beta \otimes \left(\mathbf{d\Omega}^\beta_\alpha + \mathbf{\Omega}^\beta_\lambda \wedge \mathbf{\Omega}^\lambda_\alpha\right). \tag{E.6}$$

These structure equations will be compiled in the next subsection.

E.1.4 Structure equations for calculating curvature

The equations derived in the previous sections will be combined below for convenience:

$$\mathbf{d}g_{\alpha\beta} = g_{\alpha\lambda}\mathbf{\Omega}^\lambda_\beta + \mathbf{\Omega}^\lambda_\alpha g_{\lambda\beta}, \tag{E.7}$$

$$\mathbf{d}\vec{e}_\alpha = \vec{e}_\lambda \otimes \mathbf{\Omega}^\lambda_\alpha \quad\Leftrightarrow\quad \mathbf{d}\mathbf{e}^\beta = -\mathbf{\Omega}^\beta_\lambda \wedge \mathbf{e}^\lambda, \tag{E.8}$$

$$\mathbf{R}^\beta_\alpha = \mathbf{d\Omega}^\beta_\alpha + \mathbf{\Omega}^\beta_\lambda \wedge \mathbf{\Omega}^\lambda_\alpha. \tag{E.9}$$

The structure equations for calculating curvature are often most convenient when used in an orthonormal basis. For the inner product of orthonormal basis vectors, $\vec{e}_{\hat\alpha} \cdot \vec{e}_{\hat\beta} = \eta_{\hat\alpha\hat\beta}$, so that (Eq. E.7) is conveniently satisfied. A useful calculation will be done below.

E.1.5 Calculating curvatures

The curvature components for a general static spherically symmetric geometry will be calculated here. For convenience, the metric will be expressed in the form:

$$ds^2 = -e^{2a(r)}(dct)^2 + e^{2b(r)}dr^2 + r^2 d\theta^2 + r^2 \sin^2\theta\, d\phi^2. \tag{E.10}$$

It is convenient to define an orthonormal basis $\vec{e}_{\hat\alpha} \cdot \vec{e}_{\hat\beta} = \eta_{\hat\alpha\hat\beta}$:

$$\vec{e}_t = e^a\,\vec{e}_{\hat t},\; \vec{e}_r = e^b\,\vec{e}_{\hat r},\; \vec{e}_\theta = r\,\vec{e}_{\hat\theta},\; \vec{e}_\phi = r\sin\theta\,\vec{e}_{\hat\phi}. \tag{E.11}$$

One approach involves calculating the connections, then using those connections to calculate the curvature. The connections can be calculated from Eq. E.8:

$$\begin{aligned}
&\mathbf{d}\,\vec{e}_t = a'(r)\,\vec{e}_t \otimes \mathbf{d}r \;\Rightarrow\; \Gamma^t_{tr} = a'(r)\\
&\mathbf{d}\,\vec{e}_r = b'(r)\,\vec{e}_r \otimes \mathbf{d}r \;\Rightarrow\; \Gamma^r_{rr} = b'(r)\\
&\mathbf{d}\,\vec{e}_\theta = \tfrac{1}{r}\,\vec{e}_\theta \otimes \mathbf{d}r \;\Rightarrow\; \Gamma^\theta_{\theta r} = \tfrac{1}{r}\\
&\mathbf{d}\,\vec{e}_\phi = \vec{e}_\theta \otimes \left(\tfrac{\mathbf{d}r}{r} + \cot\theta\,\mathbf{d}\theta\right) \;\Rightarrow\; \Gamma^\phi_{\phi r} = \tfrac{1}{r},\; \Gamma^\phi_{\phi\theta} = \cot\theta.
\end{aligned} \tag{E.12}$$

The remaining non-vanishing connections are obtained using Eq. E.7 or Eq. 5.18. For instance, $\partial_\theta g_{tr} = \Gamma^r_{tt}\, g_{rr} + g_{tt}\, \Gamma^t_{tr} = 0$. Thus, the remaining non-vanishing connections are given by:

$$\Gamma^r_{tt} = a'\, e^{2a-2b},\ \Gamma^r_{\theta\theta} = -r\, e^{-2b},\ \Gamma^r_{\phi\phi} = -r \sin^2\theta\, e^{-2b}. \tag{E.13}$$

The rotation forms are then substituted into Eq. E.9 to obtain the curvatures.

Rather than calculating using the original basis, the non-vanishing curvature components can be obtained more rapidly by working directly with the orthonormal basis. For that basis, the condition E.7 identically vanishes $\mathbf{d}g_{\hat{\alpha}\hat{\beta}} = 0$, reducing the number of steps needed in the calculation. Begin by constructing orthonormal dual basis forms:

$$\mathbf{e}^{\hat{t}} = e^a\, \mathbf{d}ct,\ \mathbf{e}^{\hat{r}} = e^b\, \mathbf{d}r,\ \mathbf{e}^{\hat{\theta}} = r\, \mathbf{d}\theta,\ \mathbf{e}^{\hat{\phi}} = r \sin\theta\, \mathbf{d}\phi.$$

Exterior derivatives on those basis forms construct rotation 1-forms:

$$\begin{aligned}
&\mathbf{d}\,\mathbf{e}^{\hat{t}} = a'\, e^{-b}\mathbf{e}^{\hat{r}} \wedge \mathbf{e}^{\hat{t}} \ \Rightarrow\ -\mathbf{\Omega}^{\hat{t}}_{\hat{r}} = -a'\, e^{-b}\, \mathbf{e}^{\hat{t}},\\
&\mathbf{d}\,\mathbf{e}^{\hat{r}} = 0,\\
&\mathbf{d}\,\mathbf{e}^{\hat{\theta}} = \tfrac{e^{-b}}{r}\mathbf{e}^{\hat{r}} \wedge \mathbf{e}^{\hat{\theta}} \ \Rightarrow\ -\mathbf{\Omega}^{\hat{\theta}}_{\hat{r}} = -\tfrac{e^{-b}}{r}\, \mathbf{e}^{\hat{\theta}},\\
&\mathbf{d}\,\mathbf{e}^{\hat{\phi}} = \left(\tfrac{e^{-b}}{r}\mathbf{e}^{\hat{r}} + \tfrac{\mathbf{e}^{\hat{\theta}}}{r}\right) \wedge \mathbf{e}^{\hat{\phi}} \Rightarrow -\mathbf{\Omega}^{\hat{\phi}}_{\hat{r}} = -\tfrac{e^{-b}}{r}\, \mathbf{e}^{\hat{\phi}},\ -\mathbf{\Omega}^{\hat{\phi}}_{\hat{\theta}} = -\tfrac{\cot\theta}{r}\, \mathbf{e}^{\hat{\phi}}.
\end{aligned} \tag{E.14}$$

It is straightforward to take exterior derivatives of the rotation 1-forms, giving:

$$\begin{aligned}
&\mathbf{d}\,\mathbf{\Omega}^{\hat{t}}_{\hat{r}} = -e^{-2b}\left(a'' + a'^2 - a'\, b'\right) \mathbf{e}^{\hat{t}} \wedge \mathbf{e}^{\hat{r}},\\
&\mathbf{d}\,\mathbf{\Omega}^{\hat{\theta}}_{\hat{r}} = -\tfrac{b'}{r} e^{-2b}\, \mathbf{e}^{\hat{r}} \wedge \mathbf{e}^{\hat{\theta}},\\
&\mathbf{d}\,\mathbf{\Omega}^{\hat{\phi}}_{\hat{r}} = \tfrac{e^{-b}}{r}\left(-b' e^{-b}\mathbf{e}^{\hat{r}} + \tfrac{\cot\theta}{r}\mathbf{e}^{\hat{\theta}}\right) \wedge \mathbf{e}^{\hat{\phi}},\\
&\mathbf{d}\,\mathbf{\Omega}^{\hat{\phi}}_{\hat{\theta}} = -\tfrac{1}{r^2}\, \mathbf{e}^{\hat{\theta}} \wedge \mathbf{e}^{\hat{\phi}}.
\end{aligned} \tag{E.15}$$

Using Eq. E.9, the curvature components can be obtained from the curvature two-forms using $\mathbf{R}^{\hat{\beta}}_{\hat{\alpha}} = \frac{1}{2}\mathcal{R}^{\hat{\beta}}{}_{\hat{\alpha}\hat{\mu}\hat{\nu}}\mathbf{e}^{\hat{\mu}} \wedge \mathbf{e}^{\hat{\nu}}$, where:

$$\begin{aligned}
&\mathbf{R}^{\hat{t}}_{\hat{r}} = \mathbf{d}\,\mathbf{\Omega}^{\hat{t}}_{\hat{r}} = -e^{-2b}\left(a'' + a'^2 - a'\, b'\right) \mathbf{e}^{\hat{t}} \wedge \mathbf{e}^{\hat{r}},\\
&\mathbf{R}^{\hat{t}}_{\hat{\theta}} = \mathbf{\Omega}^{\hat{t}}_{\hat{r}} \wedge \mathbf{\Omega}^{\hat{r}}_{\hat{\theta}} = a'\, \tfrac{e^{-2b}}{r}\, \mathbf{e}^{\hat{t}} \wedge \mathbf{e}^{\hat{\theta}},\\
&\mathbf{R}^{\hat{t}}_{\hat{\phi}} = \mathbf{\Omega}^{\hat{t}}_{\hat{r}} \wedge \mathbf{\Omega}^{\hat{r}}_{\hat{\phi}} = a'\, \tfrac{e^{-2b}}{r}\, \mathbf{e}^{\hat{t}} \wedge \mathbf{e}^{\hat{\phi}},\\
&\mathbf{R}^{\hat{\theta}}_{\hat{r}} = \mathbf{d}\,\mathbf{\Omega}^{\hat{\theta}}_{\hat{r}} + \mathbf{\Omega}^{\hat{\theta}}_{\hat{\phi}} \wedge \mathbf{\Omega}^{\hat{\phi}}_{\hat{r}} = -b'\, \tfrac{e^{-2b}}{r}\, \mathbf{e}^{\hat{r}} \wedge \mathbf{e}^{\hat{\theta}},\\
&\mathbf{R}^{\hat{\theta}}_{\hat{\phi}} = \mathbf{d}\,\mathbf{\Omega}^{\hat{\theta}}_{\hat{\phi}} + \mathbf{\Omega}^{\hat{\theta}}_{\hat{r}} \wedge \mathbf{\Omega}^{\hat{r}}_{\hat{\phi}} = \tfrac{1}{r^2}\left(1 - e^{-2b}\right) \mathbf{e}^{\hat{\theta}} \wedge \mathbf{e}^{\hat{\phi}},\\
&\mathbf{R}^{\hat{\phi}}_{\hat{r}} = \mathbf{d}\,\mathbf{\Omega}^{\hat{\phi}}_{\hat{r}} + \mathbf{\Omega}^{\hat{\phi}}_{\hat{\theta}} \wedge \mathbf{\Omega}^{\hat{\theta}}_{\hat{r}} = -b'\, \tfrac{e^{-2b}}{r}\, \mathbf{e}^{\hat{r}} \wedge \mathbf{e}^{\hat{\phi}}.
\end{aligned} \tag{E.16}$$

The Einstein tensor components can be calculated from the Riemann curvature tensor components $\mathcal{R}^{\hat{\beta}\hat{\lambda}}_{\hat{\mu}\hat{\nu}} \equiv g^{\hat{\lambda}\hat{\alpha}}\,\mathcal{R}^{\hat{\beta}}_{\hat{\alpha}\hat{\mu}\hat{\nu}}$ using the relation (see Reference [77], p. 344):

$$\begin{aligned}
G^{0}_{0} &= -\left(\mathcal{R}^{j-\,j}_{j-\,j} + \mathcal{R}^{j\,j+}_{j\,j+} + \mathcal{R}^{j+\,j-}_{j+\,j-}\right) \\
G^{j}_{j} &= -\left(\mathcal{R}^{0\,j-}_{0\,j-} + \mathcal{R}^{0\,j+}_{0\,j+} + \mathcal{R}^{j+\,j-}_{j+\,j-}\right) \\
G^{0}_{j} &= \mathcal{R}^{0\,j+}_{j\,j+} + \mathcal{R}^{0\,j-}_{j\,j-} \\
G^{j}_{j\pm} &= \mathcal{R}^{j\,0}_{j\pm 0} + \mathcal{R}^{j\,j\mp}_{j\pm\,j\mp}.
\end{aligned} \tag{E.17}$$

where the Latin indices $(j-,\, j,\, j+)$ are a cyclic permutation of (1,2,3), and *there are no implied sums*. Substituting, the non-vanishing components of the Einstein tensor are given by:

$$G^{\hat{t}}_{\hat{t}} = -\frac{1}{r^2}\left(1 - e^{-2b}\right) - 2b'\,\frac{e^{-2b}}{r} \tag{E.18}$$

$$G^{\hat{r}}_{\hat{r}} = -\frac{1}{r^2}\left(1 - e^{-2b}\right) + 2a'\,\frac{e^{-2b}}{r} \tag{E.19}$$

$$G^{\hat{\theta}}_{\hat{\theta}} = e^{-2b}\left(a'' + a'^2 - a'\,b'\right) - b'\,\frac{e^{-2b}}{r} + a'\,\frac{e^{-2b}}{r} = G^{\hat{\phi}}_{\hat{\phi}}. \tag{E.20}$$

These are also the correct mixed components of the Einstein tensor even for the non-orthonormal original components indexed by $(ct, r, \vartheta, \varphi)$.

Appendix F

Addendum for Chapter 7

F.1 Calculation of Schwarzschild entropy using action principle

Consider the thermodynamics of Rindler's geometry. The thermal partition function is directly related to the free energy that minimizes the Euclidean action in the form:

$$Z = e^{-\theta_R F_R} = e^{-W_{Euclidean}/\hbar} \tag{F.1}$$

where:

$$W_{Euclidean} = \frac{c^3}{16\pi\, G_N} \int d^4x_{Euclidean} \sqrt{g_{Euclidean}}\, \mathcal{R}_{Euclidean}. \tag{F.2}$$

The conical coordinate singularity at $\rho_r = 0$ is degenerate in the Rindler time parameter $\theta_R = -i\omega_t$, just as the North Pole of the Earth is degenerate in longitude. The thermodynamic parameters can be obtained in terms of derivatives with respect to θ_R, as long as it is not assigned the final constant value 2π. This angle will be evaluated between $0 \leq \theta_R \leq 2\pi + \epsilon$, where the parameter ϵ will be referred to as the *conical excess* angle. The singular nature of this geometry can be obtained by parameterizing a generalized metric, then taking the appropriate limits.

For calculational convenience, define the dimensionless parameter $\lambda_\epsilon \equiv 1 + \frac{\epsilon}{2\pi}$ that takes the value of unity in the needed limit. The thermal Euclidean metric will be parameterized as:

$$ds^2_{Euclidean} = \rho_r^2\, d\omega_t^2 + \frac{(\rho_r^2 + \delta^2)}{\lambda_\epsilon^2 \rho_r^2 + \delta^2}\, d\rho_r^2 + |\mathbf{dx}_\perp|^2 \tag{F.3}$$

This four-surface, which gives non-vanishing curvatures, becomes Rindler space with conical excess $\theta_R : 0 \to 2\pi + \epsilon$ when $\epsilon \to 0$ and $\delta \to 0$.

The Euclidean action F.2 requires the calculation of the determinant of the metric, as well as the Ricci scalar generated by that metric. The determinant of the Euclidean metric is given by $g_{Euclidean} = \frac{(\rho_r^2+\delta^2)\,\rho_r^2}{\lambda_\epsilon^2\rho_r^2+\delta^2}$. The Ricci scalar can be calculated in a straightforward manner, yielding $\mathcal{R}_{Euclidean} = 2\delta^2 \frac{\lambda_\epsilon^2 - 1}{(\rho_r^2+\delta^2)^2}$. This scalar vanishes in the limit that the conical excess vanishes (the vacuum solution), as expected. Combining these terms, the Euclidean action can be directly calculated to be given by:

$$\frac{W_{Euclidean}}{\hbar} = \frac{c^3}{16\pi\, G_N}\, 2\pi\, \frac{\text{Area}_\perp}{\hbar}\, 2\sqrt{\frac{\lambda_\epsilon^2\rho_r^2 + \delta^2}{\rho_r^2 + \delta^2}}\Bigg|_0^\infty = \frac{\epsilon\, c^3\, \text{Area}_\perp}{\hbar\, 8\pi\, G_N}, \tag{F.4}$$

since $2\pi(\lambda_\epsilon - 1) = \epsilon = \theta_R - 2\pi$. In this expression, the factor $\text{Area}_\perp$ is the surface area associated with the coordinates $\mathbf{dx}_\perp$ in the Rindler metric. The action is seen to be independent of the parameter δ.

The entropy can then be evaluated from F.1 using the standard relationships from thermodynamics:

$$S = -\frac{\partial F}{\partial T} = k_B\,\theta_T^2\,\frac{\partial}{\partial\theta_T}\frac{\log Z}{\theta_T} = k_B\,\theta_R^2\,\frac{\partial}{\partial\theta_R}\frac{\log Z}{\theta_R}, \tag{F.5}$$

where $\theta_T \equiv 1/k_B T$. The last equality is true because the Rindler and Schwarzschild temperatures are linearly related, and the dimensional factors cancel. This gives an entropy of the form:

$$S = k_B\,\theta_R^2 \frac{\partial}{\partial\theta_R}\left(\frac{(\theta_R - 2\pi)}{\theta_R}\frac{c^3}{\hbar}\frac{\text{Area}_\perp}{8\pi\,G_N}\right)\Bigg|_{\theta_R \to 2\pi} = \left(\frac{k_B c^3}{\hbar}\right)\frac{\text{Area}_\perp}{4\,G_N}. \tag{F.6}$$

This calculation bypassed any direct expression of the temperature in obtaining a form for the entropy.

Appendix G

Addendum for Chapter 8

G.1 Fluctuations and dissipations

There are general connections between the fluctuations of a system in a thermal environment, and the energy dissipation, which is the difference between energy absorbed and energy emitted. A general quantum operator translates according to:

$$\hat{A}(\tau) = e^{\frac{i}{\hbar}\tau\hat{K}}\hat{A}e^{-\frac{i}{\hbar}\tau\hat{K}}, \tag{G.1}$$

using proper-energy operators. The symmetric autocorrelation function:

$$C_A(\tau - \tau') \equiv \frac{1}{2}\langle\{\hat{A}(\tau), \hat{A}(\tau')\}\rangle, \tag{G.2}$$

is a measure of the fluctuations of the operator over time. The general susceptibility is directly related to the expectation value of the commutator of the operator at different times:

$$\chi_A(\tau - \tau') \equiv i\langle[\hat{A}(\tau), \hat{A}(\tau')]\rangle. \tag{G.3}$$

The energy dissipation can be directly related to the imaginary part of the susceptibility.

The thermal averages in these expressions will be taken at fixed temperature parameter $\beta \equiv \frac{1}{k_B T}$. By inserting a complete set of eigenstates of $\hat{K}$ to evaluate the thermal average, these expressions satisfy:

$$\begin{aligned} C_A(\tau - \tau') &= \frac{1}{2}e^{\beta F_K}\sum_{m,n}\left(e^{-\beta K_n} + e^{-\beta K_m}\right)|\langle n|\hat{A}|m\rangle|^2 e^{\frac{i}{\hbar}(K_m - K_n)(\tau' - \tau)} \\ \chi_A(\tau - \tau') &= i e^{\beta F_K}\sum_{m,n}\left(e^{-\beta K_n} - e^{-\beta K_m}\right)|\langle n|\hat{A}|m\rangle|^2 e^{\frac{i}{\hbar}(K_m - K_n)(\tau' - \tau)}. \end{aligned} \tag{G.4}$$

The Fourier transforms of the correlation $\tilde{C}_A(\omega) \equiv \int \frac{d\tau}{\sqrt{2\pi}} C_A(\tau)e^{i\omega\tau}$ are sometimes referred to as the *power spectrum* of the autocorrelations. Likewise, the Fourier transform of the susceptibility will be denoted $\tilde{\chi}_A(\omega) \equiv \int \frac{d\tau}{\sqrt{2\pi}} \chi_A(\tau)e^{i\omega\tau}$. Comparing these transforms yields the expression:

$$\tilde{C}_A(\omega) = \left(\frac{1 + e^{-\beta\hbar\omega}}{2(1 - e^{-\beta\hbar\omega})} \right) \frac{\tilde{\chi}_A(\omega)}{i} = \left(\frac{1}{e^{\beta\hbar\omega} - 1} + \frac{1}{2} \right) \frac{\tilde{\chi}_A(\omega)}{i}. \tag{G.5}$$

The quantity in the parenthesis is just the thermal Planck factor $(n(\omega) + \frac{1}{2})$. Thus, the power spectrum is proportional to the imaginary part of the transform of the susceptibility, with a proportionality constant that depends only upon the temperature.

References

[1] J.G. King, *Phys. Rev. Letters* **5**, 562 (1960); L.J. Fraser, E.R. Carlson, and V.W. Hughes, *Bull. Am. Phys. Soc.* **13**, 636 (1968).

[2] E.M. Purcell, *Electricity and Magnetism*, McGraw-Hill, New York (1965).

[3] P.A.M. Dirac, *Proc. Roy. Soc.* **A133**, 60 (1931); and *Phys. Rev.* **74**, 817 (1948).

[4] J. Lindesay and T. Gill, "Canonical proper time formulation for physical systems", *Foundations of Physics* **34**(1), 169–82 (2004). See also T. Gill and J. Lindesay, "Canonical proper time formulation of relativistic particle dynamics", *Int. J. Theoretical Phys.* **32**, 2087–98 (1993).

[5] H.B.G. Casimir, *Proc. K. Ned. Akad. Wet.* **51**, 793 (1948).

[6] T.H. Boyer, *Phys. Rev.* **174**, 1764 (1968).

[7] E.M. Lifshitz, *Soviet Phys. JETP* **2**, 73 (1956). I.D. Dzyaloshinskii, E.M. Lifshitz, and I.P. Pitaevskii, *Soviet Phys. Usp.* **4**, 153 (1961). I.D. Landau and E.M. Lifshitz, *Electrodynamics of Continuous Media*, pp. 368–76. Pergamon, Oxford (1960).

[8] E.S. Sabisky and C.H. Anderson, *Phys. Rev.* A **7**, 790 (1973).

[9] J. Schwinger, *Lett. Math. Phys.* **1**, 43 (1975).

[10] K.A. Milton, L.L. DeRaad, and J. Schwinger, "Casimir effect in dielectrics", *Ann. Phys.* **115**, 1 (1978).

[11] D. Kleppner, "With apologies to Casimir", *Phys. Today*, **43**, 10 (Oct 1990) 9–11.

[12] J.A. Wheeler and R.P. Feynman, *Rev. Mod. Phys.* **17**, 157 (1945); and *Rev. Mod. Phys.* **21**, 425 (1949).

[13] J.V. Lindesay and H.P. Noyes, Non-perturbative, unitary quantum-particle scattering amplitudes from three-particle equations, hep-th/0203262, *Found. Phys.* **34**, 1573–606 (2004).

[14] N. Bohr and L. Rosenfeld, "On the question of the measurability of electromagnetic field quantities", *Selected Papers of Leon Rosenfeld*, Cohen and Stachel, eds., Reidel, Dordrecht (1979) 357–400, translated from *Mat.-fys. Medd. Dan. Vid. Selsk.*, **12**, no. 8 (1933).

[15] A. Einstein, B. Podolsky, and N. Rosen, *Phys. Rev.* **47**, 777 (1935).

[16] J.S. Bell, *Physics* **1**, 195 (1964).

[17] A. Aspect, P. Grangier, and G. Roger, *Phys. Rev. Lett.* **49**, 91 (1982).

[18] Belle experiment collaboration, KEK, Tsukuba, Japan. http://www.kek.jp/intra-e/press/2007/BellePress9e.html. Accessed November 20, 2012.

[19] A. Leggett, *Found. Phys.* **33**, 1469–93 (2003).

[20] J. Romero, J. Leach, B. Jack, S.M. Barnett, M.J. Padgett, and S. Franke-Arnold, "Violation of Leggett inequalities in orbital angular momentum subspaces". *New J. Phys.* **12**, 123007 (2010) and cited references therein.

[21] W. Pauli (1940); "On the connection between spin and statistics", *Phys. Rev.* **58**, 716.

[22] R.F. Streater and A.S. Wightman, *PCT, Spin and Statistics, and All That.* Benjamin/Cummings, Reading, MA (1964).

[23] L. Susskind and J. Lindesay, *An Introduction to Black Holes, Information, and the String Theory Revolution: The Holographic Universe.* World Scientific, Singapore (2005).

[24] L. Susskind, *The Black Hole War: My Battle with Stephen Hawking to Make the World Safe for Quantum Mechanics*, Little, Brown & Co (2009).

[25] L. Dyson, J. Lindesay, and L. Susskind, "Is There Really a de Sitter/CFT Duality?" *J. High Energy Phys.* online, arXiv:hep-th/0202163, JHEP08(2002)045, 10 pages (2002).

[26] C.E. Shannon, "A mathematical theory of communication", *Bell System Technical Journal*, **27**: 379423–623656 (1948).

[27] M. Tribus, *Thermostatics and Thermodynamics*, D. van Nostrand Company, Inc., Princeton, NJ (1961).

[28] J. Lindesay, T.E. Mason, L. Ricks-Santi, W. Hercules, P. Kurian, and G.M. Dunston, "A new biophysical metric for interrogating the information content in human genome sequence variation: Proof of concept". arXiv:1108.3012 [physics. bio-ph] pages, *J. of Comp. Bio. and Bioinf. Res.* **4**(2), (online) (2012), http://www.academicjournals.org/jcbbr DOI: 10.5897/JCBBR11.026.

[29] W.K. Wootters and W.H. Zurek, "A single quantum cannot be cloned", *Nature* **299**. 802–3 (1982).

[30] D. Dieks, "Communication by EPR devices," *Phys. Lett. A* **92**(6), 271–2 (1982).

[31] R. Colella, A.W. Overhauser, and S.A. Werner, "Observation of gravitationally induced quantum interference", *Phys. Rev. Lett.* **34**, 1472–74 (1975). See also A.W. Overhauser and R. Colella, *Phys. Rev. Lett.* **33**, 1237 (1974).

[32] H. Muller, A. Peters, and S. Chu, "A precision measurement of the gravitational redshift by the interference of matter waves". *Nature* **463**, 926–9 (2010), doi:10:1038/nature08776.

[33] B.G. Levi, "Ultracold gases of fermionic atoms offer another path to atom interferometry". *Phys. Today*, 25 (August 2004).

[34] G. Roati, E. de Mirandes, F. Ferlaino, H. Ott, G. Modugno, and M. Inguscio, *Phys. Rev. Lett.* **92**, 230402 (2004).

[35] M.R.R. Good, *Thermalizing the Vacuum*, (2006), 6 pages available online at www.unc.edu/~mgood/research/Unruh.pdf (accessed 6 December, 2012).

[36] M. Tinkham, *Introduction to Superconductivity*, McGraw-Hill (1975) p. 18.

[37] V. Efimov, *Phys. Lett.* **33B**, 563 (1970). Also V. Efimov, *Nuclear Phys.* **A210**, 157 (1973), and V. Efimov, *Sov. J. Nucl. Phys.* **12**, 589 (1971). For experimental evidence see H.-C. Nageri, *Nature* **440**, 315 (2006).

[38] F.J. Dyson, *Phys. Rev.*, **85**, 631 (1952).

[39] S.S. Schweber, *Q.E.D. and the Men Who Made It*, Princeton University Press, NJ, (1994), p. 565.

[40] H.P. Noyes, *Prog. In Nucl. Phys.* **10**, 355 (1968).

[41] L.D. Faddeev, *Zh. Eksp. Teor. Fiz.* **39**, 1459 (1960), *Sov. Phys.-JETP* **12**, 1014 (1961). See also L.D. Faddeev, *Mathematical Aspects of the Three-Body Problem in Quantum Scattering Theory*, Davey, New York, (1965). For the extension to the four-body problem, see e.g., O.A. Yakubovsky, *Yad. Fiz.* **5**, (1967), *Sov. J. Nucl. Phys.* **5**, 937 (1967).

[42] D.Z. Freedman, C. Lovelace, and J.M. Namyslowski, *Nuovo Cimento* **43A**, 258 (1966). Also H.P. Noyes, *Czech. J. Phys.* **B24**, 1205 (1974).

[43] O.A. Yakubovsky, *Yad. Fiz.* **5**, 1312 (1967), *Sov. J. Nucl. Phys.* **5**, 937 (1967).
[44] H.P. Noyes, *Phys. Rev. Letters* **23**, 1201 (1969); H.P. Noyes in *Three Body Problem in Nuclear and Particle Physics*, J.S.C. McKee and P.M. Rolph (eds), North Holland, Amsterdam (1970).
[45] N. Levinson, *Kgl. Danske Vidensk., Mat.-fys. Medd.* **25** (9) (1947).
[46] J.V. Lindesay, PhD Thesis, available as SLAC Report No. SLAC-243 (1981).
[47] J.V. Lindesay, A.J. Markevich, H.P. Noyes, and G. Pastrana, *Phys. Rev.* **D33**, 2339 (1986).
[48] M. Alfred, P. Kwizera, J.V. Lindesay, and H.P. Noyes, "A non-perturbative, finite particle number approach to relativistic scattering theory", SLAC-PUB-8821, arXiv:hep-th/0105241 (2001), *Found. Phys.* **34**(4), 581–616 (2004).
[49] P.A.M. Dirac, *Rev. Mod. Phys.* **21**, 392 (1949). See also H. Leutwyler and J. Stern, *Ann. Phys.* **112**, 94 (1978).
[50] S.N. Sokolov and A.N. Shatnyi, *Theor. Math. Phys.* **37**; 1029 (1979).
[51] N.F. Mott and H.S.W. Massey, *The Theory of Atomic Collisions*, Oxford University Press, New York (1949), 2nd edn, Ch. III.
[52] L.I. Schiff, *Quantum Mechanics*, McGraw-Hill (1955), Sec. 20.
[53] J. Lindesay and H.P. Noyes, "Construction of non-perturbative, unitary particle-antiparticle amplitudes for finite particle number scattering formalisms", *Found. Phys.* **35**(5), 39 pages (2005), DOI 10.1007/s10701-005-4563-8. SLAC-PUB-9156, arXiv:nucl-th/0203042, 47 pages (2002).
[54] R.P. Feynman and S. Weinberg, *Elementary Particles and the Laws of Physics. The 1986 Dirac Memorial Lectures*. Cambridge University Press, Cambridge (1987).
[55] P.A.M. Dirac, *Proc. Roy. Soc.* (London), **A117**, 610 (1928); ibid, **A118**, 351 (1928).
[56] J.D. Bjorken and S.D. Drell, *Relativistic Quantum Mechanics*, McGraw-Hill, New York, 1964, p. 17.
[57] F. Belinfante, *Physica* **6**, 887 (1939).
[58] S. Weinberg, *Gravitation and Cosmology*, John Wiley & Sons, Inc., New York (1972), Ch. 12.
[59] J.V. Lindesay and H.L. Morrison, "The Geometry of Quantum Flow", in *Mathematical Analysis of Physical Systems*, R.E. Mickens (ed.), Van Nostrand Reinhold, New York, (1985) p. 135.
[60] H.L. Morrison and J.V. Lindesay, "Galilean presymmetry and the quantization of circulation", *J. Low Temp. Phys.* **26**, 899–907 (1977).
[61] H. Flanders *Differential Forms*, Academic Press (1963) p. 77.
[62] V. Schweikhard, I. Coddington, P. Engels, S. Tung, and E.A. Cornell, "Vortex lattice dynamics in rotating spinor Bose–Einstein condensates", arXiv:cond-mat/0410237v2 (2004) 4 pages. See also V. Schweikhard, I. Coddington, P. Engels, et al., "Rapidly rotating Bose–Einstein condensates in and near the lowest Landau level", *Phys. Rev. Lett.* **92**, 040404 (2004).
[63] O.J.J. d'Halloy, *Introduction a la Geologie*, Levrault, Paris (1833).
[64] J. Lindesay and H.L. Morrison, "Properties of anisotropic superfluids", arXiv:cond-mat/0203045, 12 pages (2002).
[65] N.D. Mermin and T.L. Ho, *Phys. Rev. Lett.* **36**, 594 (1976).
[66] M.C. Cross, *J. Low Temp. Phys.* **26**, 165 (1976).
[67] S. Blaha, *Phys. Rev. Lett.* **36**, 874 (1976).
[68] J. Lindesay and H.P. Noyes, "Non-perturbative, unitary quantum-particle scattering amplitudes", SLAC-PUB-9164, arXiv:hep-th/0203262 35 pages (2002), *Found. Phys.* **34** (4), 1573–606 (2004), DOI 10.1023/B:FOOP.0000044105.74284.04.

[69] J. Lindesay, "Linear Spinor Field Equations for Arbitrary Spins", arXiv:math-ph/0308003 (2003).
[70] J. Lindesay, "Group Structure of an Extended Lorentz Group", arXiv:math-ph/0309060 (2003).
[71] J. Lindesay, "An Extended Poincaré Algebra for Linear Spinor Field Equations", arXiv:math-ph/0308015 (2003).
[72] E.P. Wigner, *Ann. Math.* **40**, 149 (1939).
[73] S. Weinberg, *The Quantum Theory of Fields*, Cambridge University Press, Cambridge (1995).
[74] A. Angelopoulos et al., "First direct observation of time-reversal non-invariance in the neutral-kaon system", *Phys. Lett. B* **444**, 43–51 (1998).
[75] R.H. Dicke, "The Eotvos Experiment", *Scientific American* (December, 1961).
[76] H. Flanders, *Differential Forms*. Academic Press, New York (1963).
[77] C.W. Misner, K.S. Thorne, and J.A. Wheeler, *Gravitation*. W.H. Freeman and Company, San Francisco (1971).
[78] A.J.S. Hamilton and J.P. Lisle, "The river model of black holes", arXiv:gr-qc/0411060 (2004) 14 pages.
[79] J. Lindesay, "Coordinates with non-singular curvature for a time-dependent black hole horizon", arXiv:gr-qc/0609019 (2006), *Foundations of Physics* **37**, 1181–96 (2007), online DOI 10.1007/s10701-007-9146-4, 15 (2007), 16 pages.
[80] R.P. Kerr, "Gravitational field of a spinning mass as an example of algebraically special metrics", *Phys. Rev. Lett.* **11**, 237–8 (1963).
[81] R.H. Boyer and R.W. Lindquist, "Maximum analytic extension of the Kerr metric", *J. Math. Phys.* **8**, 265–81 (1967).
[82] J. Polchinski, *String Theory*, Cambridge University Press, Cambridge (2000).
[83] J. Lindesay, "An Exploration of the Physics of Spherically Symmetric Dynamic Horizons", arXiv:0803.3018 [gr-qc] 25 pages (2008).
[84] G.W. Gibbons and S.W. Hawking, *Phys. Rev.* **D 15**, 2738–51 (1977).
[85] B.A. Brown and J. Lindesay, "Construction of a Penrose diagram for a spatially coherent evaporating black hole". arXiv:0710.2032v1 [gr-qc] 12 pages, *Class. Quantum Grav.* **25**, 105026 doi:10.1088/0264-9381/25/10/105026 (2008).
[86] T. Clifton, "Properties of black hole radiation from tunneling", arXiv:0804.2635v2[gr-qc] 14 pages (2008).
[87] J. Lindesay, "Quantum behaviors on an excreting black hole", arXiv:0810.4515v1 [gr-qc] *Class. Quantum Grav.* **26**, 125014 24 pages (2009).
[88] T. Finch and J. Lindesay, "Global causal structure of a transient black object", arXiv:1110.6928 [gr-qc] 25 pages (2011).
[89] A.G. Riess, et al., *Astron. J.* 1009 **116**, (1998), P. Garnavich, et al., *Astrophys. J.* **509**, 74 (1998) S. Perlmutter, et al., *Astrophys. J.* **517**, 565 (1999).
[90] C.L. Bennett, et al., *Astrophys. J. Supp.* **148**, 1 (2003).
[91] Particle Data Group, *Astrophysics and Cosmology*, as posted (2009).
[92] E.W. Kolb and M.S. Turner, *The Early Universe*, Westview: Boulder, Co. (1994).
[93] J. Lindesay, "An Introduction of Multiple Scales in a Dynamical Cosmology", arXiv:gr-qc/0605007, 8 pages (2006); J. Lindesay, "Global Structure of a Multi-Fluid Cosmology", arXiv:0901.2741v1 [gr-qc] 18 pages (2006); J. Lindesay, "Diagonal Forms of a Dual Scale Cosmology". arXiv:gr-qc/0605086, 5 pages (2006).
[94] Section 19. Big Bang cosmology, review of particle physics, *Phys. Lett. B* **667**, 217–27 (2008).
[95] J. Lindesay and T. Finch, "Global Geometry of a Transient Black Hole in a Dynamic de Sitter Cosmology", in *Classical and Quantum Gravity: Theory,*

Analysis and Applications. V. Frignanni, (ed.) Nova Science Publishers, Inc. ISBN 978-1-61122-957-8 40 pages (2011). Available at www.novapublishers.com/catalog/product_info.php?products_id=23125.

[96] D.H. Lyth and A.R. Liddle, *The Primordial Density Perturbation: Cosmology, Inflation, and the Origin of Structure*. Cambridge University Press, Cambridge (2009).

[97] J.V. Lindesay, H.P. Noyes, and E.D. Jones, (2004). "CMB fluctuation amplitude from dark energy partitions", arXiv:astro-ph/0412477 v3, 7 pages; *Phys. Lett. B* **633**, 433–5 (2006).

[98] R.H. Lambert, "Density of states of a sphere and cylinder", *Am. J. Phys.* **36**, 417–20 (1968). More generally, see H. Weyl, *Math. Ann.* **71**, 441 (1912).

[99] J. Lindesay, "Consequences of a cosmological phase transition at the TeV scale", *Found. Phys.* (online March 13 2007) DOI 10.1007/s10701-007-9115-y 41 pages, print volume **37**, nos 4–5/May, 2007 pp. 491–531.

[100] P.A.M. Dirac, *General Theory of Relativity*, John Wiley & Sons Inc., New York (1975).

[101] G.B. Arfken and H.J. Weber, *Mathematical Methods for Physicists*. Academic Press, San Diego (1995).

[102] J. Lindesay, "Group Structure of an Extended Poincaré Group", arXiv:math-ph/0311044, 7 pages (2003).

[103] M. Visser, "Essential and Inessential Features of Hawking Radiation", arXiv:hep-th/0106111 (2001) 16 pages.

Index